BLOOD PARASITES
TAXONOMY, ECOPARASITISM, PATHOBIOLOGY AND CONTROL

BLOOD PARASITES
TAXONOMY, ECOPARASITISM, PATHOBIOLOGY AND CONTROL

Prof. Neelima Gupta

Delve
PUBLISHING

www.delvepublishing.com

Blood Parasites: Taxonomy, Ecoparasitism, Pathobiology and Control

Prof. Neelima Gupta

Delve Publishing
4164 Lakeshore Road
Burlington, ON L7L 1A4
Canada
www.delvepublishing.com

Email: orders@arclereducation.com

ISBN: 978-1-77956-346-0 (Hardcover)

ABOUT THE AUTHOR

Professor Neelima Gupta, a notable academician and administrator of the country is the first woman Vice Chancellor of Dr. Harisingh Gour Vishwavidyalaya Sagar (Central University). She has the experience of serving 4 government universities as a Vice-Chancellor [Ex VC Chhatrapati Shahuji Maharaj University, Kanpur, Tilka Manjhi Bhagalpur University, Bhagalpur, Munger University, Munger (Addtl charge)]. Prof. Gupta is a Gold Medalist from Aligarh Muslim University, M.Phil, and Ph.D. from AMU, Aligarh. In addition, she was also awarded the degree of D.Sc. Prof. Gupta has visited the 5 continents of the world including USA, UK, Australia, Egypt, Japan, Poland, Singapore, China, Thailand, Hong Kong, Germany, Taiwan and France. She has foreign research collaborations with Poland, Egypt and the Czech Republic. Prof. Gupta has been recognized amongst the Most Influential Women in education space in 2024 by B.W. Education. She has been honoured with more than 70 awards and distinctions, including E.K. Janaki Ammal National Award (Government of India), Saraswati Samman, the highest award in the field of education conferred by Yogi Adityanath, Chief Minister, Government of Uttar Pradesh, Vigyan Ratna Award for Excellence in Science conferred by the State Government, U.P. Dr. Sah Abidi Vishisht Krishi Vaigyanik Award by the Governor, Platinum Jubilee Award of the Indian Science Congress, ZSI Gold Medal, Indian Academy of Environmental Sciences Gold Medal, Dr. B.N. Singh Memorial Oration Award, Prof. J.S. Dutta Munshi Gold Medal, Millennium Women Scientist Award, Best Orator Gold Medal, IAES Silver Jubilee Gold Medal, Indian Medical Association Recognition, Salgare Foundation Best Scientist of the Year Gold Medal, Women of the Year-Award etc. She has featured in several programs on SANSAD TV, Delhi Doordarshan, as panelist discussing important issues related to education and research. She represents in various high powered international (Member, Governing Board of IGSTC-DST, Government of India, Member of Higher Education-U.K. British Council, Visiting Scientist at Poland DST-KBN, INSA-PAS)

and National bodies/committees (Member, Governing Council of the Association of Indian Universities, Chairperson, NAAC Peer Team/UGC Committee, Expert in UPSC, Selection Committees for the appointment of Vice-Chancellors, PSC-U.P., Uttrakhand, UPHE Commission, WB College Service Commission etc.).

She has been an excellent researcher and was awarded the "Centre of Excellence" by the Govt. of Uttar Pradesh for the development of High Tech Lab. She has been the Principal Investigator of 11 international & national research projects funded by UGC, DST, AICTE, Ministry of Environment & Forests and UP Council of Agricultural Research.

She has published 225 research papers through publishers like Taylor & Francis, USA, Springer Germany, Magnolia Press, New Zealand etc. She has participated in 424 Conferences, including 71 International and 353 National level, chaired sessions in more than 100 Conferences and organized several International and National Conferences.

Prof. Gupta is a Member of the Editorial Board of several National and International Journals from USA, Egypt, Africa, Costa Rica, India and Executive Member of various scientific societies. She is the Chairman/ Expert in NAAC, UGC, Public Service Commission, selection committees and various apex bodles of national importance.

TABLE OF CONTENTS

LIST OF FIGURES

LIST OF TABLE

FOREWORD

Blood borne parasitic infections are a major source of mortality and morbidity especially in the tropical regions of the world.

The book entitled "*Blood Parasites: Taxonomy, Ecoparasitism, Pathobiology, and Control*" highlights all the major parasites that have been discovered in fishes and mammals and it elaborates their taxonomic, environmental interactions, evolutionary trends as well pathobiological impact in fishes and mammals. The book goes on further elaborating in detail on how the presence of blood parasites can affect the health of their hosts. The book covers a detailed literature review of research work on the topics mentioned in the book. The chapters in this book bring out the knowledge gained through work done during the past decades in this field by the author and her group.

Section one of the book deals with the relevance of studying blood parasites. The impacts on human health, the history of understanding the impact of blood- borne parasites and the future challenges of research in this field.

The second section deals in detail the taxonomic classification of blood parasites, that focuses on fish and mammalian trypanosomes, as well as their evolutionary origins. The first chapter of the section gives a detailed overview of the fish parasites. Numerous fishes of the world harbor blood parasites which can be very detrimental to the health of humans as the majority of the population of the world live near the coastal areas and consume fish. Fishes are a staple diet for millions of people mainly in coastal areas and living in the river valleys, flood plains, who can be more vulnerable to these parasites. Hence the first section of the book deals with blood parasites that are nowadays commonly encountered in fishes around the world.

The chapters in the section deal with several fish *Trypanosoma*, fish *Trypanplasma*, and blood parasites. Parasites of *Cyprinus carpio* that invade fish in Poland are also highlighted in a separate chapter in quite detail. Next part of this section deals with chapters on mammalian parasites, with chapters on *Trypanosoma* and *Babesia*. The other chapters in this section deal with taxonomy and evolution of these parasites.

In the third section, the author focuses on Ecoparasitism. The section covers, how blood parasites are affected by age, length and weight of the host, prevalence is also influenced by impact of season and annual cycles. Further, chapter in the section discusses how the local ecological niche determines the parasite prevalence. Host specificity of the parasite in the ecological niche and also how the parasite interacts among themselves in the environment are the other topics of prime importance that have been covered in other chapters of the section.

The fourth section deal with the pathobiology of blood parasites. The section includes the chapters on '*Blood Parasites vis-à-vis Host Hematology and Blood Parasites vis-*

à-vis Host Biochemistry.' These chapters elaborate on the latest research findings of how parasites interact with the host by altering their hematological parameters, the chapter in the section also throws light on the findings of "Parasite physiology and Biochemistry" of glucose, protein and cholesterol metabolism that still holds challenges for researchers. The culture of Trypanosomes from fishes is one such challenge and here the author has described her research contribution on "*Cultivation of Fish Trypanosomes in Monophasic and Diphasic Media*" in the next section.

The final section of the book deals on how the immune system of the host deals with blood parasites and cites how researchers are attempting to create a long-term immunity against the blood parasites. The research gaps, challenges, and success in the field of drug development have been covered in the final chapter.

This book covers all important aspects of the *Blood Parasites* with a focus on fish and mammalian blood parasites, covering the latest studies. The studies on fish blood parasites have been rare and not well documented so far, have been covered in this book. The author has highlighted the latest research in this field. The book has meticulously documented the details of blood parasites and presented them through well-illustrated pictures and diagrams. The tables and charts are self-explanatory presenting in a well-articulated way, all the latest research findings in this field. This book is an essential read for medical fraternity and researchers working on tropical medicine and related diseases.

—Professor Nirmal Kumar Ganguly

M.B.B.S., M.D., FRCP (London), Fellow, Imperial College (London), FAMS, FNA (Ind. Med.) FNASc, FTWAS (Italy), FIACS (Canada), FIMSA

Former Director General – Indian Council of Medical Research

PREFACE

"Every book has a story; this story has a book"

The present title, *"Blood Parasites: Taxonomy, Ecoparasitism, Pathobiology and Control*" is a humble attempt to pack the information on blood parasites under a single cover. Blood is the highly sophisticated transport system that allows communication and coordination between different parts of the human body serving three main functions: transport, protection and regulation and is an indispensable tissue for the body performing vital functions. Being rich in nutrients, it also serves as an excellent medium for the growth of blood parasites that live in the blood of their animal hosts. Feeding on blood, these parasites distort the entire physiology of the host. Thus, certain questions become pertinent, what is the nature of these parasites, how do they propagate, what impacts do they cause on host blood, how do they affect the immunity of the host, in case of multi-infection, how do they react with each other, is it possible to control these parasites through drugs? These and other related questions compelled the author to carry out this piece of work and through this book, she has tried to address these questions.

Blood parasites have often been neglected specially those present in fishes and lower animals. Trypanosomes, common inhabitants of Africa have often been thought to be absent in other countries in humans due to the absence of Tsetse flies (*Glossina palpalis*) but what about fishes in which the parasite is spread through the leeches? Aroused by this interest, and looking into the scope of the work, erudite studies were undertaken on the blood of fishes and unknown species were discovered one after the other leading to discoveries of other blood parasites and their biological and physiological implications.

The author has tried to make the book unique and novel by comprehensively designing and compiling all information on blood parasites to attract a wide range of audience: students, researchers, faculty, hematologists, biochemists, parasite environmentalists, immunologists, pharmacologists, pathologists, policy makers, academicians and also to provide a step-by-step, easy-to-follow techniques in handling blood and parasites.

The entire work presented in the thesis was completed in three decades. Taking a grip on the subject after completion of work on fishes during Ph.D., the author was interested to pursue studies on blood parasites onwards on different phyla, research was executed on select host species of fishes, amphibians, reptiles, birds and mammals under different programs and projects. The work became so exhaustive that it had to be restricted on fishes and mammals those on other phyla are published elsewhere. During this period, other academic and administrative responsibilities were executed as well, being a Professor, Head of Animal Science Department, Dean of the Faculty of Applied Sciences, officiating Finance Officer, Dean Students Welfare of Mahatma Jyotiba Phule Rohilkhand University, Bareilly but even with these responsibilities, always snatched out time to continue with the research interest. Now, as Vice Chancellor of Doctor Harising Gour Sagar University (A Central University), Sagar, Madhya Pradesh, India, through this book, my work is being shared with the readers.

The over-riding goal of this book is to present essential information regarding blood parasites in a compact, readily accessible form which leads to unraveling of facts and simple understanding. The book is structured into six sections in a uniform, systematic format: Chapter 1 introduces the subject, gives an idea about the past researches done in the area and a step-wise, easily reproducible methodology for the execution of the work. Chapter 2 is dedicated to the taxonomy of the parasites in fish (*Trypanosoma, Trypanoplasma*) and mammalian (*Trypanosoma, Babesia*) hosts providing illustrated accounts of 16 species, out of which nine have been reported as new species by the author. This section also describes the multiplication pattern in trypanosomes and the evolutionary trends in fish and mammalian trypanosomes. Chapter 3 deals with the ecological aspects of parasitism, impact of variables (age, season and locality), interaction amongst parasites under concurrent infectivity and host specificity. Chapter 4 focuses on pathobiological studies (hematological and biochemical) and in Chapter 5, the methods of trypanosome culture are described. Chapter 6 describes lymphocyte culture, cellular immunity, induction of immunological memory, development of immune sera against fish trypanosomes and drug action on trypanosomes. The book concludes with paying a tribute to all researchers whose work was consulted during the entire research and acted as stepping stones to reach the ulterior goal as set forth in this book. Simple, clear figures and photographs, self-explanatory tables and graphs provided with in the book will definitely help the reader to readily grasp the contents in a lucid manner.

Sincere gratitude is expressed to her mentor, Prof. Durdana S Jairajpuri, former Chairman, Department of Zoology, Aligarh Muslim University, Aligarh who aroused interest in the author during her post-graduate program by suggesting to work on blood parasites of fishes with a comment, "Neelima, the problem is difficult, but promising, I have confidence that you will be able to work on it with good results." This was taken as a challenge and with passing time, the author probed deeper and deeper into the problem, for her MPhil, PhD, and ultimately her Doctor of Science was awarded on the topic "*Characterization of blood parasites, pathobiological manifestations and ecological variance in vertebrate hosts*." The present book brings into print the work presented in this thesis. The person who stood as a strong orientation wind was Late Prof. M. Shamim Jairajpuri, Ex-Vice Chancellor, MANU, Hyderabad who gave direction at the proper time for the proper task. All the funding agencies (University Grants Commission, Department of Science and Technology, Ministry of Environment and Forests, New Delhi and Uttar Pradesh Council of Agricultural Research, Lucknow and Uttar Pradesh State Government) that provided financial assistance on various occasions for the execution of the research work are gratefully acknowledged. Thanks are expressed to the Department of Science and Technology and Indian National Science

Academy, New Delhi for nomination to visit Poland on different occasions which made it possible to work in internationally renowned parasitological laboratories as a Visiting Scientist. The camaraderie and expertise of Prof. Irena Wita, Head, Blood Parasitology Laboratory, Institute of Parasitology, Warsaw, Poland, Prof. Andrzej Pilarczyk, Director, Institute of Ichthyobiology and Aquaculture, Chybie, Poland who were my collaborators during the INSA- PAS and DST-KBN collaboration are appreciated. The exchange of research ideas with Prof. Jamina Pashtuszko, Warsaw Veterinary University, Warsaw and Dr. Hanna Lutnicka, Scientist, Fish Disease and Research Laboratory, University of Agriculture, Lublin, went a long way in planning experiments and familiarization with research techniques. It has been an honor to have worked with Prof. N. Kingston of Wyoming University, USA, a Visiting Scientist at the Institute of Parasitology, Warsaw through whom the intricacies of trypanosome culture were familiarized.

Special thanks are expressed to my loving and encouraging husband, Dr. D.K. Gupta for his kind gestures and all-time support who has always stood beside me in all my pursuits. My worthy children (sons, Er. Dr. Mudit, USA and Dr. Ruchit, Australia and their wives, Shilpa and Ankita) have always stimulated me to pioneer ahead. I am also grateful to the Managing Editor, Delve Publications, Canada in bringing out this book.

It is to be confessed that while working on this project, each finding aroused greater interest and propelled the author to move further. The expanse of knowledge could be felt to increase to develop greater confidence on the subject matter. It is sanguinely hoped that the readers will be equally fascinated as they go through each chapter and will lure them to execute further work on these lines.

The author will remain responsible for any shortcomings of the book. She will remain grateful to the readers for their critical suggestions for the further improvement of the book.

—Neelima Gupta

Vice Chancellor

Dr. Harisingh Gour Vishwavldyalaya, Sagar, MP, India

(A Central University)

e-mail: vcneelimagupta@gmail.com

Chapter **1**

Introduction

Contents

1.1. INTRODUCTION

"*The seat of the soul*" or blood has always been a source of fascination for man. Ancient Egyptians bathed in it, aristocrats drank it, authors and playwrights used it as themes and modern man transfuses it. It has been regarded as the most important system of the body as it performs several vital functions. It has been recognized as an agent for distributing nutrients, oxygen, detoxicating compounds, hormones that control and integrate the activities of various organs and maintains a uniform body temperature. It transports antibodies and cells that protect the body from invading pathogens and also transports drugs administrated to the body to facilitate their therapeutic effects on specific sites of disease. Moreover, it collects the waste material generated by the body's metabolic activities.

Mankind has always been interested in blood, since it is likely that even primitive man realized that loss of blood, if sufficiently great was associated with death. The biblical phase "*Blood is life*" has for long irresistibly attracted man. According to the Bible, "*to shed blood*" was a term used in the sense "*to kill.*" Due to its importance to life, blood has been associated with a large number of beliefs. "*Related by blood*" is used as a symbol for family relationships. Sayings such as "*blood is thicker than water*" and "*bad blood,*" as well as "*blood brother*" are not uncommon. Not only this, blood has gained popularity around the globe. Among the German tribes (Anglo-Saxons and Norsemen), blood was used during their sacrifices, the *Blo'ts*. The act of sprinkling blood (*bleodsian*) in Old English and the blood of the Gods (*ichor*), a mineral that was poisonous to mortals by the Ancient Greek are well remembered.

Having an ancient and strong footing, blood has drawn attention of biochemists, biophysicists, cellular and molecular biologists, geneticists, immunologists, pharmacologists, oncologists and above all, parasitologists. Hematological techniques offer excellent diagnostic aids and assists in assessing the general health and evaluation of diagnosis of diseases. Blood is a useful indicator of the physiological state of the animal. It has been used in the screening procedure to evaluate the general health, as an adjunct for the diagnosis of diseases to assess the body's ability to fight against infection and to evaluate the progress of diseases.

Parasitic diseases can be called as '*Cancer*' of tropical countries as they probe into the very population causing great harm. Morbidity and mortality consequent to parasitic invasion are of major concern and pose a great challenge and strain to the community and the country at large. One cannot have experienced the war without having been impressed anew, and depressed by the amount of parasitism in world.

Parasites are a world-wide major health problem ranking amongst the greater threats to the well-being and survival of the hosts. They occur in all natural communities, sometimes as benign infections, but may also have massive economic impact by seriously damaging populations of domestic animals, fish and crop plants. A continuous effort by the world community of parasitologists is necessary to understand and further the control of parasites and their vectors.

The blood feeding mode evolved independently more than 150 million years ago in at least 6 different groups of arthropods and this habit simultaneously acted as a viable source for the transmission of blood parasites during next blood meal of the vector. These arthropods (insects and ticks), had to independently solve the problem of how to deal with their host's defense against blood loss, the hemostatic mechanisms. In addition, some annelids like sanguinivorous leeches attack a variety of invertebrate and vertebrate hosts (including fish) and act as transmitters of blood-borne parasitic diseases. 50 million years ago, dinosaurs became extinct and mammals became the main food source, bringing with it a new problem posed by blood platelets which are much more efficient than bird or reptiles thrombocytes in preventing blood loss. Blood feeding arthropods in the same family had to discover new pharmacological products to deal with this new feeding problem. Perhaps because of these evolutionary reasons, a very large diversity of pharmacological products is being discovered in the saliva of such arthropods, which comprise around 15,000 species in nearly 600 genera.

Parasitic diseases are a heterogeneous term of convenience summing up the woes of an individual and are a phenomenon which causes a population to be abnormal. The reasons for their outbreaks are varied representing complex interactions between the host and the disease-causing situation. In case of host-parasite interactions, the onset of disease represents a decrease in the resistance of the host and an increase in the numbers and/or virulence of the parasites. It is especially undesirable that any species should suffer from the rigors of disease. Unfortunately, the etiology of disease is poorly understood and warrants a deeper study. The pathological processes attributed to these infectious agents are often complicated by clinical and sub-clinical processes which are largely due to the inability of the host to cope with the health-compromising changes in the environment.

When, and in what manner blood was first examined is unknown. However, in the past 50 years, hematology has grown into a broad science, which, in seeking to understand the normal and pathologic physiology of hematopoietic system employs all the methodologies of diverse scientific disciplines. Hematological investigations have been ever increasing in importance and today, it encompasses many aspects of applied sciences and diagnosis by hematological techniques have proved to be an excellent aid.

1.1.1. Blood Parasites

Being nutritionally rich, blood serves as a favorable site for parasites. Once present, they may interfere with the blood physiology and terminate in diseased condition. Initiation of disease is a complex process interacting the set of circles of the parasite, host and environment creating a conducive condition for disease outbreaks.

Parasites are known to inhabit all groups of vertebrates commencing from fishes, through amphibia, reptiles, birds and mammals. The blood micro-niche has not escaped parasitism and has acted as a suitable environment of hemoparasitic invasion. "Parasitism in blood" has been a matter of curiosity and interest amongst parasitologists and hematologists all around the world. Being rich in nutrients, blood serves as an appropriate site for parasite habitation. The parasites include mastigophoreans *Trypanosoma,* the monoflagellates and *Trypanoplasma,* the biflagellates. These are extracellular blood parasites inhabiting the blood plasma and move actively in the blood by means of flagella. The sporozoans (*Lankesterella, Haemogregariana, Hepatozoon, Plasmodium, Haemoproteus, Dactylosoma, Babesiosoma, Thromocytozoon, Leucocytozoon, Myxosoma, Babesia*) also inhabit the blood extracellularly or intracellularly.

1.1.2. Ecology

Although inputs on parasite study are rapidly on the increase, our knowledge regarding parasites producing diseases with respect to the environment is limited which requires investigations on multidisciplinary aspects encompassing parasitology, hematology, biochemistry, immunology and environmental relationships.

Parasites have evolved numerous ways to adapt to the dramatic changes in environment experienced throughout their life cycle. An alteration in habitat is communicated to the parasites by chemical or physical signals which triggers metabolic and ultra-structural changes (Czichos et al., 1986).

Parasitic diseases can pose major threats to animal populations and have serious economic impacts on host health, therefore, research into host-parasite ecology is critical (Cleaveland et al., 2002; Galvani, 2003). However, knowledge on how hosts and parasites interact in natural systems remains limited and consequently, international bodies like the World Health Organization have advocated an increased study of host-parasite ecology in wild host populations (Real, 1996; Schall and Pearson, 2000).

Purely vertically transmitted parasites are predicted to cause low degrees of pathogenicity in their host, however, the role of stressful or diversified environmental conditions on host fitness is poorly understood and it is obvious that parasitized hosts would suffer a great reduction in fitness than uninfected hosts under adverse environmental conditions.

Parasite abundance can vary in response to numerous factors, both in terms of prevalence, defined as the proportion of the host population that is infected and the parasite load, which is the number of parasites in an individual host. Factors that can influence parasite abundance include host sex and age (Schall et al., 2000; Smallridge and Bull, 2000); host reproductive effort (Norris et al., 1994; Nordling et al., 1998; Veiga et al., 1998), host condition and physiology (Salvador et al., 1997; Appleby et al., 1999; Dowell, 2001); vector biology (Sol et al., 2000; Reardam and Norbury, 2004), and host density (Arneberg et al., 1998). Abiotic factors such as temperature and rainfall can also influence parasite abundance (Fuller, 1996; Bajer et al., 2001). Each locus is typified by its own ecological complex comprising of parasites, vector, host and various features of the host's environment forming an ecological web, and influence the existence of parasites.

1.1.3. Growth of the Host

The growth of the host is one of the biotic factors contributing to the distribution of parasitic fauna. It has been an open question whether the growth of the host has any role to play on the incidence of parasites as the latter have been reported to increase with growth of the host, decrease with growth of the host or even be independent of the growth of the host. Keeping this in mind, the incidence of blood parasites was also taken into account with respect to the growth of the host in question.

1.1.4. Seasonal Variations

The climatic factors change worldwide during the annual cycle and the climatic pattern with respect to parasite is not only a matter of scientific interest but can also help to identify mechanisms such as the demography of hosts and vectors which influence parasite transmission dynamics. It has been commonly proposed that temperature is one of the most important factor determining the existence, abundance and distribution of parasites. Temperature varies during the different months of the year and therefore its significance as a regulatory parameter for parasite density is natural. The availability and infectivity of the vector, its seasonal prevalence, the prevalence of the parasite, immunological behavioral and biology of the vector and the host are to be taken into account for determining infectivity of blood parasites during the annual cycle.

1.1.5. Habitat

The typical environment in which parasites occur is commonly referred to as habitat. The geographical division of the external environment where a parasite is found is also frequently used to denote the term. It has been regarded as an important factor playing a significant role in the infectivity of parasitic species. Ecological and biological factors may affect the incidence of parasites as the environmental factors have a controlling system on the availability of parasites in their hosts. Habitat, lifecycle and host specificity may be said to be closely interrelated as host specificity may play a role in the parasite's distribution.

1.1.6. Concurrent Infectivity

The resources available in nature are not sufficient for the survival of all individuals born and thus there is a constant competition for the available resources. Darwin (1859) drew attention to the importance of competition for resources (food or other necessities of life) which prevails even today. Two types of behavior may cause competition: exploitation where individuals of different species use some of the same resources and interference where directly damaging behavior occurs. Occurrence of concomitant infectivity amongst blood parasites is not so frequent but when present, intraspecific (within species) and interspecific (different species) competition amongst blood parasites may occur but how far they act as limiting agents for parasites is questionable.

The parasite assemblage, interaction, adaptability and survival of species is a field of parasitology deprived of due attention and little efforts have been made to study the co-existence of a species in a particular ecological niche. Darwin (1859) stressed the importance of competition for food and other resources for co-existing species which paved the way for studies in other groups of animals (Hogstedt, 1980).

1.1.7. Host Specificity

There is little experimental proof to demonstrate hosts specificity amongst blood parasites. They have been reported to be euryhostpitalic (Khan, 1977a), or may also be reported from the same host.

Parasitic species show variation in their adaptation to single host species or to multiple host species. The pivotal studies on specificity of host-parasite relationships concern the mechanisms which limit host selection and which control the ability of the parasite to invade the host and to survive within it. The emerging interest shown in the problems of specificity can be attributed to their great importance for the understanding of certain questions of evolution and phylogeny of the parasites as well as their hosts.

1.1.8. Hematology

Incidentally, blood is the most accessible component of the vertebrate body fluid system. Accordingly, it is the tissue most frequently examined in attempts to assess the physiological status. The value of the classical blood variables (erythrocyte count, hemoglobin) and their derivatives like mean corpuscular volume (MCV), mean corpuscular hemoglobin (MCH), mean corpuscular hemoglobin concentration (MCHC) are regarded as valid indicators of health which are to be assessed from the view point of measurement precision and prevention of distortion of information content by sampling procedures and storage condition.

Hemograms act as mirrors of metabolic dysfunction and are increasingly being used by biologists as bases for the evaluation of dietary adequacy for the demonstration of pathological status, the assessment of physiological response to stress conditions and for the establishment of hematological norms for various species. The proper evaluation of the hemogram of any animal involves the determination of total erythrocyte count (TEC), total leukocyte count (TLC), hemoglobin concentration (Hb), packed cell volume (PCV), differential leukocyte count (DLC), mean corpuscular volume (MCV) and a detailed study of stained peripheral blood films. The basic techniques used in mammalian hematology also apply to that of other vertebrates, such as birds, amphibians and reptiles. The reasons for the growing interest in hematological studies are mainly due to its diagnostic potential in toxicology and pathology.

Hematology plays an important role in disease diagnostics and in assessing the toxic effects of etiological agents. Fishes being poikilothermic aquatic vertebrates exhibit changes in their physiology during different stages of their life history which are reflected in their hemograms, thus its true picture is difficult to obtain. Infection with parasitic diseases make the situation worse as they are regarded to be one of the major factors responsible for changes in the blood values of fishes.

The assessment of blood parameters and morphological/cytochemical characterization of blood cells plays an important role in determining host health and can therefore reflect alteration in the physiological status according to exo and/or endogenous influences. Any kind of stress not resulting in gross changes and mortality produces certain changes in blood characteristics (Christensen et al., 1972). Apparently, the hosts are under stress due to parasitic load.

For diagnostic purposes, a set of tests are to be established followed by a more detailed hematological examination. A detailed hemogram or complete blood count (CBC) includes determinations of RBC, Hb, PCV, MCV, MCHC, MCH, icterus index, total plasma protein concentration, WBC and a 200-cell differential leukocyte count. However, the choice of tests vary according to the animal investigated, the disease suspect and further laboratory investigations may accordingly be required. It is thus essential to generate a library of information by collecting basic data related to ecological, environmental and physiological conditions in order to promote future comparison of hematological data which is important for the proper health management.

1.1.9. Biochemistry

The biochemical composition of blood reflects the state of function of every organ (Shell, 1961). There is not much deviation in the values of biochemical indices in members of the same species and are generally fixed for a particular species under similar conditions of assay. Variation in these values would reflect the impact of environmental factors, behavioral habits, metabolic activity or exposure to some sort of pathogen.

Like other physiological parameters, the biochemical constituents of blood are also influenced by the presence of parasites. Studies on chemical changes in blood due to the presence of hemoflagellates provide an important index of thc pathological condition of the host. The increasing use of lower animals for investigating the pathological conditions and the assumption that some of these conditions maybe reflected by changes in the blood picture point to the need of establishing hemograms of the host. Reports on changes in blood biochemistry are spasmodic and those on record are on hypoglycemia (Tandon and Joshi, 1974; Joshi, 1981), hypocholesterolemia (Tandon and Chandra, 1977b) and anemia (Khan, 1977; Gupta et al., 2002a).

1.1.10. Culture

The development of an *in vitro* system allows a continuous multiplication of parasites retaining their infectivity to the host. The establishment of a proper *in vitro* system has been along standing requisite for a metabolic and immunological studies and as such, a large number of hematozoans have been isolated from the host blood without altering their properties and an attempt has been made to cultivate them. Axenic culture of parasitic organisms is a valuable tool for *in vitro* analysis of properties such as infectivity, metabolic requirements and antigenic variation (Taylor and Baker, 1978). The inability to grow animal infective forms of parasites *in vitro* has been one of the major hindrances which has prevented the development of effective control strategies. Attempts to cultivate parasites *in-vitro* are aimed at the application of appropriate *in vitro* systems and have met with great difficulties as in many of the cases, the protocols are not well defined or there may be contamination of the media and have thus posed serious problems for the parasite culturist.

For the successful cultivation of blood parasites, it is important to have a defined or semi-defined medium available for biochemical, nutritional and immunological studies as well as investigations of the transformation of the blood stream form to the culture which are poorly understood. The cultivation of most of the blood parasites have been done from mammalian hosts including hamster (Berens *et. al.,* 1976); mouse (Hirumi et al., 1977; Abdallah et al., 1990); lion (Brun et al.,1981); cattle (Brun, 1982) and opossum (Thomaz and Deane, 1990). Cultivation of blood parasites from fish hosts have been attempted on *Trypanosoma striati* (Qadri, 1962c); *T. danilewskyi* (Rouskova and Kulda, 1983); *T. carassii* (Overath et al., 1998*)* and *T. catostomi* and *T. phaleri* (Jones and Woo, 1991).

There has been a remarkable progress in the area of fish immunology as fishes are one of the most ancient group in which an immune system characterized by the presence of immunoglobulin has been demonstrated. The occurrence of immunoglobulin class in most fish provides a unique experimental model for studying cellular reorganization by Ig receptors and the subsequent immune response. Relatively little is known about protective immunity in fish and the effect of parasites on piscine immune system.

1.1.11. Lymphocyte Culture and Cellular Immunity

Small mononuclear cells, the lymphocytes, constituting 20–30% of circulating leukocytes have become the focus of interest of immunological investigations due to their protective capability. They are the center of interest for immunological studies and have been credited with protective potency. Contributions on these aspects are gradually gaining momentum (Van Muiswinkel et al., 1985) and sporadic investigations on lymphocyte culture are also on the increase (Blaxhall, 1985).

1.1.12. Immunological Memory and Immune Sera

During the past few decades, interest has grown to develop vaccination schedules evoking protective immunity in fish from different quarters of the globe. With the decided advantages of immunological techniques over chemotherapy in being prophylactic, having no toxic side effects and no accumulation of toxic residues, there has been a rising interest on immunological studies at the international scenario.

The rapid and exaggerated response to secondary antigenic stimulation betokens "memory" of previous experience with a given antigen which is attributed to small lymphocytes and its activity can be demonstrated by cell transfer experiments.

The development of immune sera provides an excellent means for inculcating immunity in fish. The naive fish can be challenged with parasites and the number and nature of challenge dose required to obtain "immune" fish would facilitate control and prophylaxis against the specified pathogen.

1.1.13. Drug Action

Many parasitic infections are chronic in nature, and due to lack of apparent clinical symptoms, diagnosis is delayed. Early diagnosis and effective treatment is the key to successful intervention, and the combination of proper and timely diagnosis clubbed with more effective chemotherapy is a holy grail for an effective control program.

There is a need for treatment of parasitic infection throughout the whole spectrum and the search for better drugs is a perpetual process. Avenues have been opened up for new drugs which were unimaginable just a few decades ago with the advancement in parasite genomic research and its related technology. However, the soaring costs of drug discovery programs has made research and development specially difficult to promote and sustain the global economy of today.

Protozoans, the simple single-celled organisms can invade even the most complex organisms. This necessitates the formulation of suitable drugs for the effective control of protozoan infection. A complex interplay of activities in the field of animal health, education and intricate research is involved for parasitic eradication. The situation is grim in lower animals where not much attention is paid for their healthy survival and often the sick animals remain ignored and neglected.

The above brief introduction indicates that many avenues pertaining to blood parasites are as yet unexplored. Although we are steadily exploring hematozoan taxonomy of different groups of vertebrates, the path on hemato-biochemistry, co-existence of blood dwelling parasites, infectivity in relation to eco-variants, host susceptibility and specificity, cultivation of parasites and their control through chemotherapy still remains deserted and it is the call of the day that we focus our attention on these lines. This would enable us not only to gain insight into the physiology of the most vital tissue, blood but would also enable us to march ahead in the field of blood parasites, their eco-physiological and biochemical impacts and to devise appropriate strategies for the control of the blood pathogens. This piece of work is a humble attempt to fill-in the lacuna existing in this field and is likely to prove an asset for the hematologist, biochemist, taxonomist and parasite environmentalist.

1.2. RELEVANCE IN CONTEXT TO THE PAST

1.2.1. Fish Farming

The United Nations charter of food security for the alleviation of poverty and hunger can only be implemented by enhancing global food production. This is possible by adopting appropriate technologies as may be applicable to the diverse agro-climatic conditions of different geographical areas and by taking into consideration the ecosystem as a whole on a long-term and sustainable basis. The world human population in 1998 was estimated at 6 billion with an annual growth rate of 1.4%. It is expected to grow to 8 billion by the year 2025. The global requirement for fishery products, as projected by the FAO, will approximately be an additional 26 million tons by 2010 from that of the 1996 level, of which 7.2% would be required in Asia and 9.6% in Africa.

The trend in increased demand for fish is expected to continue worldwide, and will certainly result in mounting pressure on aquatic systems to produce more fish. Fish farming with freshwater species has become a highly productive industry in several countries.

Asia plays a major role in world fish production, accounting for more than 40% of capture fisheries and 80% of aquaculture production. Fisheries in most Asian countries are of socioeconomic importance as fish constitutes a major part of the diet of more than 1 billion people particularly those who live close to water edges along the coast and river banks. It is also a major income source for many poor fisher folk and contributes substantially to foreign exchange earnings in many developing nations. Although the total contribution of fisheries to the Gross National Product (GNP) of developing countries is not significantly high as compared to contributions from the manufacturing, servicing and agriculture sector, fisheries employs more than 50 million people, mostly from Asia and many in the post-harvest sector.

In India, we have a combined river length of about 45,000 km comprising of 113 major and minor rivers. While the inland fisheries resources are supported by most of the rivers, only five major rivers with a network of their tributaries, canals and streams crisscrossing the entire length and breadth of the country, covering an area of 3.12 million sq. km and total length of several 1,000 km contribute to the maximum. Reservoirs constructed on these rivers have an area of 2.1 million ha with a combined storage capacity of 16,247 cubic km.

1.2.2. Disease

Utilization of fish for its complete caloric output necessities solving of multifarious and intricate problems, the foremost being to develop disease-free fishes. Intensive or semi-intensive fish culture propagates artificial stress due to overcrowding, mal-nutrition, accumulation of metabolites, deterioration of water quality and all such factors magnify the risk of outbreak of disease. The gap in knowledge on fish diseases can be filled by conducting advanced studies on different pathogens, their biology, life cycle and pathogenicity.

Under natural environmental conditions, aquatic fauna succumb to infections, parasites, toxins, malnutrition and neoplasms, but predators and scavengers so quickly eliminate the weak and consume the dead that most losses are unapparent to man.

Disease, *per se*, is not an entity or an end in itself. Disease is the end result of an interaction between a noxious stimulus and biological system, and to understand disease is to understand all aspects of the biology of the species. The interdisciplinary approach which is currently characterizing many branches of scientific research is also an evidence in the field of diseases of fish.

Severe outbreaks of diseases can result from the introduction of parasite or pathogens, from malnutrition, from chemical and physical alterations of water, from the genetic makeup of the fish, and from interrelationship of none or all these factors. Parasites may be responsible for causing various ichthyopathological manifestations constituting one of the most important scourges confronting fish culturists.

1.2.3. Parasites

The science of Parasitology is one of the many new disciplines of twentieth century, as such it is a dynamic and rapidly evolving science which encompasses an increasing number of sub-disciplines and technologies.

The field of host-parasite relationships represents a challenging new direction in Parasitology which recently has attracted the attention of endocrinologists, physiologists, immunologists, neurologists, biologists and ecologists. This field has been dominated by biologists who have demonstrably intense "*feeling for the organism*" and their findings have been reported historically from time to time. The perspectives of parasitologists and epidemiologists are greatly enhancing our appreciation of the adaptive significance of the host-parasite relationship for the parasites involved and their respective hosts. It is the considered view of parasitologists and fishery biologists that micro- parasites induce increased mortality rates and cause reductions in fecundity, while macro -parasites tend to cause reductions in the condition of fish which reduces their market value (Dobson and May, 1987).

The cultured populations in aquaculture are often crowded into confined areas for extended periods. Parasites are common and dangerous among fishes living in confined space such as aquarium, hatcheries, stocking ponds and tanks (Ali, 1990). Fishes usually have mixed infections of parasites. The degree of damage by infection is influenced to a large extent by the type and the numbers of parasites present (Bauer, 1941). Parasites can affect fish population by causing mortality, reduction in growth, weight loss and suppression of reproductive activity (Bauer, 1961). The high population density facilitates the transmission of most protozoan disease agents and therefore, increases the probability of disease outbreaks. The degree of success in preventing and treating disease outbreaks is roughly proportional to the information available on the pathogen involved.

One of the major fish pathogen group is Protozoans. The Protozoa literally means first animal. It includes amazingly diverse group of organisms which are mostly microscopic, generally unicellular and sometimes appear to be simple. But these are the most complex cells known because all the biological and biochemical mechanisms for a complex life-style are contained within these single cells (Sleigh, 1991). Protozoans act as major enemies in reaching maximum fish production target as they detrimentally harm the fish and decrease its food value. These parasites have mainly been studied in higher vertebrates and fishes are largely neglected in spite of their high prospects of being an excellent nutritional food item.

Several protozoan parasites, *Trypanosoma, Trypanoplasma, Haemogregarina, Babesiosoma, Dactylosoma, Mesnilium, Haematractidium, Icthyophthirius, Costa, Chiloden, Myxosoma, Myxobolus, Trichodina, Bodomonas, Unicauda, Henneguya, Chloromyxum, Kudoa, Nosema, Glugea, Sphaerospora* and *Rudicapsula* may inhabit different organs of fish causing variable degrees of pathogenicity (Gupta, 1981). Amongst these, six of the genera are reported from the blood itself. The parasites of present concern are the hemoflagellates, *Trypanosoma* Gruby, 1843 (Family Trypanosomatidae Doflein, 1901) and *Trypanoplasma* (Laveran and Mesnil, 1901) (Family Cryptobiidae Poche, 1913). The studies on the hematozoan parasites of Indian fishes have assumed a special importance due to the vast and varied amount of fish potentialities available in the marine and fresh water sources and their huge consumption.

1.2.4. Blood

Studies on blood encompasses the science of Hematology which is a rapidly expanding field with changing concepts in understanding of pathophysiology of immuno-hematological disorders. It is now an established fact that blood acts as a mirror of eco-physiological activities and lays impact on the body of vertebrates. Blood related bio-chemical, hematological and immunological studies are increasingly being used to have a firsthand preliminary to diagnostic follow-up and a proper monitoring of the vertebrates of economic importance all over the world. The hematological picture not only reflects the intrinsic metabolic activities going on within the whole organism, their requirements, and their products but also act as a wonderful transporter. Immediate short term and long-term influences are truly and fully reflected by the blood.

1.2.5. Blood Flagellates

The class "Flagellata" includes protozoan parasites characterized by the presence of one or more flagella which originate from the anterior end directed forward during locomotion.

Studies on flagellates that parasitize fish have gained importance in recent years. This is the result of an intensified effort to cultivate a wide variety of fresh water and marine fish under semi artificial conditions so that they can eventually be harvested by man. Fish hemoflagellates are less extensively studied in comparison to those infecting man and domestic animals.

Hemoflagellates are Kinetoplastids having one or two flagella, an axoneme and a paraxial rod which arises from a flagellar pit. The single mitochondrion usually extends along the length of the body and the kinetoplast, which is part of the mitochondrial system, stains with Feulgen, Janus green or Romanovsky stains. The mitochondrion is believed to be concerned with respiratory adaptations to different environmental conditions (Vickerman, 1976). Most species are extra- cellular endoparasites in the intestinal tract, internal organs, blood and/or tissue fluids of their vertebrate host. The ectoparasitic flagellates and those associated with the intestinal tract have direct transmissions while the hematozoic parasites are usually transmitted by blood sucking invertebrates.

The research strategies being framed for the study of protozoan parasites, calls for a detailed literature survey and this chapter caters in fulfilling the advances made in this progressive area of research.

1.2.6. Fish Trypanosoma

The monoflagellates, *Trypanosoma* Gruby, 1843 are blood parasites existing in all groups of vertebrates: fish, amphibians, reptiles, birds and mammals. They have played an interesting role in the development of certain concepts in biology and medicine (Chandler and Read, 1970). A survey of literature indicates that Valentin (1841) was the first to discover a trypanosome (or probably a trypanoplasm) in the blood of *Salmo fario*. In the fish blood, trypanosomes usually occur as trypomastigotes and may be monomorphic, dimorphic or polymorphic in nature. Depending upon the species and state of development, body lengths range from 15 μm (*T. remarki* from pike) to 130 μm (*T. giganticum* from ray), average being 50 μm long and 2–5 μm wide. Amongst the early workers, *danilewskyi* (1885) discovered a trypanosome from *Cyprinus carpio*, Lingard (1904) was the first to record *Trypanosoma* from India. Castellani and Willey (1905) found *T. saccobranchi* from *Saccobranchus fossilis*; Montel (1905) reported *T. clariae* from *Clarias macrocephalus*; Neave (1906) reported forms from *Polypterus* sp.; Robertson (1906, 1909, 1911) studied transmission; Dutton et al. (1907) recorded trypanosomes from *Clarias angolensis* while Woodcock (1909) published a treatise on hemoflagellates and allied forms. Mathis and Leger (1911) reported unnamed trypanosomes. Doflein (1916) discussed their pathogenic role; Pearse (1933) found *T. ophiocephali* from *Ophicephalus straitus*; De-Mello and Valles (1936) observed *Trypanosoma* in *Clarias batrachus*; Rohdain (1942) reported *T. gandei* (H-*Labeo macrostoma*) and *T. malopteruri* (host: *Malapterurus electricus*). Bacigalupo and De la Plaza (1948) described *T. marplatensis* (host: *Psammotica microps*); Laird (1948) reported *T. hepatatretae* (host: *Hepataturetus cirrhatus*); Dias (1955) described 3 sp. of hemoflagellates from *Tilapia mossambica*; Noble (1955) contributed towards the morphology and life cycle of trypanosomes; Qadri (1955) studied *T. striati* (host: *Ophicephalus straitus*) and further (Qadri, 1962 a,b) described *T. batrachi* (H-*Clarias batrachus,*) and *T. danilewskyi saccobranchi* (host: *Saccobranchus fossilis*). Saunders (1959) reported *T. balistes* (host: *Balistes capriscus*); Baker (1960; 1961) recorded *T. mukasai* in 11 sp. of fresh water fishes. Hasan and Qasim (1962) described *T. punctati* (host: *Ophicephalus punctatus*); Chen and Hsieh (1964) recorded parasitic flagellates of fishes from Hwama lake; Becker (1967) found *T. occidentalis* from 3 sp. of fresh water teleosts; Smirnova (1970) described *T. lotai*; Daly and Degiusti (1971) reported T. *catostomi*, (host: *Catostomus C. commersoni*); Burreson and Pratt (1972) described *T. pacifica* (host: *Parophrys vetulus*).

From India, trypanosome infectivity is apparent from the infection of *T. gachuii* (host: *Ophicephalus gachua*) (Misra et al., 1973) and *T. mukundi*, *T. elongatus* (host: *H. fossilis*) (Raychaudhari and Misra, 1973). Mandal (1975, 1977, 1978a,b, 1979, 1980) recorded trypanosomes from different fish hosts and compiled a monograph on hematozoa of commercially important Indian fishes (Mandal, 1984). Rao (1975) proposed that the quality and quantity of food intake of fishes resulted in fluctuation of *T. saccobranchi* in *Saccobranchus fossilis*. Nandi et al. (1983) listed 20 *Trypanosome* sp. in a catalogue of protozoan parasites from Indian fishes *T. notopterae* was described from *Notopterus notopterus* collected from Ranaghat, West Bengal, India (Kundu and Chatterjee, 2005).

Hill stream and fresh water fish trypanosomes from India have been reported by Joshi (1976a, 1978, 1979a, 1983, 1985) who described *T. maguri* (host:

Clarias batrachus); *T. vittati* (host: *Mystus vittatus*); *T. seenghali* (host: *Mystus seenghala*); *T. aori* (host: *Mystus aor*); *T. mrigali* (host: *Cirrhinus mrigala*); *T. batai* (host: *Labeo bata*); *T. stigmai* (host: *Puntius stigma*) and *T. rupicola* (host: *Nemacheilus rupicola*) amongst others.

From Andhra Pradesh, India, Narasimhamurti and Saratchandra (1980) reported *T channai* (host: *Channa punctata*) and *T. qadrii* (host: *Clarias batrachus*) from the fishes of Visakhapatnam and Srikakulum. *T. godavariensis* (host: *Channa punctata*) was reported by Saratchandra and Jayaramaraju (1981). A heavy infection (about 75%) of *T. danilewskyi* and *T. saccobranchi* from the blood of *Heteropeneustes fossilis* from Hyderabad was described by Rao et al. (1984).

From Uttar Pradesh, India, Gupta and Jairajpuri (1981a,b; 1982a,b, 1985), Gupta (1986 a,b) and Gupta and Gupta (1988) described monomorphic, dimorphic and polymorphic trypanosomes from fresh water fishes of northern India. The validity of *T. batrachi* (Qadri, 1962a) was confirmed by Gupta and Jairajpuri (1983) who reported dimorphism with 24% incidence. *T. bareilliana* was recorded from the blood of *Channa punctatus* from river Ramganga, Bareilly (Gupta, Gupta, and Yadav, 1987) having 65.2% infection. Eight new species of trypanosomes and five new species of trypanoplasms were discovered by Yadav (1990). Gupta and Saraswat (1991, 1993) described trypanosomes and *Myxosoma* from some fresh water fishes. Gupta, Saraswat et al. (1998) and Gupta, Gupta, and Yadav (2000) described new varieties of trypanosomes from the blood of *Colisa fasciatus* and *Catla catla* respectively. Gupta et al. (1998) recorded some hemoparasites affecting *Puntius ticto* in India. *T. karelensis* was recorded at an infectivity of 12.5% in *Heteropneustes fossilis* (Gupta, Gupta, and Saraswat, 2000). Dimorphism was reported in *T. artii* at an average infectivity of 15.56% (Gupta et al., 2002b). *T. piscidium* (host: *Colisa fasciatus*) was monomorphic (Gupta et al., 2003) whereas two dimorphic species, *T. saulii* and *T. heteropneusti* were described from the blood of *Channa punctatus* and *Heteropneustes fossilis* respectively (Gupta et al., 2006). Gupta (2006) reviewed piscine trypanosomiasis with a focus on Indian *Trypanosoma*.

Other species recorded from India include *T. mukasai* (host: *Tilapia mossambica*) (Sinha, 1986); *T. seenghali* var. nov. *sophorae* (host: *Punctius sophore*) (Yatindra and Mathur, 1986). *T. puntii*, *T. rayi* and *T. marathwadensis* from the genus *Puntius* (Wahul, 1986a); *T. aurangabadensis*, *T. mastacembeli* and *T. murtii* (Wahul, 1987) from *Mastacembalus armatus* were described from Maharashtra, state of India.

From Iraq, *T. acanthobramae* (host: *Acanthobrama marmid*) (Warsi and Fattohy, 1976); *T. neinavana* (host: *Barbus grypus*) (Fattohy, 1978) and *T. mystaii* (host: *Mystus pelatius*) (Al-Jafery and Rahemo, 1982) were reported.

Cottrell (1977) recorded a trypanosome from the plaice, *Pleuronectes platessa* and Haldar and Mukherjee (1979) listed 14 protozoan parasites from *Ophicephalus punctatus*. Grogl et al. (1980) was the first to describe *T. magdulenae* from the fresh water teleost, *Petenia karussii*. Trypanosomes from the Atlantic Cod (*Gadus morhua*) were found to be prevalent in Newfoundland (Khan et al., 1980).

From Brazilian fishes, *T. pintoi* (Ribeiro et al., 1990) and *T. vialeri* (Lopes et al., 1991) were recorded. From Poland, trypanosomes from the carp and perch were observed by Jastrzebski (1984); Wita and Ovcharenko (1997) reported *T. carassii* from the blood of Rutilus *rutilus* and *Tinca tinca* and Wita et al. (2001) recorded the presence of *T. abramidis* from *Abramis brama*.

Reports on fish trypanosomes followed: *T. carasobarbi* (host: *Carasobarbus luteus*) (Al-Salim, 1985) and *T. humboldti* (H-*Schroederichthyes chilensis*) (Morillas et al., 1987) were added to trypanosome literature. Eiras (1988) described *Trypanosoma granulosum* in *Anguilla anguilla* from Portugal.

Jones and Woo (1993) reported 62% infection of *T. burresoni* in the American eel. Zajicek and Peckova (1995) studied the morphological and growth characteristics of seven selected enzymes and isoenzymes of alcohol dehydrogenase in trypanosome strains from perch pike, carp and tench. The development of *T. mukasai* in its biological vector, *Bactracobdelloides tricarinata* in relation to its life cycle pattern was investigated by Negm-Eldin (1997, 1998).

Trypanosomes have also been recorded from Chinese fishes. Zhao (1998) recorded flagellates of fresh water fishes from Sichuan province of China and Gu et al. (2006, 2007) redescribed *Trypanosoma ophicephali* and *Trypanosoma siniperca* from *Odontobutis obscura* and *Siniperca chuatsi* respectively.

The phylogeny of fish trypanosomes by assessing kinetoplastid sequences (Figueroa et al., 1999) as evidenced from genes (Hamilton et al., 2004) and by using ssu rRNA sequences and random amplification of polymorphic DNA (Gibson et al., 2005) has also been worked upon during the recent years. From Okavango Delta, Botswana, Smit et al. (2004) and Nico et al. (2004) reported trypanosomes from the silver cat fish and morphometric variability in trypanosomes from fish blood smears. Karlsbakk and Nylund (2006) reviewed trypanosomes from North Atlantic and Mediterranean teleosts while dealing with trypanosomes from *Gadus morhua*. Hassam et al. (2007) sampled 270 fish specimens and recorded higher prevalence of trypanosomes in *Syndontis clarias* (35.24%) as compared to that of *Clarias gariepinus* (10.78%). Gu et al. (2010) discussed the phylogenetic position of *Trypanosoma ophiocephali* from the complete small subunit ribosomal RNA gene sequence.

1.2.7. Rodent Trypanosoma

Members of the order Rodentia are one of the key mammalian group, more abundant and successfully varied in form and function than are mammals of any other order. Rodents occupy virtually every terrestrial habitat capable of supporting mammalian life (Mackenzie, 1972). They have successfully thrived in many environments across the globe.

In tropical and sub-tropical countries, at least 20 species of rodents have been recognized as pests to agricultural crops, including *Rattus norvegicus* (Norwegian rat/brown rat/sewer rat) and *Rattus Rattus* (house rat/roof rat/black rat/ship rat), two of the most successful and widespread species in the world (Fernando et al., 1985).

The excellent adaptability, small size and prolific reproduction rate of rodents have enabled them to occupy different ecological niches as well as to exploit a variety of food materials (Mafiana et al., 1997).

In addition to the tremendous economic losses to agriculture owing to their pestiferous nature, rats survive and proliferate in close association with humans in households, agricultural and commercial places (Benigno and Marges, 1978) thus making those interesting subjects for research. In some parts of the world, rodents are even consumed as a source of protein (Nmorsi and Egwunyenga, 1999). All these factors make rodents an interesting model for research purposes.

Biological invasions are global threats to biodiversity and parasites might play a role in determining invasion outcomes. Transmission of parasites from invading to native species can occur, aiding the invasion process, whilst the 'release' of invaders from parasites can also facilitate invasions. Parasites might also have indirect effects on the outcomes of invasions by mediating a range of competitive and predatory interactions among native and invading species.

Being in close proximation with man and animal habitats, they can transmit diseases efficiently thus acting as reservoir hosts for many zoonotic pathogens, posing a health risk to them and causing public health nuisances (Cochrane et al., 1975).

A rodent disease outbreak is accompanied with high morbidity and mortality. These animals are known to harbor pathogens and parasites (Ugbomoiko and Obimiwe, 1991) which primarily affect the health and well- being of the animals. They can harbor many different protozoan and helminthic endoparasites (Davoust et al., 1997). Endoparasites of rodents play an important role in the zoonotic cycles of many diseases, some being more important than others.

In order to characterize and accommodate, morphologically similar but biologically divergent forms, members of the genus *Trypanosoma* (Protozoa;

Kinetoplastida) infecting mammals were separated into two sections by Hoare (1964), viz., the Stercoraria and the Salivaria. Members of the Stercoraria are so designated primarily because they show a pattern of discontinuous multiplication in the vertebrate host and development to the infective stage in the vectors occurs in the posterior station; infection is contaminative. Certain morphological features also characterize this section. With some exceptions (*T. lewisi* at times, *T. cruzi*) stercorarian trypanosomes are considered non-pathogenic to their hosts. No members of the Salivaria show a pattern of continuous multiplication in the vertebrate host and development to the infective stage in the vector in the anterior station with infection being inoculative (with the exception of *T. equiperdum*). The majority of the salivarian species are transmitted by tsetse flies and, with some exceptions e.g., *T. evansi, T. vivax,* are to be found only on the African continent. Some salivarian species are highly pathogenic to their vertebrate hosts (including man) causing conditions referred to as surra, nagana, African sleeping sickness and other diseases; often, there is high mortality among their vertebrate hosts.

Trypanosomes from *Cervus canadensis* (Kingston and Morton, 1973, 1975) and ruminants (Kingston, 1991) have been reported from Wyoming, USA. From Poland, trypanosomes have been reported from red deer (*Cervus elaphus*) and elk, *Alces alces* (Kingston et al., 1985), ruminants (Kingston, 1991) and the European bison, *Bison bonasus* (Kingston Drozdz et al., 1992). Kingston, Bobek et al. (1992) described *Trypanosoma* (*Megatrypanum*) *stefanskii* from *Capreolus capreolus*, Wita

and Kingston (1993) and Wita et al. (1998) reported *Trypanosoma* (*Megatrypanum*) infections in wild ruminants, Karbowiak and Sinski (1996a); Karbowiak and Wita (1998) and Karbowiak et al. (1998) reported the occurrence and morphological features of *Herpetosoma* and Wita and Kingston (1999) recorded *T. cervi* in redder.

Herpetsoma is a subgenus of the genus *Trypanosoma* and the section *Stecoraria*. It comprises about 90 or so described species and a considerable number of sub species and strains. Representatives of *Herpetosoma* are parasites of rodents of the families Muridae, *microti*dae, Zapodidae and of some other mammals, mainly of the orders Lagomorpha, Insectivora, as well as of primates (Molyneux, 1970; Hoare,1972).

Rodents are susceptible to trypanosomes as most of them are capable of infecting mice and rats and these animals serve widely as the experimental hosts. Such experimental infections are used as models of the diseases. Elton et al. (1931) discussed the health and parasitic infection of a wild mouse population. Galuzo and Novinskaya (1961) described trypanosomes of rodents and Baker et al. (1963) reported blood parasites of wild voles, *Microtus agrestis* from

England. Fay and Rausch (1969) surveyed parasitic organisms in the blood of arvicoline rodents in Alaska.

The rodent trypanosome, *Trypanosoma (Herpetsoma) lewisi* was first reported by Kent in 1880 and subsequently by Laveran and Mesnil in 1901. It is a trypanosome of a sub genus of *Herpetsoma* (*Sterocoraria* Section), a parasite of rats (*Rattus Rattus* and *Rattus norvegicus)* that is transmitted by fleas: *Nosophyllus fasciatus* in temperate countries, and *Xenopsylla cheopis* in inter tropical areas. *T. lewisi* can also develop in *Ctenocephalides canis* (dog flea), *Leptopsylla segnis* (mouse flea) and *Pulex irritans* (human flea) which can act as reservoirs. As *T. lewisi* is specific to rats, it cannot grow in mice. Rats are infected principally by oral route, through contamination by flea faeces or ingestion of fleas. Some *Herpetosoma* like the mouse trypanosome *Trypanosoma musculi,* have a similar strict host specificity, but others like *Trypanosoma rangeli* can infect a wide range of hosts. All trypanosomes of the sub-genus *Herpetostoma* are non-pathogenic to their vertebrate host, it has even been reported that rats infected by *T. lewisi* and mice infected by *T. musculi* have a quicker growth rate and reach a heavier weight than those which are not infected. However, it has been suggested that infection of rats by *T. lewisi* can increase their sensitivity to other parasites (Desquesnes et al. (2002).

Studies on prevalence of *T. lewisi* in murine populations were mainly made in the middle of the 19th century by Laird (1951) in New Zealand, Callero (1952) in Panama, Eyles (1952) in USA, Kartman (1954) in Hawaii, Asiatic or African countries (Linxian et al., 1994), Ugbomoiko (1997) in Nigeria, Abdel-Aal and Abou-Eisha (1997) in Egypt and Laha et al. (1997) in India. South American records include Chile (Franjola et al., 1995) and Venezuela (Herrera and Urdaneta- Morales, 1997).

The hematozoan parasites have also been studied from Californian (Davis, 1952) and Australian (Mackerras, 1959) mammals. Cox (1970) reported parasitic protozoa of British wild mammals. Wilson et al. (1971) reported the fine structure of *T. musculi* in the blood stream phase. Hoare (1972) published a monumental volume of trypanosomes of mammals. Ewers (1973) provided a host parasite list of the protozoan and helminth parasites of new Guinea animals. Ashman and Seed (1974) observed the effects of *Trypanosoma brucei gambiense* infection on the reproductive organs of male *Microtus montanus* and Herbert and Lumsden (1976) proposed a rapid matching method for estimating the host's parasitemia. Molyneux (1976) made elaborate investigations on the biology of *Herpetosoma* comprising of isolation techniques, its cycle in the mammalian and invertebrate hosts, cultivation of the parasite and the factors affecting the epizootiology of infections. Nielsen et al. (1978) observed that *T. lewisi*-infected rats were more prone to *Salmonella typhimurium* infection.

Mohamed and Molyneux (1987) cultivated *Herpetosoma* and Mohamed et al. (1987) performed isoenzyme characterization of *Herpetosoma*. Maraghi et al. (1989) differentiated rodent trypanosomes by lectins.

Anderson (1990) could identify trypanosome infections in three genera of rodent hosts. Oral transmissions of *Herpetosoma* from small mammals (Maraghi et al., 1995) and congenital transmission of *Trypanosoma brucei* in mice (Ijagbone and Agbede, 2000) have also been reported.

From Polish rodents, contributions include the findings of Karbowiak and Wita (2001a) who recorded *T. lewisi* infection in *Rattus norvegicus* in two out of three rats examined and the same authors (Karbowiak and Wita, 2001b) detected *Herpetosoma evotomys* infection using microhematocrit centrifugation technique and in blood smears. Wita et al. (2003) reported the light and electron microscopic examination of *Herpetosoma microti* from *Microtus socialias* from Ukraine. Karbowiak and Wita (2004) recorded a new sub-species, *Herpetosoma grosi kosewiense* from the yellow necked mouse, *Apodemus flavicollis.*

Other workers include Laha et al. (1997) who observed the prevalence of *T. lewisi* infection in wild rats and attempted to infect the parasite in an unnatural host. Bajer et al. (1998) recorded blood parasites in a wild rodent community of Mazury lakes District, Poland whereas Davila et al. (1998) discussed biometrical alterations of *Trypanosoma evansi* isolate in laboratory rodents. Monroy et al. (1998) reported the ultra-structure, surface characteristics and serological interactions of *T. musculi.*

Desquesnes et al. (2002) showed that the use of ITSI sequence of ribosomal DNA provided an efficient and sensitive assay for detection and identification of *Trypanosoma lewisi* infection in rats. Linardi and Botelho (2002) reported a prevalence of 21.7% of *T. lewisi* in 429 *Rattus norvegicus* from Brazil. Ochiogu et al. (2008) examined the effects of *Trypanosoma brucei* infection on different stages of gestation in female *Rattus norvegicus.*

1.2.8. Trypanoplasma

The genus *Cryptobia* (Family Bodonidae) was initially proposed for biflagellated organisms in the reproductive systems of snails (Leidy, 1846). The genus was later expanded to include other morphologically similar flagellates infecting other invertebrates and vertebrates (e.g., ectoparasites on gills/ body surface and endoparasites in blood/ digestive tract of fishes). The genus *Trypanoplasma* was proposed by Laveran and Mesnil (1901) for a biflagellated hemoflagellate in freshwater fish. Crawley (1909) synonymized *Trypanoplasma* with *Cryptobia* because of their morphological similarities. However, some workers (Lom, 1979) retained the genus *Trypanoplasma* for hematozoic species and

Cryptobia for non-hematozoic species because they believed the differences in life cycles (aematozoic species are transmitted indirectly by leeches, while non-hematozoic species have direct transmissions) and the presence of a contractile vacuole (in ectoparasitic species) were sufficient for the separation of the two species.

Cryptobia Leidy, 1846 and *Trypanoplasma* (Laveran and Mesnil, 1901) (Kinetoplastida; Cryptobiidae) were described in literature to be parasitic in marine and fresh water fishes, salamanders, frogs, heteropods, planarians, siphenophores, leeches, mole crickets, lizards, snails and also act as free-living flagellates (Noble, 1968).

According to Burreson (1983), kinetoplastid biflagellates that inhabit the blood and require a vector for transmission and are biologically distinct should be placed in *Trypanoplasma*; those that inhabit the gut, gills or other sites and are apparently transmitted directly should be placed in *Cryptobia*.

Failure of establishment of the parasite in the intestine of the same host was held as a criterion to reserve the genus *Cryptobia* for the biflagellates of the seminal vesicles of snails and digestive tract, gills and skin of fish and thc genus *Trypanoplasma* be confined to the blood biflagellates of piscine hosts (Gupta and Gupta, 1987a).

Woo (1994) divided the *Cryptobia* into two subgenera. The hematozoic and normally digenetic species (i.e., indirect life cycle) were placed in the subgenus *Trypanoplasma*, while the subgenus *Cryptobia* includes the non-hematozoic monogenetic (i.e., one host) parasites. This proposal still indicates the very close phylogenetic relationships between the groups but shows that there are biological differences between them. He further suggested that non-hematozoic species (ectoparasites on gills/ body surface and endoparasitic in intestine) may warrant division into additional subgenera because of the distinct physiological conditions under which they live after careful experimental studies.

These biflagellates were observed from all parts of the globe by various workers during the past viz. Mavor (1915), Swezy (1919) and Strout (1965). *Cryptobia salmositica* was initially described from coho salmon (*Oncorhynchus kisutch*) in Washington, U.S.A. (Katz, 1951). Putz (1972a,b) described *C. cataractae* from the cyprinid fishes of West Virgina and also observed clinical symptoms of the disease in its natural hosts, *Onchorhynchus kisutch.* Bower and Woo (1977a) studied the mechanism of resistance of rainbow trout to *C. salmositica* in laboratory raised fish using the *in vitro* plasma incubation test. Lom (1979) suggested that the poorly described *Cryptobia* (*Trypanoplasma*) *makeevi* from *Oncorhynchus gorbuocha* and *O. keta* from the Amur River in East Asia may be *C.* (*T.*) *salmositica*. The parasite causes mortality in experimentally infected gold fish (*Carassius auratus*) and the common carp (*Cyprinus carpio*) (Dykova and Lom, 1979; Lom, 1979), respectively. Woo (1979) reported *Tpl. salmositica* experimentally in the rainbow trout, *Salmo giardneri.* Ecto-commensal function of *Cryptobia branchialis* from Czechoslovakia and lllinois was reported by Lom (1980). 100% ocurrence of *T. bullocki* was recorded in *Paralichtys dentalus* in the lower Chesapeake Bay region (Burreson and Zwerner, 1982).

The water temperature has been shown to influence *Trypanoplasma* infectivity. Burreson (1982) and Burreson and Zwerner (1982) traced the life cycle of *T. bullocki* in *Caliobdella vivida* contributing significantly on the role of host and vector biology and temperature in the distribution of *T. bullocki* infections in the lower Chesapeake Bay, Virginia. Sypek and Burreson (1983) reported *Trypanoplasma*cidal lysis in experimental infections with *T. bullocki* in *Paralichthys dentatus* maintained at 24°C and in natural infections over a range of 10°C to 22°C. Woo et al. (1983) also stressed the role of temperature on the susceptibility of trout to *Cryptobia salmositica.* Bower and Margolis (1985) observed the effect of temperature and salinity on the course of infection of *C. salmositica* in juvenile *Onchorhynchus* by intra peritoneal inoculation.

Woo and Wehnert (1983) observed that though hemoflagellates of fishes are transmitted by leeches, but under crowded conditions in hatcheries, transmission between fish may occur in the absence of leeches suggesting that the vascular species of *Cryptobia* were originally ectoparasitic on fishes which in turn descended from the free living *Procryptobia.* Rehulka (1985) investigated carp yearling and older fish for *T. borreli* invasion and their pathogenic role in the blood of the host. Woo and Wehnert (1986) recommended residual oxygen bioassay for studies in the pathogenesis of *C. salmositica.*

Trypanoplasms have also been reported from fishes of India. Krishnamurthy and Wahul (1986) and Wahul (1985, 1986a,b,c) reported trypanoplasm species from Maharashatra state India: *Trypanoplasma qadrii* from *Labeo calbasu* and *L. dyocheilus; T. krishnamurthy* and *T. cavacii* from *Mystus cavacius; T. vidyai* from *M. seenghala; T. saranae* from *Puntius sarana; T. lomi* from *P. hexastichus; T. solapurensis* from *P. jerdoni* and *T. wallagoi* from *Wallago attu* and three species from the genus *Puntius. T. ompoki* from fresh water fish *Ompok bimaculatus* (Bloch) and *Clarias dussumieri* was reported by Shanavas et al. (1989) from Kerala.

Amongst Indian workers from Uttar Pradesh, India, monomorphism in *Trypanoplasma tengari* (Gupta, Gupta, and Yadav, 1988) was reported from blood samples of *Mystus vittatus*. Gupta, Yadav and Gupta (1990) experimentally verified the pathogenecity of *T. tengari* and *T. maguri* in *Mystus vittatus* and *Clarias batrachus* respectively by evaluating the K factor and kidney and gonad histopathology. Gupta, Gupta, and Yadav (1998) reported a dimorphic species, *T. cyprinoides* from the blood of *Puntius ticto.*

Woo (1987a) reviewed the genus *Cryptobia* and cryptobiosis in fishes. Laidley et al. (1988) monitored the physiological responses of rainbow trout, *Salmo gairdneri* to experimental infection with *C. salmositica* and observed that the parasitaemia elicited physiological duress as apparent by progressive anaemia, hepatomegaly, splenomegaly and depressed plasma T3 and T4 levels.

A redescription of *T. borreli* (Laveran and Mesnil, 1901) was given by Kruse, Steinhagen and Korting et al. (1989) infecting the common carp *Cyprinus carpio* who observed high variability in cell dimensions of *T. borreli* during experimental infections in common carp. Kruse, Steinhagen and Korting (1989) and Steinhagen et al. (1989) made observations on the development and parasitaemia of *T. borreli* in *Cyprinus carpio* and monitored the effect of immuno-suppressive agents on *Cyprinus carpio* experimentally infected with *T. borreli.*

Diamant (1990) reported ectoparasitic species of *Cryptobia eilatica* from the gills of two marine teleosts in the northern Red Sea. Jones and Woo (1990) recorded *Cryptobia cataractae* from the blood of *Semotilus atromaculatus*. Forward et al. (1995) reported that all *Salvelinas fontinalis* were susceptible to *Cryptobia salmositica* infection, however, the severity of disease varied. Wiegertjes et al. (1995) suggested genetic variation in susceptibility to *Trypanoplasma borreli* infection in common carp (*Cyprinus carpio*). The pathogenecity of *T. atraria* was observed from three species of fishes: Utah chub (*Gila ataria),* redside shiner *(Richardsonius balteatus)* and speckled dace *(Rhinichthys osculus)* by Cranney and Heckmann (1996).

Bunnajirakul (1998) suggested the association of *T. borreli* with sleeping sickness in carp. Woo (1998) described *C.* (*T.*) *salmositica,* its mode of transmission and clinical signs of disease caused by the parasite. Hughes and Smith (2003) considered trypanoplasm infection to be one of the causes of losses in commercial aquaculture for all life stages of fish production. It has also been proposed that carp subjected to daily handling stress were much more susceptible to *T. borreli* infection than control fish (Saeij Verbung-van-Kemenade et al., 2003). Woo (2003) reported salmonid cryptobiosis from North America and also investigated the clinical signs and strategies against the disease (Woo, 2005). Burreson (2007) recorded three distinct trypanoplasms: *T. beckeri, T. bobolsoni* and an unnamed trypanoplasm from marine fishes. Currie and Woo (2007) observed that spawning female trout were more susceptible (higher parasitemia and mortality) than sexually mature males. Ruszczyk et al. (2008) considered papain-like cysteine proteinases expressed by trypanoplasm parasites to be an important target for the development of new drugs as these were responsible for causing pathogenicity in the hosts. Early reports in literature on the division of trypanosomes of fishes include those of *danilewskyi* (1885) who reported unequal binary division, the young forms appearing like monads. Chalachnikov (1888) reported multiplication by longitudinal fission in the blood of *Cyprinus carpio* and *Esox lucius in vitro* which he regarded as multiplication forms of trypanosomes. Equal or sub-equal binary fission was reported by Laveran and Mesnil (1907) and Baker (1960) observed small forms of *T. mukasai* which he presumed to be dividing.

Longitudinal binary fission in the culture form of *T. striati* was reported by Qadri (1955). Occasional records of binucleated trypanosomes were observed in *T. batrachi* (Qadri, 1962a); *T maguri* (Tandon and Joshi, 1973) and *T. attii* (Gupta and Jairajpuri, 1982b) from the peripheral blood of fishes.

Lewis and Ball (1981) reported the divisional pattern in *Trypanosoma cobitis* in the leech vector and in culture which divided by binary fission, kinetoplast division occurring prior to nuclear division. Woo (1981) recorded transverse cytokinensis in *T. danilewskyi.* Flagellar division was reported to occur prior to cytokinesis by Reilly and Woo (1982). Longitudinal as well as transverse kinetoplast division has been reported in *Colisa fasciata* (Gupta, Gupta, and Sharma, 1987). Burreson and Karlsbakk (2007) observed unequal cell division in *T. pacifica* parasitizing *Parophrys vetulus* from Oregon coastal waters.

Division in trypanoplasms has not been reported frequently. *Cryptobia salmositica* divided by unequal longitudinal binary fission (Woo, 1978); *C. cataractae* by equal binary fission producing elongate, slender daughter cells (Jones and Woo, 1990). Peckova and Lom (1990) cultivated nine strains of trypanoplasms and observed the transformational changes in them. Kruse et al. (1991) described the division process of *Trypanoplasma borreli* in infected common carp (*Cyprinus carpio*). Woo (2003) reported that *Cryptobia salmositica* divides rapidly by binary fission in salmons and is responsible for causing disease, the severity of which is directly related to parasitaemia.

1.2.9. Babesia

Most CBC's in the industrialized world are done by machines which are not able to "see" *Babesia* and catch the subtle parasite forms. Species of *Babesia* are intra- erythrocytic protozoal parasites of many species of mammals in all parts of the world where the tick-vectors are found. The *Babesia* organism was named after Viktor Babes, who first recognized it in blood cells of cattle in the late 19th century. Babesiosis was also the first vector-borne disease to be described in literature, as the etiologic agent of Texas Cattle Fever. Babesiosis or piroplasmosis causes considerable damage to the livestock and other animals. Human babesiosis have been documented in many countries. Consequently, babesiosis is a zoonotic disease of medical, veterinary and economic importance. Human cases are often caused by *Babesia microti* of rodents and they become infected when they accidentally intrude into a natural disease cycle (Morsy et al., 1994).

Babesia and its close relatives are members of a group of organisms called piroplasms, a name which comes from their pear-shaped (piroplasms) outlines as seen in the infected erythrocytes (piroplasmosis). Long associated with blood diseases of cattle and other mammals, members of the genus *Babesia* have been recognized since 1950's as infectious agents in humans. Species of this protozoan blood parasite that have routinely been isolated from mice (*B. microti*) or cattle (*B. divergens*) have also been isolated from humans. The parasite represents a threat to the safety of the blood supply in that blood from asymptomatic humans can transmit *Babesia* to blood recipients.

In recent years, there has been a developing interest in this protozoan parasite with recognition of its role as a zoonotic agent of human disease, i.e., a disease communicable from animals to humans. Babesiosis, as the disease is known, is found in a wide variety of mammals but is perhaps most prevalent in rodents, carnivores and cattle.

Cases of human piroplasmosis caused by *B. microti* were recorded since 1969 in USA (Kreier, 1977; Anderson et al., 1991). In Europe, *B microti* piroplasmosis was found in Bulgaria, Yugoslavia, Czeschoslovakia (Sebek et al., 1977) and Germany (Krampitz et al., 1986; Brasseur and Gorenflot, 1992).

B. microti was recorded in *Clethrionomys glareolus* and *Apodemus sylvaticus* in England (Turner and Cox, 1985; Turner, 1986), *Microtus agrestis* in Austria (Sebek et al., 1980) and Germany (Walter and Liebisch, 1980) and in all three species in Great Britain (Healing, 1981). Morsy et al. (1994) examined 310 rodents from Saudi Arabia and detected natural *Babesia* infection in 30.65% rodents.

However, in Poland, study on the animal reservoir of *B. microti* was done in Mazury (Karbowiak and Sinski, 1994). One case of human piroplasmosis was recorded in Poland (Humniczewska and Kuzna-Grygiel, 1995). Sinski and Karbowiak (1995) discussed the role of Ixodidae in the epidemiology of human piroplasmosis with special emphasis on *Babesia*. Karbowiak and Sinski (1996b) reported the first case of *Babesia microti* infection in *Clethrionomys glareolus*. Karbowiak et al. (1999) examined seven species of rodents for the presence of piroplasms and showed the competence of *Microtus agrestis* and *M. oeconomus* as zoonotic reservoir of *B. microti* in south and north Poland. Karbowiak (2004) observed the prevalence of *Babesia* infection in *M. arvalis* (9–33%); *M. agrestis* (50%) and *M. oeconomus* (7.7–50.0%). Karbowiak,

Wita and Czaplinska (2004) investigated blood parasites in the social vole (*M. socialis*) and recorded *B. microti* and *Trypanosoma microti* at a prevalence of 50% each. Karbowiak, Slemenda et al. (2004) observed a rise in the incidence of canine babesiosis and recommended the combined use of morphological and molecular diagnosis for monitoring the infection. The authors further reported the occurrence of *B. microti* in the root vole, *Microtus oeconomus* (Karbowiak et al., 2005).

Shih et al. (1997) detected asymptomatic *Babesia* infection in humans from Taiwan which was serologically related to *B. microti* that infects rodents recording a high prevalence (83%) of babesial infection in the latter trapped from areas around the neighbourhood. Schuster (2002) cultured *Babesia* likc blood parasites and discussed the culture- related factors, Yamasaki et al. (2003) examined four enzymes in *B. gibsoni* by polyacrylamide gel electrophoresis (PAGE). Dantrakool et al. (2004) identified a new type of *Babesia* species in wild rats (*Bandicota indica*) in Thialand resembling *B. microti* but were phylogenetically placed close to *B. canis*. Schaller (2008) provided an illustrated laboratory guide for the diagnosis of *Babesia* in human blood for the first time.

1.2.10. Environmental Parasitology

1.2.10.1. Incidence

Many workers have surveyed host species and mentioned about the incidence and parasitic load of the parasites. Surveys have been conducted from different parts of world to identify the hematozoan parasites. Baker (1960) and Becker (1970) reported parasites from East Africa and North America respectively. Molnar et al. (1974) discovered 84 species of parasites from Ontario, Canada and Lom and Laird (1976) surveyed parasitic protozoa from marine and euryhaline fish of Newfoundland and New Brunswick reporting microsporidian species. Letch and Ball (1979) sampled fishes from river Lee for 33 months and recorded the incidence of parasites. Grogl et al. (1980) surveyed fishes from Colombia. Khan and Newman (1982) made reports from fishes of north Atlantic waters.

Masoumian and Pazooki (1999) examined myxosporean parasites of freshwater fishes from Mesopotamian part of Iran. Puentes-Zambrano et al. (2003) studied the parasitic fauna of 70 grey snapper juveniles from Venezuela. Pazooki and Masoumian (2004) investigated blood parasites of 60 acipensarid fishes from Caspian Sea (Iran). Svobodova and Kolarova (2004) reviewed the parasites, diseases and poisonings of *Tinca tinca*. Baska and Gonda (2006) surveyed common protozoal diseases of pet fish.

Indian fishes have attracted attention of Haldar and Mukherjee (1979) who listed the protozoan parasites from *Ophiocephalus punctatus.* Joshi (1979a) examined 2150 fresh water teleosts recording 3.48% prevalence. Nandi et al. (1983) listed twenty species of trypanosomes from Indian fishes.

1.2.10.2. Parasite Mix

The parasite mix, interaction amongst concurrent species and their survival is a field of parasitology which has been deprived of due attention. Darwin (1859) stressed the importance of competition for food and other necessities of life for co-existing species and subsequently, the phenomenon was studied in some other groups of animals (Crombie, 1946; Pontin, 1961; Dhandt, 1977; Hogstedt, 1980). Pontin (1982) made an attempt to consolidate the earlier reports. Khan and Lacey (1986) recorded concurrent infection of *Lernaeocera branchialis* and *T. murmanensis* and Gupta and Gupta (1990c) reported *T. vittati* and *Trypanoplasma tengari* as the competing species in *Mystus vitattus.* Jansen and Burreson (1990) collected 38 species of parasites from 341 *Paralichthys dentatus* in the Chesapeake Bay which included 8 species of protozoans, 1 monogenean, 11 digeneans, 2 acanthocephalans, 1 copepod, 1 branchiuran, 1 leech, 10 cestodes and 4 nematodes. Jalali and Barzegar (2006) isolated 24 ecto and endoparasites species in different organs of 8 native and exotic fish species from Iran including five protozoan species. Sobecka and Sominsha (2007) conducted parasitological studies on bream and perch and reported the occurrence of 21 and 13 species of parasites respectively.

1.2.10.3. Ecology

An ecological perspective of disease outbreaks and their causative organisms are the essential aspects of fish culture and are the basis for rational management.

Fish, however, themselves being poikilothermic, exhibit marked behavioral and ecophysiological changes within the body in fluctuating environmental conditions. Thus, the intrinsic factors in fish environment may be identified as components of the micro-environment for the endoparasites of fish and extrinsic factors including various abiotic and biotic factors may be segregated as components of macro-environment influencing fish as well as parasite biology under different bio-ecological situations.

Parasitization index values depicting correlation of infection with fish weight were effectively used in ichthyo-parasitology by Kennedy (1970) who assessed environmental modeling in parasitology in British freshwater fishes.

Computation of k values was expressed as a measure of over dispersion in ecological interactions between species of parasites by Pennycuick (1971a,b,c,d

) after parasite frequency exhibited a poor agreement on Poisson series (Kennedy, 1968). Circular distribution (Zar, 1974) was introduced in ichthyo-parasitology for the first time by Davis and Huffman (1978). Halvorsen (1976) gave the basic concepts of negative interactions amongst parasites stressing their ecological implications. The importance of r- and k- selection in parasitism was further elaborated by Esch et al. (1977). Walter and Liebisch (1980) studied the ecology of some blood protozoa of wild small mammals in North Germany. Similar situations in ichthyo-parasitology in India have been attempted only by a few authors (Malhotra, 1981, 1985).

Concurrent infectivity of parasites have been reported. Freeman et al. (1973) recorded the effects of *Trypanosoma lewisi* on concurrent infections with *Hymenolepsis diminuta.* Hoffmeyer (1973) reported interaction and habitat selection of parasites in *Apodemus flavicollis* and *A. sylvaticus*. The interactions of *Trypanosoma lewisi* and *Toxoplasma gondii* under concurrent conditions were studied by Guerrero et al. (1997). Arrea et al. (1998) described an apparent *Trypanosoma lewisi*-induced immunosuppression which enhanced *Toxoplasma gondii* multiplication using tachyzoites to infect the animals.

Turner (1986) described the seasonal and age distributions of *Babesia, Hepatozoon, Trypanosoma* and *Grahamella* species in *Clethrionomys glareolus* and *Apodemus sylvaticus* populations. Asin and Catala (1995) quantified *T. cruzi* development within *Triatoma infestans* maintained at two different temperatures thereby observing the role of temperature on *Trypanosoma* development.

Ryce et al. (2004, 2005) determined the ages at which juvenile rainbow trout *Oncorhynchus mykiss* became resistant to the effects of whirling disease following exposure to a range of parasite doses and also examined the effects of both fish age and size on the development of resistance to the disease in Erwin strain rainbow trout.

1.2.11. Host Specificity

Experimental proof on host specificity amongst fish parasites includes the findings of Baker (1960), Becker (1967) and Khan (1977a) who reported fish trypanosomes to be euryhostpitalic. Contrarily, various trypanosomes species have been recorded from the same host. Khan (1972) demonstrated host specificity of trypanosomes from *Gadus morhua* and Mukherjee and Haldar (1982) from *Nandus nandus*.

Parasites may show affinity not only for the final host but also for the vector (Mandal, 1979). Khan (1991) studied the distribution, abundance, host selection and prevalence of trypanosomes in the marine leech, *Johannssonia arctica* in north western Atlantic. Jones and Woo (1992) observed that *T. catostomi* could be infected only by the leech *Actinobdella inequiannulata* although two other species of leeches also occurred in the same environment (*Hemilepsis marginata* and *Desserobdella phalera*). Noyes et al. (2002) observed that bank voles carry only one whereas wood mice carry at least two polyphylectic parasites. Baska and Gonda (2006) observed that fish parasitic coccidians are highly host specific thus they cause only limited damages in aquarium fish.

1.2.12. Hematology

Study of hematological parameters of fishes is gaining recognition in recent years as a useful tool for regular monitoring of health of stocked fish. The importance of hematological techniques for the diagnosis of fish disease has been widely accepted and act as the easiest tools for diagnostic physiology. It cannot be over emphasized that the application of these techniques will prove as important diagnostic tools for fishery biologists in assessing fish health.

Hematological values for higher mammals are available (Schalm, 1967; Wintrobe, 1981) but very little is known regarding piscine hematology. Normal hematological indices of fishes have been investigated (Joshi, 1980; Sharma and Joshi, 1984). The hypotonic fragility of fish erythrocytes (Khan et al., 1972), the impact of starvation (Joshi, 1979b), partial hepatectomy (Joshi et al., 1982), sex related variations (Joshi and Sharma, 1982; Kapila et al., 2000), asphyxiation, ambient water salinity and body size (Sharma and Joshi, 1985, 1986a,b) and pollutants (Sawhney and Johal, 2000) on the hematological parameters have been evaluated in different fish species. However, reports on hematological indices due to trypanosome infectivity are comparatively few. Attempts have been made to observe such changes in higher mammals (Nayler, 1971; Facer, 1976; Ikede et al., 1977; Raina et al., 1986) by various workers.

Flagellate infections in fish induce physiological and pathological changes. A perusal of literature reports changes in hematological parameters of infected fish. *Trypanosoma lotai* was responsible for the alteration in blood values of *Lota lota* (Smirnova, 1970), *T. maguri* (*H: Clarias batrachus*) and *T. vittati (H: Mystus vittatus)* affected the hemato-physiology of freshwater fishes (Tandon and Joshi, 1973). Khan (1977b, 1985) held similar views and added that hematocrit and Hb levels decreased and ESR increased following infection and anemia was a frequent outcome. Seasonal variation in the hematological values *Heteropneustes fossilis* and *M. vittatus* was reported by Joshi and Tandon (1977) and the physiology of host-parasite relationship parasitized by trypanosomes was studied by Tandon and Chandra (1977a,b).

The erythrocytes of infected fishes may show a great distortion in their structure and values (Joshi, 1979c, 1985, 1989a,b). Joshi and Tandon (1980) further observed hematological changes in ten species of freshwater teleosts.

Trypanosoma murmanensis altered the blood values of adult *Myxocephalus octadecemspinosus* (Khan et al., 1980). The changes in hematological values of *H. fossilis* due to *T. maguri* was observed by Joshi and Dabral (1981). A report on the hematological studies on a hill-stream fish, *Schizothorax richardsoni* was given by Bhatt and Singh (1981). Raizada and Singh (1982) observed hematological values of *Cirrhina mrigala.*

Rao et al. (1984) made observations on Hb concentration of parasitized hosts. The patho-physiological effect of trypanosome infection on the plasma electrolytes was studied by Dubey and Pandey (1985). TEC, TLC, DLC and Hb% of trypanosome-infected *Garra gotyla* (Sharma and Joshi, 1991) have been found to show marked changes as compared to uninfected fishes.

Trypanosomes affecting the blood values of *Punctius ticto*, *Channa gachua, Nemacheilus rupicola* and *Garra gotyla* were studied by Joshi (1985). *T. murmanensis* from four species of marine fishes were responsible for the alteration in blood values (Khan, 1985). The parasitemia and other factors are also responsible for changes in blood values of fishes (Joshi, 1986).

Khan (1987) observed that the marine fish, winter flounder infected with trypanosomes and exposed by petroleum hydrocarbons can not only cause mortality but can also jeopardize health and reproduction in surviving fish.

Ford (1990) observed trypanosomes lying beside erythrocytes in lethargic eels and presumed that the trypanosomes were deriving nourishment from them. Murad et al. (1990) reported that anemia was induced at low temperatures (5°C) to reduce mortality and erythropoiesis can be readily initiated by gradual warming. Sahlab et al. (1990) reported that trypanosome-infected Nile catfish, *Clarias lazera* had decreased number of erythrocyte counts, neutrophils and Hb content, while a marked increase in monocytes and polymorphocytes percentage were observed in the diseased fish. Hematocrit, hemoglobin and total plasma proteins reduced significantly in trypanosome-infected fishes (Steinhagen et al., 1990) where *Cyprinus carpio* infected experimentally with cloned *Trypanoplasma borreli* showed reduction in hematocrit and total erythrocyte counts. These counts were lowest at 29 days post inoculation (p.i.) when the infection was at its peak, whereas the total leukocyte number markedly increased and peaked at day 44 p.i.

Anemia as an outcome of trypanosome infection has been confirmed by Islam and Woo (1991a) in *Carassius auratus*. The latter workers (Islam and Woo, 1991b) also reported the occurrence of anorexia in the same fish.

Erondu et al. (1993) studied erythrocyte and leukocyte numbers, hematocrit, hemoglobin, blood group and genotype of four catfish groups and found that erythrocyte and leukocyte counts showed intra-species variability whereas fish size was inversely correlated to hemoglobin concentration and hematocrit values. Hamers (1995) reported that the lymphoid cells mainly depend on the parasitemia in fish. Rauthan et al. (1995) studied some hematological changes in a hill stream fish *Barilius bendelisis* infected with trypanosomes.

Saha et al. (1997) observed that in *H. fossilis,* lowest infection of *T. mukundi* occurred in pre-monsoon period (20%) while it increased in monsoon season (81.8%) and it was nil during December and January. Lymphocytes and eosinophils increased whereas hemoglobin values decreased significantly in heavily infected fishes. Akmirza and Tepecik (2006) assessed and compared hematological parameters of Rudd, *Scardinius erythrophthalmus* parasitized and unparasitized by *Cryptobia tincae.*

Hemoglobin is a respiratory pigment, made up of 95% globin (a colorless protein) and 5% hematin (a red compound of iron). In fishes, amphibians, reptiles and birds, their percentage in erythrocytes vary from group to group.

More active the fish, the higher will be the Hb values as was recorded for *Colisa* (Engel and Davis, 1964). Estimation of Hb% values is very limited specially pertaining to the changes in response to flagellate infection.

The low Hb content of infected *Colisa* blood is indicative of anemia which also corresponds to the lethargic and motionless *Colisa* which were parasitized by trypanosomes. Fishes may become anemic and suffer from asphyxiation and suffocation and the metabolic activities decrease during trypanosomiasis which might be responsible for low Hb concentrations.

Smirnova (1970) reported pathological alteration in certain blood values of *Lota lota* infected with trypanosomes. Joshi (1973), Tandon and Joshi (1973) and Dabral (1980) opined that Hb concentration and RBC decreased due to trypanosome infection in different species of fresh water fishes.

Joshi (1982) found that the Hb values decreased when the fish was motionless and compared the blood glucose and liver glycogen contents of uninfected and trypanosome-infected *Clarias batrachus* subsequent to intramuscular glucose injection. Alteration in hematological values due to concomitant parasite infestation of helminthic cyst was observed in *Nemacheilus rupicola* (Joshi et al., 1983) where the fish became erythropenic and anemic. Sharma and Joshi (1991) observed a decline in Hb% in a hillstream fish, *Garra gotyla.*

1.2.13. Biochemistry

Serum biochemistry of fishes has been investigated for various purposes as also changes caused therein due to divergent factors. Factors such as age and growth, diet and disease have been reported to affect the biochemical composition of fish blood (Elliot et al., 1966).

Parasites are known to influence the blood constituents of the host similar to other physiological parameters. Studies on chemical changes in blood due to hemoflagellate infection provide an important index of the pathological condition of fish. Linton (1930) studied the blood chemistry in acute trypanosome infection.

The normal blood glucose level of teleosts has been studied by different workers with varying results. Studies on the blood glucose level of fish infected with trypanosomes have been comparatively few.

Pathogenic trypanosomes consume sugar in the blood stream to an extent so as to cause exhaustion of the carbohydrate reserves of the host. It may cause strain on the liver leading to its functional breakdown and even death of the animal (von Brand, 1973). Tandon and Joshi (1974) observed changes in blood glucose in *Clarias batrachus, Heteropneustes. fossilis, Channa punctatus* and *Mystus seenghala* parasitized by trypanosomes reporting maximum loss in *C. punctatus* (47.0%). Glucose is by far the most important exogenous substrate used by blood- stream forms of trypanosomes *in vivo* (Gutteridge and Coombs, 1977). Joshi (1981, 1982a, 1985) emphasized the physiological needs of the parasite indicating a highly significant ($P<0.001$) fall in glucose content of parasitized fish.

Gupta and Jairajpuri (1983) observed changes in glucose content of *C. batrachus* parasitized by *T. batrachi.* Gupta, Gupta, and Saraswat (2000) reported hypoglycemia in *H. fossilis* under concurrent infectivity of *T. karelensis* and *Myxosoma fossilii.* Gupta (2003) reported that hypoglycemia due to *Trypanoplasma* infectivity was more severe as compared to *Trypanosoma.* Laidley et al. (1988) observed changes in glucose content of rainbow trout in *Cryptobia*-infected fish which appeared to be under physiological duress.

Protein contents appear to be related with increase in fish size. The increase is correspondingly found to be associated with the changing dietary habit of the fish. The protein content of fish blood may change under varying conditions such as season, stage of maturity, spawning, type of food consumed and feeding habits. Plasma proteins also vary widely from species to species and are found to vary with respect to age, diet and water temperature.

High values of protein are found in fishes containing ripe gonads and during intensive feeding periods. Brooke (1964) found very little variation in protein content of *Salvelinus fontinalis* in various months. Elliot et al. (1966) recorded low values of protein from April to June in Chinook salmon. Calcium and protein contents of freshwater fishes were estimated by Siddiqui and Naseem (1971). A reduction in biochemical parameters like blood glucose and protein appears to be too obvious which may occur under infected conditions. Protein metabolism is considered to be one of the most sensitive physiological system responding to environmental stresses (Gordon, 1972; Thompson, 1977). A decrease in the amount of protein in fish tissues with advancing days of starvation support the assumption that caloric demand of the body to maintain basal metabolic rate and body temperature causes tissue protein to be lost (Borah and Yadav, 1996).

According to Joshi (1989a), studies related to the diagnostic aspects of fish diseases and disease-related physiobiochemical alterations in fish tissues, including blood are yet to be made. Parasites are known to influence, alter and interfere with the normal physiology of the host in order to adjust to their environment. Trypanosome infection altered the serum phophatase level in white mice (Chatterji, 1960). Other contributors include Tandon and Joshi (1973, 1974), Joshi (1981), Woo (1979), Joshi and Dabral (1981) and Rai et al. (1984). Moreover, trypanosomes have their own enzyme activity (Harvey, 1949; Grant and Sargent, 1960; Seed et al., 1967). Gupta et al. (2001) recorded a change of 16.04% in the protein content of *Clarias batrachus* infected with *T. qadrii.*

1.2.14. Trypanosome Culture

Progress in the techniques of cultivation of parasites envisages the development of an *in vitro* system allowing a continuous multiplication and retaining their infectivity to the host. The establishment of proper *in vitro* systems have been a long-standing requirement for metabolic and immunological studies. Large numbers of trypanosomes have been isolated from host blood without altering their properties. Literature on the cultivation of trypanosomes from higher animals is significant (Hirumi et al., 1977; Brun et al., 1981; Simpson et al., 1985; Czichos et al., 1986). Mehlhorn (1988) defined the optimum conditions for growth of the parasites.

Efforts to culture fish trypanosomes date back to the beginning of the century when Petrie (1905) isolated *T. danilewskyi* but failed to establish them in sub-culture. Thomson (1908) observed *T. carassii* from gold fish on 6th and 7th days p.i. (post inoculation) in cultures. Tanabe (1925) cultured the parasites in NNN agar which showed full growth in about two weeks. Qadri (1962c) cultivated *T. striati* in Shortt's modification of NNN. Berens et al. (1976) devised a simple monophasic medium for the cultivation of hemoflagellates. Taylor and Baker (1978) described chemically defined media for parasite culture. Trypomastigote rich suspensions of *T. danilewskyi* were obtained by Rouskova and Kulda (1983) in a complex medium L4NHS. Fish et al. (1987) observed cell adhesion of *Trypanosoma* in *in vitro* studies and Peckova and Lom (1990) observed divisional stages of flagellates *in vitro.* The culture characteristics of *T. catostomi* and *T. phaleri* were compared (Jones and Woo, 1991) and the role of temperature in the *in vitro* multiplication of *T. danilewskyi* has also been worked upon (Islam and Woo, 1992).

Davis et al. (1995) cultured *T. granulosum* of the eel in semi-defined and defined media. Jones et al. (1995) observed that cultures of *Trypanoplasma borreli* occasionally contained small numbers of peripheral blood leukocytes (PBL) from the fish. Bienek and Belosevic (1997) proposed the requirement of fish serum for growth of *Trypanosoma carassii*, presumably as bloodstream forms. Overath et al. (1998) have shown that adaptation of methods in use for *T. brucei* allows purification of *T. carassii* from fish blood and its long-term cultivation as infective trypomastigotes *in vitro*. Lubega et al. (2002) reported that anti-nTbTub serum strongly inhibited trypanosome proliferation in culture, and immunoagglutinated trypanosomes even after heat inactivation of complement. The effects of insulin on the *in vitro* growth characteristics of *Trypanosoma granulosum* were investigated by Mackintosh et al. (2003). The authors observed that insulin growth stimulus had a rapid onset with little or no lag phase but was relatively short lived, growth peaking typically on day 3 or 4 of culture. This contrasted with the medium containing 10% fetal calf serum, normally used for *in vitro* culture of this trypanosome, which after a 3-day lag stimulated sustained log-phase growth. Insulin demonstrated a biphasic dose response with maximum growth typically occurring at 3–10 microg/ml and a much-reduced effect at higher concentrations. Vassella et al. (2009) reported that when *T. brucei* was cultured *in vitro*, the null mutant and wild type trypanosomes (tagged with cyan fluorescent protein) maintained a near-constant equilibrium.

1.2.15. Immunology

A remarkable progress has been made in the area of fish immunology due to the fact that fishes are the most ancient group in which an immune system characterized by the presence of immunoglobulin has been demonstrated. It acts as a unique experimental model for studying cellular recognition by Ig receptors and the subsequent immune response. Although much is known about the taxonomy and biology of fish parasites, relatively little is known about protective immunity in fish and the effect of parasites on the piscine immune system. The concept of vaccinating fish has been realized (Ellis, 1985) and this technique has been proposed as a viable control strategy. The need for a coordinated global research program on fish diseases has been duly stressed (Woo, 1987b).

The protective capability of lymphocytes have rendered them to become the focus of interest of immunological investigations. Contributions on the seaspects are gradually increasing (Parker and Lukes, 1971; Kang and Park, 1975; Anderson and Hennessen, 1981; Van Muiswinkel et al., 1985) and the sporadic investigations on lymphocyte culture are also rising (Hadden, 1978; Blaxhall, 1985). Grace and Manning (1980) documented the lymphoid organs of fish from the ontogenic viewpoint. The impact of *Yersinia ruckeri* O-antigen on the immune response and antigen localization was reported by Lamers and Pilarczyk (1982). Fijan (1984) discussed the prospects and constraints of vaccination of fish in European pond culture. Blaxhall (1985) cultured lymphocytes by using four separation techniques for isolation. Sudarto et al. (1990) demonstrated *Trypanosoma* immuno-histochemically in tissues of experimentally infected rats and naturally infected water buffaloes. Woo (1999) used monoclonal antibody probes in salmonid cryptobiosis. Mehta and Woo (2002) regarded humoral and cell-mediated immunity to be important components of protection against *C. salmositica* in both vaccinated and recovered fish. *Cyprinus carpio* macrophages depleted by intraperitoneal injection of clodronate-liposomes for the *in vivo* study of the effect of macrophage depletion on the resistance of carp to infection with blood parasite flagellates (Saeij, Grroeneveld et al., 2003). Scharsack et al. (2003a,b) studied the responses of *Cyprinus carpio* head-kidney derived neutrophils to *Trypanoplasma borreli*, the consequences of these responses for parasite survival and also examined the carp leukocytes obtained during *T. borreli* infection for their capacity to generate nitric oxide (NO). Scharsack et al. (2004) studied the specific immune response of carp to *T. borreli* infection and observed that the sera of resistant carp contained antibodies which agglutinated the flagellates *in vitro*. Wiegertjes and Joerink (2004) discussed the role of inducible nitric oxide synthase (iNOS) and arginase activity as potential markers of type I or type II immune response in fish in relation to macrophage polarization. Recently, Woo (2010) adopted the principle of adaptive immunity for protecting *Oncorhynchus* spp. from cryptobiosis. He was able to protect the fish against cryptobiosis by using live attenuated *Cryptobia* vaccine.

1.2.16. Drug Action

Piscine infection and infestations have been duly classified and investigated but the therapeutic and preventive measures dealing with them are still demanding. Studies dealing with fishes as research models should be able to evaluate the potential impact of "disease" and define the state of health and subsequently propose suitable therapeutic measures for the diseased conditions. Pharmacological data on fish disease is deprived of due attention and calls for greater research inputs.

Parasitic infections maybe chronic in nature, characterized by a comparatively late onset of clinical symptoms. Quick diagnosis coupled with more effective chemotherapy is recommended for control programs. The action of any anti-parasitic drug is crucially based on the efficiency by which it is taken up by the parasite. Drug selectivity can depend on biochemical differences between the host and the parasite.

Berenil (diminazene aceturate), an aromatic compound was discovered in 1944 and was developed from a drug called "Congasin" and other aminoquineldines that were found active against trypanosomes and babesias (Fussanger and Bauer, 1962). Chemically, diminazine is an aromatic diamidine. The drug has been used for the treatment of babesiosis in mammals but its effect on fish blood parasites like *Trypanosoma* and *Trypanoplasma* are less documented in literature.

All over the world, diminazene is sold as Berenil (Intervet), Dimisol (Virbac), Babeze (Milborrow. Bayer AH), Berenil RTU (Intervet), Crede Bab-Minazene (Experto vet), Dizene Cattle (Virbac) and Veriben (Sanofi AH).

Therapeutic success has been obtained during the past years for the treatment of protozoal disease in domestic animals using a novel drug, Samorin (Isometamidium chloride) developed in the research laboratories of Fabwerke Hoechst. The life of mice infected with *T. equiperdum* could be prolonged by streptothricin obtained from *Actinomyces lavendula* (Waksman et al., 1941) and Tobie (1954) described the effect of the antibiotic stylomycin on the same parasite. Hewitt et al. (1956) reported the effect of Nucleocidin, an antibiotic injected intraperitoneally or orally to the mice infected with *T. equiperdum* and Lynch et al. (1956) noticed the action of complexes of antimony trichloride with tetracycline antibiotics against the parasite.

Jaffe (1967) demonstrated that antibiotics like Philomycin, Profiromycin, Mctomycin C and Actinomycin D. could significantly reduce parasitaemia in mice. The tissue retention properties of Berenil were indicated by Williamson (1970) who also recorded its greater activity *in vivo* against *T. brucei* tissue form. Delain et al. (1971) also investigated the effect of trypanocidal drugs on the fine structure of trypanosomes.

Ikede and Losos (1972) treated dogs with Berenil at a dose of 3.5 mg/kg body weight and observed hematological changes during trypanosomiasis caused by *T. brucei.* Trypanosomiasis and other hemoprotozoan infections could be effectively controlled by Berenil in cattle, buffalo, sheep and several other domestic and wild animals (Verma et al., 1973; Pandey and Mishra, 1978). Woodruff (1975) described trypanocidal activity of anti-tumor antibiotics and other metabolic inhibitors.

Klatt and Hajdu (1976) conducted pharmocokinetic investigations on diminazene and rolitetracycline and made a comparative analysis of both the drugs. The susceptibility of a rodent adapted stain of *T. vivax* (Lee flangstain Y58) to Berenil, Samorin, and Novidium was tested in mice and described by

Arowolo and Ikede (1977). Ferrante et al. (1982) described the intermediate products of the oxidation of polyamines by polyamine oxidase for the killing of intracellular protozoan parasites.

The disposition of Berenil in cattle was reported by Kellner et al. (1985) and in pregnant and lactating cows by Mdachi et al. (1995). Mercado et al. (1986) observed the structural changes in *T. cruzi* produced by an anti*Trypanosoma*l factor from *Pseudomonas fluorescens*. Mc-Donald and Roberson (1988) also described anti-protozoan drugs. Mamman et al. (1996) explained the pharmocokinetics of diminazene in plasma and lymph of goats. Therapeutic and chemophropylactic efficacy of various drugs like Suramin, Quinapyramine, Berenil, Samorin against clinical trypanosomiasis (surra) in buffaloes were evaluated by Joshi and Singh (2000). Plasma protein binding activity of Berenil in black Bengal goats was described by Mallick et al. (2000).

The hematological changes due to theileriasis and trypanosomiasis in cross bred cows treated with chloroquin phosphate (10 ml, 1/m), Berenil (3.5 g, 1/m), Belamyl (10 ml, 1/m), 20% Dextrose (500 ml 1/v) and externally with Deltamethrin (3 ml/litre of water) were reported by Nalinikumari et al. (2000). Bhikane et al. (2001) tested the efficacy of Berenil on the basis of survival rate improvement in clinical signs and blood parameters in cattle against babesiosis. Othman and Ahmed (2004) recorded the cytogenic effect of Berenil in the lymphocyte cultures of river buffalo at three final concentrations of 0.25, 0.50 and 1.00 mg ml^{-1}. Van den Bossche and Doran (2004) discussed the use and misuse of trypanocidal drugs.

Babigumira et al. (2009) inoculated domestic pigs with *T. brucei* to study the clinical course of the disease and response to trypanocidal treatment. Kumar et al. (2009) observed the effect of isometamidium treatment on *T. evansi*- infected rats. The recent preliminary studies conducted by Woo (2010) indicated that the drug-antibody conjugate (polyclonal antibodies-conjugated drug: PAIC) were effective when injected into infected juvenile chinook salmon. The above brief review focuses on the advances made in heme-parasitology over the years up to the recent times. Looking into the scope and academic and applied significance of the work, diligent attempts have been made by the researcher to fill-in the gaps in the area and are presented herein through this piece of work.

1.3. TECHNIQUES

1.3.1. Collection and Maintenance of Fish Samples

The work presented in the thesis was conducted in India and at Poland under the nomination as INSA Scientist. The procedure adopted was as follows:

- India: Live specimens of local fish fauna viz. *Heteropneustes fossilis, Clarias batrachus, Mystus vittatus, Wallago attu, Channa punctatus, Notopterus notopterus, Catla catla, Cyprinus carpio, Colisa fasciatus, Barbus barbus,* and *Puntius ticto* were collected from the local ponds, fish market, river Ramganga and its tributaries, Deorania, Nakatia and Sankha depending on their availability (Plate I).

  They were transported alive to the laboratory and maintained separately in aquaria under proper conditions of food and water. They were fed on commercial fish food capable of providing the fish with fat (for energy) and amino acids (building blocks of proteins) and could be speedily digested in order to prevent buildup of intestinal gas, renal failure and to avoid aquarium pollution due to excessive ammonia. The aquaria were kept well aerated throughout the experimentation.

- Poland: The fishes examined at the Institute of Ichthyobiology and Aquaculture (Plate IIA) were collected from ponds Golysz, Wyszni II and Baginiec III (Plate IIB,C) and maintained under recirculating water system at the station (Plate IIIA).

1.3.2. Taxonomical Studies

1.3.2.1. Collection of Blood

Blood was collected either from live fish from the caudal vein or from the heart after autopsy. Kidney smears were also made to detect flagellates.

1. Caudal Vein: For collection of blood from the caudal vein, large fish were anesthetized with 0.2 mg/ml of tricaine methanesulfonate (MS-222, Sigma) (Poland) or CIFECALM (2 drops in 1l of water) and whole blood was collected from the caudal vein using 5-ml heparinized syringes.

   At the time of examination, fish were taken out from the aquaria, numbered (Plate IIIB) and placed carefully for blood extraction. The caudal vein of the fish was located (Plate IVA) and blood withdrawn Plate IVB) in a heparinized syringe for examination. The tested fish were reverted back into the aquaria.

2. Heart (Cardiac Puncture): The sample of blood was taken from the heart by gentle cardial puncture or by using a thin Pasteur pipette introduced through the ventral body wall or below the pectoral fins. A heparinized pipette or one containing a droplet of heparin saline

solution was used occasionally for large fish to avoid clotting. The fish was then killed by severing the vertebral column just behind the skull. For large fish, use of a Pasteur pipette to collect blood was not convenient, and when the fish were autopsied, the blood was also taken directly from the heart after the body cavity opening. Heparin was used as an anticoagulant for trypanosomes and EDTA was used for biochemical estimations.

The collected blood was examined by hanging drop preparations and microhematocrit for the presence of parasites.

3. Kidney: The kidney of carps collected from the experimental ponds of Institute of Ichthyobiology and Aquaculture were also examined for the presence of flagellates. The autopsied fish were used for this purpose. The kidney was extracted from the fish and minced tissue of kidney suspended in saline was used for parasite detection ["kidney stroke" according to Snedecor and Cochran (1989)].

1.3.2.2. Examination of Blood

The blood was examined either by hanging drop preparations to observe live parasites or permanent stained smears were prepared for making studies on prevalence, intensity, taxonomy, hematology and drug effects.

1. Hanging Drop Preparation: This technique was used for first hand detection of trypanosomes in the blood. The technique requires a coverslip, a microscope, and a special slide having a central concavity. The blood sample suspected of containing the trypanosomes was diluted with a sterile isotonic solution. A drop of this fluid mixture was placed on a glass coverslip, which was then inverted carefully and placed over the slide so that the drop hung from the coverslip into the concavity in the slide. The motile and delicate structure and its characteristic movement was viewed through the microscope for diagnostic purposes. A positive infection with hemoflagellates could be confirmed by the vigorous movement of the flagellates in fresh blood mounts, observed under the compound microscope at 40x magnification. The live flagellates could be identified not only by their constant motion but also by the undulating movement of the body continuously. The periodic contractions and elongations of the body changing the orientation of the anterior-posterior axis of the flagellates also aided in diagnosis.
2. Concentration of Parasites
   - **i. Clotting method:** When the infections were of extremely low

intensity, the blood was allowed to clot in a centrifuge tube placed overnight in a refrigerator. The following day, it was spinned at 13,000 g for 10 mins and the serum was discarded. Flagellates were revealed in the sediment even if parasitemia was extremely low. Conventional thin films were made of the pellet which contained trypanosomes and RBC. This is a modification of the method by Strout (1962) who originally used just a small clot of blood from which the flagellates escaped into the serum, without later centrifugation. However, this method was not a sure shot method of confirmation as subpatent chronic infection could not be confirmed by this method.

ii. **Micro-centrifugation:** Low intensity infections were also confirmed by the hematocrit centrifuge technique (Woo, 1969). REMI microcentrifuge RM 12C was used for the purpose having maximum rpm of 16,000. A heparinized capillary tube was filled with blood (about 0.06–0.7 ml) and one end was sealed with plasticine. The sealed tube was centrifuged in a hematocrit centrifuge for approximately 4 min at 12,000 rpm. The lid of the centrifuge was not completely closed in order to prevent heat buildup. The microcentrifuge was operated under appropriate safety precautions. After centrifugation, the tube was carefully placed horizontally in a drop of immersion oil on a microscope slide and examined under the compound microscope at 1,000× magnification. Infections were confirmed by detecting the parasites writhing at the junction of the buffy layer (the layer of packed blood cells) and the clear plasma. The tube was cut immediately above this layer, the suspension layered on slides, and stained with any of the techniques given below.

The two step centrifugation method was also experimented: low speed spinning removed most of the blood corpuscles (and some flagellates) while the following high speed (2,000g) centrifugation of the supernatant sediments, the flagellates. Cottrell (1977) used centrifugation of heparinized blood in hematocrit tubes, which after spinning were broken at the erythrocyte-leukocyte interface and the buffy layer of white cells examined for live trypanosomes.

3. Blood Film Preparation:

i. **Microscope slides:** Clear and colorless slides which were corrosion- resistant, flat, free from distortions and ripples were used for preparing the blood films. It was also ensured that the slides were scratch-free, free of dust, lint and fingerprints and dry at the time of use.

New slides were placed in potassium dichromate cleaning solution (20 g $Cr_2K_2O_7$ in 100 ml water with 900 ml concentrated H_2SO_4) for 48 hours followed by a thorough rinse in running tap water. Slides were stored in 95% methanol and carefully wiped clean and dry before use.

Used slides were soaked in a detergent at 60°C for 15–20 mins, rinsed in hot water before drying. Other cleaning agents used were chromic acid or 70% alcohol + 30% ether. The microscope lens and slides after being used with oil immersion were also cleaned in the latter solution.

Used slides (when used) were cleaned by soaking in a detergent at 60°C for 15 to 20 mins and rinsed in hot water before drying.

Subsequent to collection/concentration of blood, the smears were made and processed. The "Push" method was conveniently used for preparing the blood films and it was the preferred method as sufficiently large area is available for microscopic examination: this area shows all cells barely touching or separated from each other (monolayer part). The blood volume used was sufficient to draw a film 2.5 to 4.0 cm in length. The blood drop was placed at a distance of 1 cm from the slide end. In cases of small volume of blood, the film was prepared by the cover glass method. Care was taken to use clean spreaders having smooth ends in order to ensure even thickness for the entire width of the blood film. After placing the blood drop on the slide, the drop was allowed to spread quickly along the spreader's edge. The spreader was now moved forward immediately at a steady rate, at a fairly fast speed and at 45° angle until all the blood was consumed into the film. The films were immediately air dried.

ii. **Fixation and Staining:** The air-dried films were immediately fixed in methanol (10 mins) in order to avoid gray/blue background effects. Slides were stained by immersion in reagent filled Coplin jars. The blood films were stained by Romanowsky dyes and the following methods were employed:

a. **Giemsa/ Leishman Staining:** The films were stained in

Giemsa/ Leishman + buffer (pH 6.5) in the ratio of 1:7 for 40 mins as per standard techniques.

b. JSB Staining: Preparation of Solutions

- **JSB I:** In 1 liter of round bottom flask, 500 ml of distilled water and 500 mg of methylene blue (medicinal) were dissolved. 1% of dilute H_2SO_4 drop wise was added and 3 cc of $K_2Cr_2O_7$ and shaken properly. A purple precipitate was observed. 3.5 gm of Na_2HPO_4 $2H_2O$ was added and mixed thoroughly till the precipitate dissolved. The solution was boiled for 1 hour at low temperature till it turned deep blue with a pinkish shade (end point). It was cooled, filtered and stored in a bottle and allowed to mature for 48 hours. The stain was filtered prior to use.
- **JSB II:** 1 gm of Eosin powder was mixed with 500 ml of distilled water. It was stored in bottle and filtered when used.
- **Buffer:** Na_2 HPO_4–0.417 gm. KH_2PO_4–0.752 gm were mixed in 2,000 ml of distilled water (pH 5.5–6.5).
- **Procedure**: The methanol fixed slide was dipped in JSB II for 5 sec, washed in buffer 10–20 times, dipped in JSB I for 30–40 sec, washed in buffer, dried and examined.

The positive slides were mounted in xylene-DPX having low viscosity so that there were no air bubbles and covered with cover glass to ensure safety of the slides. The slides were properly labeled and viewed under oil immersion in LEICA DMLB Microscope with photoautomat or OLYMPUS CX 21 TR trinocular microscope with digital photographic unit and also photographed therein.

c. **Hemacolor® (Merck) Rapid Blood Smear Staining Technique:** Hemacolor® is a staining set for rapid manual staining of blood smears. This method was used for staining smears at the Institute of Parasitology, Warsaw.

Solutions

- Fixing solution (Soln. 1)
- Two buffered staining solutions (Solns. 2 and 3)
- Buffered tablets.

The staining pattern of blood smear corresponds to classical staining patterns, cell elements and structures being stained distinctly. Morphological alterations can thus be easily recognized.

The air-dried blood smears were fixed and stained by immersion into solutions 1 to 3 for 30 seconds.

Procedure

1. Each solution was transferred into a separate cuvette with a lid or into a staining trough.
2. Blood smears were prepared and dried in air.
3. The blood smear was immersed five times to one second each time into solution 1. The solution was allowed to drip off.
4. The blood smear was immersed three times for one second each time into solution 2.
5. The blood smear was immersed six times for one second each time into solution 3. The solution was allowed to drip off.
6. The smear was rinsed with buffer solution (pH 7.2). The buffer solution was prepared by dissolving 1 buffer tablet in 1 liter of freshly prepared distilled water.

The slides were dried and examined under the microscope.

d. Wright Stain Reagents

1. Absolute methanol.
2. Staining solution (0.3 g of Wright stain powder dissolved in 100 ml absolute methanol) and left in a closed container at room temperature for 24 hours. It was filtered before use.
3. Sorensen's buffer solution at pH 6.4.

- **Stock Solution A:** 0.1 M KH_2PO_4 –(13.61 g KH_2PO_4 dissolved in 1,000 ml of distilled water).
- **Stock Solution B:** 0.1 M Na_2HPO_4 (17.8 g $Na_2HPO_4.2H_2O$, dissolved in 1,000 ml of distilled water).

Stock solution A (70 ml) and Stock solution B (30 ml) were mixed to give Sorensen's buffer solution (pH 6.47).

Method

- Step 1: The slides were fixed for 30 seconds in absolute methanol.
- Step 2: Methanol was removed by tilting the slide.
- Step 3: The slides were stained for 2 minutes on a horizontally positioned slide.

- Step 4: The aliquot of the buffer solution was added without any of the stain running off the slide. The buffer and stain was mixed gently without touching the surface of the blood film on the slide. A metallic sheen appeared on the surface of the staining solution mixture.
- Step 5: The slide was allowed to stand for 3 minutes.
- Step 6: It was rinsed with (distilled) water for 30 seconds.
- Step 7: It was dried in tilted position.
- Step 8: The positive slides were mounted in DPX.

However, this method did not give good results and was therefore not adopted for routine preparations after a few trials.

e. Giemsa Azur Eosin Methylene Blue: This stain was either used readymade from MERCK or prepared as follows:

   1 g of Giemsa Azur eosin methylene blue was dissolved in 54 ml of glycerol by heating. After cooling, 84 ml methanol GR was added and the solution was filtered.

f. May-Grunwald-Giemsa Stain Reagents

- Absolute methanol.
- **Staining Solution I:** 0.3 g May-Grunwald powder in 100 ml absolute methanol was left in closed container at room temperature for 24 hours. It was filtered before use. Staining solution II (Giemsa stain): 1 g Giemsa stain powder was dissolved in 66 ml glycerol and heated to 56°C for 90 to 120 mins. After addition of 66 ml absolute methanol and thorough mixing, the solution was left at room temperature in a closed container. It was filtered before use.
- **Buffer:** Sorensen's buffer solution. The pH was 6.8 for the May-Grunwald-Giemsa stain instead of 6.4 as in the Wright stain. For pH 6.81, Stock solution A (50 ml) and B (50 ml) (prepared as given above for Wright stain) were mixed to give Sorensen's buffer solution.

Method

- Step 1: The slides were fixed for 30 seconds in absolute methanol.
- Step 2: Methanol was removed by tilting the slide or by simply
- removing from the fixing jar.
- Step 3: Staining solution I was applied freshly diluted with an

equal part of buffer for 5 minutes on a horizontally positioned slide or in a jar.

- Step 4: The slide was transferred from jar without washing (or the stain was removed by holding slide vertically) into staining solution II that was freshly diluted with 9 parts buffer for 10 to 15 minutes.
- Step 5: The slide was transferred to a jar with buffer for 1 rinse after removing stain.
- Step 6: The slide was washed with ample water.
- Step 7: The slide was transferred to a jar containing water for 2 to 5 minutes.
- Step 8: It was dried in tilted position.
- Step 9: The slide was mounted in DPX with a cover glass.

g. Gomori Trichrome Staining: This technique demonstrated the nucleus and kinetoplast of parasitic flagellates clearly with good results as it was more easily obtained than with the use of Heidenhain's iron hematoxylin staining. This technique was employed for staining rodent blood parasites at the Institute of Parasitology, Warsaw, Poland.

1. Smears were fixed in Schaudinn's fluid for 15 min.
2. They were then washed for 2 min in 70% ethanol.
3. Thereafter washed for 3 min in distilled water.
4. The smears were stained for 5–8 min in Harris's hematoxylin.

Composition of Harris's Hematoxylin:

Ammonium potassium alum	20.0 g
Hematoxylin	1.0 g
Mercuric oxide	0.5 g
Distilled water	100.0 ml
Absolute ethyl alcohol	10.0 ml

Hematoxylin was dissolved in alcohol and the alum in the water, the two solutions were combined, rapidly boiled, mercuric oxide was added and the solution was cooled. This solution is stable for many years.

5. The slides were then differentiated in acid alcohol (1 ml conc. HCl in 99 ml 70% ethanol) until only the nuclei were stained.
6. The slides were stained for 10 min (range of 5–20 min) in the following solution: 0.6 g phosphotungstic acid was dissolved in 100

ml distilled water, then 0.6 Chromotrope 2R, 0.3 g Fast green FCF and 1 ml of glacial acetic acid was added.

7. The slides were rinsed in a solution of 0.2 ml glacial acetic acid in 100 ml of distilled water for 1 min.
8. The slides were dehydrated through an alcohol series, cleared in xylene, and mounted using a permanent mounting medium (non-viscous Canada balsam or DPX). Xylene was used as a diluent if required.

The permanent preparations made as above were mounted in clear, non-viscous DPX (xylene was used as a diluent) and sealed. They were properly labeled and stacked in slide boxes.

h. Calculation of Parasitemia (Intensity): In addition to determination of parasitemia (intensity) based on number of parasites counted per 15/20 mins rigourous scan on stained smears, it was also determined as follows:

Number of Parasites/100 RBC: The number of parasites and RBC's were counted simultaneously and at 100 RBC count, the number of parasites were noted by Manual Blood Cell Counter (Model Photocon with 8 unit display).

Parasitemia was determined in a hemocytometer according to the standard procedure of Archer (1965). Blood was diluted with 0.6% cold-blooded Ringer's solution. Pipettes containing blood were shaken thoroughly for 45 secs before filling the hemocytometer. Intensity was enumerated in the white cell count chambers according to Woo and Wehnert (1986).

i. Diagrams, Measurements and Photography: The stained slides were screened for blood parasites using a 100x oil immersion objective in a LEICA microscope. The microscopic images of the parasites were drawn to scale with the aid of a camera lucida, using a 20x hyperplane ocular and a 100x oil immersion objective. A scale of the μMeter was also projected on the drawing surface and calibrated. The midline of the parasite body was traced with a piece of thread and then compared with the divisions of a μMeter scale projected through the camera lucida onto the drawing surface. The flagella and other structures were likewise measured (Strout, 1965). The cell organelles were measured with a divider set at ½ μm and measured on the projected scale to maintain accuracy. The measurements of the parasites are expressed in μm, as mean and standard deviation (+ SD), with ranges and numbers of observations (n) in parentheses. The parasites were studied and photographed under LEICA DMLB photoautomat with computerized photo system and database MPS 28 and OLYMPUS CX21 TR trinocular microscope with digital photographic unit.

At the Institute of Parasitology, Warsaw, Poland, the parasites were measured by light microscopy by the "Analysis" Pro 2.11 software in combination with a video camera and Olympus BX50F4 microscope at a magnification of 1200X or 1500X. This method allowed obtaining results accurate to 0.01 μm. The nomenclature of morphometrics parameters of trypanosomes was that commonly used by Hoare (1972); Matthews et al. (1977) and Kingston Bobek et al. (1992) and Kingston Drozdz et al. (1992).

The body parts of the parasites have been abbreviated as follows:

AF	Anterior flagellum
CB	Parasite maximum width
CF	Total length including free flagellum
CL	Body length
FL	Flagellar length
KB	Kinetoplast breadth
KL	Kinetoplast length
KN	Distance from kinetoplast to nucleus center
NA	Distance from nucleus center to anterior end of body
NB	Nuclear breadth
NL	Nuclear length
NP	Distance from posterior end of body to nucleus
PK	Distance from posterior end of body to kinetoplast
UM	Width of undulating membrane

Indices

Flagellar Index	FF: BL
Kinetoplast Index	PN/ KN
Nuclear Index	PN/NA

Holotypes and Paratypes: All holotypes and paratypes are deposited in the Section of Parasitology, Department of Animal Science, MJP Rohilkhand University, Bareilly, U.P., India.

1.3.3. Hematological Parameters

The following hematological parameters were taken into consideration for hematological assays:

TEC	Total Erythrocyte Counts
TLC	Total Leukocyte Counts
DLC	Differentiate Leukocyte Counts
PCV	Packed Cell Volume
ESR	Erythrocyte Sedimentation Rate
Hb	Hemoglobin

The hematological parameters were estimated as follows:

1.3.3.1. Total Erythrocyte Counts (TEC)

1. **Apparatus**: Neubauer chamber, cover slip, RBC pipette, RBC diluting fluid (Hayem's fluid), microscope.

Composition of Hayem's Fluid

Mercuric chloride	0.25 g
Sodium chloride	0.50 g
Sodium sulphate	2.50 g
Distilled water	100.00 ml

2. **Procedure:** The preparations for total erythrocyte counts were made by standard clinical methods using Neubauer's Hemocytometer. Blood was collected from the caudal vein, the RBC pipette was filled up to 0.5 mark, the diluting fluid was added upto mark 101 and mixed with RBC diluting fluid by gentle rotation. The first few drops were discarded from the pipette to remove the fluid in the stem which had escaped mixing with blood. The diluted blood was charged in the chamber by holding the RBC pipette at 45° angles at the space between the chamber and coverslip. The drop of the mixture was allowed to run under the cover glass by capillary action, two minutes were given for settling of the cells. A coverslip was placed over the chamber covering both platforms and the number of erythrocytes were counted in 80 small squares comprising of 5 medium sized squares consisting of 4 corners and one central square.

3. Calculation:

The total area of one small square = 1/400

Depth of the small square = 1/10

The volume one small square = 1/4000 cu mm.

Therefore, total volume of 80 small square =

80 1

4000̄ 50 cu mm

R.B.C. Counts = Dilution × 1 × No. of cells

Volume Volume

= 200 × 50 × No. of cells

Units:/cu mm of blood.

= No. of cells × 10000

1.3.3.2. Total Leukocyte Counts (TLC)

TLC was also counted by using Neubauer's hemocytometer but Turk's fluid was used as the diluting fluid.

1. **Apparatus:** Neubauer counting chamber, cover slip, WBC pipette, WBC diluting fluid and microscope.
   - Composition of Turk's Fluid

     Glacial acetic acid: 3.0 ml

     1% aqueous gentian violet: 2.0 ml

     Distilled water: 195.0 ml
2. Procedure: Blood was filled in a clean dry pipette up to mark 0.5, diluting fluid was added in the pipette up to mark 11, the contents of the pipette were mixed gently by rotation. Care was taken to discard the first four drops from the pipette. The chamber was charged by holding the pipette at angle of 45° and an appropriate drop of the mixture was gradually poured under the coverglass (large coverslip covering the whole platform) by capillary action. The leukocytes were counted at the four large squares of the corners subsequent to 2 mins of standing.
3. Calculation

   TLC = No. of cells × Dilution × 10

   No. of cells counted in 1 sq. mm.

   Units: /cu mm of blood

1.3.3.3. Differential Leukocyte Counts (DLC)

1. Apparatus: Manual Blood Cell Counter: Model Photocon with 8-unit display. Electronic Blood Cell Counter: Yorco BCC (Digital Display) model YSI 153 with 12-unit display.
2. Preparation of blood film: Thin blood smears were prepared on clean slides, fixed and stained by any of the methods given above.
3. Counting: The cells were counted using high power microscope in a strip running along the whole length of the film by means of manual/ electronic blood cell counter. At least 100 cells were counted and the leukocytes differentiated by the morphological characteristics of the nucleus and characterized as follows based on the figures given below:

    i. **Lymphocytes**: Majority of lymphocytes in the blood are 'small' but 'large' lymphocytes are also found. It is believed that they produce antibodies and destroy the toxic products of protein metabolism. The cytoplasm of a lymphocyte is clear sky blue, scanty, with few unevenly distributed, azurophilic granules with a halo around them. The nucleus is generally round, oval, or slightly indented, and the chromatin (a network of fibers within the nucleus) is lumpy and condensed at the periphery.

    ii. **Monocytes:** The monocyte, the largest of the normal white blood cells, destroys bacteria, foreign particles, and protozoa. Its color resembles that of a lymphocyte, but its cytoplasm is a muddy gray-blue. The nucleus is lobulated, deeply indented or horseshoe-shaped, and has a relatively fine chromatin structure. Occasionally, the cytoplasm is more abundant than in the lymphocyte.

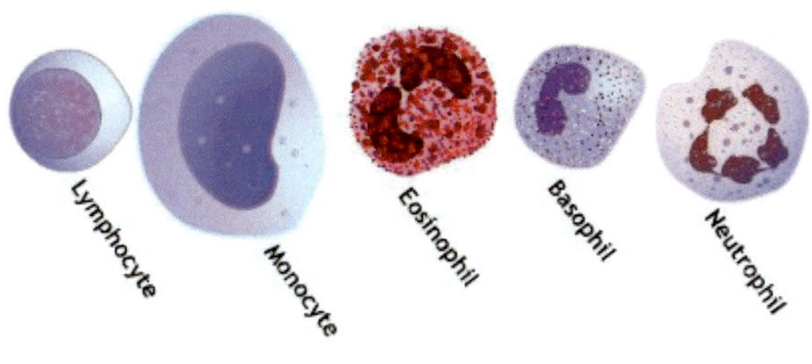

Figure 1.1. *Identification of different types of white blood cells.*

   iii. **Neutrophils:** Neutrophils account for the largest percentage of leukocytes found in a normal blood sample, and function by ingesting invading bacteria. On a stained blood smear, the cytoplasm of a neutrophil has numerous fine, barely visible lilac- colored granules and a dark purple or reddish-purple nucleus. The nucleus may be oval, horseshoe, or "S"-shaped, or segmented (lobulated). Neutrophils are sub-classified according to their age or maturity, which is indicated by changes in the nucleus. The sub-classifications for neutrophilic cells are metamyelocyte, band, segmented, and hypersegmented.

iv. **Eosinophils:** Eosinophils aid in detoxification. They also break down and remove protein material. The cytoplasm of an eosinophil contains numerous coarse, reddish-orange granules, which are lighter colored than the nucleus.

v. **Basophils:** The function of basophilic cells is unknown. It is believed, however, that basophilic cells prevent the blood from clotting in inflamed tissue. Scattered large, dark-blue granules that are darker than the nucleus, characterize the cell as basophil granules may overlay the nucleus as well as the cytoplasm.

Units: Percentage of White Blood Cells

1.3.3.4. Packed Cell Volume (PCV)/Hematocrit (Hct)

The blood was anticoagulated with EDTA and the sample was processed within 2 hrs after collection.

1. **Apparatus:** Heparinized capillary, micro centrifuge, reading device
2. Procedure: PCV was estimated by the Micro method. A capillary tube 70 mm long with an internal bore of 1 mm was used and blood was taken directly into heparized capillary tube.
   - A heparinized capillary tube was filled with blood by capillary action leaving 10 mm unfilled.
   - The empty end was sealed by plasticine.
   - The tubes were centrifuged in microhematocrit centrifuge at 10,000 g for 5 mins.
   - The blood column was measured by using a reading devise.

Units: The results are expressed as percentage

1.3.3.5. Erythrocyte Sedimentation Rate (ESR)

When well mixed anti-coagulated blood is placed in a vertical tube, the erythrocytes tend to fall towards the bottom of the tube/pipette till they form a packed column in the lower part of the tube in a given time.

ESR takes place in the following 3 phases carried out in one hour:

- **Rouleaux Formation**: In the initial 10 mins, the process of rouleaux formation occurs and there is little sedimentation.
- **Settling**: In the next 40 mins, settling of RBC occurs at a constant phase.
- **Packing**: In the last 10 mins, sedimentation slows and packing of the RBC's to the bottom occurs.
- Due to the above, ESR is expressed as mm first hour rather than per hour (Mohan, 2000).

Fresh anti-coagulated blood was allowed to stand in a Wintrobe's tube, the red cells sedimented slowly. The rate of sedimentation was estimated under standard conditions.

1. **Apparatus:** Pipette, sedimentation rack.
2. Procedure: Due to the scarcity of blood in fish, the micro ESR method was used for ESR determination. A capillary 160 mm long with an internal bore diameter of 1 mm was used. The capillary was graduated 1 mm apart for 50 mm with two red lines on it. Alternatively, non-graduated heparinized capillary was used and the reading taken by measurement of length of column.
   - The micro-sedimentation pipette was filled up to first red mark with anticoagulant (sodium citrate and EDTA).
   - The pipette was filled with blood up to second red mark.
   - The pipette was inverted several times and allowed to stand for one hour in the sedimentation rack.
   - The reading was taken at the upper meniscus of the RBC's after one hour.

Units: The results are expressed in mm/hr.

1.3.3.6. Hemoglobin Percentage (Hb%)

Several methods have been proposed for the estimation of hemoglobin in the blood (Wintrobe, 1981).

1. Direct reading electric hemoglobinometer method
2. Electronic counter method
3. Oxyhemoglobin method
4. Cyanmethemoglobin method
5. Colorimetric method

6. Sahli's method

Hemoglobin was estimated by using standard Sahli's Hemoglobinometer employing acid hematin method under proper light.

1. **Apparatus:** Sahli's graduated tube, N/10 HCl, stirrer, distilled water.
2. Principle: HCl (N/10 HCl) converts hemoglobin into acid hematin, the brown color is matched against a brown glass standard in a comparator called Sahli's Hemoglobinometer. It is based on the principle that hemoglobin is converted into acid hematin with the action of dilute HCl (N/10). The brown color of acid hematin is matched with the standard brown glass comparator in a visual colorimeter called Sahli's colorimeter. This method was adopted due to the ready availability of the reagents and its simplicity.
3. Procedure: N/10 HCl was Filled in the Hemoglobin Pipette up to Mark 20
   - Fresh blood (0.2 ml) was sucked in the pipette carefully, introduced into the hemoglobin tube, mixed with N/10 HCl by a glass stirrer and distilled water was added, drop by drop and matched with standard glass.
   - The exact matched color at the upper meniscus was recorded as the hemoglobin value from Sahli's graduated tube.

Units: gm/100 ml of blood.

1.3.4. Biochemical Studies

1.3.4.1. Glucose

Glucose ($C_6H_{12}O_6$), also known as D-glucose, dextrose, or grape sugar is a simple sugar (monosaccharide) and the most important carbohydrate fuel in the body. It is used by the cells as the primary source of energy and a metabolic intermediate.

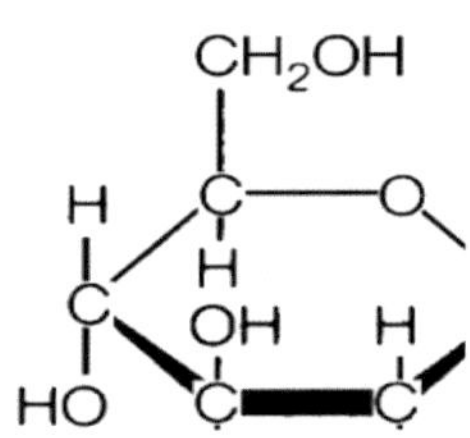

Glucose

Glucose is classified as a monosaccharide because it cannot be broken down further by hydrolysis. It is further classified as a hexose because of its six-carbon skeleton and as an aldose, because of the presence of an aldehyde group on carbon 1. The aldehyde group condenses with a hydroxyl group so that glucose exists as a hemiacetal ring structure. This ring structure explains many of the reactions of glucose.

1.3.4.2. Serum Glucose Estimation

Serum glucose was estimated by:

i. O-Toludine Method (Cooper and Mc Daniel, 1970)

ii. Ruby Auto-analyzer

1.3.4.2.1. O-Toludine Method (Cooper and Mc Daniel, 1970)

- Reagents Used:
  - 6% O- Toludine in glacial acetic acid.
  - Glucose- standard 10 mg/ 10 ml.
- Methodology: In a test tube labeled 't,' 5 μl of serum was taken. In other tube, 50 μl standard was taken and for blank, 50 μl of distilled water was taken. 5 ml O- Toludine reagent was added, mixed properly and the solutions were kept in a boiling water bath exactly for 10 min. It was cooled under running tap water and OD was measured at 625 nm.
- Calculation

$$\text{Serum glucose (mg/100 ml)} = \frac{\text{OD of test} \times \text{conc. of glucose} \times 100\ \text{ml}}{\text{OD of standard}}$$

$$= \text{OD of test} \times 100\ \text{OD of standard}$$

Auto-Analyzer

Blood samples were collected as above by sterile syringe in sterile tubes. Serum was obtained by centrifuging blood in REMI Centrifuge at 3,000 rpm for 15 minutes at room temperature and collected carefully in another vial. This serum was used to determine concentrations of glucose, by using "Ruby" Brand of Snijders autoanalyzer with the help of commercially available laboratory kits.

AUTOSPAN Glucose GOD-POD, End Point Assay and Kinetic Assay kit was used for glucose analysis by the autoanalyzer (SPAN Diagnostics Ltd., marketed by COGENT).

- Principle for Assay by the Glucose Kit

Glucose oxidase (GOD) oxidizes glucose to gluconic acid and hydrogen peroxide. In the presence of enzyme peroxidase, released hydrogen peroxide is coupled with phenol and 4-aminoantipyrine (4-AAP) to form colored quinoneimine dye. Absorbance of colored dye is measured at 505 nm and is directly proportional to glucose concentration present in the sample.

$$\text{Glucose} + O_2 + H_2O \xrightarrow{\text{Glucose oxidase}} \text{Gluconic acid} + H_2O_2$$

$$H_2O_2 + \text{Phenol} + 4\text{-AAP} \xrightarrow{\text{Peroxidase}} \text{Quinoneimine dye} + H_2O$$

Composition		
1. Glucose Reagent:	Phosphate Buffer	100 mM/L
	Glucose oxidase	> 15,000 U/L
	Peroxidase	> 1,600 U/L
	4 -AAP	0.28 mM/L
	Stabilizers	qs
2. Glucose Diluent:	Phenol	10 mM/L
	Preservative	qs
3. Glucose Standard:	Dextrose Preservative	100 mg/dL qs
Glucose Standard:	Dextrose Preservative	400 mg/dL qs

Working Reagent: 1 vial of Reagent 1 (R1) was mixed with 500 ml of Reagent 2 (R2) so that the glucose reagent dissolved completely.

Storage and Stability

- **Prior to Reconstitution:** Reagent 1 (Glucose reagent), 3 and 4 (Glucose standard) are stable at 2–8°C and Reagent 2 (Glucose diluent) is stable at room temperature (15–30°C). All reagents were used prior to the expiry date mentioned on the label.
- **After Reconstitution:** Working reagent is stable for 12 months from the date of reconstitution till the expiry date, whichever is less. It was stored at 2–8°C in dark.

Assay Parameters (According to the Kit)

Mode	End Point
Wavelength	505 nm (490–550 nm)
Flow-cell temperature	37°C
Optical path length	1 cm.
Blanking	Reagent blank
Sample volume	10 µl
Working reagent volume	1,000 µl
Incubation time	10 min at 37°C/30 min at room temperature (15–30°C).
Concentration of Standard	100 mg/dL
Stability of color	1 hour
Permissible reagent blank absorbance	<0.2 AU
Linearity	Up to 500 mg/dl
Unit	mg/dl

Methodology

Tube	Blank	Standard	Test
Serum/Plasma	–	–	10 µl
Glucose standard (Reagent 3)	–	10 µl	–
Working Glucose reagent	1,000 µl	1,000 µl	1,000 µl

Serum and standard were mixed with working reagent and incubated at 37°C for 10 minutes or at room temperature for 30 minutes.

The analyzer was programmed as follows:

- The analyzer was blanked with reagent blank
- Absorbance was measured
- Results were calculated

The experiment was conducted following the same precautions as given for glucose.

The following precautions were taken while performing the test:

1. Clean and dry glassware was used.
2. Reagents were used at selected assay temperatures.
3. Only the required amount of reagent was taken for assay and immediately, the reagent bottle was tightly capped and stored in a refrigerator to avoid reagent deterioration.
4. Time between sample mixing and aspiration was minimum.

Serum Cholesterol

Cholesterol is a waxy steroid of fat that is produced in the liver or intestines. It is an important component for the manufacture of bile acids, steroid hormones, and Vitamin D. It is the principal sterol synthesized by animals.

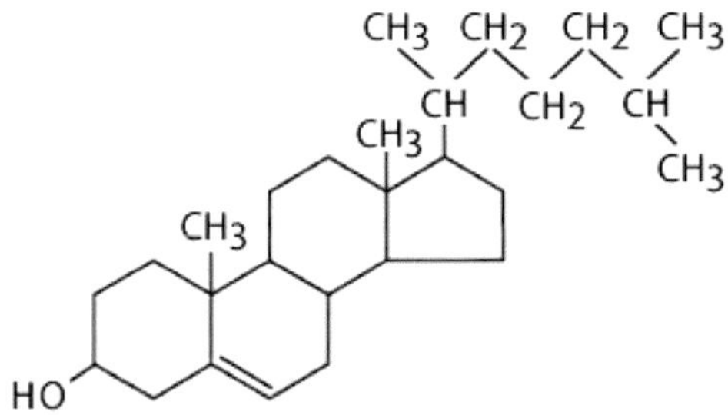

Cholesterol

Serum Cholesterol Estimation

Serum cholesterol was estimated by two methods:

i. Method of Zatkis et al. (1953):

- Principle: Blood serum was treated with glacial acetic acid, and cholesterol was determined by means of modified Liebermann-Burchard reaction. The estimation (Zatkis et al., 1953) is based on the principle that red color is produced when acetic acid solutions of certain steroids are treated with ferric chloride and sulfuric acid.
- Solutions:
    - Aldehyde free acetic acid (refuming over chromium trioxide for three hours, then distilling twice. Acetic acid from each distillation were accepted and the residue was discarded).
    - Aldehyde free, by:
    - o 10 % Ferric chloride solution ($FeCl_3 6H_2O$)
    - o 10 gm ferric chloride was dissolved in 100 ml of acetic acid.

- Color Reagent: 10 % ferric chloride solution (0.5 ml) was added to 7.5 ml conc. H_2SO_4 in a dry flask. The solution was mixed and diluted to 50 ml with conc. H_2SO_4. The original clear and light yellow gradually became colorless. Care was taken to prepare fresh reagent daily.
- Cholesterol Standard: The stock standard (1 mg/ml) cholesterol solution was prepared by dissolving 10 mg of pure dry cholesterol in 10 ml acetic acid. 0.2 ml of stock cholesterol standard was diluted to 10 ml with acetic acid for working standard which contained 0.2 mg/ml.
- Methodology: 0.1 ml of serum was added to 6 ml of glacial acetic acid in 1 × 6" boiling tube. To this tube, 4 ml of ferric chloride-sulfuric acid color reagent was added slowly along the side of the tube which formed a layer below the acetic acid. The two layers were mixed immediately thoroughly to ensure even and smooth distribution. The tubes were then allowed to cool.
- **Standard:** 1 ml of standard cholesterol (0.2 mg of cholesterol) was added to 5 ml of glacial acetic acid and 0.1 ml of distilled water. 4 ml of ferric chloride- sulfuric acid color reagent was added in a similar manner as the test system.
- **Blank:** The Blank comprised of 0.1 ml of distilled water, 6 ml of acetic acid and 4 ml of color reagent. The O.D of the different samples was observed at 570 nm.
- Calculation

Total cholesterol (mg/ ml) = Reading of test × 0.2 × 100/ 0.1

Reading of standard

Reading of test × 200

Reading of standard

ii. Auto-Analyzer: Serum was obtained by centrifuging blood as mentioned for "glucose." This serum was used to determine concentrations of cholesterol, by using "Ruby" Brand of Snijders Autoanalyzer with the help of Liquid Gold Cholesterol kit CHOD-PAP (SPAN Diagnostics Ltd., marketed by COGENT).

- Principle for Assay by the Cholesterol Kit: Cholesterol esters are hydrolyzed by Cholesterol Esterase (CE) to give free cholesterol and fatty acids. In subsequent reaction, Cholesterol Oxidase (CHOD) oxidizes the 3-OH group of free cholesterol to liberate Cholest-4-

en- 3-one and Hydrogen Peroxide. In presence of Peroxidase (POD), Hydrogen Peroxide couples with 4-Aminoantipyrine (4 -AAP) and Phenol to produce Red Quinoneimine dye. Absorbance of colored dye is proportional to the amount of total cholesterol concentration in the sample.

$$\text{Cholesterol esters} \xrightarrow{\text{CE}} \text{Cholesterol + Fatty acids}$$

$$\text{Cholesterol} + O_2 \xrightarrow{\text{CHOD}} \text{Cholesten- 3- one} + H_2O_2$$

$$H_2O_2 + \text{Phenol} + 4\text{–AAP} \xrightarrow{\text{POD}} \text{Quinoneimine dye} + H_2O$$

Composition:

1. Cholesterol Reagent: Good's Buffer (Ph 6.7)	50 mmol/L Cholesterol Esterase Cholesterol Oxidase Peroxidase 4 – Aminoantipyrine Stabilizer	>100 U/L >50 U/L >3 KU/L 0.4 mmol/L qs
2. Cholesterol Standard: Cholesterol	200 mg/dl Preservative Stabilizer	qs qs
3. Precipitating Reagent: PEG 6000	200 mM/L Stabilizer Preservative	qs qs

- **Storage and Stability:** All the reagents are stable at 2–8°C. Care was taken to use the reagents before the expiry date after which they were discarded.
- Assay Parameters (According to the Kit)

Mode	**End Point**
Wavelength	505 nm (490–510 nm)
Flow-cell temperature	37°C
Optical path length	1 cm
Blanking	Reagent Blank
Sample volume	10 µl
Standard volume	10 µl
Reagent volume	1,000 µl

Incubation time	10 mins at 37°C
Concentration of Standard	200 mg/dl
Stability of final color	1 hr
Permissible reagent blank	< 0.2 AU
Linearity	750 mg/dl
Unit	mg/dl

- **Methodology:**

Tube	Blank	Standard	Test
Serum/Plasma		–	10 μl
Cholesterol standard (Reagent 2)	–	10 μl	
Working Cholesterol reagent (Reagent 1)	1,000 μl	1,000 μl	1,000 μl

Serum and standard were mixed with working reagent and incubated at 37°C for 10 mins or at room temperature for 30 mins.

The analyzer was programmed as follows:

- The analyzer was blanked with reagent blank
- Absorbance was measured
- Results were calculated

The experiment was conducted following the same precautions as given for glucose.

1.3.4.4. Total Serum Protein

Total Serum Protein (TSP) was estimated by two methods:

1. According to Lowry et al. (1951)

2. Ruby Auto-analyzer

1. Method of Lowry et al. (1951)

Reagents Used

- Reagent A: 2% sodium carbonate in 0.1 NaOH
- Reagent B: 2% sodium potassium tartrate, 1%, $CuSO_4$ which were mixed in the ratio of 1:1 at the time of experiment.
- Reagent C: 50 parts of reagent A + 1 part of reagent B.
- Reagent D: Folin-Ciocalteu reagent. This was diluted to 1:2 ratio with distilled water at the time of experiment.

- Bovine Serum Albumin (BSA): 50 mg of BSA was dissolved in 50 ml of distilled water and used as standard.
- Methodology: 5.0 ml of freshly prepared alkaline copper sulphate solution (Reagent C) was added to 1 ml of serum and left at room temperature for 10–15 min. 0.5 ml of Folin-Ciocalteu (Reagent D) was added to it, and mixed thoroughly. This was incubated at room temperature for 30 to 40 mins, blue color developed which was measured at 690 nm or 500 nm. For BSA standard, 100 µl was added to 900 µl of distilled water. For blank, 1,000 µl of distilled water was taken. Other solutions were added according to the sample.
- Calculation

Protein (mg/1 ml) = OD of test × conc. of standard in µl × DF

OD of standard

2. Auto Analyzer: Serum was again obtained by centrifuging blood as mentioned for "glucose." This serum was used to determine concentrations of protein by using "Ruby" Brand of Snijders Autoanalyzer with thc help of Liquid Gold Total Protein Kit Modified Biuret, End Point Assay (SPAN Diagnostics Ltd., marketed by COGENT).

- Principle for Assay by the Protein Kit: The peptide bonds of proteins react with Cupric ions in alkaline solution to form a colored chelate, the absorbance of which is measured at 578 nm. The Biuret reagent contains Sodium Potassium Tartrate, which helps in maintaining solubility of this complex at alkaline pH. The absorbance of final color is proportional to the concentration of Total Protein in the sample.

Protein + Cu++ **Alkaline pH** Cu- Protein complex

Composition

Biuret reagent	Copper Sulphate Sodium Hydroxide Sodium-Potassium Tartrate Surfactant	7 mM/L 200 mM/L 20 mM/L qs
Protein Standard	BSA Preservative	6.5 g/dl qs

Assay Parameters (According to the Kit):

Mode	End Point
Wavelength	578 nm (550–580 nm)
Flow-cell temperature	37°C
Optical path length	1 cm
Blanking	Reagent blank
Sample volume	10 µl
Standard volume	10 µl
Working reagent volume	1,000 µl
Incubation time	5 mins
Concentration of standard	6.5 gm/dl
Stability of final color	2 hr
Permissible reagent blank absorbance	< 0.2 AU
Linearity	Upto 20 g/dl
Units	g/dl

Methodology:

Tube	Blank	Standard	Test
Serum/Plasma		–	10 µl
Glucose standard (Reagent 2)	–	10 µl	–
Biuret reagent (Reagent 1)	1,000 µl	1,000 µl	1,000 µl

Serum and standard were mixed with working reagent and incubated at 37°C for 10 mins or at room temperature for 30 mins.

The analyzer was programmed as follows:

- The analyzer was blanked with reagent blank
- Absorbance was measured
- Results were calculated

The experiment was conducted following the same precautions as given for glucose.

1.3.5. Statistical Analyses

- **The Data Were Statistically Analyzed:** The mean + SE and the values of co-variance were calculated for different experimental groups to determine the inter-group and intra-group significance. The one-way analysis of variance (ANOVA) was employed for statistical analysis and P values of 0.001 were considered to be significant. The significant difference between each group were determined and analyzed separately using Critical Difference (CD) and Duncan Multiple Range (DMR) test as post-hoc tests.
- **The SPSS** (originally, Statistical Package for the Social Sciences) was used for the analyses. The first version of the package was released in 1968 after being developed by Norman H. Nie and C. Hadlai Hull at Stanford University, USA and the formal manual was published shortly afterwards (Nie et al., 1970). It is now among the most widely used programs for statistical analysis. For the present purposes, SPSS version 16 was used.

1.3.6. Other Experimental Work

Studies on multiplication patterns of parasites were conducted on permanent stained preparations. The formulation of groups/sub-groups and the experimental protocol adopted for eco-parasitological studies, trypanosome culture, lymphocyte culture, development of immune sera and drug action are mentioned in the respective chapters for the ease of reference and coordination of the procedural techniques with the findings.

Chapter 2

Taxonomy of Blood Parasites

Contents

The assignment of a name to an organism provides the only key to all the information available about that species. Careful and accurate identification and classification are of vital importance. The taxonomist is faced with the difficult task of characterizing a species and to arrange it in hierarchy of higher categories. The aim of modern taxonomy is not only to describe, identify and arrange organisms in convenient categories but also to understand their evolutionary histories and mechanisms (Kapoor, 1998).

The classification of Protozoans prepared by the Society of Protozoologists published by Levine et al. (1980) has been followed. The flagellate parasites discussed in the thesis fall under the phylum Sarcomastigophora and can be subcategorized as follows:

- **Phylum *Sarcomastigophora***: Simple type of nucleus, sexuality when present, essentially syngamy, flagella, pseudopodia or both types of locomotor organs.
- **Subphylum Mastigophora**: Flagellates. One or more flagella typically present in trophozoites, asexual reproduction basically by intrakinetal (symmetrogenic) binary fission; sexual reproduction known in some groups.
- **Class Zoomastigophorea**: "Animal-like" flagellates. Chloroplasts absent. One to many flagella; amoeboid forms, with or without flagella; sexuality known in a few groups; a polyphylectic group.
- **Order Kinetoplastida**: One or two flagella arising from a depression; flagella typically with paraxial rod in addition to axoneme; single mitochondrion (sometimes non-functional) extending length of body. Feulgin positive (DNA containing) kinetoplast.

The existing taxonomy of kinetoplastids is based on morphology and life cycles. Kinetoplastids are a group of protozoa characterized by the presence of a characteristic organelle, the kinetoplast (Vickerman, 1976). This structure was observed by early researchers in light microscope as basophilic granules located near the basal body of a flagellum. This association was thought to be meaningful, and the term `kinetoplast' was coined by Alexeieff in 1917 (Alexeieff, 1917; Vickerman and Preston, 1976). Only in the 1960's, after the first electron microscopic studies of this organelle were conducted it became clear that the kinetoplast represents a portion of a single branched mitochondrion of the cell containing a large mass of mitochondrial DNA. The presence of the kinetoplast is easy to determine by fluorescence with a DNA binding dye DAPI or by Giemsa staining.

This group of organisms is usually assigned the rank of an order-Kinetoplastida, which, together with two other groups: Euglenida and

Diplonemida – were placed in the phylum Euglenozoa (Cavalier-Smith, 1993; Corliss, 1994). This assemblage has been confirmed by a number of works, mainly based on the ribosomal SSU RNA phylogenetic analysis, which showed Euglenozoa as a monophyletic group, well separated from the eukaryotic crown (Sogin and Silberman, 1998; Maslov et al., 1999). However, the classification proposed by Levine et al. (1980) is still followed and Kinetoplastida is placed in the phylum *Sarcomastigophora*.

The order Kinetoplastida was created to incorporate the various flagellated protozoans containing organelles that, on cytological and cytochemical evidence, are classified as kinetoplasts. Protozoans having kinetoplasts include the Trypanosomatids, the cryptobids, the bodonids, and a few other types of flagellates. Kinetoplasts, in addition to being Feulgen-positive and staining selectively with Janus green B, are also stained by acridine orange, are sensitive to DNase digestion, and, in a number of species, can incorporate tritiated thymidine. Electron microscopy has confirmed that all kinetoplasts are basically similar. They consist of a modified enlarged region of the mitochondrion that contains all or almost all of the mitochondrial DNA (Clark and Wallance, 1960; Simpson, 1972).

Lom (1976) and Vickerman (1976) subdivided the order Kinetoplastida into two suborders, Bodonina and Trypanosomatina.

2.1. SUBORDER BODONINA

It includes two families:

1. **Bodonidae and Cryptobiidae,** members of which have a relatively large kinetoplast and two flagella: anterior (always remains free) and recurrent (free in bodonids and attached in cryptobiids). Bodonids show a variety of life styles, ranging from free living, such as *Bodo, Parabodo, Dimastigella, Rhynchobodo, Cruzella*, to parasitic. The latter are represented by ectoparasites of the fish skin and gills such as *Ichthyobodo*, endoparasites of the reproductive tract of snails (some *Cryptobia* species) and gastro-intestinal tract of fish (some other *Cryptobia*) and also by digenetic parasites of fish blood (*Trypanoplasma*) transmitted by leeches.
2. **Family Cryptobiidae:** The family Cryptobiidae belongs to the order (Kinetoplastida Honigberg, 1963), class Zoomastigophorea. Cryptobiids are allied with the suborder Bodonina. All have a kinetoplast, many species are parasitic, and an undulating membrane (often difficult to detect) may be present or absent. Until recently, some authorities placed the Cryptobiidae along with

the Trypanosomatidae, Bodonidae, and related forms in the order Protomonadina.

Cryptobiids differ from trypanosomes chiefly in the presence of two flagella rather than one-hence, the terms "biflagellate" and "monoflagellate." Members of the family Bodonidae also have two flagella and a Feulgen-positive kinetoplast. However, the posterior flagellum is usually free of the body in the Bodonidae, whereas it is partially or completely adherent in the Cryptobiidae. Moreover, contractile vacuoles may occur in the Bodonidae, which are mostly free-living flagellates, but not in the Cryptobiidae, which are mostly parasitic.

Cryptobiids were, until recently, believed to occur in three main situations:

(i) the blood of freshwater and marine fish; (ii) the digestive tract of marine fish; and (iii) the internal organs of various invertebrates including pulmonate snails, chaetognaths, and siphonophores. All these biflagellates are structurally very similar so that many workers believe them to belong to a single genus, *Cryptobia* Leidy 1846, in which *C. helicis*, a parasite of the seminal vesicles of snails (genus *Helix)* is the type species. The first cryptobiids reported from fish blood were placed in a new genus, *Trypanoplasma* Laveran and Mesnil 1901, and *T. borreli* from a European freshwater fish became the type species. Because of close similarities between the two forms, the claim was soon made that *Trypanoplasma* was morphologically identical with *Cryptobia* and that the latter held taxonomic priority. However, a rational case was made to retain the biflagellates of fish blood in the genus *Trypanoplasma* on the basis that, although the organisms from snails and fish appeared to be superficially similar, they were biologically distinct (Laveran and Mesnil, 1912). Moreover, the parasites in the reproductive organs of snails were known to be transmitted directly during copulation, whereas those in the blood of fish were transmitted indirectly by leeches.

There is no valid reason to reject the use to *Trypanoplasma* for those cryptobiids parasitizing the blood of fish and transmitted by leech vectors. Not only is taxonomic differentiation determined largely by isolation, but biological and physiological attributes are valuable aids in establishing taxonomic categories among the parasitic protozoa (Becker, 1970). It has also been shown that morphologically the two genera are not exactly similar, differences in shape, undulating membrane (Kozloff, 1948; Becker, 1977); aciculum (Hall, 1953) exist and the movement of the two is also different in live condition (Burreson and Sypek, 1981). Hence the genus *Cryptobia* be reserved for the biflagellates of the seminal vesicles of snails and digestive tract, gill and skin of fish where they are naturally known to occur and the genus *Trypanoplasma* should designate the blood biflagellates of piscine hosts. Accordingly, the cryptobiids discovered from the fish blood and transmitted by leeches are placed in the genus *Trypanoplasma* in the thesis. The other types of cryptobiids associated with fish should be retained in the genus *Cryptobia* (Becker, 1977). The early taxonomy of biflagellates from the digestive tract of marine fish is interwoven and conflicting (Wenyon, 1926), but ectoparasitic biflagellates from freshwater fish have usually been named *Cryptobia.*

Numerous species of *Trypanoplasma* have been reported from freshwater fish in the Northern Hemisphere. Due to inadequate descriptions and lack of contrasting diagnostic features, the validity of many species remains open to question. The differentiation of species is just as problematical as for fish trypanosomes.

2.1.1. Genus Trypanoplasma (Laveran and Mesnil, 1901)

The genus *Cryptobia* was proposed to describe a biflagellated protozoan isolated from the reproductive system of snails (Leidy, 1846). It now includes morphologically similar parasites infecting other invertebrates and vertebrates. Laveran and Mesnil (1901) proposed the genus *Trypanoplasma* for a biflagellated protozoan in the blood of freshwater fish.

Trypanoplasma are parasites of bloodstream and tissues of fish. They have a wavy undulating membrane, much wider than in *Cryptobia*, but differ mainly in being transmitted by a vector, the leech and in undergoing a series of regular morphological changes in their life cycle, both in the fish and in the leech.

2.1.2. Suborder Trypanosomatina (Kent, 1880)

This group of kinetoplastids, the Trypanosomatids, forms a single family, Trypanosomatidae, which includes obligatory parasitic organisms with a single flagellum and a small kinetoplast. Trypanosomatids parasitize all classes of vertebrates, as well as some invertebrates, preferentially insects from the orders Diptera and Hemiptera, and also plants. The invertebrates can either represent hosts, as in monogenetic parasites, such as *Crithidia, Leptomonas, Herpetomonas* and *Blastocrithidia*, or serve as vectors, as in digenetic genera *Trypanosoma, Leishmania* and in bug-transmitted parasites of plants, *Phytomonas.*

The single flagellum usually arises from a small, narrow flagellar pocket either extending free or attached to the body as the undulating membrane with its end extending free anteriorly ("free flagellum"). A single, small disc- shaped kinetoplast, microtubules subtend the cell membrane around the whole circumference of the cell. All are parasitic, some are monoxenous (in invertebrates, mostly insects), some are dixenous, live in vertebrate blood or tissues, and are transmitted by an invertebrate vector (arthropod or leech). Dixenous species include some very serious pathogens (e.g., the well-known agent of a sleeping sickness of man, *Trypanosoma brucei gambiense*) and their life cycle in their vertebrate hosts and vectors consists of a series of morphological (and also physiological) stages comprising shifts in position of the kinetosome and kinetoplast in relation to the nucleus and the anterior end of the cell. Some Trypanosomatid genera exist in the form of only one of these stages, in others, the life cycle includes the sequence of many if not all of these stages.

The system of the Trypanosomatid genus assignment was largely developed by Hoare and Wallace (1966) on the basis of the observed morphotypes (Molyneux and Ashford, 1983). The morphotypes are defined by the relative position of the nucleus and the kinetoplast and the overall body shape. At least two morphotypes are observed in each genus, depending on the life cycle stage and physiological conditions of the cells. Some morphotypes are unique to a certain genus, while others, such as promastigotes, are shared by many genera (Maslov et al., 2001).

2.1.3. Family Trypanosomatidae (Doflein, 1911)

With characters of the suborder. The Trypanosomatids occurring in fish are currently assigned to a sole genus *Trypanosoma* Gruby, 1843.

2.1.4. Genus Trypanosoma (Gruby, 1843)

Dixenous (or digenetic) Trypanosomatids; their life cycle takes place partly in the blood and tissues of vertebrates and partly in the digestive tract of an invertebrate vector (mostly insects and leeches; fish trypanosomes are transmitted exclusively by leeches) and involves trypomastigote, epimastigote, spheromastigote and promastigote stages.

2.2. FISH TRYPANOSOMES

All piscine trypanosomes belong to the genus *Trypanosoma* Gruby, 1843. No sub-classification based on morphological and biological characters has been developed within the Trypanosomatidae for these flagellates. The information available is incomplete and inadequate.

Two sections, Stercoraria and Salivaria, were established for mammalian trypanosomes by Hoare (1964). Neither section holds taxonomic status; they are merely convenient grouping for subgenera. Within this scheme, fish trypanosomes might to go in the Salivaria (Becker, 1977) because they are transmitted via the proboscis sheath- the "anterior station" of leeches – by inoculation, the kinetoplast is largely terminal or subterminal, and multiplication in the fish vertebrate host occurs in the trypomastigote form

(Burreson, 1982; Burreson and Karlsbakk, 2007). Other characteristics of the Salivaria are inconsistent when applied to various piscine trypanosomes. However, trypanosomes of other host-groups were not considered. Hoare (1964) pointed out that the name *Hematomonas* Mitrophanow, 1883 is available for those who decide that fish trypanosomes are subgenerically distinct. This can be accepted. Since Mitrophanow named as representative of this taxon the species *T. carassii*, the subgenus *Hematomonas* should be restricted to include polymorphic species of trypanosomes infecting freshwater fishes, with a long, slender body, terminal or subterminal kinetoplast, a long free flagellum and with development in the leech predominantly in epimastigote stages, with a low proportion of spheromastigotes. Marine fish trypanosomes differ in having a leech sequence starting with dividing amastigotes. This group of freshwater trypanosomes, without a free flagellum, should evidently have their separate subgenera; however, such an assignment, to be well-founded, warrants further study (Lom and Dykova, 1992).

Numerous species of *Trypanosoma* have been reported from freshwater and marine fish throughout the world. At present, more than 200 named species of fish trypanosomes are, for the greatest part, ill-defined and many of them may be synonymous, relics of the wrong assumption of strict host specificity. Due to inadequate descriptions and lack of contrasting diagnostic features, the validity of many species remains open to question. Their exact differentiation is quite difficult. The following criteria have proved useful for the characterization of fish trypanosomes:

2.2.1. Characterization of a Fish Trypanosome Species

1. A detailed morphological analysis (shape, structure and position of the nucleus, shape and position of kinetoplast, sub-surface striation in adult forms, arrangement of the undulating membrane and free flagellum, cytoplasmic granules),
2. Morphometric evaluation,
3. Ascertainment of pleomorphism in the course of infection,
4. Observation on host-specificity based on cross infection experiments,
5. Observation on the sequence of stages in the leech vector.

The above criteria maybe considered as a prerequisite to make identification an exacting procedure. Letch (1979) and Zajicek (1991) could find no patterns supporting specific differentiation in trypanosomes from various freshwater fishes. Zajicek (1991) failed to differentiate species even using restriction analysis of kinetoplast DNA. Observations on behavior in culture and in the host and pathogenicity may prove to be useful.

2.3. MAMMALIAN TRYPANOSOMES

Since the publication of the comprehensive 'pre-molecular' texts on trypanosomes and trypanosomiasis by Mulligan (1970) and Hoare (1972), much new biochemical and molecular data relating to the systematics and phylogeny of trypanosomes has become available.

Previous attempts to dismember the genus *Trypanosoma*, by creating additional genera or subdividing it into subgenera, failed to gain acceptance, but, in the case of mammalian trypanosomes, the arrangement of species into informal groupings found more general recognition. However, as subordinate groups tend to multiply with advances in our knowledge, the classification of these trypanosomes threatens to become too unwieldy. It was therefore proposed to give assemblages of allied species a formal taxonomic status, by subdividing the genus *Trypanosoma* into subgenera, corresponding to previous minor groupings, but retaining the two major sections (Stevens and Brisse, 2004). The subgenera and their type-species are as follows:

1. Section Stercoraria, with 4 Subgenera:
   - i. *Megatrypanum* Hoare, 1964: *Trypanosoma* (*Megatrypanum*) *theileri*;
   - ii. *Herpetosoma* Doflein, 1901: *Trypanosoma* (*Herpetosoma*) *lewisi*;
   - iii. *Schizotrypanum Chagas, 1909*: *Trypanosoma* (*Schizotrypanum*) *cruzi*;
   - iv. *Endotrypanum* Mesnil and Brimont, 1908: *Trypanosoma* (*Endotrypanum) schaudinni.*
2. Section Salivaria, with 4 Subgenera:
   - i. *Duttonella* Chalmers, 1918: *Trypanosoma (Duttonella) vivax*;
   - ii. *Nannomonas* Hoare, 1964: *Trypanosoma (Nannomonas*) *congolense*;
   - iii. *Pycnomonas* Hoare, 1964: *Trypanosoma* (*Pycnomonas*) *suis*;
   - iv. *Trypanozoon* Lühe, 1906: *Trypanosoma (Trypanozoon*) *brucei.*

The systematic position of *Trypanosoma* among the protozoa and the revised classification of the mammalian trypanosomes, according to Levine et al. (1980) and Corliss (1994), is as follows:

- **Subkingdom**: *Protozoa* (Goldfuss, 1818); emend. (Levine et al., 1980)
- **Phylum**: *Sarcomastigophora* (Honigberg & Balaniuth, 1963); emend. (Levine et al., 1980)

- **Subphylum**: *Mastigophora* (Diesing, 1866); emend. (Levine et al., 1980)
- **Class**: *Zoomastigophorea* (Calkins, 1909)
- **Order**: *Kinetoplastida* (Honigberg, 1963)
- **Family**: Trypanosomatidae (Doflein, 1901); emend. (Grobben, 1905)
- **Genus**: *Trypanosoma* Gruby (1843)

2.3.1. Subgenus Herpetosoma (Doflein, 1901)

The great majority of trypanosomes of subgenus *Herpetosoma* parasitize rodents and are not frequently implicated as agents of trypanosomiasis of medical and veterinary importance. *Trypanosoma* (*Herpetosoma*) *lewisi* (Kent, 1880; (Laveran and Mesnil, 1901)) is the type species of subgenus *Herpetosoma* and is parasitic in rats throughout the world.

2.3.2. Subgenus Schizotrypanum (Chagas, 1909)

The subgenus *Schizotrypanum* comprises small trypanosomes that are very difficult to distinguish morphologically, with a mean size range of bloodstream forms, including flagellum, from 14 to 24 μ (Hoare, 1972). A voluminous kinetoplast and typically curved bloodstream forms constitute distinctive morphological characters of this homogeneous subgenus (Hoare, 1972). Recognized *Schizotrypanum* taxa include *T.* (*S.*) *cruzi*, the agent of Chagas disease,

T. (*S.*) *rangeli* and species (*T. vespertilionis, T. dionisii, T. hedricki* and *T. myoti*) which appear to be restricted to bats (Chiroptera). The trypanosomes discovered from the blood of Polish rodents have been placed in the sub-genus *Herpetosoma* Doflein, 1901.

2.3.3. Trypanosoma colisi (Gupta, 1986)

Host Profile	
Kingdom	Animalia
Phylum	Chordata
Superclass:	Pisces
Subclass	Actinopterygii
Order	Perciformes
Family	Anabantidae
Genus	*Colisa*

Species	*Fasciatus*
Parasite Profile	
Phylum	Protozoa
Class	Flagellata
Order	Kinetoplastida
Suborder:	Trypanosomatina
Family	Trypanosomatidae
Genus :	*Trypanosoma*
Species	*Colisi* n. sp.

When a hanging drop preparation was observed under the microscope, the trypanosomes wriggled very actively amongst the blood corpuscles in a snake – like manner. These monoflagellates moved very rapidly in one place for a few seconds then repeatedly jumped and changed their position. Trypanosomes ranging from small, stumpy forms to long, slender forms could clearly be distinguished and the latter were comparatively more motile.

In a hanging drop preparation, these parasites remained alive for about 100–110 mins. Then they gradually slowed down their movement, extended themselves and died out. While slowing down, the flagellates attached themselves to a blood corpuscle by its posterior end. The anterior end retained its motility till the last. Stained preparations showed three distinct forms (Plate V), short and broad 'stumpy' forms; long and narrow 'slender' forms and forms somewhat in between the two, 'intermediate' forms (Plate VI A-C). All the forms could be observed in a single slide but the relative abundance varied from slide to slide. A very heavy infection of 77% was recorded and the intensity was also very high. In an average about 150–200 trypanosomes could be seen in a single slide but one preparation exhibited gross infection where they outnumbered 500.

2.3.3.1. Description

- **Shape:** Mostly 'C' shaped but 'S' shape also encountered. Extremities of the body pointed, anterior end being more acute.
- **Cytoplasm**: Granular cytoplasm, metachromatic granules typically arranged in a linear manner along both the margins of the entire cell body (CL) which are darker in the anterior half of the body. Granulation pattern similar in all the forms. Myonemes absent.
- **Nucleus**: Oval or bean shaped nucleus situated slightly in the anterior half of the body i.e., PN > NA. Contains darkly stained chromatin granules arranged in patches. Karyosome not observed. Does not occupy the entire body width.

- **Kinetoplast**: Spherical or oval kinetoplast is terminal or sub-terminal in position. Two conditions are found, KL = PK or KL > PK in the different forms (KL is the length of the kinetoplast and PK the distance from posterior end to kinetoplast). But in all cases, its position is nearer to the posterior end of the body than the nucleus (PK< KN).
- **Blepharoplast**: As in other fish trypanosomes, the blepharoplast is lacking. The axoneme apparently arises directly from the kinetoplast.
- **Flagellum**: The flagellum originates from the kinetoplast, borders the undulating membrane and finally extends beyond as a free flagellum. The latter increases in size with increase in size of the flagellate. (CL > FL in all forms).
- **Undulating Membrane**: The undulating membrane has 2–6 folds and is comparatively broader in the stumpy forms than the intermediate and slender forms. It is attached to the cell body at 2–5 places.
- **Dividing forms**: The dividing form possesses 2 kinetoplasts lying adjacent to each and perhaps represents an early stage of division.
- **Measurements**: The mean and range of component parts of the parasite and nuclear, kinetoplast and flagellar indices are depicted in Table 2.1. The values support polymorphism of the parasite.

Type host	*Colisa fasciata* Bloch.
Type Locality	Muddy ponds of Bareilly, Shahjehanpur (U.P) India.
Site	Circulating blood of fish.
Vector	Unknown.

2.3.3.2. Discussion

The prevalence and intensity of trypanosome infection was very high. Polymorphism among piscine trypanosomes has been reported by Qadri (1955), Becker (1967), Fattohy (1978), Becker and Overstreet (1979), Gupta and Jairajpuri (1981b, 1982b). *Trypanosoma striati* (Qadri, 1955) and *T. occidentalis* (Becker, 1967) are large flagellates, their smallest form being larger than long slender form of the species under discussion. The present species has shorter flagellar and body lengths than *T. neinavana* (Fattohy, 1978) and the nuclear position also varies, N.I. of the former being much less than the latter. The present species is somewhat comparable to *T. mugilicola* (Becker and Overstreet, 1979) in its size but again differs drastically in N.I. which is 1.1–

1.2 in the trypanosome under discussion whereas 3.8 μm the aforementioned species. The form under study is separable form *T. singhii* and *T. attii* (Gupta and Jairajpuri, 1981b, 1982b) due to its peculiar granulation pattern, nuclear position and differences in statistical measurements.

T. trichogasteri (Gupta and Jairajpuri, 1981a) reported from the same host is a monomorphic species. A close examination of the two (Table 2.2) suggests that no form of the present species conforms completely with the earlier described species. The present slender form is comparable to *T. trichogasteri* in its body and flagellar length but is broader than the latter. On the other hand, contrasting differences exist in kinetoplast size, granulation of the cytoplasm and nuclear position which lies in the anterior half in the present case but in the posterior in *T. trichogasteri*. The dissimilarities between the two rules out the possibility of the two being the same species, in case somehow the two other forms were overlooked by the workers earlier.

The identification of trypanosomes from fish involves the host, geographical locations, measurements and structural characteristics (Becker, 1967; Daly and Degiusti, 1971; Fattohy, 1978), hence the species under discussion is being given a new species status and is designated as *Trypanosoma colisi* n sp. taking all the foregoing facts into account.

Table 2.1. *Morphometric parameters of Trypanosoma colisi n. sp.*

Component Parts of Parasite*	Stumpy	Intermediate	Slender
CF	22.9 (22.7–23.4)	30.4 (29.5–31.4)	36.6 (36.0–38.5)
CL	14.2 (12.9–15.6)	19.5 (17.0–22.3)	23.3 (20.2–25.5)
CB	1.8 (1.3–2.2)	1.6 (1.2–2.0)	1.5 (1.2–1.8)
FL	9.0 (8.0–10.3)	10.8 (9.0–12.5)	13.2 (10.5–16.0)
NL	2.5 (2.0–3.1)	3.0 (2.8–3.2)	2.5 (2.1–2.8)
NB	1.2 (0.8–1.4)	1.2 (1.0–1.5)	1.3 (1.1–1.5)
NA	5.4 (4.1–5.9)	7.6 (6.5–10.2)	9.0 (8.0–12.2)

KN	4.8 (4.4–6.3)	7.3 (6.0–8.4)	9.9 (9.3–11.5)
KL	0.7 (0.5–0.8)	0.7 (0.5–1.0)	0.8 (0.6–1.0)
PK	0.5 (0.0–1.1)	0.6 (0.0–1.2)	0.6 (0.0–1.4)
NI	1.1	1.1	1.2
KI	1.2	1.1	1.1
FI	1.6	1.7	1.7

* *Abbreviations as given in Materials and Methods.*

Note: All values are in µm. Figures in parentheses give the range values.

Table 2.2. *Comparative Table of Trypanosoma trichogasteri Gupta and Jairajpuri, 1981 and the Slender form of Trypanosoma colisi n. sp*

Component Parts of Parasite*	***T. trichogasteri* Gupta and Jairajpuri, 1981**	**Slender form *T. colisi* n.sp.**
CF	–	36.0–38.5
CL	24.0–27.4	20.2–25.5
CB 1.2–1.4		1.2–1.8
FL 11.5–12.9	10.5–16.0	
NL 3.0–3.2		2.1–2.8
NB 1.0–1.1	1.1–1.5	
NA	–	8.0–12.2
KN	–	9.3–11.5
KL	0.7–1.5	0.6–1.0
PK	–	0.0–1.4

Note: All values are in µm.

2.3.4. Trypanosoma kargenensis (Gupta and Gupta, 1994)

Host Profile	
Kingdom	Animalia
Phylum	Chordata
Superclass :	Pisces
Subclass	Actinopterygii
Order	Cypriniformes
Family	Heteropneustidae
Genus	*Heteropneustes*
Species	*Fossilis*
Parasite Profile	
Phylum	Protozoa
Class	Flagellata
Order	Kinetoplastida
Suborder	Trypanosomatina
Family	Trypanosomatidae
Genus	*Trypanosoma*
Species	*Kargenensis* n. sp.

The parasites showed an incidence of 32% and parasitemia ranged from a meager 5 to even 120 or more forms/20 mins scan.

2.3.4.1. Description

- **Shape**: The body maybe elongated but usually takes a twisted 'C' or 'S'-shaped conformation upon fixation (Plate VIIA) being either extended or contorted in appearance.
- **Cytoplasm**: The granulation pattern is typical in being conspicuous towards the curvature of the parasite.
- **Nucleus**: The nucleus is oval or bean-shaped occupying the major width of the cell body ranging in length from 3.0 to 5.2 μm and 1.2 to 3.0 μm in width. The karyosome is usually not differentiated but may be faintly outlined in rare cases as a spherical mass on one side, the other being occupied by a small vacuole.
- **Kinetoplast**: The kinetoplast isprominent, spherical in shape occupying the entire width or even protruding out of the cell body (Plate VIIB). It is always placed slightly away from the posterior tip.
- **Flagellum and Undulating Membrane**: The flagellum is usually ½

of the total body length. The undulating membrane has 4–8 folds, at times, 2–4 minor folds may also be present.

- **Measurements**: The measurements of the parasite and their component parts of 48 undistorted forms selected at random are given in Table 2.3.
- **Indices:**

NI = 0.992

FI= 2.329

KI = 1.190

Table 2.3. *Statistical Comparison of Trypanosoma kargenensis n. sp. with Existing Species from Heteropneustes Fossilis*

Parasite Author (s) Locality Component Parts	*T. saccobranchi* Castellani and Willey, 1905 Colombo No. Measurement	*T. danilewskyi* var. *saccobronchi* Qadri, 1962 India Monomorphic	*T. mukundi* Raychudhuri and Mishra. 1973 India Dimorphic		*T. singhii* Gupta and Jairajpuri 1981 India				*T. kargenensis* n. sp. Present Study India Monomorphic
				Polymorphic					
			Slender	Stumpy	Developmental	Adult			
						Small	Inter-Mediate	Large	
CF	–	38.0–47.0	32.5–43.5	22.0–28.5	8.5–10.2	24.4–29.9	35.2–40.1	44.5–47.1	38.0–55.0
CL	–	27.0–32.0	24.0–28.5	18.0–23.5	8.5–10.2	13.0–19.1	21.6–27.9	30.0–33.6	22.0–39.0
CB	–	1.0–1.5	2.0–2.5	1.9–2.5	1.0–1.4	1.1–2.4	1.2–2.9	1.5–2.0	1.0–2.5
FL	–	11.0–15.0	9.0–15.5	4.0–6.5	Absent	10.3–13.4	11.3–16.2	13.3–16.4	9.0–21.0
NL	–	2.7–3.2	2.2–3.7	2.2–3.7	1.2–2.0	2.0–4.0	3.1–4.2	3.2–5.1	3.0–5.2
NB	–	0.7–1.0	1.7–2.2	1.7–2.2	0.8–1.2	1.3–1.6	0.9–1.8	1.1–1.5	1.2–3.0
KB	–	0.7–1.0	0.7	0.7	0.6–0.7	0.4–0.7	0.8–1.0	0.8–1.1	0.8–1.5

(All values are in μm).

2.3.4.2. Discussion

A comparison of all the species reported from *H. fossilis* with the new species is shown in Table 2.3 which suggests that the latter does not fit-in adequately with any of the described species. A critical examination with *T. danilewskyi var. saccobranchi* Qadri, 1962 shows that the form under discussion not only has wider size variability (38.0–55.0 μm as compared to 38.0–47.0 μm *T. danilewskyi var. saccobranchi*) but is also broader (1.0–2.5 μm vs 1.0–1.5 μm), has a longer free flagellum (9.0–21.0 vs 11.0–15.0 μm) and a larger nucleus (3.0–5.2 μm vs 2.7–3.2 μm). Vacuoles were always found to be present in the cytoplasm at the ends of the nucleus which were lacking in the present species. The latter also differs from the former in having a shorter free flagellum in relation to the body length (usually 1/3rd) as compared to the latter (usually 1/2). The present form does not correspond to any of the forms of *T. mukundi* Raychaudhuri and Misra, 1973. It is somewhat comparable to the large form of *T. singhii* Gupta and Jairajpuri, 1981 in the nuclear and kinetoplast width but again differs in being monomorphic, in reaching a greater total and cell body length, in having a broader nucleus and wider undulating membrane.

The above discussion justifies the new status of the trypanosome from *Heteropneustes fossilis* from Rohilkhand region. The said parasite is therefore, named as *Trypanosoma kargenensis* n. sp.

2.3.5. Trypanosoma bagroides (Gupta and Gupta, 1996)

Host Profile	
Kingdom	Animalia
Phylum	Chordata
Superclass:	Pisces
Subclass	Actinopterygii
Order	Cypriniformes
Family	Bagridae
Genus	*Mystus*
Species	*Vittatus*
Parasite Profile	
Phylum	Protozoa
Class	Flagellata
Order	Kinetoplastida
Suborder	Trypanosomatina
Family	Trypanosomatidae

Genus	*Trypanosoma*
Species	*Bagroides* n. sp.

2.3.5.1. Description

- **Shape:** Typically curved sinuous body where the curvature may lie within the "C" shape.
- **Cytoplasm:** The cytoplasmic granulation is more prominent in the anterior portion and is variably distributed in the other parts of the cell.
- **Nucleus:** The dense nucleus is long, almost twice as long (NL: 4.46 μm) as wide (NB: 1.96) (Plate VIIIA). The karyosome was not observed.
- **Kinetoplast:** The kinetoplast is sub-terminal occupying the entire posterior width of the cell body and is 1.34 μm in length. The blepharoplast was not observed, the axoneme apparently arises directly from the kinetoplast.
- **Flagellum:** The free flagellum is about 1/3rd of the cell body length (FL ~1/3 CL) and ranges in size from 10.0–18.0 μm (Table 2.4).
- **Undulating Membrane:** The undulating membrane is imperceptible from the cell body in most cases and is 1.08 μm wide in average. The number of folds varies from 4–8 (Plate VIIIB).
- **Indices**:

  NI = 0.49

  KI = 1.26

  FI = 2.87

2.3.5.2. Discussion

Over the past few decades, various species of trypanosomes have been discovered, *T. vittati* Tandon and Joshi, 1973 was described as a polymorphic species. Joshi (1976b) reported *T. seenghali* from *M. seenghala* from the river Gomti. The form being discussed herein differs in all statistical measurements from the earlier reported species (Table 2.4). Mandal (1979) described *T. bengalensis* from *M. bleckeri* from W. Bengal. This monomorphic species is conspicuously smaller than the compared species not only in its length but also in its width and nuclear dimensions. *T. aori* Joshi, 1982 was described as a polymorphic species. The present form can be distinguished in being monomorphic, narrow and in having a longer nuclear and kinetoplast statistical build up. Wahul

(1985) reported *T. joshii* (H-*M. cavacius*), *T. mandali* and *T. purnensis* (host: *M. seenghala*). All the forms are monomorphic and the present form is thus comparable. However, it differs from *T. joshii* in being larger, broader and also possessing a large nucleus. The total body length is fairly comparable to *T. mandali* and *T. purnensis* but it can be distinguished from the former in being broader and also in nuclear and kinetoplast morphometrics. It again differs from *T. purnensis* in nuclear and kinetoplast size, distinct granulation and also in vacuolar distribution. Differences also exist in the shape of the nucleus and NI which is just half of that of *T. purnensis*.

The above discussion clearly indicates that the form under analysis is definitely unlike any of the forms described so far. Considering these facts, the form is considered to be a new and the species *Trypanosoma bagroides* n. sp. is created for its reception.

Table 2.4. Trypanosomes from the Genus *Mystus*

Host Parasite Author (s) Forms	***M. vittatus T. vittati*** **Tandon & Joshi,1973 Polymorphic**	***M. seenghala T.seenghali*** **Joshi, 1976 Monomorphic**	***M. bleekeri T. bengalensis*** **Mandal, 1979 Monomorphic**	***M. aor T. aori*** **Joshi, 1982 Polymorphic**		***M. cavacius T. joshii*** **Wahul, 1986 Monomorphic**	***M. seenghala T. mandali*** **Wahul, 1986 Monomorphic**	***M. seenghala T. purnensis*** **Wahul, 1986 Monomorphic**	***M. vittatus M. bagroides*** **n.sp. Gupta & Gupta, 1996 – Monomorphic**
Component part				A	B				
CF	29.9–47.4	33.00	–	35.8	43.6	–	–	–	40.51 (31.0–50.0)
CL	18.2–34.8	24–50	15.50	17.6	32.6	22.03	29.95	34.60	30.50 (21.0–35.0)
CB	1.3–2.3	1.71	1.75	3.0	4.6	1.87	1.49	2.42	2.85 (2.0–4.0)
FL	6.5–14.0	8.50	10.50	18.2	11.0	17.41	15.69	9.53	11.93(10.0–18.0)
NL	1.9–4.0	2.76	2.50	1.9	3.1	28.20	3.25	3.22	4.46
NB	0.7–1.8	1.00	0.50	1.5	1.4	1.04	0.90	1.86	1.96

KL	0.9–1.7	0.95	–	0.9	1.0	1.06	0.74	1.05	1.34
KB	0.8–1.2	0.56	0.75	–	–	0.71	0.55	0.73	1.30

Note: (All values in µm).

2.3.6. Trypanoplasma maguri (Gupta and Gupta, 1987)

Host Profile	
Kingdom	Animalia
Phylum	Chordata
Superclass:	Pisces
Subclass	Actinopterygii
Order	Siluriformes
Family	Clariidae
Genus	*Clarias*
Species	*Batrachus*
Parasite Profile	
Phylum	Protozoa
Class	Flagellata
Order	Kinetoplastida
Suborder	Bodonina
Family	Cryptobiidae
Genus	*Trypanoplasma*
Species	*Maguri* n. sp

Out of 80 fish examined, only 10 (12.5%) were found to be infected. The intensity was also quite low- 2 to 5 parasites/20 mins scan. Thirty-six forms were drawn to scale and measured.

The parasites displayed a wriggling snake-like movement in hanging drop preparations and remained alive for 30–55 mins.

2.3.6.1. Description

The fixed and stained parasites are dimorphic in nature and exist as "bulky" (22.2–36.4 µm) and "slender" (36.5–48.7 µm) forms. The smaller forms are comparatively broader (5.0 µm) than the larger ones (3.2 µm) (Plate IX, X).

- **Shape:** The shape of the parasite is elongated, the anterior end being broader than the posterior. The fixed forms assume a curved appearance.
- **Cytoplasm:** The cytoplasm is granular in nature. The myonomes were not observed.
- **Nucleus:** The nucleus is oblong in shape and is situated opposite and posterior to the kinetoplast. Its size ranges from 5.2–6.8 μm × 0.8–1.7 μm in the bulky form and 5.6–7.1 μm × 1.0–2.1 μm in the slender form. The chromatin granules are densely packed together. The karyosome was not observed.
- **Kinetoplast:** The rod shaped kinetoplast occupies an anterior position. The average size in the bulky form is 4.9 × 1.1 μm and 5.3 × 1.3 μm in the slender form. The blepharoplast appears to be absent as it was not observed in any of the specimens.
- **Flagella:** Two flagella arise from the kinetoplast- the anterior flagellum (bulky form 18.0 μm; slender form 26.2 μm) which is oriented anteriorly and the posterior flagellum (bulky form 7.8 μm; slender form 12.0 μm), usually less than half of the anterior flagellum which trails behind posteriorly.
- **Undulating membrane**: The undulating membrane is poorly outlined.
- **Measurements:** The biometric parameters of the species are presented in Table 2.5.
    - Type host: *Clarias batrachus* Linn.
    - Type locality: Nakatia pond of Bareilly (U.P.), India Site: Blood
    - Vector: Unknown

2.3.6.2. Discussion

Different species of *Trypanoplasma (Cryptobia)* have been reported from piscine hosts during the past years (Strout, 1965; Putz, 1972a; Joshi, 1982a). However, most of the earlier workers had observed uniformity in the size of the biflagellate i.e., exhibiting monomorphism.

Becker (1977) referred to trypanoplasms as '*organisms fairly uniform in size and configuration.*' Contrarily, Joshi (1982b) mentioned polymorphism in *T. Atti* which occurred as "long" and "ovoid" forms ranging from 18.0–50.0 μm. As the names of the species under discussion suggest, difference exists in the shape of the parasite, the present parasite being much narrower than *T.*

Atti. Moreover, the lengths of both the flagella in both the forms of *T. Atti* are comparable to each other, whereas in the species under speculation, the length of the anterior flagellum is more than twice of the posterior flagellum (Plate IX). The nuclear and kinetoplast dimensions of the present species also differ from *T. Atti.* Taking the foregoing facts into account, the trypanoplasm from the blood of *Clarias batrachus* is assigned the name of *Trypanoplasma maguri* n. sp. with the specific characters as given in this account.

Table 2.5. Morphometric Dimensions of *Trypanoplasma maguri* n.sp.

Component Parts of Parasite*	**Bulky Form**	**Slender Form**
CL	28.6 (22.2–36.4)	40.5 (36.5–48.7)
CB	5.0 (4.2–6.5)	3.2 (2.6–3.9)
AF	18.0 (15.3–21.6)	26.2 (22.4–29.5)
PF	7.8 (4.8–9.6)	12.0 (10.3–15.4)
NL	6.0 (5.2–6.8)	6.4 (5.6–7.1)
NB	1.1 (0.8–1.7)	1.5 (1.0–2.1)
KL	4.9 (4.2–5.6)	5.3 (4.6–5.9)
KB	1.1 (0.6–1.8)	1.3 (0.8–1.9)

Note: All values are in µm. Figures in parentheses give the range values.

2.3.7. Trypanoplasma guptii (Gupta and Gupta, 1987)

Host Profile	
Kingdom	Animalia
Phylum	Chordata
Superclass :	Pisces
Subclass	Actinopterygii
Order	Cypriniformes

Family	Bagridae
Genus	*Mystus*
Species	*Vittatus*
Parasite Profile	
Phylum	Protozoa
Class	Flagellata
Order	Kinetoplastida
Suborder	Bodonina
Family	Cryptobiidae
Genus	*Trypanoplasma*
Species	*Guptii* n. sp.

A meager infection of 11% was recorded with a low intensity of 1–5 parasites/20 mins scan. 56 forms were measured. No external pathological signs were apparent.

Live specimens could be readily observed in hanging drop preparations and whole blood. They displayed wriggling, worm-like and somewhat jerky movement, the nucleus was not discernable and the flagella could be observed with difficulty under optimum light. Size variability could be clearly observed in live specimens.

2.3.7.1. Description

The stained biflagellates took a bluish color and ranged from 15 μm –51 μm in body length. They are arbitrarily grouped into the following categories for the sake of convenience-'minute' (15–26 μm); 'intermediate' (28–38 μm) and 'bulky' (40–51 μm) forms (Plate XI).

- **Shape**: The minute forms, small in size were either pointed at the posterior end (Plate XIIA, XIVA) or almost uniformly broad (Plate XIIB) and were most abundant (Plate XIIC), followed by the intermediate forms and these two forms could be seen lying in close proximity to each other (Plate XIIIA). The intermediate forms could take a straight (Plate XIIIB), notched (Plate XIIIC) or even curved (Plate XIVB) posture on fixation. The bulky form (Plate XIC) was least abundant, elongated in shape, and was never encountered in a straight configuration. Sometimes the posterior end may be flattened disc-like or funnel shaped (Plate XIVC). The fixed forms assume a curved appearance.
- **Cytoplasm**: The cytoplasm does not take a uniform stain thereby suggesting variability in thickness of the parasite. It is mildly alveolar

but densely granulated, more so in the bulky forms and appears to increase with the increase in size of the body. Myonemes were not observed.

- **Nucleus:** The nucleus is oval in shape and takes a pink stain. It is always situated opposite to the kinetoplast and posterior or adjacent to it. In some cases, especially the minute forms, the nuclear membrane may not be well demarcated, instead, dense granules may only be present. The karyosome is not observed in the minute forms but some intermediate and bulky forms possess a distinct karyosome which takes a uniform, deeper stain and is less than 1/3 of the nuclear length.
- **Kinetoplast**: The kinetoplast is usually rod-shaped, takes a deep red stain and is situated very near to the anterior end. A distinct blepharoplast could not be observed and two flagella appear to arise from the kinetoplast: one moves anteriorly, is larger being usually more than twice of the posterior flagellum which trails from the hind end of the body.
- **Flagella**: The flagella take a very light blue stain and occasionally they may be barely visible.
- **Undulating membrane**: The undulating membrane may not be well outlined though the bulky forms do clearly possess it but it is not wider than 1 μm in any case.
- **Division**: Division in *Trypanoplasma* was not observed.
- **Measurements**: The morphometric parameters of the species are given in Table 2.6.
    - Type host: *Mystus vittatus* Bloch
    - Type locality: Freshwater ponds of Bareilly, U.P. India.
    - Site: Blood
    - Vector: Unknown

2.3.7.2. Discussion

Controversy regarding the applicability of the genus *Cryptobia* and *Trypanoplasma* has posed problems for the flagellate taxonomist during the past. Leidy (1846) described *Cryptobia helicis* from the seminal vesicula and spermatophores of pulmonate snails, *Helix albolabris*, *H. tridenta* and *H. alternata* as a new genus and species. The genus *Trypanoplasma* was established by Laveran and Mesnil (1901) for biflagellates from the blood of fresh water rudd, *Leuciscus erythrophthalmus* with *T. borelli* as the type species. Crawley (1909) examined the snail biflagellates (but *not* fish-blood biflagellates) and proved morphological identity of *Cryptobia* Leidy, 1846 and *Trypanoplasma* (Laveran and Mesnil, 1901) and opined that the former genus held taxonomic priority and should thus be established. According to Putz (1972b) "morphology together with comparative biological studies between similar types are necessary before a firm taxonomic classification can be made." Differences do exist, firstly from the habitat point of view- *Cryptobia,* inhabits the seminal vesicle of the snail and *Trypanoplasma,* fish blood, secondly, in the life cycle pattern where the former is a monoxenous parasite and has a different life cycle and the latter is heteroxenous parasite having two types of hosts and is transmitted via aquatic leeches; thirdly, morphologically they are not exactly similar, differences in shape, undulating membrane (Kozloff, 1948; Becker, 1977); aciculum (Hall, 1953) exist; fourthly, the movement of the two is different in live condition (Burreson and Sypek, 1981). Our transmission experiments to establish the parasite in the intestine of the same host failed to give positive results. Had the two genera been similar and habitat not been a criterion for differentiation, the parasite should have successfully survived in the intestine. Hence the genus *Cryptobia* should be reserved for the biflagellates of the seminal vascular of snails and digestive tract, gill and skin of fish where they are naturally known to occur and the genus *Trypanoplasma* should confine itself to the blood biflagellates of piscine hosts on which assumption the present new species is described.

Table 2.6. Biometrical Data of *Trypanoplasma guptii* n.sp.

Component Parts of Parasites	Minute	Intermediate	Bulky
CL	20.5 (15.2–26.4)	32.2 (28.3–38.1)	43.0 (40.1–51.0)
CB	3.5 (3.2–3.8)	4.6 (3.6–5.5)	4.9 (4.2–6.2)
AF	6.9 (5.0–8.8)	17.9 (13.4–20.0)	23.0 (17.6–26.2)
PF	3.1 (2.2–4.1)	8.2 (6.1–10.0)	10.5 (7.0–12.1)
NL	4.9 (4.5–5.3)	6.6 (6.0–8.4)	9.3 (8.0–11.5)
NB	1.2 (1.0–1.5)	1.5 (1.1–1.8)	1.5 (1.2–1.8)
NA	3.4 (2.8–4.0)	5.2 (5.1–6.2)	5.7 (4.0–6.5)
NP	11.5 (9.0–14.0)	22.8 (17.0–32.0)	30.5 (27.8–33.2)
KL	5.0 (4.5–6.1)	6.1 (5.2–9.0)	6.5 (5.2–9.8)
KB	1.1 (0.8–1.3)	1.1 (1.0–1.3)	1.2 (1.0–1.4)

Note: All values are in µm. Figures in parentheses give the range values.

Table 2.7. *Morphometric Comparison of the Allied and New Species of Trypanoplasma* **(*Cryptobia*)**

Component Parts of Parasite	**Parasite Author**	***C.borreli* Laveran & Mesnil 1901**	***C.gurneyorum* Minchin, 909 Scattered giants**	***T.sp* Tanabe, 1925**	***C. salmostica* Katz,1951**	***C. bullocki* Strout, 1965**	***C. cataractae* Putz., 1972**	***C. indica* Mandal., 1979**	***T. mysti* Joshi, 1982**	***T. atti* Joshi, 1982**			***T. guptii* n.sp. Present species**	
										Long	**Ovoid**	**Minute**	**Intermediate**	**Bulky**
Length of cell body (CL)		20–25	25.1 (19.2–28.2) 38.8	25.0 (17.40)	- 14.94	16.5 (10.9–21.7)	17.0 (14.7–18.9)	28.5 (25.0–30.5)	28.2 (24.0–34.2)	35.3 (26.0–50.0)	22. 2 (18.0–27.0)	20.5 (15.2–26.4)	32.2 (28.3–38.1)	43.0 40.1–51.0)
Breadth of cell body (CB)		3–4	6.7 (3.1–7.1) 9.1	- 5.0	- 2.46	4.3 (2.9–6.0)	2.0 (1.54–2.2))	8.5 (6–10.5)	10.9 (8.8–13.0)	6.8 (5.0–8.0)	12.0 (9.0–14.0)	3.5 (3.2–3.8))	4.6 (3.6–5.5)	4.9 (4.2–6.2)
Anterior flagellum (AF)	-	-	19.0	15.0	16.05	13.1 (5.8–17.5)	11.0 (9.6–13.2)	25.0 (21–30)	13.8 (10.0–22.0)	9.6 (7.5–12.0)	7.5 (6.0–9.0)	6.9 (5.0–8.8)	17.9 (13.4–20.0)	23.0 (17.5–26.2)
Posterior Flagellum (PF)	-	-	10.0	-	8.96	8.7 (4.6–18.2)	14.0 (11–16.4))	10.5 (9–12)	11.2 (8.0–15.0)	8.3 (7.0–10.5)	9.0 (6.0–15.0)	3.1 (2.2–4.1)	8.2 (6.1–10.0)	10.5 (7.0–12.1)
Nuclear length (NL)				2.0	3.50	3.6 (1.9–5.6)	1.5	7.5	6.5 (5.0–5.0)	5.5 (3.5–7.0)	4.9 (4.0–6.0)	4.9 (4.5–5.3)	6.6 (6.0–8.4)	9.3 (8.0–11.5)
Nuclear breadth (NB)				1.5	1.50	1.8 (1.0–2.6)	1.0	5.0	3.4 (2.3–4.0)	1.9 (1.0–3.0)	2.9 (1.5–3.5)	1.2 (1.0–1.5)	1.5 (1.1–1.8))	1.5 (1.12–1.8))

Nucleus to anterior end (NA)						4.1 (1.2–6.2))	-	-	-	-	-	3.4 (2.8–4.0)	5.2 (5.1–6.2)	5.7 (4.0–6.5)
Nucleus to posterior end (NP)						1.4 (0.1–3.0))						11.5 (9.5–14.0)	22.8 (17.0–32.0)	30.5 (27.8–33.2)
Kineto-plast length (KL)				5.6 (4–9)	4.58	3.6 (1.3–5.7)	3.1	5.0	1.3 (0.5–2.5)	4.0 (3.0–5.0)	5.5 (4.5–9.4)	5.0 (4.5–6.1)	6.1 (5.2–9.0)	6.5 (5.2–9.8)
Kineto-plast breadth (KB)					0.78	1.6 (0.8–3.6)	2.6	1.5	0.9 (0.4–1.5)	1.4 (0.9–2.0)	2.5 (1.8–3.7)	1.1 (0.8–3.7)	1.1 (1.0–1.3)	1.2 (1.0–1.4)

Mavor (1915), Becker (1970, 1977), Mandal (1979) reported that blood trypanoplasms are more or less uniform in shape and do not exhibit polymorphism. Contrarily, Tanabe (1925) had reported polymorphism in the trypanoplasm of *Misqurnus anquillicaudatus* and mentioned three 'types' ranging from 17–40 µm in length. Unfortunately, this unnamed parasite lacks statistical measurements of the different forms so that the present species cannot be more crucially compared to it. Apart from size difference, the parasites described by him were much broader being 1/5th of the body length whereas in the present species it was never found to be so. Differences also exist in the cytoplasmic granulation and kinetoplast size which decreases with the increase in size of the parasite in the former but is just reverse in our findings. Joshi (1982b) reported polymorphism in *T. atti* mentioning long and ovoid forms ranging from 18 µm–50 µm. However, the present species differs from the above in the body width which decreases with the increase in size of the body in the latter, contrary to the former. Apart from this, differences also exist in the size of both flagella, nuclear and kinetoplast dimensions (Table 2.7). These dissimilarities set the present species apart from the ones described by Tanabe (1925) and Joshi (1982b). Polymorphism in fish trypanoplasms may be either due to sexual or due to differences in the developmental stages.

Mandal (1979) described *Cryptobia indica* from the blood of *Mystus vittatus*. The species under consideration differs from the above in its polymorphic nature, narrower body width, nuclear and kinetoplast size. This is the first report of such giant *Trypanoplasma* (51 µm) inhabiting fish blood. The foregoing facts make the present species distinct from the species described so far and is being given a new species status, *Trypanoplasma guptii* n. sp.

2.3.8. Trypanoplasma catli (Gupta and Gupta, 1998)

Host Profile	
Kingdom	Animalia
Phylum	Chordata
Superclass:	Pisces
Subclass	Actinopterygii
Order	Cypriniformes
Family	Cyprinidae
Genus	*Catla*
Species	*Catla*
Parasite Profile	
Phylum	Protozoa

Class	Flagellata
Order	Kinetoplastida
Suborder	Bodonina
Family	Cryptobiidae
Genus	*Trypanoplasma*
Species	*Catli* n. sp.

Examination of 100 specimens of *Catla catla* for hemoflagellate infection proved positive only for 4 samples. The parasitaemia was meager (5–8 forms /15 mins scan) and the parasite existed as a monomorphic from.

*This part of the thesis has been published in Bull. NRC, Egypt 24 (1): 75–83 entitled "A new record of *Trypanoplasma* (Kinetoplastida: Cryptobiidae) inhabiting the blood of carp, *Catla catla* in India"

2.3.8.1. Description

- **Shape:** The body is typically curved though variation exists among different parasites. Some maybe irregularly shaped (Plate XVA,B), U-shaped (Plate XVC), ovoid (Plate XVD), notched (Plate XVE,F) or pointed posteriorly (Plate XVIA).
- **Cytoplasm:** Coarse chromatophilic granules are found along the undulating membrane, on the lateral margins or may even be present in bunches (Plate XVB). Sometimes, a clear, nongranulated area may also be visible (Plate XVD).
- **Nucleus:** Nucleus is a conspicuously large and deeply stained structure. L:W may be 2:1. but is 7.5 μm in length and 4.5 μm in width on an average. The organelle is extremely variable in shape, situated along the dorsal border of the parasite. Its position is opposed to the kinetoplast (Plate XVIB).
- **Kinetoplast:** The kinetoplast is an elongated structure which may be marginal (Plate XV) or partially marginal (Plate XV) but in many cases, it lies well within the cytoplasmic mass (Plate XVD). In all cases, it is a deeply stained and well differentiated structure and its size is 4.0 μm X 1.5 μm on an average.
- **Flagella and Undulating Membrane:** The two flagella arise just anterior to the kinetoplast; the anterior flagellum runs freely whereas the other traverses along the dorsal surface, bordering the undulating membrane, then protrudes freely. The anterior flagellum is larger (11.7 μm) than the posterior one (10.0 μm).

- **Measurements**: Table 2.8 provides the statistical composition of the parasite under investigation and its comparison with other monomorphic trypanoplasms.
  - Type host: *Catla catla* Ham.
  - Type locality: Kargana pond, Bareilly. U.P. India Site of infection: Blood
  - Vector: Unknown

2.3.8.2. Discussion

The question of synonymity between *Cryptobia* (Leidy, 1846) and *Trypanoplasma* (Laveran and Mesnil, 1901) has been a topic of great controversy and has puzzled the flagellate taxonomist since long. The validity of the genus *Trypanoplasma* is supported by a complex of taxonomically important characters such as profound physiological adaptation needed to cope with change of hosts (vector ↔ vertebrate) plus living media (leech digestive tract ↔ fish blood system and tissues) which are reflected in cyclical polymorphism both in the fish and in the leech (Lom, 1979). The present report accordingly describes the flagellate from the blood of *Catla* as a species of the genus *Trypanoplasma*.

The species discovered from the blood of carp is a monomorphic species and is thus being compared with the monomorphic forms. Seven species have been recorded from the genus *Mystus* from India; *T. indica* from *M. vittatus* (Mandal, 1979), *T. mysti* from *M.aor* (Joshi, 1982b), *T. krishnamurthyi*, *T. cavacii* from *M. cavacius* (Wahul, 1985), *T. vidhyai, T. seenghali* from *M. seenghala* (Wahul, 1985) and *T. tengari* from *M. vittatus* (Gupta et al., 1988). Three other monomorphic species have been described from the genus *Puntius*:*T. saranae, T. lomi* and *T. solapurensis* from *P. sarana, P. hexastichus* and *P. jerdoni*, respectively. One species has been described from the fresh water shark, *Wallago attu, T. wallagoi* (Wahul, 1986b) (Table 2.8).

Based on the size of the parasite, it can be differentiated from *T. indica, T. mysti, T. seenghali, T. solapurensis* and *T. tengari* which are all distinctly larger than the form under discussion. On the other hand, *T. cavacii* and *T. lomi* are definitely smaller. The species is therefore comparable to *T. krishnamurthyi, T. vidhyai, T. saranae* and *T. wallagoi.* The first mentioned species in the fore-mentioned line differs in having both their flagella twice a long of the species under speculation and in lacking a kinetoplast. The present species is comparable to *T. vidhyai* and *T. saranae* in their cell body dimensions but it differs from the former in having both of its flagella remarkably shorter and varied nuclear width; from the latter in having a shorter anterior flagellum which approximates half its total length whereas in the compared species, it is much more than that.

Differences also exist in the nuclear and kinetoplast size. When compared with *T. wallagoi*, it is again observed that the present species cannot be treated at par with it due to the following reasons: firstly, the anterior flagellum of the present form is not only considerably smaller than *T. wallagoi* but it is also smaller than its body length whereas in the trypanoplasm from *W. Attu*, it normally exceeds the said component. Secondly, the form from *Catla catla* is broader in cell body, nuclear and kinetoplast statistics.

The identification of the status of a discovered species has frequently puzzled the taxonomist. The description of a new species of flagellate from fish involves consideration of several problems: the host, its geographical location, measurements and structural and cytomorphological characteristics must all be given adequate weightage in order to designate a new flagellate. In view of these facts, it is felt that a combination of these factors involving the trypanoplasm of the carp, *Catla catla* differentiates it sufficiently from other reported fish trypanoplasms so as to treat the debated form as a new species. It is therefore proposed to name this organism as *Trypanoplasma catli* n. sp.

Table 2.8. *Comparative Table Showing Statistical Comparison of Trypanoplasma catli n. sp. with Other Monomorphic Trypanoplasma*

Host Parasite Author	*Mystus vittatus* *T. indica* **Mandal, 1979**	*M. aor* *T. mysti* **Joshi, 1982**	*M. cavacius* *T. krishnamurthyi* **Wahul, 1985**	*M.cavacius* *T.cavacii* **Wahul, 1985**	*M. seenghala* *T. vidyai* **Wahul, 1985**	*M. seenghala* *T. seenghali* **Wahul, 1985**	*Puntius sarana* *T. saranae* **Wahul, 1986**	*P. hexastichus* *T. lomi* **Wahul, 1986**	*P. jerdoni* *T. solapurensis* **Wahul, 1986**	*Wallago attu* *T. wallagoi* **Wahul, 1986**	*M. vittatus* *T. tengari* **Gupta et al., 1988**	*Catla catla* *T. catli* **Gupta et al., 1999**
CL	(25.0–30.5) (28.5)	28.2	19.89	17.63	19.16	25.27	18.88	14.67	23.42	18.74	26.9	20.2(15.2–28.5)
CB	(6.0–10.5) (8.0)	10.9	10.30	9.58	11.84	8.00	9.00	8.11	5.64	9.76	9.5	12.0(7.5–20.0)
AF	(21.0–30.0) (25.0)	13.8	22.52	20.41	25.05	23.69	16.80	17.36	16.53	21.63	9.8	11.7(8.0–17.0)
PF	(9.0–12.0) (10.5)	11.2	18.38	16.94	18.06	7.84	10.54	8.97	8.47	10.16	12.3	10.0(6.0–14.0)
NL	(5.0–8.5) (7.5)	6.5	6.91	6.25	6.76	6.93	6.34	5.64	6.66	6.15	6.6	7.5(6.5–10.0)
NB	(2.5–3.5) (3.0)	3.4	3.08	2.65	3.15	2.80	2.67	2.19	1.84	2.61	3.8	4.5(2.0–8.0)
KL	(5.0–6.0) (5.5)	1.3	-	4.33	4.78	5.70	4.84	2.75	6.66	3.92	6.3	4.0(2.0–8.0)
KB	(0.75–2.00) (1.5)	0.9	-	1.43	1.43	0.98	0.93	0.79	0.68	0.97	1.8	1.5(1.0–2.0)

Note: All values are in µm.

2.3.9. Trypanoplasma hematalis (Gupta and Gupta, 2010)

Host Profile	
Kingdom	Animalia
Phylum	Chordata
Superclass:	Pisces
Subclass	Actinopterygii
Order	Cypriniformes
Family	Clariidae
Genus	*Clarias*
Species	*Batrachus*
Parasite Profile	
Phylum	Protozoa
Class	Flagellata
Order	Kinetoplastida
Suborder	Bodonina
Family	Trypanoplasmiidae
Genus	*Trypanoplasma*
Species	*Haematalis* n. sp.

The blood of *C. batrachus* revealed 29.5% trypanosome infectivity out of which 8.1% of the infected samples were concomitant and under such conditions, the biflagellates, *Trypanoplasma* (10–30 forms/20 min scan outnumbered the monoflagellates, *Trypanosoma* (4–20 forms/20 min scan). Trypanoplasms were always found concomitantly with trypanosomes (Plate XVIIA).

2.3.9.1. Description

1. ***Trypanosoma batrachi* Qadri, 1962 (Redescription, (Gupta and Jairajpuri, 1983) (Plate XVIIB):** The dimorphic trypanosome observed at an intensity of 4–40 forms/20 min scan was found to conform to *Trypanosoma batrachi* Qadri, 1962 and is thus identified. However, as it is reported for the first time from Bareilly, it is being reported as a new locality record (Bareilly, Uttar Pradesh India).
2. ***Trypanoplasma hematalis* n.sp. (Plate XVIIC)**

   (Description based on 30 intact forms)

Parasite Profile	
Type host	*Clarias batrachus* Linnaeus
Type locality	Bareilly, Uttar Pradesh, India, (28.35° N, 79.42° E) Site of Infection – Blood
Etymology	The parasite is named after its site of infection Prevalence–3.7%
Intensity	10–30 forms/20 min scan
Holotype	- RU/Tpl/Cl-1
Paratype	RU/Tpl/Cl-2

The parasite is a Zoomastigid protistan typified by the presence of two flagella with proximal rod in addition to axoneme.

2.3.9.2. Description

- **Shape:** The parasite has a short (average 22.1 μm; range 15.8–27.8 μm) and broad (average 9.8 μm; range 6.1–14.1 μm) body which may sometimes be irregular in shape.
- **Cytoplasm:** The cytoplasm is densely vacuolated with deep staining chromatin granules.
- ***Nucleus*:** The nucleus (4.7–9.6 × 2.6–6.9 μm) is typically elongated or spherical in shape occupying an anterior position in the cell body. The karyosome was not observed.
- **Kinetoplast:** The kinetoplast is elongated and generally rod shaped (6.6 × 1.8 μm). This organelle is longer than those of other trypanoplasms recorded.
- **Flagella:** The anterior flagellum (AF, 12.4 μm) is slightly shorter than the
- posterior one (PF, 13.9 μm) and approximates half the cell body length.
- **Undulating Membrane:** The undulating membrane is poorly developed and often not demarcated from the body of the parasite.

2.3.9.3. Discussion

C. batrachus has served as a host for *Trypanoplasma maguri* n. sp. as given in the preceeding chapter. But the species is dimorphic and the present parasite lacks conformity with any of its forms as evident by their gross appearance, shape, statistical data and cytomorphological features. The parasite can also be differentiated from *T. guptii* n. sp. (preceeding chapter) (polymorphic) and

T. atti (Joshi, 1982), *T. qadrii* (Krishnamurthy and Wahul, 1986) (dimorphic) based on the same criteria.

Amongst monomorphic species described from India, the genus *Mystus* serves as a host for *T. indica* (Mandal, 1979) and *T. tengari* (Gupta et al., 1988). The present species differs from them in having a differential AF length and AF: PF ratio. The form under consideration can be differentiated from *T. mysti* (Joshi, 1982) (Host: *M. aor*) in again being smaller in size (*T. mysti* 28.2 μm, present species 22.1 μm), *T. mysti* having a much-reduced nucleus 6.5 × 3.4 μm) than the present species (4.7–9.6 μm × 2.6–6.9 μm) and a more prominent kinetoplast (*T. mysti* 1.3 μm × 0.9 μm, present species 6.6 × 1.8 μm). The present form differs from *T. krishnamurthyi* and *T. cavacii* (Host: *Mystus cavacius*) (Wahul, 1985) in CL and AF/PF lengths and ratio (CL: *T. krishnamurthyi* 19.89 μm, *T. cavacii* 17.63 μm, present species 22.1 μm; AF: *T. krishnamurthyi* 22.52 μm, *T. cavacii* 20.41 μm, present species 12.4 μm; PF: *T. krishnamurthyi* 18.38 μm, *T. cavacii* 16.94 μm, present species 13.9 μm).

T. vidyai and *T. seenghali* (Wahul, 1985) have been reported from *M. seenghala*. The form under question stands midway between the two forms in CL and CB (cell breadth) but has far shorter AF and PF lengths.

Monomorphic trypanoplasms have also been recorded from the genus *Puntius*. The present form can be differentiated from *T. saranae* (Host *P. sarana*) and *T. lomi* (Host- *P. hexastichus*) on statistical grounds; however, it is comparable to *T. solapurensis* (Host- *P. jerdoni* Wahul, 1986) in CL and nuclear and kinetoplast longitudinal axis. But the present species is broader in CB, nuclear and kinetoplast components and also differs in AF and PF.

A comparison of this species with *T. wallagoi* Wahul, 1986 (Host- *Wallago attu*) shows that it has a significantly shorter AF and is also contrasted by its relative sizes of AF and PF.

In view of the above critical discussion and comparison with the described monomorphic species, the form under speculation from *Clarias batrachus* is considered to be new to science and is named as *Trypanoplasma hematalis* n. sp. to accommodate the new identity with its specific characters as given in this account.

2.4. BLOOD PARASITES FROM *CYPRINUS CARPIO* FROM POLAND

During the Scientist Exchange Program between Indian National Science Academy (INSA) and Polish Academy of Sciences (PAS), *Cpyrinus carpio* at the Fish Culture Experimental Station, Golysz, Chybie, Poland were being harvested. Table 2.9 shows the pond and fish characteristics of the experimental set- up. Fishes were collected from four ponds: Golysz, Wyszni II, Wyszni III and Baginiec III. The stock material was Chylinski W.I. footling from the first three ponds whereas it was mixed in the last one.

Fresh examination revealed a meager 6.25% *Trypanosoma* infection in blood and the kidney was free of monoflagellate infection, whereas infectivity was 12.50% in blood and 42.15% in the kidney due to the biflagellate, *Trypanoplasma*. Interestingly, all infection was observed in fish showing stunted growth where infectivity reached 70%.

The degree of parasitemia was variable reaching a maximum of 20 forms/100 RBC in fish examined from pond Baginiec III. All the infected fish were the small sized ones collected from ponds Golysz, Wyszni II and Baginiec III which had smaller pond surfaces and a higher stock density.

Table 2.9. Pond and Fish Characteristics of the Experimental Set-Up at the Fish Culture Experimental Station, Golysz, Poland

Pond	Pond Surface (ha)	Origin of Stock Material and Initial Weight	Stock Number Per Ha	Harvest		Average Weight (kg)	Survival %	% of S* fish
				Kg	Number			
Golyszi	26.00	Chylinski W.I. footling	220 1,500	27,000	35,526	0.76	91	40
Wyszni II	8.24	Same as above	1,900	12,800	15,058	0.85	96	35
Wyszni III	6.40	Same as above	1,900	11,300	12,021	0.94	98	40
Baginiec III	23.70	Mixed	1,900	40,500	44,020	0.92	97	15

*Stunted.

2.4.1. Trypanosoma danilewskyi (Laveran and Mesnil, 1904)

Host Profile	
Kingdom	Animalia
Phylum	Chordata
Superclass:	Pisces
Subclass	Actinopterygii
Order	Cypriniformes
Family	Cyprinidae
Genus	*Cyprinus*
Species	*Carpio*
Parasite Profile	
Phylum	Protozoa
Class	Flagellata
Order	Kinetoplastida
Suborder	Trypanosomatina
Family	Trypanosomatidae
Genus	*Trypanosoma*
Species	*Danilewskyi* (Laveran and Mesnil, 1904)

Trypanosoma prevalence (6.25%) and intensity (2–3 forms/100 RBC) were both very low.

2.4.1.1. Description

- **Shape and Size:** The parasite is monomorphic and slender in shape. It has a sinuous body and usually takes 'C' (Plate XVIIIA) a 'S' (Plate XVIIIB) shape on fixation. It has a more or less undulated body (Plate XVIIIC). The length of the body is 24.0–35.0 μm and the maximum width, 1.6–3.1 μm.
- **Cytoplasm:** The protoplasm contains chromatic granules varying in size and number. The granules are more concentrated near the nucleus and towards the curvature of the body.
- **Nucleus:** The nucleus is elongated, nearer to the anterior end. It is less deeply stained than the kinetoplast. It is 2.4 × 4.3 μm in size.
- **Kinetoplast:** The kinetoplast is fairly large and close to the posterior extremity. It is elongated or spherical in shape and placed longitudinally.
- **Flagella and Undulating Membrane:** The undulating membrane

is broad with many folds. The flagellum is well developed usually more than half of the body length (9.2–16.2 μm).

- **Measurements:** The measurements of the parasite are given in Table 2.10 and are compared to those of earlier reported species.

2.4.1.2. Discussion

Trypanosoma danilewskyi was first described from the blood of the common carp (*Cyprinus carpio*) in Europe (Laveran and Mesnil, 1904). It has since then been found in the common carp, goldfish (*Carassius auratus*), tench (*Tinca tinca*) and eel (*Anguilla* sp.) in Europe (Thompson, 1908; Pavlovskii, 1964; Lom, 1973) and in *Saccobranchus fossilis* in India (Qadri, 1962a). The parasite is not host specific and is infective when inoculated experimentally into *Barbus conhus*, *Danio malabaricus*, *Catostomus commersoni*, *Notropis cornutus*, *Etheostoma caeruleum*, and *Ictalurus nebulosus* (Woo and Black, 1984).

In fact, *T. carassii* Mitrophanow, 1883 first observed in *Carassius carassius* is perhaps the most common and most studied trypanosome in freshwater fish. *T. danilewskyi* (Laveran and Mesnil, 1904), described from common carp (*Cyprinus carpio*) is obviously its synonym; there is no essential morphological difference between the two and cross infections are invariably positive (Lom and Dykova, 1992). *T. wincheniense* (Qadri, 1962), described in England from carp imported from Germany, is quite obviously identical and hence a synonym too; several other species described under different names are probably also conspecific, as for example, *T. tincae* (Laveran and Mesnil, 1904) of an undistinguishable morphology which can be easily established in carps and goldfish. *T. carassii* isolated from crucian carp or common carp could be transmitted by syringe passage into heterologous hosts including those from non-cyprinid families (Lom, 1979; Woo and Black, 1984) but other hosts, e.g., rainbow trout, *Perca fluviatilis* or *Lepomis gibbosus* were refractile. Morphometric values were consistent in isolates from crucian carp passaged in the same host, but were significantly different after the trypanosome was inoculated in other fishes, so that less than 7% of such populations could be classified by their measurements as the original type from the crucian carp (Lom and Dykova, 1992). The authors observed a prevalence of 100% in some carp stocks, the prevalence and intensity peaks occurring in the spring in fish debilitated by over wintering but also in the summer in densely stocked ponds. Vectors of the parasite were *Piscicola geometra and Hemiclepsis marginata*. In peak parasitemia, the average size of the parasite was about 21 × 2.3 μm, free flagellum, about 14 μm long. Body was slender, with pointed anterior and posterior ends. Oval nucleus 2.5 × 1.6 μm lacked the endosome and was located in the anterior half of the body. Young, slender forms were as short as 15.6

µm; stout, rarely occurring forms of the chronic phase of infection were 26–32 (sometimes even 40 µm) × 3–5.3 µm, with a relatively short flagellum (Table 2.10). The leech stages were described by Qadri (1962b); the life cycle was outlined by Lom (1979); division process was studied by Woo (1981a); culture stages were recorded by Wang and Belosevic (1994). The best model for studies of *T. carassii* infection was proposed to be goldfish (Lom, 1979). In laboratory infections, the survivors could be protected from re-infection. Trypanosomes incubated *in vitro* in immune goldfish plasma, for two hours at 20°C, were not infective.

T. carassii is pathogenic for some of its hosts (Lom and Dykova, 1992). Approximately 70% of goldfish succumbed to intraperitoneally administered infections soon after the strain had been isolated; the virulence of a strain increased with time over which it had been passaged in goldfish, and, after about 50 passages, the mortality was about 90%. The most pronounced histopathological changes were located in the hemopoietic organs, spleen, liver and kidney, although other organs were also affected. There was a striking absence of inflammatory reaction (Dykova and Lom, 1979).

T. danilewskyi is also pathogenic for juvenile carp, which may die, and which reveal significant histopathological changes (Lom and Machackova, 1986). The parasite can easily be cultivated in biphasic and liquid media (Lom, 1979) yielding up to 1.6×10^7 cells /ml, with a succession of epimastigotes, broad and then slender trypomastigotes.

Woo and Black (1984) showed that the host fish significantly affects the morphometrics of the trypanosome. The PK, AF and area of the nucleus are similar among samples from the same species of fish host but differ significantly between samples from five different host species.

The parasite was compared to earlier reported species from carps by Szulman (1984), Jastrzebski (1984), Lom and Dykova (1992) and Wita and Ovchrenko (1997) (Table 2.10). The parasite resembles *T. danilewskyi* recorded from *Cyprinus carpio* and fits well with the description given by Jastrzebski (1984) and is therefore identified as *Trypanosoma danilewskyi* (Laveran and Mesnil, 1904).

Table 2.10. *Comparison of Morphometrical Values (μm) of T. carassii (T. danilewskyi) According to Different Authors and the Present Species*

Component Parts of the Parasite	**Authors**						
	Szulman 1984	**Jastrzebski 1984** ***Cyprinus carpio***	**Lom & Dykova 1992** ***Cyprinus carpio***			**Wita & Ovcharenko 1997** ***Rutilus rutilus***	**Present sp. *Cyprinus carpio***
			a	**b**	**c***		
Length of the body (CL)	27.0–60.0	23.5–32.8	15.6	21.0	26–32 (max-40)	24.4 (12.0–44.1)	24.0–35.0
Width of the body (CB)	1.4–6.0	0.9–5.3	–	2.3	3.0–5.3	2.8 (0.8–5.7)	(1.6–3.1)
Length of free flagellum (AF)	7.0–22.0	7.2–10.9	–	14.0	relatively short	12.6 (8.4–19.8)	(9.2–16.2)
Size of nucleus (NL X NB)	2.2 × 4.7	2.3 × 4.4	–	1.6×2.5	–	2.8 × 4.0	2.4 × 4.3
Distance from posterior tip to kinetoplast (PK)	0.2–2.2	0.6–1.9				0.5–1.9	0.5–2.2
kinetoplast to middle of nucleus (KN)	–	–				–	7.8–15.0

*a- early phase of infection, b-peak of infection, c- chronic phase of infection.

2.4.2. Trypanoplasma golysiana (Gupta and Pilarczyk, 1994)

Peculiarly, trypanoplasm parasites from pond Wyszni II and Baginiec III differed from each other conspicuously. The blood and kidney forms from the former were rather uniform in size and thus monomorphic, whereas those from the latter were highly variable in size and polymorphic in nature. It was further interesting that the kidney from of both the species of *Trypanoplasma* were devoid of flagella, contrarily present in the blood forms of Wyszni II carp.

***Trypanoplasma golysiana* n. Sp.**

Host Profile	
Kingdom	Animalia
Phylum	Chordata
Superclass	Pisces
Subclass	Actinopterygii
Order	Cypriniformes
Family	Cyprinidae
Genus	*Cyprinus*
Species	*Carpio*
Parasite Profile	
Phylum	Protozoa
Class	Flagellata
Order	Kinetoplastida
Suborder	Bodonina
Family	Cryptobiidae
Genus	*Trypanoplasma*
Species	*Golysiana* n. sp.

2.4.2.1. Description

- **Shape and Size:** The body of the parasite is elongated or slightly curved in most of the forms (Plate XIX). The anterior end is broader whereas the posterior was bluntly pointed. Sometimes, the body assumes a wavy appearance being broader at some place. The length of the cell body in the blood form is 17.31 μm and that of the kidney form, 19.50 μm. The width of the two forms is 5.03 μm and 3.69 μm respectively indicating that the blood form (Plate XX) is shorter and broader as compared to the kidney form (Plate XXI).
- **Cytoplasm:** Parasites of both the sites had conspicuous granules in

the posterior half of the body which stained deeply but the anterior half had relatively few granules.

- **Nucleus:** Variously shaped nucleus occupies a position opposite to that of the kinetoplast. It contains deep staining chromatin material which is occasionally arranged in the form of a ring. The average length of the nucleus (3.99 μm) is usually more than twice its breadth (1.39 μm).
- **Kinetoplast:** The kinetoplast is conspicuously elongated and protrudes into the interior of the cell body being 9–11 times longer than its width (average KL, 7.07 μm; KB, 0.78 μm). It is the most deeply stained cell organelle.
- **Flagella:** The flagella were seen only in the blood forms and the anterior flagellum (12.6 μm) is usually twice as long as the posterior one (6.0 μm).
- **Measurements:** The measurements of the blood and kidney forms of the parasite along with their mean values are given in Table 2.11.

Table 2.11. *Biometric data of Trypanoplasma golysiana n. sp. and Trypanoplasma polygolysiana n. sp.*

Com-ponent parts of Parasite	***Trypanoplasma golysiana* n. sp.**			***Trypanoplasma polygolysiana* n. sp.**		
	Blood from	**Kidney from**	**Mean**	**Stumpy**	**Inter- Me-diate**	**Large**
CL	17.31 (14.6–19.0)	19.50 (16.3–24.5)	18.40	15.00 (10.2–21.2)	27.18 (23.5–29.5)	31.94 (30.0–34.8)
CB	5.03 (4.0–6.1)	3.69 (2.5–5.0)	4.36	10.55 (7.0–13.0)	7.48 (4.6–12.8)	6.41 (4.3–10.7)
AF	12.60 (7.0–17.0)	-	-	-	-	-
PF	6.00 (5.5–6.8)	-	-	-	-	-
NL	3.93 (3.8–4.2)	4.50 (3.0–5.0)	3.99	4.61 (2.0–6.4)	5.85 (3.6–8.8)	6.48 (5.0–9.7)
NB	1.35 (1.2–1.6)	1.43 (1.0–2.0)	1.39	2.47 (1.7–4.5)	2.38 (0.8–3.9)	2.27 (1.5–3.0)

KL	6.83 (5.3–7.5)	7.31 (6.3–8.8)	7.07	8.87 (6.3–12.2)	9.96 (8.5–12.0)	10.10 (8.4–11.8)
KB	0.86 (0.8–1.0)	0.70 (0.5–1.0)	0.78	0.95 (0.7–1.2)	0.92 (0.5–1.3)	0.97 (0.8–1.2)

Note: All values are in μm.

2.4.2.2. Trypanoplasma Polygolysiana Gupta and Pilarczyk, 1994

Host Profile	
Kingdom	Animalia
Phylum	Chordata
Superclass	Pisces
Subclass	Actinopterygii
Order	Cypriniformes
Family	Cyprinidae
Genus	*Cyprinus*
Species	*Carpio*
Parasite Profile	
Phylum	Protozoa
Class	Flagellata
Order	Kinetoplastida
Suborder	Bodonina
Family	Cryptobiidae
Genus	*Trypanoplasma*
Species	*Polygolysiana* n. sp.

Examination of the blood and kidney of Baginiec III carp revealed hemoflagellates amongst the cells of kidney. This polymorphic parasite (Plate XXII) existed as short, stumpy forms (Plate XXIIIA) and elongated forms discernible as intermediate (Plate XXIIIA) and large (Plate XXIIIB). 18 stumpy forms/100 *Trypanoplasma* could be observed and the intermediate forms outnumbered the others. The infection was intense and all the polymorphic forms could be visible in a single field under oil immersion (Plate XXIV).

2.4.2.3. Description

- **Shape and Size:** The shape of the parasite is highly variable. Almost spherical, triangular, 'C'-shaped, comma, irregular, shoe-shaped and typically elongated forms could be observed. The stumpy forms were the widest. The size of the parasite is 15.0 × 10.55 µm (stumpy form), 27.18 × 7.48 µm (intermediate form) and 31.94 × 6.41 µm (large form).
- **Cytoplasm:** Typically, the posterior part of the forms was most densely granulated and the vacuoles were irregularly scattered.
- **Nucleus:** There was a corresponding increase in the length of the nucleus with an increase in size of parasite though the width did not increase. It ranges in length from 4.61 µm in the stumpy form to 6.48 µm in the large form. Chromatin material is densely placed in the form of a ring under the outer margin of the nucleus.
- **Kinetoplast:** The deeply stained kinetoplast is an elongated structure increasing proportionally with the size the parasite (6.3 µm – 11.8 µm). Its position ranges from sub terminal, oblique or internally placed. One end takes a deeper stain and is usually swollen, in a few instances it arises from a pocket-like structure and it rarely protrudes from the body of the parasite.
- **Flagella and Undulating Membrane**: Neither the flagella were discernable nor was a definite undulating membrane seen in any of the forms.
- **Measurements:** The measurements of the 'stumpy,' 'intermediate' and 'large' forms of the parasite along with their mean values are given in Table 2.11.

2.4.2.4. Discussion

The monomorphic nature of the *Trypanoplasma* from pond Wyszni II marks it off from various di- and polymorphic species of *Trypanoplasma.* Compared with various other described monomorphic trypanosomes, it has an overlapping length range with *Cryptobia bullocki* Strout, 1965; *T. cavacii* Wahul, 1985 and *T. wallagoi* Wahul, 1985. However, the present form differs from *T. bullocki* in size of flagella and kinetoplast and from *T. cavacii* and *T. wallagoi* in shape, kinetoplast and nuclear dimensions. In view of its distinctness, this species is considered to be new to science and is designated as *Trypanoplasma golysiana* n. sp.

Becker (1977) stated that nearly all species of *Trypanoplasma* are monomorphic while in the blood of fish. However, the definite existence of the sequence of different morphological types (polymorphism) of *Trypanoplasma* during the course of infection in carp was firmly established by Keysselitz (1906). A clear proof of trimorphism has been observed in *T. willoughbii* (Qadri, 1962d), *T. catostomi* (Bower and Woo, 1977b) and *T. guptii* (Gupta and Gupta, 1987a).

The biflagellates from carp of pond Wyszni III having a pond surface of 6.4 ha demonstrate similar polymorphism. Interestingly, this parasite showed a diversification in contour which has not been recorded from any carp of European waters or elsewhere. The other diagnostic features which warrant creation of new species are orientation and characterization of the kinetoplast, nuclear dimensions and differential cytomorphological characterization as compared to other species of polymorphic trypanoplasms. At this extensive carp rearing center, the blood parasites were identified for the first time. These facts conclusively establish the need for creation of a new species and it is named *Trypanoplasma polygolysiana* n. sp.

2.4.3. Trypanoplasma borreli (Laveran and Mesnil, 1901)

Host Profile	
Kingdom	Animalia
Phylum	Chordata
Superclass	Pisces
Subclass	Actinopterygii
Order	Cypriniformes
Family	Cyprinidae
Genus	*Cyprinus*
Species	*Carpio*
Parasite Profile	
Phylum	Protozoa
Class	Flagellata
Order	Kinetoplastida
Suborder	Bodonina
Family	Cryptobiidae
Genus	*Trypanoplasma*
Species	*Borreli* (Laveran and Mesnil, 1901)

2.4.3.1. Description

- Shape and Size: The parasite is elongated with considerable variations in size and shape. Upon fixation, they could assume horse-shoe-shape (Plate XXVA) C-shape (Plate XXVB) stumpy (Plate XXVIA) question-mark-shaped (Plate XXVIB) or elongated (XXVIIA). Some forms were more or less compressed and assumed a triangular shape. During acute infections, small and slender forms dominate whereas under chronic conditions, stumpy parasites are the principal forms. As stumpy forms dominated, the infection appeared to be chronic.

The body length is approximately 5-fold of its width being 25.3 + 2.8 μm long (CL) and 5.2 + 1.4 μm wide (CB). As compared to trypanosomes, the parasite was slower in movement under live condition with wavy movements of the undulating membrane.

- Cytoplasm: The cytoplasm is irregularly granular, the margins and posterior end of the parasite usually taking a deeper stain. Clear, non-granulated areas are also visible.
- Nucleus: Nucleus is usually oval in shape situated in the anterior part of the parasite (near the convex margin) opposite to the kinetoplast. In some forms it is almost circular (Plate XXVIB) or maybe pointed at one end (Plate XXVIIA). L:W may be about 2:1. and is 4.6 + 0.75 μm in length (NL) and 2.2 + 0.33 μm in width (NB) on an average.
- Kinetoplast: The kinetoplast is the most diversified organelle. The shape ranges from elongated (Plate XXVIIA) to curved (Plate XXVIIB) and is occasionally notched (Plate XXVB). The position of the elongated kinetoplast may be marginal (Plate XXVIIB) or partially marginal (Plate XXVIIIA) but may also be obliquely placed (Plate XXVIB). In all cases, it is a deeply stained and well differentiated structure and its length (KL) is 8.1 + 0.91 μm and width (KB) 0.7 + 0.12 μm.
- Flagella and Undulating Membrane: The two flagella arise just anterior to the kinetoplast (Plate XXVIIIA); the anterior flagellum runs anteriorly whereas the other may border the undulating membrane, then protrude freely (Plate XXVIIIB). In some cases, the posterior flagellum does not traverse the undulating membrane, instead, it is inconspicuous in parasites having bulging body (Plate XXVIIIB). The anterior flagellum is always larger (13.9 + 3.1 μm) than the posterior one (8.4 + 2.4 μm).
- Measurements: The measurements of the trypanoplasms discovered from the blood of *Cyprinus carpio* and their comparison to *Trypanoplasma borreli* (Laveran and Mesnil, 1901) as described by Kruse et al. (1989) are shown in Table 2.12.

Table 2.12. Biometrical Data of the Present Species Compared with *Trypanoplasma borreli* (Laveran and Mesnil, 1901) as described by Kruse et al. (1989)

Component Parts of Parasite	***Trypanoplasma borreli* Kruse et al. (1989)**	***Trypanoplasma borreli* Present Species**
CL	24.4±3.0	25.3±2.80
CB	4.9±1.2	5.2±1.40
AF	14.6±3.6	13.9±3.10
PF	7.9±2.7	8.4±2.40
NL	4.1±0.86	4.6±0.75
NB	2.1±0.47	2.2±0.33
KL	7.6±1.6	8.1±0.91
KB	–	0.7±0.12

Note: All values are in μm.

Trypanoplasm division was rare. Occasionally, *T. borreli* with two kinetoplasts could be observed (Plate XXVA) but the sequential pattern of division could not be traced.

Type host	*Cyprinus carpio*
Type locality	Pond Wyzni II, Golysz, Poland
Site of infection	Blood
Vector	*Hemiclepsis marginata* (Europe and Asia); *Piscicola geometra* (Eurasia) (Woo, 1995)

2.4.3.2. Discussion

Trypanoplasma borreli (Laveran and Mesnil, 1901) is a pathogenic blood flagellate from cyprinid fishes in Europe. It is transmitted by the blood-sucking leeches *Piscicola geometra* and *Hemiclepsis marginata* and causes 'sleeping sickness' of *Cyprinus carpio* and *Tinca tinca* (Lom and Dykova, 1992; Woo and Poynton, 1995). Infections of carp with *T. borreli* can easily be induced and monitored and therefore are good models for the study of interactions of the parasite with the host immune system (Jones et al., 1995; Wiegertjes et al., 1995) and mechanisms of pathogenesis (Lom and Dykova, 1992; Bunnajirakul et al., 2000). *Trypanoplasma borreli* was described from red eye (*Leuciscus erythrophthalmus*) caught in ponds from Garches (France). Since then, the parasite has been reported in the blood and tissues of several freshwater cyprinids in Europe. Sometimes, it has caused heavy losses to cultured carps, Tench and crucian carp (Lom, 1979). The parasite is not host specific and causes mortality in experimentally infected goldfish (*Carassius auratus*) and *Cyprinus carpio* (Dykova and Lom, 1979; Lom, 1979). Extensive morphological and cross transmission studies by Lom (1979) concluded that the parasites from carp, goldfish and crucian carp in Europe are all *T. borreli* and that the pathogenic *T. cyprini* is a junior synonym of *Cryptobia* (*Trypanoplasma*) *borreli*. During its life cycle, it develops in two freshwater leeches (*Hemiclepsis marginata* and *Piscicola geometra*).

Keysselitz (1906) indicated that post spawning cyprinids seemed to be more susceptible to the parasite than sexually mature fish. The parasite multiplies rapidly after infection and this is followed by their elimination from peripheral blood (Steinhagen et al., 1989; Jones et al., 1993). The prepatent period, intensity and durations of infection are affected by water temperature (Steinhagen et al., 1989) as was also shown earlier for salmonid cryptobiasis (Woo et al., 1983). *T. borreli* infection lasted for 20 weeks in fish kept at 20°C which shortened to 12 weeks at 30°C (Steinhagen et al., 1989).

The course of spontaneous trypanoplasm infections in the wild very much resembles that of trypanosomes with the difference that many *Trypanoplasma* species are serious pathogens and the infections are fatal. The large "adult" forms are most often found in symptomless hosts. Successive relapses of the acute phase of infection caused neither by reinfection nor environmental factors, has been shown to date to occur only in *T. salmositica*. In other species, it may be due to constant contact with infective leeches or due to seasonal effects, such as early spring stresses in European fish polyculture, exacerbating the latent chronic infections.

T. borreli appears in the peripheral blood of goldfish in 4–47 days depending primarily on the number of flagellates introduced. Parasitaemias peak in next 7–18 days depending on the individual resistance of fish and water temperature. The maximum infection intensity may reach 120,000 flagellates per c mm of blood. Erythrocyte counts are reduced to about 9,50,000 from a normal 1–2 million and eventually mortality is about 80%. Surviving goldfish are free from detectable infection after about 90 days and the fish are at least temporarily immune to re-infection (Lom, 1973). At temperatures of 10°C and below, patent infections are not reported to occur (Lom, 1973).

Stress is known to promote infections of *T. borreli* infections in carps and Tench. Under natural conditions, trypanoplasms and trypanosomes can exist simultaneously in carp, but one or the other, usually the trypanoplasms gradually disappear from goldfish (Krier, 1977).

Experimental induction of the parasite induced anaemia (Lom, 1979; Jones et al., 1993). This is manifested at higher intensities and is absent from recovered fish (Woo, 1995). The parasite was responsible for killing 56–80% of experimentally infected goldfish (Lom, 1979), mortality being related to the size of the inoculum and water temperature.

The morphological development of cloned *Trypanoplasma borreli* during the course of experimental infections in common carp was studied by Kruse, Steinhagen and Korting et al. (1989). During the course of infection, the cell dimensions of *T. borreli* showed high variability. At the beginning of parasitemia at day 8 p.i., small and slender *T. borreli* predominated: surface area 48.9 ± 13.1 mm^2; shape index (SI, length/width) 7.8 + 2.1 (n = 110). During the exponential growth phase until day 13 p.i. the size further decreased significantly *(P* < 1%): SA 29.9 ± 9.5 µm: (n = 109).

During the transition of parasitemia from exponential growth to the chronic phase, the size of *T. borreli* enlarged significantly *(P* < 1%) and it became stumpy in shape. Day 15 p.i.: SA 63.2 ± 15.8 μm^2; SI 4.2 ± 1.0 (n = 51). During the chronic phase significantly *(P* < 1%) larger trypanoplasms predominated: day 41 p.i. SA 90.6 ± 22.7 μm^2, SI 4.9 ± 1.1 (n = 221). Further samples showed only moderate variability of cell dimensions during this phase of parasitemia. Measurements of dimensions of kinetoplast and nucleus and length of flagella varied in a similar manner. The position of the kinetoplast remained unchanged, while the nucleus had a more anterior position in the chronic phase. Experimental infections of various cyprinids were successful, perch and rainbow trout were not susceptible to experimental infections with *T. borreli.*

The trypanoplasm species reported from freshwater fishes are shown in Table 2.13. The trypanoplasm recovered from the blood of *Cyprinus carpio* from pond Wyzni II, Golysz, Poland was compared to the earlier reported species and resembles *Trypanoplasma borreli* (Laveran and Mesnil, 1901) as described by Kruse et al. (1989) (Table 2.12) and is identified as such.

Table 2.13. *Trypanoplasma spp. of Freshwater Fish*

Species	**Principal Hosts**	**Geographic Area**
T. abramidis (Brumpt, 1906)	*Abramis brama*	Europe, Asia
T. acipenseris Ioff, (Levashov & Boschenko, 1926)	*Huso huso, Acipenser nudiventris, A. ruthenus, A. guldenstadti, A. stellatus, Pseudoscaphirhynchus* sp.	Europe, Asia

T. barbi (Brumpt, 1906)	*Barbus barbus, B. fluviatilis, B. vulgaris*	Europe
T. beckeri (Burreson, 1979)	*Scorpaenichthys marmoratus*	Oregon coast, USA
T. bliccae (Nikitin, 1929)	*Blicca bjoerkna*	Asia
T. borreli (Laveran & Mesnil, 1901)	*Scardinius erythrophthalmus, Phoxinus phoxinus,* possibly other cyprinids	Europe, Asia
T. bullocki (Strout, 1965)	*Opsanus tau, Fundulus heteroclitus, F. majalis, Liopsetta putnami, Pseudopleuronectes americanus, Paralichthys dentatus, Trinectes maculates*	North America
T. cataractae (Putz, 1972)	*Rhinichthys cataractae, R. atratulus; Exoglossum maxillingua; Campostoma anomalum*	West Virginia
T. catostomi (Bower & Woo, 1977)	*Catostomus commersoni*	USA
T. capoetobramae (Osmanov, 1963)	*Capoetobrama kuschakewitschi*	USSR
T. carassii (Kaschkovsky, 1974)	*Carassius auratus*	Siberia
T. clariae (Mathis & Leger, 1910)	*Clarias macrocephalus*	Vietnam
T. cyprini (Plehn, 1903)	*Cyprinus carpio, Carassius auratus, C. vulgaris, Ctenopharyngodon idella,* other cyprinids	Europe, Asia, Germany
T. erythroculteri (Chen & Hsieh, 1964)	*Erythroculter dabryi*	China
T. gandei (Rodhain, 1942)	*Labeo macrostomata*	Africa
T. guernei (Brumpt, 1906)	*Cottus gobio*	France
T. gurneyorum (Minchin, 1909)	*Esox lucius, Coregonus clupeiformis, Salvelinus namaycush*	Europe, Asia, northern Canada, England

T. keysselitzi (Minchin, 1909)	*Tinca vulgaris*	England
T. lomakini (Khaybuly-aev, 1971)	*Gobius melanostomus affinis, G. kessleri gorlap, G. fluviatilis pallasi, Benthophilus macrocephalus*	Caspian Sea
T. makeevi (Achmerov, 1959)	*Oncorhynchus gorbuscha, O. keta*	Amur river, East Asia
T. markewitchi (Schapow-al, 1953)	*Anguilla anguilla*	European part of USSR
T. megalobrami (Chen & Hsieh, 1964)	*Megalobrama amblycephala*	China
T. mirabilis (Dzhalilov, 1965)	*Varicorhinus heratensis steindachneri*	Tadzhikistan, USSR
T. misgurni (Chen, 1964)	*Misgurnus anguillicaudatus*	China
T. ninae kohl-yakimovi (Yakimov & Schochcr, 1917)	*Siluris* spp.	Turkestan, USSR
T. pseudobagri (Chang, 1964)	*Pseudobagrus fulvidraco*	China
T. pseudocaphirhynchi (Ostroumov, 1949)	*Pseudocaphirhynchus kaufmanni*	Uzbekistan, USSR
T. rutili (Osmanov, 1963)	*Rutilus rutilus caspicus*	Uzbekistan, USSR
T. salmositica (Katz, 1951)	*Oncorhynchus kisutch, Salmo trutta, Prosopium williamsoni, Coyyus spp. Rhinichthys cataractae, Catostomus snyderi, Gasterosteus aculeatus*	Washington, California, British Columbia, southwestern Alaska
T. sarcocheilichthysi (Chen & Hsieh, 1964)	*Sarcocheilichthys nigripinnis*	(US Pacific coast)
T. tincae (Schaperclaus, 1954)	*Tinca tinca*	China
T. truttae (Brumpt, 1906)	*Salmo trutta fario*	Germany
T. valentine (Gauthier, 1920)	*Salmo trutta fario*	France
T. varium (Leger, 1904)	*Noemacheilus barbatula*	France

T. willoughbii (Qadri, 1962)	*Salvelinus willoughbii*	France
T. keysselitzi (Minchin, 1909)	*Misgurnus anguillicaudatus*	England

2.4.4. Trypanosoma lewisi (Laveran and Mesnil, 1901)

Host Profile	
Kingdom	Animalia
Phylum	Chordata
Class	Mammalia
Order	Rodentia
Family	Muridae
Genus	*Rattus*
Species	*Rattus* (Linnaeus, 1758)
Parasite Profile	
Phylum	Protozoa
Class	Flagellata
Order	Kinetoplastida
Suborder	Trypanosomatina
Family	Trypanosomatidae
Genus	*Trypanosoma* (*Herpetosoma*)
Species	*Lewisi* (Laveran and Mesnil, 1901)

2.4.5. The Host: Rattus (Linnaeus, 1758)

The black rat (*Rattus rattus*) is a common long-tailed rodent of the genus *Rattus* (rats) in the subfamily Murinae (murine rodents). The species originated in tropical Asia and spread through the Near East in Roman times before reaching Europe by the Ist century and spreading with Europeans across the world.

A typical adult black rat is 12.75–18.25" (32.4–46.4 cm) long, including a 6.5–10 in (17–25 cm) tail and weighs 4–12 oz (110–340 g). Despite its name, the black rat exhibits several color forms. It is usually black to light brown in color with a lighter underside. In the 1920's in England, several variations were bred and shown alongside domesticated brown rats. This included an unusual green tinted variety.

It is nocturnal and omnivorous, with a preference for grains and fruit. Compared to the brown rat, it is a poor swimmer, but more agile and a better climber, tending even to flee upwards. In a suitable environment, it breeds throughout the year, with a female producing three to six litters of up to ten young. Females may regulate their production of offspring during times when food is scarce, producing as few as only one litter a year. *R. Rattus* lives for about 2–3 years. Social groups of up to sixty can be formed.

Rats serve as outstanding vectors for transmittance of diseases because they have the ability to carry bacteria, viruses and parasites in their systems. A few parasites that are common to rats include *Streptococcus pnuemoniae*, *Corynebacterium kutsheri, Bacillus piliformis*, *Pasteurella pneumotropica*, *Streptobacillus moniliformis* and *Trypanosoma lewisi* to name a few. In some cases, these diseases are incurable.

2.4.5.1. Methodology

2.4.5.1.1. Maintenance of Host

Rattus Rattus were trapped alive in cages from storehouses, godowns in and around Bareilly, U.P. India. They were maintained in cages and fed on food grains, bread crumbs and water *ad libitum*. They were maintained in a hygienic environment. Uneaten left-over foods were removed and dishes and water cans were cleaned daily when new food and fresh water was given. The cage and accessories were cleaned weekly with hot soapy water to remove any hardened material and were sanitized using disinfecting solutions to kill germs. The cages and accessories were rinsed thoroughly when new animals were introduced and air dried. Care was taken to dry the cage and accessories before resetting the animals.

2.4.5.1.2. Examination of Blood

Tail vein sampling was used for blood collection. Blood was collected from the lateral tail vein. The tail was washed with alcohol. Finger pressure, 5 cm from the tail tip was applied to enhance visibility of the tail vessels. Blood samples were taken from the base of the tail with the help of a 2 ml disposable syringe.

After collection of the sample, the blood flow was stopped by applying finger pressure on soft tissue placed at the blood sampling site for approximately 30 seconds before the animal was returned to its cage.

The collected blood was transferred to tubes containing heparin. The samples were tested for trypomastigotes by examination in micro hematocrit tubes, hanging drop and permanent smear preparation.

When centrifuged, the trypanosomes seen swimming above or sometimes in the buffy coat were placed on microscope slides and thin films prepared, air dried, fixed in absolute methanol and stained with Giemsa. Thin smears were prepared for permanent slides, air dried, fixed in methyl alcohol and stained for 1 hour in Giemsa, diluted (1:7) in phosphate buffer (pH 7.2). Slides were rinsed in buffer, dried and examined under a light microscope.

The stained blood films were examined at a magnification of 1000×. Morphometric observations followed the nomenclature and methodology of Hoare (1972).

2.4.5.1.3. Description

Trypanosoma (*Herpetosoma*) *lewisi* was recovered from the blood of five black rats, *Rattus rattus.* The rats apparently did not show any clinical symptoms. In fresh preparations, the adult form of *T. lewisi* could be seen dashing at high speed through the microscopic field. The posterior end of the parasite is very thin, long and rigid (Plate XXIXA).

The parasite was more distinct in the microscopic observation of Giemsa-stained blood smears, where the parasites appeared as described by Hoare (1972). Depending on the incubation period, parasites appeared as immature forms (Plate XXIXB) or classical adult slender forms (Plate XXIXC).

The intensity of infection ranged from 10–15 parasites/100 RBC. All trypanosomes observed in the bloodstream of the black rats were trypomastigotes. Thirty-five undistorted trypomastigotes, typical *Herpetosoma* forms were measured.

The adult form is monomorphic, and can easily be distinguished from *Trypanosoma vivax* (large and circular kinetoplast situated at the round posterior end of the body, nucleus in central position). The thin straight posterior end of the parasite in adult forms takes lighter stain and may not be clearly visible and therefore, the parasites were over-stained.

- **Shape and size:** *Trypanosoma lewisi* is a long and slender parasite, both extremities of the body tapering to a fine point. It usually appears C-shaped (Plate XXXA) or S-shaped (Plate XXXB) in permanent preparations. The length of the cell body is 23.5 + 2.1 μm whereas the total length of the parasite including free flagellum is 31.4 + 1.1 μm.
- **Cytoplasm:** The alveolar cytoplasm stains light blue with Giemsa. Some of the larger forms of *T. lewisi* have from one-to-many extranuclear chromatin granules in the cytoplasm. The cytoplasm often has large vacuoles and an irregular border on the convex side. It is likely that such trypanosomes are degenerative forms which are undergoing necrosis.

- **Nucleus:** The nucleus is anterior in position, oval in shape, its long axis being parallel to the body (Plate XXXC). Both the nuclear membrane and the small central karyosome stain deep red, and the space between them lighter red. The nuclei of forms having extranuclear chromatin granules frequently contain several densely staining granules of similar nature. The size of the nucleus is 1.2 + 0.6 × 1.6 + 0.8 µm.
- **Kinetoplast:** The kinetoplast is a well developed, deeply stained structure oval in shape, situated far from the posterior end (Plate XXXIA). The kinetoplast as a whole stains bright pink, and appears circular to rod-shaped. It measures 0.82 × 0.73 µm.
- **Flagellum:** The flagellum varies in apparent thickness according to the degree of extraction of the stain. It originates from a blepharoplast which lies immediately anterior to the parabasal body but which is not always differentiated from the latter structure in Giemsa-stained material and runs anteriorly as a free flagellum (Plate XXXIB). It is approximately one third of the body length (7.9 + 1.0 µm).
- **Undulating Membrane:** The prominent undulating membrane very rarely appears at the concavities of the curves. It usually keeps to the convex side, crossing over the body at each curve (Plate XXXIC). The maximum width of the undulating membrane is 0.9 + 0.01 µm.
- **Dividing Forms**: A few forms undergoing division were observed which are described in the chapter "Division in Trypanosomes."
- **Measurements:** The statistical measurements of the parasite have been provided in Table 2.14 and is compared with some of the earlier reported species as mentioned by other authors.

2.4.6. Key for the Diagnosis of Mammalian Trypanosomes in Stained thin Blood Films (Kreier and Baker, 1987)

1	Trypanosomes more than 40 µm long Trypanosomes less than 40 µm long	*T. theileri* 2
2	All individuals with free flagellum	3
	Some or all individuals without free flagellum	11
3	Kinetoplast not terminal, medium sized to small (or absent)	4
	Kinetoplast terminal, large	9

4	Posterior end sharply pointed	5
	Posterior end bluntly pointed or rounded	10
5	Kinetoplast medium-sized or large	6
	Kinetoplast small	8
6	Trypanosomes curved (C-shaped); kineto-plast large, round Not as above	*T.cruzi* 7
7	Kinetoplast medium-sized, large Kinetoplast large, rod shaped	*T. rangeli* *T. lewisi*
8	Long, slender trypanosomes (rare Central African species)	*T. suis*
9	Trypanosomes more than 20 μm long Trypanosomes less than 18 μm long	*T. vivax* *T. uniforme*
10	Kinetoplast present, small	*T. evansi, T.*
		equiperdum and old
		laboratory stains of
		T. brucei (*sensu lato*)
	Kinetoplast absent	*T. equinum*
11	All individuals without free flagellum Some individuals have, or appear to have, a free flagellum	*T. congolense*
12	Differentiated into long, slender forms with free flagellum, short stumpy forms without free flagellum, and intermedi-ates; kinetoplast not markedly marginal in position Not as above; kinetoplast marginal	*T. brucei* (*sensu lato*) *T.simiae*

2.4.6.1. Discussion

The trypanosome from *Rattus Rattus* is placed in the genus *Herpetosoma* Doflein (Stercoraria section) and is identified as *Trypanosoma lewisi* (Laveran and Mesnil, 1901) according to the above key.

Trypanosoma (Herpetosoma) lewisi Kent, 1880 is generally regarded as non- pathogenic in adult animals but it can cause lethal infections in suckling rats. It has a stringent species specificity and cannot grow in other rodents such as mice. Rats are infected principally by oral route, through contamination by flea feces or ingestion of fleas. *T. lewisi* infections in rat colonies can interfere with research protocols and fleas of wild rats are often the source of such infections.

Doré (1918) was the first to record *T. lewisi* from New Zealand, stating that this flagellate was first found in rats from the Auckland area in 1912. Of "several hundred" rats from various parts of the North Island examined by Doré, 30% of those captured in the neighborhood of sewers and 12% of those from wharves and grain stores were infected. *T. lewisi* is also widespread in Australia (Pound, 1905; Johnston, 1909).

Three out of 65 specimens of *Rattus norvegicus* were reported to be parasitized by *T. lewisi*, two being lightly infected and one very heavily infected, with an average of one trypanosome to every eight host erythrocytes (Laird, 1951).

Saxena and Miyata (1993) reported an unusual morphological type of *T. lewisi* which in *Rattus norvegicus* from India was different from known types of *T. lewisi* in that the kinetoplast was at a greater distance from the posterior end (PK) than the usual type. The total length, including the free flagellum, is 35.3–39.3 μm and PK is 11.9–15.0 as against total length of 21.0 to 36.5 μm in the normal form (Table 2.14).

Karbowiak and Wita (2001) recorded *T. lewisi* from two out of three brown rats (*R. norvegicus*) caught in Milanwek near Warsaw, and one caught in the center of Warsaw for the first time from free living rats in Poland.

Allam et al. (2002) examined Damietta and Qualyobia governorates in Egypt for the seasonal activities of *Rattus* species and subspecies and their natural infection with *T. lewisi.* The common domestic rodents identified were *Rattus r. alexandrinus, R. r. frugivorus* and *R. norvegicus*. The three rodent species were naturally infected with *T. lewisi*.

Karbowiak et al. (2009) reported the light and electron microscopic examination of *T. lewisi* isolated from *R. norvegicus* from Poland. Bloodstream trypomastigotes were identified morphometrically from 100 specimens collected from three naturally infected rats *Rattus norvegicus*. Body length ranged from 15.45–23.64 μm and width from 1.3–2.32 μm while the free flagellum was 8.1 μm long. Electron microscopic study of bloodstream trypomastigotes exhibited typical ultrastructural features similar to those of other stercorarian trypanosomes.

The parasite discovered from the blood of *Rattus Rattus* from Bareilly, U.P., India conforms to the basic description of *Trypanosoma lewisi* (Laveran and Mesnil, 1901) and is identified as such.

Table 2.14. Measurements of Present Species of Trypanosoma lewisi (μm) Compared with Some of the Earlier Reported Ones

Author(s) Year Component Parts of Parasite	**Kent (1880)**	**Laveran & Mesnil (1901)**	**Minchin (1908)**	**Taliferro (1921)**	**Laird (1951)**	**Hoare (1972)**	**Saxena & Miyata (1993) (Unusual form)**	**Sorata phan (2007)**	**Karbo-wiak et.al, (2009)**	**Present species**
Length of free flagellum (FL)	-	-	6.9	6.61	7.1	-	-	7.8	8.1	7.9 ± 1.0
Length of the body (CL)	21.0–37.0	21.0–37.0	23.7	-	22.9	-	-	-	15.45–23.64	23.5 ± 2.1
Total length (CF)	-	-	30.6	31.25	30.0	30.6	35.3–39.3	31.8	-	31.4 ± 1.1
Width of the body at center of nucleus (CB)	-	-	-	1.59	1.1	-	(21–36.5) normal	-	1.3–2.32	1.2 ± 0.6
Length of nucleus (NL)	-	-	-	-	1.7	-	-	2.7	-	1.6 ± 0.8
Width of nucleus (NB)	-	-	-	-	1.1	-	-	-	-	1.0 ± 0.4
Width of undulating membrane (UM)	-	-	-	-	1.0	-	-	-	-	0.9 ± 0.01
Distance from posterior end of the body to Kinetoplast (PK)	-	-	-	4.26	-	-	-	4.2	-	3.51
Distance from Kinetoplast to center of nucleus (KN)	-	-	-	10.85	-	-	11.9–15.0	9.5	-	11.52
Distance from posterior end of the body to nucleus (NP)	-	-	-	-	-	-	-	-	-	15.03

Distance from nucleus center to anterior end of body (NA)	-	-	-	9.51	-	-	-	9.7	-	8.47
Length of Kineto-plast (KL)	-	-	-	-	-	-	-	-	-	0.82
Breadth of Kineto-plast (KB)	-	-	-	-	-	-	-	-	-	0.73
KI = PN/KN	-	-	-	-	-	-	-	-	-	1.30
NI =PN/NA	-	-	-	-	-	-	-	1.4	-	1.77
FI =FF/BL	-	-	-	-	-	-	-	-	-	0.33

2.4.7. The Host: Apodemus flavicollis Melchior, 1834

The Yellow-necked mouse, *Apodemus flavicollis* is closely related to the wood mouse, with which it was for long confused, only being recognized as a separate species in 1894. It differs in its band of yellow fur around the neck and in having slightly larger ears and usually being slightly larger overall. The head-body length is 95–120 mm, males are 95–120 mm (mean 103.8 mm), females 96–114 mm (mean 103.6 mm). The dorsal surface of females is rich orange-brown or yellow-brown in color whereas the ventral is pale grey-white, with chest collar yellow, reaching up to meet darker dorsal fur giving it the characteristic name of yellow-necked mouse.

The habitat and range of distribution is mainly mature deciduous woods, particularly drier areas long-established woodlands. It can also be found in other deciduous forests, mixed forest, and (marginal habitats) hedges, rural gardens, rural buildings. It can climb trees and sometimes overwinters in houses. It is found mostly in mountainous areas of southern Europe, but extends north into parts of Scandinavia and Britain. It is a critical host in the maintenance of tick- borne encephalitis. Other common names are *Yellow-necked Field Mouse*, *Yellow- necked Wood Mouse*, and *South China Field Mouse.*

2.4.8. Trypanosoma (Herpetosoma) Grosi Kosewiense

Until recently *Herpetosoma* trypanosomes were not investigated in Poland. Only in the last few years records have appeared about trypanosome infections of the bank-vole, *Clethtrionomys glareolus,* and voles from genus *Microtus* (Karbowiak and Sinski, 1996b; Bajer et al., 1998; Karbowiak, Wita and Rychlik et al., 1998). Trypanosomes infecting yellow-necked mice *Apodemus flavicollis* are practically unknown; however, *Trypanosoma grosi*, infecting the wood mouse, *Apodenmus sylvaticus*, is well described in England and Germany

(Krampitz, 1959; Hoare, 1972; Noyes et al., 2002). The yellow-necked mice *Apodemus flavicollis* were investigated for the presence of blood parasites and *Trypanosoma* were identified from its blood samples.

Host Profile	
Kingdom	Animalia
Phylum:	Chordata
Class:	Mammalia
Order:	Rodentia
Family:	Muridae
Genus:	*Apodemus*
Species:	*flavicollis* (Melchior, 1834)
Parasite Profile	
Phylum:	Protozoa
Class:	Flagellata
Order:	Kinetoplastida
Suborder:	Trypanosomatina
Family:	Trypanosomatidae
Genus:	*Trypanosoma* (*Herpetosoma*)
Species	*grosi kosewiense* (Karbowiak and Wita, 2004)

2.4.8.1. Methodology

2.4.8.1.1. Animals

Yellow-necked mice *Apodemus flavicollis* were collected from Kosewo Gorne, Mazurian Lakeland, Poland. The animals were trapped monthly, using live traps containing oats as bait. After collection of material, the animals were numbered by toe clipping and released where they were trapped.

2.4.8.1.2. Detection of Trypanosomes

Blood samples were obtained from the tail tip of investigated animals. The trypanosomes were detected by centrifugation in microhematocrit tubes (8 mins, 6,200 rpm). They accumulated above the WBC's fraction, and their movements were observed using a light microscopy, using magnifications 10 × 10 and 10 × 20 (eyepiece x objective). Smears were made from the fraction with trypanosomes, as well as from non-centrifuged blood. The smears were fixed in absolute methanol (10 min) and stained with Giemsa's reagent stain, diluted 1:15 in 0.2 M phosphate buffer, pH 7.2 for 60 mins. Giemsa's reagent stained the cytoplasm of *Herpetosoma* trypanosomes light blue, the nucleus and kinetoplast a violet or purple-blue. Because the quality of staining with Giemsa's reagent was poor in the case of smears made from the centrifuged blood, the commercial kit Hemacolor (Kruszewicz, 1994) was used.

For the measurements of parasites, the "Analysis" software in combination with a video camera and Olympus BX50F4 microscopy was used. This method allowed obtaining results accurate to 0.01 µm. Stained blood smears were examined under oil immersion. The nomenclature of morphometrics parameters of trypanosomes was that commonly used by Hoare, (1972); Matthews et al. (1977) and Kingston, Bobek et al. (1992b).

Typical trypomastigote forms were selected for measurements. Poorly stained, damaged or deformed individuals were ignored. The data obtained were compared with similar data given by Krampitz (1959), for *Trypanosoma grosi* Laveran et (Pettit, 1909), parasitizing *Apodemus sylvaticus*, related to *Apodemus flavicollis* and *Trypanosoma grosi kosewiense* (Karbowiak and Wita, 2004).

2.4.8.1.3. Description

Infection with *Trypanosoma grosi kosewiense* (Karbowiak and Wita, 2004) was discovered in 2 yellow-necked mice of 32 caught. Morphometric measurements were made on 30 undistorted specimens of trypomastigotes, obtained from 3 heavily infected hosts.

The trypanosomes in the blood of the yellow-necked mouse occurred as trypomastigote forms. Two different morphological forms were observed indicating that the parasite was dimorphic.

2.4.9. "Slender" Form

- **Shape and Size**: The "slender" form (Plate XXXIIA,B), characteristic of the *Trypanosoma lewisi* group has a narrow, long body, with a sharpened posterior end. The length of the body is 22.64 + 1.92 µm (range 17.20–27.5 µm), width 1.81 + 0.41 µm (range 1.2–2. 8 µm).
- **Cytoplasm**: The cytoplasm is pale in color; the metachromatic granules being concentrated along the undulating membrane in the post-nuclear region.
- **Nucleus**: The nucleus is 2.20 + 0.24 µm (range 1.7–3.2 µm) long (Plate XXXIIC), oval in shape and located in the middle or in the

anterior part of the body, parallel to the long axis of the parasite. It is situated 11.2 + 0.98 μm from the anterior and 11.80 + 0.98 μm from the posterior end of the body.

- **Kinetoplast**: The kinetoplast, unusually large is visible on the blood smears as oval shaped, 0.85 μm (range 0.33–1.40 μm) in diameter or rod shaped, 0.40 × 0.72 up to 1.15 × 2.10 μm (mean 0.66 × 1.08 μm) in size. It is located near the posterior end of the body.
- **Free Flagellum**: The free flagellum is 6.85 + 2.2 μm (range 4.1–14.3 μm) in length which originates from the kinetoplast, and extends as a free flagellum beyond the undulating membrane. It is usually less than one third of the body length.
- **Undulating Membrane**: The undulating membrane has only 3–4 folds. It is weakly developed but is always visible.

2.4.10. "Stout" Forms

- **Shape and Size**: The "stout" forms (Plate XXXIIIA-C), have a characteristic broad and short body, the posterior end being more or less sharp or blunt. The body cell is 14.29 + 1.81 μm (range 9.5–19.8 μm) long, and 1.92 + 0.57 μm (range 1.3–2.9 μm) in width. In contrast to the slender form of *T. grosi kosewiense* and many other trypanosomes, the parasite is broadest in the middle of the PN distance, not at the nuclear level.
- **Cytoplasm:** The cytoplasm of the "stout" form is similar to the "slender" form of the parasite as mentioned above.
- **Nucleus:** The nucleus, slightly oval (Plate XXXIIIA) or round in shape, shifts towards the anterior end of the body. The nucleus is oval, placed parallel, oblique, or in extremely broad individuals, perpendicular to the long axis of the body. The nucleus is 2.12 + 0.27 μm (range 1.6–2.8 μm) long situated 5.21 + 0.95 μm from the anterior and 9.08 + 1.02 μm from the posterior end indicating that its position is more anterior as compared to the "slender" form.
- **Kinetoplast:** The kinetoplast is oval (Plate XXXIIIB), or rod-shaped, 0.24 × 0.33 μm to 1.25 × 1.35 μm in size. It is located nearer to the posterior end of the body in stout forms as compared to the slender forms. Sometimes, its position is perpendicular (Plate XXXIIIC).
- **Free Flagellum**: The free flagellum is 7.60 + 1.8 μm (range 4.5–10.7 μm) long. The relative proportion of the free flagellum (FF) approximates half of the body length (BL) in contrast to one third in the "slender" form.

- **Undulating Membrane**: Contrary to the slender form, the undulating membrane is well developed.
- **Measurements**: The morphometric parameters of the trypanosomes isolated from the yellow-necked mouse *Apodemus flavicollis* were compared to *T. grosi* (Laveran and Pettitt, 1909), description by Krampitz (1959) and *T. grosi kosewiense* (Karbowiak and Wita, 2004) and are given in Table 2.15.

2.4.10.1. Discussion

Herpetosoma trypanosomes occurring in West and Middle Europe are not very well known. The first records about trypanosomes parasitizing European small rodents were made in the beginning of the 20th century (Thiroux, 1905; Laveran and Pettit, 1909). Later, these trypanosomes were investigated only in the U.S.A. and in some Western-European countries (Davis, 1952–53; Krampitz, 1959; Sebek, 1975; Molyneux, 1976; Sebek et al., 1980; Noyes et al., 2002). The studies regarded morphology, primarily; few publications concerned the results of ecological and biochemical investigations. In Europe, new species of *Herpetosoma* trypanosomes, parasitizing rodents and insectivores are constantly being described, as well as new localities of known trypanosomes (Karbowiak and Sinski, 1996a; Macdonald et al., 1999) are being added. For the description of trypanosomes, Hoare (1972) recommended the following criteria: the host species, the vector species, the morphology and the biology of the trypanosome. Since then, ultrastructural and biochemical criteria were added. On account of the great variability within the *Herpetosoma* subspecies, and incomplete knowledge about these parasites, their recognition is difficult on the basis of morphological features only. Presently, for biochemical description, isoenzyme electrophoresis (Mohamed et al., 1987), lectin affinity (Maraghi et al., 1989) as well as 18S rRNA genotyping (Noyes et al., 2002) have been applied. Because a single criterion is insufficient for a majority of trypanosomes (Maraghi et al., 1989), the combination of various features (morphological, biological, biochemical) is recommended for the description and recognition of the parasites (Karbowiak and Wita, 2004).

The *Herpetosoma* trypanosomes are not very pathogenic and make no visual pathological effects to their hosts. The metacyclic forms of trypanosomes infecting mammals penetrate through the skin, or *per os* – when the host swallows the infected vector (Maraghi et al., 1995). This way of infection with trypanosomes was showed experimentally by Calvo et al. (1992) and Maraghi et al. (1995).

So far, the trypanosomes parasitizing the yellow-necked mouse, *A. flavicollis*, were seldom found and were recognized originally as *T. grosi* (Laveran et

Pettit, 1909; Krampitz, 1959; Sebek, 1960, 1975; Hoare, 1972; Sebek et al., 1980). However, these recognitions were supported by the similarity of hosts, because the peculiar host of *T. grosi*, the wood mouse *A. sylvaticus* is closely related to *A. flavicollis.* The other criteria were not considered at that time. The results of the present investigations show that trypanosomes parasitizing the yellow-necked mouse are bigger than *T. grosi.* This phenomenon was evident in every investigated parameter. Indices NI and KI show that the nucleus is located closer to the body center, and the distance between the kinetoplast and the posterior end of the body is longer Karbowiak and Wita, 2004). The morphological differences between trypanosomes parasitizing *A. sylvaticus* and *A. flavicollis* also were noted by Sebek (1960). Moreover, the polyphyletic character of *Trypanosoma grosi* parasitizing wood mice *(Apodemus sylvaticus)* in Great Britain was demonstrated by Noyes et al. (2002).

The characteristic feature of this parasite is the pleomorphism of trypomastigote forms living in the host's blood, which manifests itself with two morphological forms; "slender" and "stout." Pleomorphism was not observed in other rodent trypanosomes, but has been reported in *T.* (*Herpetosoma*) *rangeli* and in many trypanosomes from the Salivaria section, especially from the *Trypanozoon* subgenus (Hoare, 1972; Urdaneta Morales, 1983). The reason for pleomorphism of *T. grosi kosewiense* is not fully known. In many cases, the morphological differentiation is in correlation with the age of trypanosomes However, the dividing forms, described in *T. grosi* (Kramptiz, 1959) and in other trypanosomes parasitizing rodents from the family Muridae, were absent in the blood of trypanosome infected yellow-necked mice.

Table 2.15. Dimensions of Slender and Stout Forms of the Present Species Compared with *Trypanosoma grosi* (Laveran & Pettit, 1909), Description by Krampitz (1959) and *T. grosi kosewiense* (Karbowiak and Wita, 2004)

Parameter		**PK**	**KN**	**NP**	**NA**	**CL**	**FL**	**CF**	**NL**	**CB**	**NI**	**KI**	**FL:CL**
***T. grosi* (Krampitz 1959)**	Mean	1.90	7.30	9.20	12.30	21.5	4.55	23.80	-	-	0.74	1.26	4.72
	Range	1.00–2.50	6.00–9.00	-	10.00–17.50	-	1.5–7.6	21.00 ±28.00	1.50–3.00	-	-	-	-
***T. grosi kosewiense* f. slender (Karbowiak & Wita, 2004)**	Mean	3.87 ± 0.65	8.34 ±0.79	12.21 ±1.97	10.79 ±1.98	23.00 ±2.31	7.59 ±1.35	30.59 ±2.23	2.68 ±0.36	1.43 ±0.30	1.17 ±0.25	1.47 ±0.09	3.15 ± 0.74
	Range	2.32–5.78	5.73–11.10	8.73–15.57	5.84–15.42	16.27–28.05	3.66–12.79	23.69–36.17	1.42–3.71	0.79–2.56	0.70–1.95	1.23–1.75	1.54–6.04

***T. grosi kosewiense* f. stout (Karbowiak & Wita 2004)**	Mean	2.08 ± 0.66	7.30 ± 1.42	9.37 ± 1.67	5.37 ± 1.14	14.74 ± 2.29	8.56 ±1.51	23.33 ± 2.66	2.38 ± 0.34	1.76 ± 0.30	1.80 ± 0.39	1.30 ± 0.11	1.77 ± 0.44
	Range	0.93–4.01	3.98–11.12	5.71–15.06	3.49–8.88	10.78–23.54	5.06–11.86	17.88–31.85	1.64–3.35	1.18–2.78	0.94–2.68	1.14–1.74	1.04–3.11
***T. grosi kosewiense* f. slender (present species)**	Mean	3.51 ± 0.73	7.92 ± 0.58	11.43 ± 0.98	11.21 ± 1.75	22.64 ± 1.92	6.85 ±2.2	29.49 ± 3.42	2.20 ±0.24	1.81 ± 0.41	1.05 ± 0.29	1.48 ± 0.04	3.31 ± 0.52
	Range	2.7–5.3	6.2–9.9	9.5–14.02	7.2–16.3	17.2–27.5	4.1–14.3	21.5–35.8	1.7–3.2	1.2–2.8	1.1–2.7	1.18–2.0	1.61–5.2
***T. grosi kosewiense* f. stout (present species)**	Mean	1.98 ±0.36	7.10 ±1.22	9.08 ±1.02	5.21 ±0.95	14.29 ±1.81	7.60 ±1.8	21.28 ±3.1	2.12 ±0.27	1.92 ±0.57	1.69 ±0.13	1.24 ±0.08	1.88 ± 0.28
	Range	0.98–3.7	3.5–10.5	6.7–12.4	3.5–7.9	9.5–19.8	4.5–10.7	16.0–28.2	1.6–2.8	1.3–2.9	1.2–2.2	0.94–1.8	1.2–2.8

All values are in µm.

A comparison of the present form with the earlier recorded species indicates that similar to the parasite reported by Karbowiak and Wita (2004), the present parasite also showed dimorphism and the morphometrics of the "slender" and "stout" forms are also comparable as shown in Table 2.15. Therefore, the dimorphic parasite discovered from the blood of *Apodemus flavicollis* (Melchior, 1834) is identified as *Trypanosoma* (*Herpetosoma*) *grosi kosewiense* (Karbowiak and Wita, 2004).

2.4.11. Trypanosoma (Herpetosoma) Evotomys (Molyneux, 1969)

Host Profile	
Kingdom	Animalia
Phylum	Chordata
Class	Mammalia
Order	Rodentia
Family	Muridae
Genus	*Myodes* (formerly *Clethrionomys*)
Species	*Glareolus* (Schreber, 1780)
Parasite Profile	
Phylum	Protozoa
Class	Flagellata
Order	Kinetoplastida
Suborder	Trypanosomatina

Family	Trypanosomatidae
Genus	*Trypanosoma* (*Herpetosoma*)
Species	*Evotomys* (Molyneux, 1969)

2.4.12. The Host: Myodes (Formerly Clethrionomys) glareolus (Schreber, 1780)

The bank vole, *Myodes glareolus* (formerly *Clethrionomys glareolus*) is a small vole with red-brown fur and some grey patches, with a tail about half as long as its body. A rodent, it lives in woodland areas and is around 100 mm (3.9 in) in length. The bank vole is found in western Europe and northern Asia. It is native to Great Britain but not to Ireland, where it has been accidentally introduced.

The bank vole lives in woodland, hedge rows and other dense vegetation such as bracken and bramble. Its underground chamber is lined with moss, feathers and vegetable fiber and contains a store of food. It can live for 18 months and is omnivorous, eating insects, leaves and fruits such as raspberries and hazel nuts. It readily climbs into scrub and low branches of trees.

In areas such as mainland Great Britain, where the only other small vole is the short-tailed vole (*Microtus agrestis*), the bank vole is distinguished by its more prominent ears, chestnut-brown fur and longer tail.

2.4.12.1. Methodology

The present study was conducted on coexisting populations of small mammals living east of Luknajno Lake and north of Sniardwy Lake, in the Mazury Lake District of Poland. The material consisted of blood smears which were air dried, fixed in methanol and stained with Giemsa stain. Thirty-five undistorted and stained trypanosomes were examined microscopically under oil immersion. *M. glareolus* (n= 45) were examined for the presence of parasites. These species were dominant in the small mammal community.

2.4.12.2. Description

Infection with *T. evotomys* was found in 9.50% *M. glareolus*. The parasites occurred in the trypomastigote form.

In order to exactly characterize and compare morphological features of trypanosomes described from *M. glareous* from other regions, terminology adopted by other researchers (Molyneux,1969; Hoare, 1972; Kingston Bobek et al., 1992) was employed in the present study. Measurements were done on 30 specimens.

- **Shape and Size**: The parasites are usually 'C'-shaped but also elongated (Plate XXXIVA) at times. The body length of trypanosomes varies between 23.5 and 32.0 μm (mean length 28.2 + 2.05 μm).
- **Cytoplasm**: The cytoplasm takes a pale color. Cytoplasmic granules are more prominent in the posterior (post nuclear) part of the body.
- **Nucleus:** The nucleus is slightly elongated (Plate XXXIVB) and located in the middle or nearer to the anterior body end. It is situated 8.5 + 1.04 μm from the anterior end and 11.0 + 1.21 μm from the posterior end of the body. The length of the nucleus is 3.3 + 0.32 μm.
- **Kinetoplast:** The kinetoplast is generally spherical (Plate XXXIVC) located 2.6–5.4 μm (mean 3.7 + 0.51 μm) from the posterior end of the body and 4.2–10.6 μm (mean 7.20 + 0.93 μm) from the center of nucleus.
- **Flagellum:** Length of the free flagellum is 6.0–14.0 μm (mean 8.8 + 0.76 μm). It extends freely beyond the anterior end of the body as a free flagellum.
- **Undulating Membrane**: The undulating membrane is well developed having 5–7 folds.
- **Measurements:** The results of the present studies and the data from the literature for *T. evotomys* and *T. grosi* are presented in Table 2.16.

2.4.12.3. Discussion

There are practically no data on the occurrence of *Herpetosoma* from small mammals in Poland, and other countries in central and Eastern Europe. Until recently, the species *Trypanosoma evotomys* was described from Austria (Sebek et al., 1980), Germany (Walter and Liebisch, 1980) and England (Turner, 1986) from the bank vole (*Clethrionomys glareolus*). *Trypanosoma grosi* has been described from the wood-mouse (*Apodemus sylvaticus*) from France (Laveran and Pettit, 1909), also from Germany (Krampitz, 1959) and England (Sebek et al., 1980).

The species, *Trypanosoma grosi* and *T. evotomys*, are morphologically very similar; however, there are some differences which may be useful for species identification. The morphological features of both species given by Krampitz et al., 1961 and Molyneux, 1969 are presented in Table 2.16.

The life cycles of the two parasites are completely different. *Trypanosoma evotomys* reproduces in the lymphoid tissues of the appendix and spleen in the amastigote phase. No division forms occur in the peripheral blood (Molyneux, 1969, 1970). *Trypanosoma grosi* divides in the epimastigote phase into 8–12 specimens. Because the dividing individuals occur in the peripheral blood, it is possible to find them on blood smears (Hoare, 1972).

Fleas are vectors of most representatives of *Herpetosoma*. The most appropriate flea species for *T. grosi* are *Ctenophthalmus agyrtes* (*C. nobilis*) and *C. congerer* (Krampitz, 1959), and those for *T. evotomys* are *C. agyrtes* and *Malareus (Ceratophyllus) penicilliger* (Hoare, 1972). Molyneux (1969) described the development of *T. evotomys* in *Nosopsyllus fasciatus.*

Morphological features of *Trypanosoma* detected on rodents are in accordance with the description presented in the literature (Krampitz, 1961; Molyneux, 1969). The position of kinetoplast is nearer to the posterior body end in comparison to that described by the earlier authors.

Although the representatives of the subgenus *Herpetosoma* are parasites of many non-related mammals, individual species show very high specificity towards their hosts. Natural cases of infecting other host species, though possible, occur only sporadically. Infections, conducted under laboratory conditions, are characterized by a low intensity and most often have a latent course (Hoare, 1972; Maraghi and Molyneux, 1989; Maraghi et al., 1989).

The parasite discovered from the blood of *Myodes* (=*Clethrionomys*) *glareolus* is characteristic of the subgenus *Herpetosoma.* Its morphological features and statistical measurements are similar to *T. evotomys* described by Molyneux (1969) also described from the same host and therefore, the parasite is identified as *Trypanosoma evotomys* Molyneux, 1969.

Table 2.16. *Dimensions (in μm) of Trypanosoma evotomys (present species) Compared with T. evotomys and T. grosi Described by Other Authors*

Morpho-Logical Param-eters	***T. evotomys***		***T. grosi***	***T. evotomys* Strain**	**Present Species**
	Krampitz, 1961c	**Moly-neux, 1969**	**Krampitz, 1961c**	**Karbowiak and Sinski, 1996**	
PK	- 1.5–4.5	3.9 2.7–5.0	1.0–2.5	3.5 ± 0.78a 2.0–5.0b	3.7 ± 0.51 (2.6–5.4)
KN	- 3–5.5	7.1 3.5–8.5	6.0–9.0	7.36 ± 1.46 4.5–10.5	7.2 ± 0.93 (4.2–10.6)
NP	-	-	-	10.87 ± 1.67 7.5–14.0	10.9 ± 1.21 8.1–13.3

NA	-	8.1 3.0–12.0	-	8.71 ± 1.54 6.5–11.5	8.5 ± 1.04 6.8–11.0
CL	-	-	-	19.58 ± 2.63 15.0–24.0	19.4 ± 1.73 16.0–23.1
CF	- 21–28	28.1 24.3–31.8	23.80 21–28	27.72 ± 2.83 22.5–33.0	28.2 ± 2.05 (23.5–32.0)
FL	-	9.1 6–15.8	- 1.5–7.6d	8.14 ± 1.49 6.0–12.0	8.8 ± 0.76 (6.0–14.0)
CB	-	1.3	-	1.58 ± 0.34 1.0–2.0	1.3 ± 0.15 (1.1–1.8)
NL	-	3.5	-	2.92 ± 0.43 2.0–4.0	3.3 ± 0.32 (2.4–4.2)
KI	-	-	-	1.56 ± 0.17 1.26 ± 2.1	1.52 ± 0.21 (1.05–2.3)
NI	-	-	-	1.27 ± 0.23 0.74–1.77	1.29 ± 0.12 (1.0–1.8)

[a] *mean size;* [b] *range;* [c] *in Hoare (1972);* [d] *by Yakimoff and Korssak (1910) in Hoare (1972).*

2.4.13. Trypanosoma (Herpetosoma) microti (Laveran and Pettitt, 1909)

Host Profile	
Kingdom	Animalia
Phylum	Chordata
Class	Mammalia
Order	Rodentia
Family	Cricetidae
Genus	*Microtus*
Species	*Oeconomus* (Pallas, 1776)
Parasite Profile	
Phylum	Protozoa
Class	Flagellata

Order	Kinetoplastida
Suborder	Trypanosomatina
Family	Trypanosomatidae
Genus	*Trypanosoma(Herpetosoma)*
Species	*Microti* (Laveran and Pettitt, 1909)

2.4.14. The Host: Microtus oeconomus (Pallas, 1776)

Microtus oeconomus (Pallas, 1776) commonly known as Root Vole, Tundra Vole (English), Campagnol Nordique (French) and Topillo Nordico (Spanish) is a Holarctic species, with a wide range extending from north-west Europe in the west to Alaska in the East, in North America, restricted to Alaska and northwest Canada. In Europe, its main range extends from eastern Germany and northern Fennoscandia through Poland, Belarus, and northern and central European Russia; isolated relict populations are found in the Netherlands, southern Norway and northern Sweden, the Finnish coast, and Austria, Slovakia and Hungary. In China, it is found in northern Xinjiang and northern Nei Mongol. In Mongolia, it is found in northern areas including Mongol Altai, Hangai, Hövsgöl Hentii and Ikh Hyangan mountain ranges, and Mongol Daguur Steppe (Sokolov et al., 1985). It is a widespread and abundant species, although in some areas (e.g., parts of Europe, some North American islands) its status is less favorable. It typically inhabits damp, densely-vegetated areas along the edges of lakes, streams and marshes; may be found in tundra, taiga, forest-steppe, and even semi-desert. Wet meadows, bogs, fens, riverbanks and flooded shores are all important habitats. It feeds on green vegetation.

2.4.14.1. Methodology

2.4.14.1.1 Maintenance of Host

Microtus oeconomus were caught in live traps placed at trap stations arranged in a rectangular grid or line. Traps were placed and set (without pre-baiting) in the afternoon, checked four times per day and blocked after a final check.

Captured mammals were individually identified and released after collection of blood samples. Animals previously caught in the same session were immediately released.

2.4.14.1.2. Examination of Blood

Blood samples were taken from the tail tip. Blood was collected in tubes containing heparin from the tip of the tail of ten voles. Samples were tested for trypomastigotes by examination in micro hematocrit tubes (Bennett, 1962) and by blood culture in veal infusion medium (VIM, ca 0.5 ml of blood) (Kingston and Morton, 1973).

Trypanosomes seen swimming above or sometimes in the buffy coat were placed on microscope slides and thin films prepared, air dried, fixed in absolute methanol and stained with Giemsa. Thin smears were prepared for permanent slides, air dried, fixed in methyl alcohol and stained for 1 hour in Giemsa, diluted (1:7) in phosphate buffer (pH 7.2). Slides were rinsed in buffer, dried and examined under a light microscope.

Blood films were scanned at a magnification of 1500× by light microscopy combined with a computer and video camera, using an Analysis Pro 2.11 program. Morphometric observations followed the nomenclature and methodology of Hoare (1972). Measurements and calculated indices were compared with those of *T. microti* from *M. agrestis* given by Fay and Rausch (1969) (Table 2.17).

2.4.14.1.3. Description

Trypanosoma (*Herpetosoma*) *microti* was recovered from the blood of four root voles, *Microtus oeconomus.* No clinical signs were observed in root voles infected with trypanosomes. The intensity of infection estimated from the approximate number of trypanosomes seen in slides made from 1 µl of blood, ranged from 1.6×10^4 to 1.8×10^5 per ml. All trypanosomes observed in the bloodstream of the root voles were trypomastigotes.

- **Shape and size:** The parasite has a long, slender and sinuous body, sharply pointed at either end, more so at the anterior end. The parasite usually takes a C-shape (Plate XXXVA,B) but are also S-shaped (Plate XXXVC). The body maybe sinuous (Plate XXXVIA) and is generally large in size (Plate XXXVIB). It is 22.57 + 1.81 (range 17.49–25.90 µm) in length and has a width of 1.25 + 0.19 µm (range 0.75–2.89 µm).
- **Cytoplasm**: The cytoplasm takes a light stain. Cytoplasmic granules are not very conspicuous.
- **Nucleus:** The nucleus is oval in shape and its size ranges from 1.55–3.60 µm (mean 2.50 + 0.30 µm). It is situated 10.26 µm from the anterior end and 12.31 µm from the posterior end of the parasite.
- **Kinetoplast:** The kinetoplast is a well-developed, deeply stained structure situated 5.05 + 0.32 µm from the posterior end and 7.26 + 0.48 µm from the nucleus. It is generally round in shape (Plate XXXVIC) but can also be dorso- ventrally compressed (Plate XXXVIIA).

- **Flagellum:** The length of the flagellum is approximately one third of the body length. It originates from the kinetoplast, after bordering the undulating membrane, it extends freely as a flagellum from the anterior end. It is 7.50 + 2.51 µm (range 4.25–9.23 µm) long.
- **Undulating Membrane**: The undulating membrane has 3–6 folds (Plate XXXVIIB,C).
- **Dividing Forms:** Only one dividing stage was observed in the peripheral blood and has been mentioned in the chapter "Division in trypanosomes."
- **Measurements**: Measurements were taken from 48 undistorted trypomastigotes on slides, suitable for examination and analysis. They were the typical *Herpetosoma* forms. The statistical measurements of the parasite have been provided in Table 2.17 and the parasite is compared with *T. microti* from *Microtus arvalis*, *M. oeconomus*, *M. agrestis* and *M. socialis* as reported by other authors.

2.4.14.1.4. Discussion

Data on the occurrence of *Herpatosoma* trypanosomes in small rodents are scarce. Trypanosomes parasitizing the common vole, *Microtus arvalis*, were recorded, for the first time in Europe, in France by Laveran and Pettit (1909) and named *Trypanosoma microti*. They were subsequently confirmed by Lavier (1921). In England, *T. microti*, was described from *M. agrestis* by Elton et al. (1931), Baker et al. (1963) and Molyneux (1969). In Germany, *T. microti* was found in *M. arvalis* by Krampitz (1961) and in the field vole, *M. agrestis* by Walter and Liebisch (1980). In the former Czechoslovakia, *T. microti* was found by Sebek (1975) in *M. agrestis* and, in Austria, in the root vole, *M. oeconomus* by Frank (1978). Outside Europe, this species was recovered in Kazakstan by Galuzo and Novinskaya (1961) in *M. arvalis* and in Alaska in *M. oeconomus* by Fay and Rausch (1969).

A detailed morphological description of *T. microti* was given by Molyneux (1969). The vectors of *T. microti* have been found to include rat fleas *Nosopsyllus fasciatus* (Molyneux, 1969) and *Ctenophthalmus agyrtes* (Krampitz, 1961). Taking into consideration differences in host and geographic distribution, Hoare (1972) proposed *T. microti* be differentiated into three subspecies *T. microti* Laveran and Pettit, 1909 in Europe; *T. microti arvali* Galuzo and Novinskaya, 1961 in Asia and *T. microti nearcticum* Hoare, 1972 in North America.

The results demonstrated that the intensity of infections with *T. microti* in *M. oeconomus* was higher (1.6×10^4 to 1.8×10^5 per ml) in comparison with results given by Molyneux (1969) for *T. microti* infections in *M. agrestis* (4 ×

10^3–1×10^4/ ml). However, the values were lower than reported in voles by other authors – infections with *T. musculi* in *Mus musculus* were 5×10^7/ ml and those with *T. evotomys* in *Clethrionomys glareolus* were 1.5×10^5–5.5×10^7/ ml (Molyneux, 1969; Hoare, 1972; Karbowiak and Wita, 2001b).

The comparison of morphometric values of bloodstream trypomastigotes from *M. oeconomus* with those data obtained by other authors for *T. microti* from *M. agrestis* from England, *M. oeconomus* from Alaska and from *M. arvalis* from Germany and France revealed many similarities with most parameters having overlapping ranges (Laveran and Pettit, 1909; Krampitz, 1961; Fay and Rausch, 1969; Molyneux, 1969) (Table 2.17).

The mean values of PK, KN, NP, NA, CL, CF, CB, KI and NI parameters in trypanosomes from *M. socialis* were higher than the mean values given by Molyneux (1969) for *T. microti* from *M. agrestis*. Only the average free flagellar length (FL) was slightly lower and FL:CL ratio higher in trypanosomes from *M. socialis*. The mean length of the nucleus was similar for trypanosomes from both host species. The value of the nuclear index (NI = 1.33) showed that the nuclei in the present specimens were located more anteriorly than nuclei in *T. microti* from *M. agrestis* (NI = 1.13). Kinetoplastic indices (KI) derived from trypanosomes from *M. socialis* and *M. agrestis* were similar (1.70 and 1.63 respectively) and showed that all the specimens had a kinetoplast located far from the nucleus, near the posterior end of the body.

Fay and Rausch (1969) reported the trypanosomes from the root vole, *M. oeconomus*, in Alaska but gave only the ranges of measurements, without mean values (except for body length). Generally, the recorded measurements for each parameter for trypanosomes from root voles correspond to those from Alaska. However, the upper ranges given by Fay and Rausch (1969) are lower than the highest values of PK, KN, NP, CL, FL, CF, NL and CB parameters for the present *T. microti*.

T. microti has its kinetoplast located closer to the posterior end than to the nucleus. The value of nuclear index (NI= 1.19) showed that the nuclei were located more anteriorly than nuclei in *T. microti* from *M. oeconomus* (NI= 0.83) as reported by Fay and Rausch (1969) but was almost similar to that reported from *M. agrestis* (NI = 1.13). The FL:CL ratios in bloodstream trypomastigotes of *T. microti* from *M. oeconomus* (Fay and Rausch, 1969) and from *M. agrestis* were similar (2.4 and 2.02 respectively) but slightly higher (2.71) in trypanosomes from *M. socialis* (Wita et al., 2003). In the present case, this ratio was the highest (3.00 + 0.40) as compared to the species mentioned earlier. When *T. microti* from the root vole were compared with *T. microti* from *M. arvalis* in Germany (Kramptiz, 1961) and France (Laveran and Pettit, 1909), differences in the range of length measurements were not evident.

Table 2.17. Measurement (in µm) of Trypomastigotes from the Bloodstream of the Root Vole *Microtus oeconomus* and Those Given by Other Authors for *Trypanosoma microti*

Parasite Host Author (s)		***T. microti M.arvalis* Other authors**	***T. microti nearcticum M.oeconomus* Fay & Rausch (1969)**	***T. microti M.agrestis* Molyneux (1969)**	***T. microti M. socialis* Wita et al., 2003**	***T. microti M. oeconomus* Present species**
PK	mean ± SD Range	-	- 3.00–3.50	3.90 2.60–5.60	5.30 ± 0.70 2.59–7.44	5.05 ± 0.32 3.25–7.12
KN	mean ± SD Range	-	- 6.00–8.00	6.2 3.40–8.00	7.60 ± 0.66 4.78–8.81	7.26 ± 0.48 4.52–9.1
NP	mean ± SD Range	-	- 9.00–11.50	10.10 -	12.92 ± 1.00 7.61–14.78	12.31 ± 0.95 9.21–15.60
NA	mean ± SD Range	-	- 8.00–10.00	8.90 4.00–11.60	9.93 ± 1.59 6.00–14.83	10.26 ± 1.27 8.23–13.52
CL	mean ± SD Range	-	- 17.00–21.50	19.00 -	22.84 ± 2.12 13.88–27.79	22.57 ± 1.81 17.47–25.90
FL	mean ± SD Range	-	- 7.00–9.00	9.40 6.50–14.00	8.71 ± 1.45 5.12–11.69	7.50 ± 2.51 4.25–9.23

CF	mean ± SD Range	23.61, 20.0–30.0 25.0–30.02	24.84 ± 0.23 22.90 – 27.10	27.40 8.00–32.20	31.55 ± 2.23 19.00–37.32	30.07 ± 2.85 21.72–34.50
NL	mean ± SD Range	-	- 1.50–3.00	2.8 0.00–3.90	2.70 ± 0.42 1.64–3.88	2.50 ± 0.30 1.55–3.60
CB	mean ± SD Range	1.52	- 1.50–2.00	1.30 0.90–1.70	1.44 ± 0.28 0.53–2.06	1.25 ± 0.19 0.75–2.89
Indices						
KI	mean ± SD Range	-	1.63 -	1.63 -	1.70 ± 0.11 1.35–2.06	1.69 ± 0.09 1.27–2.18
NI	mean ± SD Range	-	0.83 -	- 1.13	1.33 ± 0.21 0.87–2.00	1.19 ± 0.18 0.75–2.16
FL:CL	mean ± SD Range	-	1:2. 40 -	1:2.02 -	1:2.71 ± 0.58 1.66–4.71	3.00 ± 0.40 1.89–4.11

[1]*Krampitz 1961,* [2]*Laveran and Pettit (1909).*

The above observations indicate that the trypanosomes discovered from the blood of *Microtus oeconomus* resemble *Trypanosoma* (*Herpetosoma*) *microti* reported from the social vole, *Microtus socialis* and are identified as *Trypanosoma* (*Herpetosoma*) *microti* (Laveran and Pettit, 1909).

2.4.15. Babesia microti (Franca, 1912)

Host Profile	
Kingdom	Animalia
Phylum	Chordata
Class	Mammalia
Order	Rodentia
Family	Cricetidae
Genus:	*Microtus*
Species	*Oeconomus (*Pallas, 1776)
Parasite Profile	
Phylum	Protozoa
Class:	Piroplasmea
Order	Piroplasmida
Family	Babesiidae
Genus:	*Babesia*
Species:	*Microti (*Franca, 1912)

The genus *Babesia* contains parasitic, intra-erythrocytic protozoans infecting mammals. Together with related genus *Theileria, Babesia* is included in the family Babesiidae, class Piroplasmea, phylum Protozoa. These are most often parasites of large mammals belonging to the orders Ruminantia and Carnivora. There are, however, species, like *Babesia microti*, parasitizing other mammals (Kreier, 1977). However, the systematic position of *Babesia* is not well established and some authors have proposed different classifications of this genus (Nowell, 1969; Mahnert, 1972; Krylov, 1981; Mehlhorn and Schein, 1984).

The occurrence of *Babesia* piroplasms in European voles was described adequately in Great Britain, Germany and Austria (Baker, 1974; Walter and Liebisch, 1980). Individual findings were also reported from the former Czechoslovakia, Hungary and Poland (Sebek et al., 1977; Karbowiak et al., 1999).

2.4.15.1. Methodology

Microtus oeconomus were collected in live traps. Blood samples from *Microtus oeconomus* were taken from the tip of the tail from eight voles, and thin smears were prepared. After air drying, the smears were fixed in methyl alcohol and stained for 1 hr in Giemsa stain diluted (1:15) in phosphate buffer, (pH 7.2). The slides were rinsed in water, dried and examined under light microscope. For each animal, 30 fields were examined for the presence of *Babesia* at X 450 and X 1200 magnifications. The morphological characteristics of *Babesia* under light microscopy were analyzed in the blood smears stained with Giemsa. The forms were identified according to Schaller (2008). The measurements of the parasites were made using "Analysis" program combined with video camera and Olympus BX50F4 microscope at a magnification of 1200X. The obtained results were accurate to 0.01 µm.

2.4.15.2. Description

Babesia microti was found in two out of the eight voles examined. The percent of erythrocyte infection was 1.5. The parasites were generally 1.5–3.0 µm in diameter. The morphological features of the parasite are according to the data given by Schall (2008). Several forms were encountered in the blood: the sporozoite (Plate XXXVIIIA) was the infective stage and could be seen invading the erythrocytes. During the very early development, the parasite appeared as a dot-like structure (Plate XXXVIIIB) probably representing the chromatin or DNA dot with very little cytoplasm. The crescent form was poorly developed (Plate XXXVIIIC). The micro ring stage was visible either at the tip of the cell (Plate XXXIXA) or even inside it (Plate XXXIXB). The ring form (Plate XXXIXC) characteristic of *Babesia* was also observed. Some smears revealed the tear drop form and in such cell, the eye form was also visible (Plate XLA). Occasionally, double eye was observed (Plate XLB). The trophozoite (Plate XLC) was well developed with cytoplasm around the nucleus. Tetrad occurred in cluster (Plate XLIA). The thick long worm string pattern (Plate XLIB was comparatively rare. Merozoites could be seen escaping from the erthrocytes (Plate XLIC) to infect additional blood cells and reinitiate the replicative cycle. The characteristic "maltese cross" was not noticed. The infected animals did not show any diseased symptoms superficially, but on autopsy, splenomegaly was evident which is typical of *Babesia* infection (Fay and Rausch, 1969; Watkins et al., 1991; Karbowiak et al., 1999).

The results indicate that in Poland, voles of the genus *Microtus oeconomus* may play an important role as zoonotic reservoirs of the piroplasms belonging to the genus, *Babesia*.

2.4.15.3. Manifestations

The infection of *Microtus* voles with *Babesia microti* resulted in a dramatically enlarged spleen. Splenomegaly was not observed with other common hemoparasite infections (*Trypanosoma* or *Trypanoplasma*). However, apart from splenomegaly symptoms, natural *Babesia* infections do not have any visible signs, therefore it is evident that piroplasms cause chronic avirulent infections in their natural hosts (Baker et al., 1963; Krampitz and Baumler, 1978; Turner, 1986).

2.4.15.4. Family Babesiidae

Most members of this family parasitize mammals; a few infect birds or reptiles. Intra-erythrocytic stages reproduce by binary fission or budding into two (rarely four) individuals. Until relatively recently, it was thought that babesiids infect only the erythrocytes of their vertebrate hosts. Two species, (*B. microti, B. equi*) have a single pre-erythrocytic generation of meronts within lymphocytes. The Babisiidae of veterinary importance are shown in Table 2.18 (Krier and Baker, 1977).

2.4.16. Genus Babesia (Starcovici, 1893)

Some of the most common species of *Babesia* are of veterinary importance, their main hosts and geograpical distribution in Europe are listed in Tables 2.19 and 2.20. A large number of species of *Babesia* have been described from wild rodents and other wild animals. Many of these appear to be quite benign in their hosts. Some *Babesia* species produce disease (babesiosis) in domestic animals. Babesiosis of cattle, horses, and dogs is often particularly severe. *Babesia rodhaini*, a natural parasite of wild rodents in Africa, and *B. microti*, a world-wide parasite of rodents, are often maintained in laboratory mice for use in experimental work. *Babesia microti* has been a cause of human infection, mainly in New England and New York coastal regions of the United States; it is one of the species with a pre- erythrocytic cycle of merogony within lymphocytes.

Most species of *Babesia* are host specific. *B. microti*, however, is a parasite of numerous wild and domestic mammals, especially small rodents. *B. microti* is characterized by relatively low specificity towards its host and can infect many non-related animals. Several species of insectivores belonging to the family Soricidae also can be infected by this pathogen (Baker, 1974). In Europe, *B. microti* infects rodents of Muridae and *microti*dae families (Mehlhorn and Schein, 1984), but this species may occur in Insectivora and even in man.

The presence of two species of *Babesia* piroplasms in ticks *Ixodes ricinus* in Poland was documented using PCR method in Szczecin area and in the forests near the Tri-City agglomeration by Karbowiak (2004). Different authors have reported parasitization by *B. microti* (infection rate 6.2–13.3%), *B. divergens* (3.0%) in Szczecin (Skotarczak et al., 2003), and *B. microti* in Tri-City (infection rate in ticks 2.3%) (Stanczak et al., 2004). These demonstrations confirm the findings that *I. ricinus* can be also involved in circulation of *B. microti* in Europe; tick infection rates with babesiae have been calculated at 7.4% in Slovenia (Duh et al., 2001) and 3.5% in Hungary (Kalman et al., 2003). The most common infections were found in adult females up to 14.6%, other developmental forms were lighter in infection where 11.1% of nymphs were infected (Skotarczak et al., 2002). The percentage of infection was changeable, depending on the season (Skotarczak and Cichocka, 2001a,b, Skotarczak et al., 2003).

Table 2.18. Babesiidae of Veterinary Importance (According to Krier and Baker, 1997)

			Main Hosts			
Species	**Size***	**Geographical Distribution**	**Vertebrate**	**Invertebrate**	**Pathogenicity**	**Common Name of Disease**
Babesia bigemina	large	Central & South America, Europe, Africa, Australia	cattle, deer	*Boophilus, Haemaphysalis, Rhipicephalus*	moderate	redwater fever
***B. bovis* (*B. argentina*)**	small	Europe, Russia, Africa, Australia, South & Central America	cattle, deer	*Ixodes, Boophilus, Rhipicephalus*	high	redwater fever
B. divergens	small	Western & Central Europe including England	cattle	*Ixodes*	moderate	redwater fever
B. major	large	Europe, Russia	cattle	*Boophilus*	low	-

B. caballi	large	Southern Europe, Asia, Russia, Africa	Equidae (domestic)	*Dermacentor, Hyalomma, Rhipicephalus*	Moderate	-
B. equi	small	Southern Europe, Asia, Russia, Africa, South America	Equidae (domestic and zebra)	*Dermacentor, Hyalomma, Rhipicephalus*	high	biliary fever
B. motasi	large	Southern Europe, Russia, Africa, Far East, tropical America	sheep, goats	*Rhipicephalus, Haemaphysalis, Dermacentor*	moderate	-
B. ovis	small	As for *B. motasi*	sheep, goats	*Rhipicephalus, Ixodes*	low	-
B. trautmanni	large	Southern Europe, Africa, Russia	pig	*Rhipicephalus*	moderate	-
B. canis	large	North, Central, South America, Southern Europe, Russia, Africa, Asia	dogs and wild canids	*Rhipicephalus, Dermacentor, Haemaphysalis*	moderate-high	tick fever
B. gibsoni	small	India, Ceylon, China	dogs and wild canids	*Rhipicephalus, Haemaphysalis*	high (in domestic dog)	tick fever

B. felis	small	Africa, India	domestic cat, lion, leopard	*Haemaphysalis*	moderate	-
#B. microti	small	World-wide	rodents	*Ixodes*	low	-
#B. rodhaini	small	Africa	rodents	unknown	low	-
** "Large": 2–5 μ m long; "small": 1–2 μ m long. # These species, while not strictly speaking of veterinary importance, have been domesticated and are maintained in hamsters (B. microti), rats and mice (B. rodhaini) in laboratories around the world. They are widely used for preliminary screening of drugs for antibabesial activity and have thus assumed veterinary importance (Krier and Baker, 1987).*						

2.4.16.1. Babesiosis: Geographical Distribution

Endemic regions in the USA include Massachusetts and New York State with Nantucket Island, Long Island, the coast of Connecticut as well as foci in Georgia, California and Wisconsin. Cases have also been reported from various European countries such as Ireland, Scotland, Sweden, former Yugoslavia, France and Russia. There have been isolated case reports from Africa, Asia and Latin America.

2.4.16.2. Babesiamicroti (Apicomplexa: Piroplasmida)

Babesiamicroti (Apicomplexa: Piroplasmida) has historically been considered a common parasite of Holarctic rodents. However, human babesiosis due to this species has generally been limited to the northeastern seaboard of the United States and Minnesota and Wisconsin. The absence of reports of *B. microti* babesiosis from sites where the agent is enzootic, such as in western Europe, remains unexplained.

Studies on the occurrence of *B. microti* in rodents were performed in West Europe. There are reports on the occurrence of this species in rodents and insectivores in Yugoslavia, Czechoslovakia, Austria, Hungary and Bulgaria (Sebek et al., 1977). This species was found in *Microtus agrestis* in Austria (Sebek et al., 1980) and Germany (Walter and Liebisch, 1980), in *Clethrionomys glareolus* and *Apodemus sylvaticus* in England (Turner, 1986), and in all three species in Great Britain (Healing, 1981). However, in Poland, study on the animal reservoir of *B. microti* was done in Mazury (Karbowiak and Sinski, 1994).

Since the early 1960's, *Babesia microti* has been considered to be a medically important parasite. First human infections with *B. microti* were reported in the USA (Grunwaldt, 1977; Parry et al., 1977; Scharfman and Taft, 1977; Steketee et al., 1985). Later, babesiosis caused by *B. microti* infection was also recorded in Europe (Sebek et al., 1977; Krampitz et al., 1986; Brasseur and Gorenflot, 1996). A case of imported babesiosis was also reported in Poland (Humiczewska and Kuzna-Grygiel, 1997).

Babesia develop in the vertebrate host in erythrocytes, in some species in lymphocytes also. In the arthropod vector, *Babesia* develop in a variety of cells and tissues. *Babesia* are classed as large (2–5 μm long) and small (1–2 μm long) types. Many hosts may be infected with both a large and a small type. The large and small forms differ to some extent in susceptibility to drugs (e.g., large species are susceptible to trypan blue; small are not) and in mode of reproduction in erythrocytes (e.g., large species produce one or two merozoites at a time, and small ones may produce four yielding a "Maltese cross" form). To date, development in lymphocytes has been reported for a few species only.

2.4.16.3. Human Babesiosis

Although not normally infective to man, human infections with babesiids have been recorded. The first of these was in 1956, in a Jugoslav farmer whose spleen had been removed after an accident 11 years earlier: he developed a fatal infection with *B. divergens*. The second case was an American hunter, who in 1966 became infected with an unidentified babesiid. Subsequent infections in Americans have been shown to be due to *B. microti* and are usually self-limiting. Those in Europe were all caused by *B. divergens*; they occurred exclusively (unlike the American cases) in splenectomized persons and most ended fatally.

2.4.16.4. The Disease: Babesiosis

Babesiosis is a typical zoonotic disease which is caused by *Babesia* spp., a tick- borne intra-erythrocytic protozoan parasite (Gorenflot et al., 1998; Homer et al., 2000).

Some piroplasms infect fish, birds, reptiles or mammals. The rodent parasite *Babesia microti* (USA) and the bovine parasites *B. divergens* and *B. bovis* (Europe) cause most infections in humans. Newly emerging species, referred to as WA1 and CA1 have been found to cause human infection in the western United States (Persing et al., 1995).

2.4.16.5. Babesiosis: Transmission

Voles form the reservoir. Transmission is via the bite of hard ticks such as *Ixodes scapularis* and *Ixodes ricinus*. The transmission of *B. bovis* is via *Boophilus microplus*. In USA, larval nymphs of *Ixodes scapularis* feed chiefly on *Peromyscus maniculatus* ("the white-footed deer mouse"). The adult ticks suck blood from deer (Lyme disease). Strangely enough, the deer are not infected with *B. microti*. Transmission occurs via ticks. The parasite passes from larva to nymph to adult tick. There is no transovarian transmission. Infections in humans are accidental occurrences. After injection of saliva of the tick, the micro-organisms penetrate red blood cells and mature. *Babesia microti* trophozoites undergo asexual reproduction in human blood and divide into two or four merozoites. To date, no exo-erythrocytic cycle, as exists in *Plasmodium* species, has been described. Infected red blood cells undergo hemolysis. This releases the protozoa which can then penetrate new red blood cells. Infections via blood transfusions have been described. Transplacental infection may occur. When a tick sucks blood, the parasites are in the blood meal. This blood meal is enclosed in a peritrophic membrane in the tick's gut. The parasite changes shape and it is assumed that it produces gametes. Once fertilization is complete, the zygote penetrates the peritrophic membrane and passes through the intestinal epithelium by endocytosis. Once in the hemolymph, it forms an ookinete, which penetrates the salivary glands. There, the parasite forms sporoblasts, a stage which can over winter. Thousands of sporozoites are formed. At the time of the tick's next feeding, the parasite is mature and ready to be injected with the saliva. The part of the cycle between sporozoites and intra-erythrocytic merozoites is as yet unclear.

2.4.16.6. Babesia and Zoonosis

Investigations of the zoonotic reservoirs of *B. microti* in west and central Europe indicate that the following species of rodents can be infected by this parasite: bank vole (*Clethrionomys glareolus*), field vole (*Microtus arvalis*), wood mouse (*Apodemus sylvaticus*), black-strip-ped mouse (*Apodemus agrarius*), snow vole (*Microtus nivalis*), pine vole (*Pitymys subterraneus*), Mediterranean water shrew (*N. anomalus*) (Cox, 1970; Mahnert, 1972; Sebek et al., 1977, 1980; Healing, 1981; Turner, 1986). In Poland, *B. microti* was found in bank vole and yellow-necked mouse (*Apodemus flavicollis*) (Karbowiak and Sinski, 1996b; Bajer et al., 1998).

Karbowiak et al. (1999) reported that in Poland, beside *C. glareolus*, two species of voles of the genus *Microtus* (*M. agrestis* and *M. oeconomus*) may play important roles as zoonotic reservoirs of the piroplasms of the genus *Babesia*.

Because densities of both vole species are very high during outbreaks (Pucek, 1981), the finding of *Babesia microti* infection in *M. oeconomus* appears to be of importance.

Voles belonging to the genus *Microtus* are still considered to be the main reservoir of *Babesia microti* in the natural environment of Europe. The analysis of data collected from various research centers shows that the prevalence rate of infection in *Microtus* voles is much higher than in other rodents (Table 2.19). Similar zoonotic situation was observed in Poland (Karbowiak, 2004; Table 2.20). The prevalence of infection was highest in *Microtus* voles: in common vole, *Microtus arvalis* from Mazurian Lakeland, the prevalence was 9–33%; in field vole *Microtus agrestis* in Katowice agglomeration reached almost 50%, in root vole *Microtus oeconomus* in Bialowieza, 7.7–50%. *Clethrionomys* voles, *Apodemus* mice and shrews play a lesser role as zoonotic reservoirs; the prevalence of infections in these mammals not exceeding 2%.

Karbowiak and Wita (2001) reported *B. microti* from the same host, *Microtus agrestis* and Karbowiak et al. (2004) recorded the same parasite from *M. arvalis*. The present piroplasm discovered from the blood of *Microtus agrestis* resembles the *Babesia microti* as reported by the above authors and is identified as *Babesia microti* Franca, 1912.

Table 2.19. *Host-Wise Prevalence (in %) of Babesia microti Infections in Europe (Excluding Poland)*

Host Species	Prevalence	Localization	Authors
Microtus arvalis	8.3	Austria, Steiermark	Sebek et al. (1980)
	0.7	Czech, S. Moravia	Sebek (1975)
	0.6	Czech	Sebek et al. (1977)
Microtus agrestis	25.5	Southern England	Baker et al. (1963)
	4.1–22.2	England Germany, Bavaria	Healing (1981)
	25.2	Austria, Steiermark	Cox (1970)
	38.0	Austria, North Tyrol	Kramptiz and Baumler (1978)
	30.5	Czech, S. Bohemia	Sebek et al. (1980)
	6.5	Czech	Mahnert (1972)
	0.5		Sebek (1975)
	0.5		Sebek et al. (1977)

Microtus nivalis	2.3	Austria, North Tyrol	Mahnert (1972)
Microtus socialis	50.0	Ukraine, Askania Nova	Karbowiak et al. (2002)
Clethri-onomys glareolus	9.6– 13.0 21.3 16.3 1.8 1.0 0.4 0.3	England, Sussex England Austria, Steiermark Austria, North Tyrol Czech, S. Moravia Czech	Healing (1981) Turner (1986) Cox (1970) Sebek et al. (1980) Mahnert (1972) Sebek (1975) Sebek et al. (1977)
Pitymys subterra-neus	11.1 18.1	Austria, Steiermark Austria, North Tyrol	Sebek et al. (1980) Mahnert (1972)
Apodemus flavicollis	1.6 0.1	Austria, Steiermark Bosnia-Herzegovina	Sebek et al. (1980) Sebek et al. (1977)
Apodemus sylvaticus	2.1– 10.0 8.8 1.6 0.1	England, Sussex Eng-land Bulgaria	Healing (1981) Turner (1986) Cox (1970) Sebek et al. (1977)
Apodemus agrarius	1.1 0.8	Bosnia-Herzegovina Eastern Slovakia	Sebek et al. (1977) Karbowiak et al. (2003)
Mus mus-culus	0.2	Macedonia	Sebek et al. (1977)
Sorex ara-neus	6.8 1.9	England Austria, North Tyrol	Cox (1970) Mahnert (1972)
Sorex minutus	5.4	England	Cox (1970)
Neomys anomalus	1.3	Macedonia	Sebek et al. (1977)

Table 2.20. *Babesia microti Infections in Small Mammals in Poland (in %)*

Rodent Species	Prevalence	Localization	Authors
Microtus arvalis	13.8 9.0	Mazurian Districta	Sinski (1999); Bajer et al. (2001); Pawelczyk et al. (2004)
Microtus arvalis	33.3	Mazurian Districtb	Karbowiak (personal communication)
Microtus agrestis	50.0	Katowice	Karbowiak et al. (1999)
Microtus eoconomus	17.6 7.7 50.0	Bialowieza	Karbowiak et al. (1999) Karbowiak et al. (2002) Karbowiak (personal communication)
Microtus sp.	14.3	Mazurian Districta	Bajer et al. (1998)
Clethrionomys	-	Mazurian Districtb	Karbowiak and Sinski (1996)
Glareolus	0.6 1.0	Mazurian Districta	Sinski (1999); Bajer et al. (2001)
Apodemus flavicollis	2.1	Mazurian Districta	Bajer et al. (1998)
Apodemus sp.	0.7	Mazurian Districta	Sinski (1999)
Sorex minutus	20.0	Bialowieza	Karbowiak (personal communication)

Abbreviations: [a]*Urwitalt near Mikolajki; bKosewo Gorne near Mragow*

2.5. DIVISION IN TRYPANOSOMES

Trypanosoma are truly hematozoic extracellular parasites spending substantial part of their life cycle in the plasma. Fish trypanosomes use blood as a route of entry to their vector, hematophagous leeches. After re-entry into the fish, they may either directly invade (or be injected by the vector into) the blood or they may enter it after an initial period of multiplication in tissue fluid local to the point of entry (Baker, 1976).

During investigations on fish blood, *Colisa fasciata* (Perciformes) was found to be infected with trypanosomes which were dividing. Examinations were conducted on permanent smears of the fish in order to trace out the sequential pattern of division *in vivo* and is being described herein.

2.5.1. Methodology

A survey of 250 fishes belonging to 5 genera, *Channa punctatus* (Perciformes: Channidae), *Heteropneustes fossilis* (Siluriformes: Heteropneustidae), *Clarias batrachus* (Siluriformes: Clariidae), *Wallago attu* (Siluriformes: Siluridae) and *Colisa fasciata* (Perciformes: Belontiidae) was carried out which were procured from the local ponds of Bareilly and maintained in aquaria. Blood was taken directly from the heart as well as from the caudal vein. Thin blood films were prepared, fixed in methanol and stained in Giemsa.

2.5.2. Results

2.5.2.1. Division in Trypanosoma Trichogasteri (Host: Colisa fasciata)

Out of all the specimens examined, only one fish (*Colisa fasciata*) infected with *Trypanosoma trichogasteri* Gupta and Jairajpuri, 1981 revealed all the divisional stages. They are shown diagrammatically in Plate XLII. The infected host exhibited exorbitant parasitemia. The trypanosome multiplied by longitudinal binary fission. Multiple fission was not observed.

The first indications of division were seen in the kinetoplast which showed several types of configurations. The kinetoplast elongated longitudinally/transversely; a constriction appeared in the middle and finally two daughter kinetoplasts were produced either lying one above the other (transverse fission, Plate XLIIIA) or side by side (longitudinal fission, Plate XLIIIB). The daughter kinetoplasts were nearly of the same size.

The nucleus divided only by transverse binary fission. The karyosome elongated longitudinally and divided within the nuclear membrane. The divided karyosome represented the polar caps between which strands of spindle fibers could be seen. The polar caps became the karyosomes of the daughter nuclei when the parent nuclear body had completely splitted into two. The two nuclei were found to be one above the other (Plate XLIIIC).

It was surprising to observe that parasites with two kinetoplasts as well as two nuclei were both observed indicating that division could be initiated either with the kinetoplast, or nucleus. Forms with two nuclei and two kinetoplasts showing divided condition of both organelles were also evident (Plate XLIVA).

The sequence of flagellar division was also not fixed. Forms with two flagella and two nuclei were visible in which the kinetoplast had not divided as also forms with distinct two kinetoplasts, two nuclei and two flagella. Rarely, flagellar division commenced prior to nuclear division and such forms had a single nucleus, 2 kinetoplasts and 2 flagella (Plate XLIVB). Clear cut forms with two kinetoplasts, two nuclei and two flagella represented an advanced stage of division.

Now the cytoplasm started to divide from the posterior end towards the anterior (Plate XLIVB). Gradually, each cytoplasmic arm contained one kinetoplast. The mother kinetoplast retained the undulating membrane and the free flagellum while a new flagellum arose from the daughter kinetoplast cither simultaneous or subsequent to cytokinesis. As cytokinesis proceeded, each body contained one nucleus. Complete undulating membranes were the last to acquire perfection. Finally, the two completed trypanosomes could be seen attached to each other at their anterior ends before becoming free of each other to exist independently as two separate identities.

2.5.2.2. Division in Trypanosoma kargenensis (Host: Heteropneustes fossilis)

A few forms undergoing division were also visible in *T. kargenensis*. The pattern of division was almost similar to *T. trichogasteri* except that cytokinesis commenced from the anterior end (Plate XLIVC) instead of posterior. However, the pattern of multiplication in the kinetoplast (transverse: Plate XLVA; longitudinal: Plate XLVB) and nucleus (Plate XLVC) was similar.

2.5.2.3. Division in Trypanosoma lewisi (Host: Rattus rattus)

A few forms of *T. lewisi* undergoing multiplication were observed in the blood of *Rattus rattus*.

- **Kinetoplast Division**: Kinetoplast division could be observed in the immature form (Plate XLVIA) as well as in the classical slender adult forms (Plate XLVIB) indicating that kinetoplast division commences at an early stage of trypanosome maturation.
- **Nuclear Division**: In the early stage of nuclear division, the two nuclei could be seen lying in close proximity (Plate XLVIIA)) to each other which later migrated in opposite directions ((Plate XLVIIB).

  Other divisional stages could not be observed.

2.5.2.4. Division in Microtus microti (Host: Microtus oeconomus)

In *Microtus microti*, only flagellar division was observed (Plate XLVIII). Peculiarly, in this form, the nucleus and kinetoplast were undivided, apparently, the flagellum was the first organelle to divide in *Trypanosoma microti*.

2.5.3. Discussion

Qadri (1955), Daly and Degiusti (1971), Khan (1972), Grogl et al. (1980) and Mukherjee and Haldar (1982) reported absence of divisional stages in fish blood during their studies.

It was presumed by some authors that the parasites did not multiply in fish blood (Letch, 1977). Occasional reports of divisional stages sprung up without providing clear evidence of the sequential pattern of division.

Lom and Dykova (1992) divided the development of the parasite in the fish host into four different phases:

- **Phase I:** Prepatent period (2–9 days) characterized by the absence of flagellates in the peripheral blood. It is not exactly known where the organisms are located during this phase and no tissue stages have been reported.
- **Phase II**: Appearance of slender forms of the flagellates in the peripheral blood – patent and increasing parasitemia due to the division of the trypanosomes.

Division of trypomastigote stages of piscine trypanosomes has been investigated rarely. The division phase lasts from several days to several weeks and a heavy parasitemia may be achieved. Sometimes, heavy infections may result in the death of the host, specially in young fish or in experimental hosts such as goldfish. However, due to immune phenomenon, the infection passes on to the next phase.

Laveran and Mesnil (1907) reported that fish trypanosomes multiply by longitudinal binary fission and protoplasmic masses undergoing division and young trypanosomes were observed in the blood of *Cyprinus carpio* and *Esox lucius*. Tanabe (1925) observed the dividing forms only once in the blood of leeches but these forms were not described. Qadri (1955) reported longitudinal binary fission in the culture forms of *T. straiti*. Baker (1960) mentioned the presence of two small trypanosomes possessing two nuclei and two kinetoplasts. Two opposite nuclei and a central kinetoplast has also been seen (Raychaudhuri and Misra, 1973). Dividing nuclei in fish trypanosomes, *T. batrachi* (Qadri, 1962a), *T. maguri* and *T. vittati* (Tandon and Joshi, 1973), *T. mrigali* (Joshi,

1976a), *T. attii* (Gupta and Jairajpuri, 1982b) and *T. mukasai* (Sinha, 1986) have been reported. Nuclear division has been described in other trypanosomes as well, *T. raiae* and *T. rhodesiense* (Vickerman and Preston, 1970); *T. vivax* (Vickerman, 1973); *T. cruzi* (De Souza and Meyer, 1974) and *T. equiperdum* (Paulin, 1975). Gupta and Saraswat (1991) observed dividing forms in *T. rohilkhandae* and binucleated stages were the most common, flagellar division was rare.

In fresh water fish, *Trypanosoma carassii* (=*T. danilewskyi), T. granulosum, T. remarki, T. tincae* and unidentified trypanosomes have been observed to divide (Needham, 1969; Lom, 1979; Woo, 1981; Paterson and Woo, 1984) or to undergo changes suggestive of division (Laveran and Mesnil, 1902; Ogawa and Uegaki, 1927; Orecka-Grabda and Wierzbicka, 1996). The latter include duplication of the kinetoplast, flagellum, and nucleus. Division in fish trypanosomes is best known in detail in some European cyprinid trypanosomes, identified as *T. carassii* and *T. tincae,* that show a particularly high frequency of division stages in highly susceptible hosts (Needham, 1969; Woo, 1981; Paterson and Woo, 1984). In dividing bloodstream forms of these trypanosomes, appearance of a short, new, anteriorly directed flagellum precedes (*T. carassii*), or directly follows (*T. tincae*) the division of the kinetoplast.

The complete sequential pattern of division of *T. danilewskyi* in *Carassius auratus* was explained by Woo (1981). The author divided the stages into four phases arbitrarily: Stage I (production of a new anteriorly directed flagellum and kinetoplast division), Stage II (enlargement and rounding of posterior end and flagellar shift to posterior end. The gradual extension of the new flagellum was accompanied by a swelling of the posterior end of the body and a transition to a posterior direction of the flagellum as a new trypanosome body gradually formed its alongside. The new flagellum flipped posteriorly and along it, new cytoplasm was produced or accumulated), Stage III (weakly formed undulating membrane. As this process produces a daughter individual of a similar size as the parental cell, the nucleus commences division, with karyokinesis where one nucleus migrates past the two kinetoplasts to the daughter trypanosome) and Stage IV (two fully formed trypanosomes joined at their posterior end, transverse constriction between the two kinetoplasts, formation of young parasites, at this stage now recognized as mother and daughter trypanosome with weakly developed undulating membrane and the nucleus situated close to the kinetoplast through the respective well-developed and poorly developed undulating membranes (Needham, 1969; Woo, 1981). The latter author also reported two abnormal forms, one with a kinetoplast and no nucleus and other with a kinetoplast and two nuclei which did not fit into the sequence of division mentioned by him. The division was thus transverse.

Yadav (1990) described the divisional pattern of trypanosomes at day 15–25 p.i. in *Heteropneutes fossils* by inoculating trypanosomes in the fish. During the first stage, multiplication in majority of the forms initiated with the division of the kinetoplast (transverse + longitudinal). The posterior end of the parasite about to undergo division widened and the kinetoplast divided. The author also reported the nucleus to divide prior to the kinetoplast in some cases, the daughter nuclei remaining enclosed in the nuclear membrane till final separation (Figure 2.1).

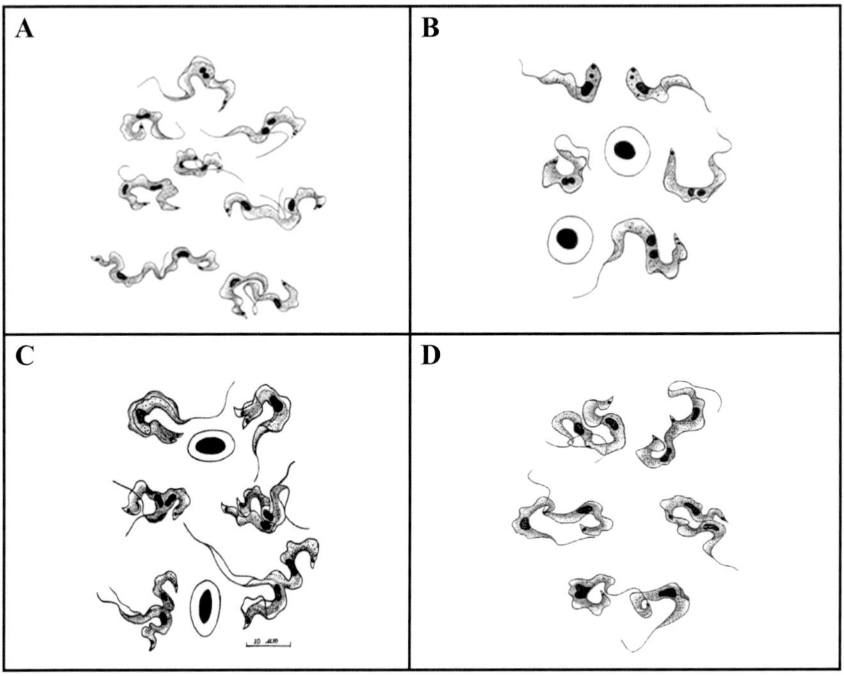

Figure 2.1. Division in *Trypanosoma kargenensis* from *Heteropneustes fossilis.*

Note: A: kinetoplast and nuclear division; B: nuclear and kinetoplast division and cytokinesis.

C: cytokinesis. D: longitudinal kinetoplast division and cytokinesis (Yadav, 1990).

During the second stage, the parasite multiplied either by transverse or longitudinal division. During the former, the division of the nucleus occurred with a simultaneous upward migration of the kinetoplast and the relative time sequence of the two varied: in most cases the migration of the kinetoplast was more swift as compared to nuclear division. The new kinetoplast gave rise to a new flagellum which increased in progression with subsequent changes and the undulating membrane was ill defined. During the process, the parasite elongated and the two nuclei also separated from each other and the trypanosome became more undulated. In majority of the divisional forms, the author reported that the kinetoplast migrated towards the opposite extremity completely in which case both the transforming forms remained attached to each other at their anterior ends before their final separation. However, the author also observed that in some forms, the new kinetoplast migrated only up to the middle and transverse cytoplasmic division occurred where one form remained attached by its anterior whereas the other by its posterior before final distinction into two forms.

The other pattern of division followed by the trypanosomes of *H. fossilis* was the longitudinal sequence of multiplication. Longitudinal cytoplasmic division commenced just after kinetoplast division. Sometimes this plan of division continued so that each organelle longitudinally splitted into two resulting, into the formation of two new trypanosomes conversely, a constriction at both the ends of the parasites appeared which progressed towards the middle and the latter was the last to separate.

In addition to the above, the authors also observed two forms in which the parasite splitted longitudinally simultaneously into two equal halves. One form which was apparently undergoing transverse division was a step ahead where division and separation of one form began even before the first division was complete.

Lom and Dykova (1992) reported division in fish trypanosomes to start with production of a new flagellum, followed by a kinetoplast and then nuclear division. The cell divided longitudinally starting anteriorly (e.g., *T. remarki*). In *T. carassii*, the new flagellum flipped posteriorly to extend in a direction opposite to that of the old flagellum and new cytoplasm grouped along it. After nuclear division, one of the nuclei moved posteriorly past the two kinetoplasts. Finally, a transverse, unequal cell cleavage occurred between the kinetoplasts; the posterior daughter individual being smaller. The vesicular nucleus divided by a modified mitosis ("closed" pleuromitosis).

Ultrastructural studies on the division of the epimastigotes of *T. cobitis* were conducted by Lewis and Ball (1981). The investigations were based on electron microscopy of the epimastigotes in the leech vector, *Hemiclepsis marginata* and in culture. The division followed the Trypanosomatid pattern of duplication of basal body, kinetoplast and then nucleus followed by cytokinesis with the exception that the dividing epimastigotes within the vector had multi-axonemal flagella whereas in culture, the new flagellum grew separately alongside the original. The kinetoplast duplicated by both lateral and equatorial enlargement. Multiplication of *Trypanosoma pacifica* in English sole, *Parophrys vetulus* from Oregon coastal waters was reported by Burreson and Karlsbakk (2007) based on observations of live flagellates and Giemsa-stained blood smears. The authors reported that multiplication commenced with the elongation of the kinetoplast, thickening of the posterior portion of the body, and appearance of a new flagellum near the kinetoplast. The new flagellum was very rigid when less than 3 μm in length, but it became flexible as it elongated. When the new flagellum was approximately 12 μm in length, cell division commenced and the kinetoplast also began to divide. The timing of nuclear division was variable. Generally, it did not occur until division of kinetoplast had begun, but occasionally binucleate individuals were observed before cell or kinetoplast division was apparent. As division continued, one nucleus migrated past the dividing kinetoplast into the future daughter trypanosome. Finally, the kinetoplast completed division and the trypanosomes separated. Cell division was unequal, with the daughter trypanosome being smaller than the parent and with a more weakly developed undulating membrane.

Among marine fish trypanosomes, evidence for division is scantly. *Trypanosoma myoxocephali* was observed dividing in the kidney of a marine sculpin (Fantham et al., 1942), but no details of figures were provided. Structural changes suggestive of division have been observed in bloodstream forms of *T. rajae* (=*T. variabile*) and *T. haploblephari* from elasmobranchs and in *Trypanosoma yakimovi* from a pipefish (Neumann, 1909; Yakimoff, 1912; Yeld and Smit, 2006), but the complete division process has not been observed.

Studies of cell division in trypanoplasms are very scarce. Findings of Keysselitz (1906) claimed sequence of division of nucleus and kinetoplast and Ponselle (1913) made it clear that it follows a pattern found in bodonids in general (Hollande, 1952) and is known also in *Cryptobia congeri* (Martin, 1910, 1913). Bower and Woo (1977b) have proved the contrary. Longitudinally-dividing bloodstream forms of *T. catostomi* start the division in their flagellar apparatus and nucleus, followed by kinetoplast division.

Woo (1978) reported unequal longitudinal binary fission during the division of *Cryptobia salmositica* in coho salmon, *Oncorhynchus kisutch*. Multiplication was initiated by the production of two flagella (one long and one short) followed by nuclear division which did not complete until kinetoplast division ended. Body division commenced from the posterior end soon after the long flagellum attached to the body. Following this, the nucleus, the kinetoplast and the blepharoplast migrated into the newly divided part of the organism. Final body division completed after the migration of these organelles.

Burreson (1982) observed division in *Trypanoplasma bullocki* inhabiting the blood of the hogchoker, *Trinectes maculatus*. *Trypanoplasma salmositica* multiplied rapidly by binary fission in fish. Intensities were significantly higher in fish with high (e.g., 3.5g dl^{-1}) than low (e.g., 1.0 gdl^{-1}) plasma proteins

(Thomas and Woo, 1990). Serum supplement was an important component under *in vitro* conditions. The parasite did not multiply in minimum essential medium unless FBS was present. Glucose depleted from the medium and its depletion increased corresponding to the number of parasites.

Cryptobia cataractae in *Semotilus atromaculatus* divided by equal binary fission that produced elongate, slender daughter cells (Jones and Woo, 1990). This was in contrast to that reported by Woo (1978) who observed unequal longitudinal binary fission during the division of *C. salmositica* in coho salmon, *O. kisutch*. Division was initiated by production of a second pair of flagella, daughter basal bodies remained near the anterior extremity. The nucleus divided and daughter nuclei were joined by a slender strand. Equal, longitudinal cell division was initiated at the posterior extremity and progressed anteriorly. The kinetoplast divided prior to disappearance of the nuclear strand and completion of cytokinesis.

Peckova and Lom (1990) grew different strains of trypanoplasms in axenic culture. Cultures of *T. borreli* from *Blicca bjoerkna, Cyprinus carpio, Scardinius erythrophthalmus* and *Tinca tinca*), *T. guerneyorum* from *Esox lucius* and *Trypanoplasma* spp. from the leech, *Piscicola geometra* were maintained in biphasic blood agar medium, SNB-9 supplemented with vitamins and antibiotics. In culture, the flagellates transformed into smaller, elongated stages with a poorly developed undulating membrane and into short forms that were morphologically similar to stages in the leech vector. The authors passaged the cultures weekly at 17–20°C but they also grew at 4°C. The flagellates divided by binary fission, being initiated by the formation of two new anterior flagella. The original anterior flagellum of the mother trypanoplasm became the recurrent flagellum of one of the daughter trypanoplasm. Nuclear division took place followed by transverse cleavage of the kinetoplast. The division was completed by longitudinal fission of the mother trypanoplasm into two offsprings. The authors also observed multiple fission resulting in rosette formation which then cleaved into several daughter cells as well as some dyskinetoplastic and other anomalous forms, not observed by other authors. In cultures isolated from Tenches with high parasitemias, non- dividing, long, filiform stages were observed. However, the culture stages were not infective for susceptible fish (Peckova and Lom, 1990).

Dividing *T. borreli* were mainly found in the muscle of infected carp near the inoculation site during the prepatent period and the exponential growth of parasitemia by Kruse et al. (1991). There was no evidence of multiple division in *T. borreli*. The division process of the hemoflagellate initiated with the formation of an additional pair of basal bodies and flagella (similar to that reported by Woo, 1978 and Jones and Woo, 1990), followed by duplication of kinetoplast and nucleus. During the duplication of organelles, the cells were stumpy shaped. After division of organelles was completed, the division of the cytoplasm started with a constriction near the origin of one pair of flagella.

- **Phase III:** Chronic phase characterized by a reducing number of flagellates in the blood and lasts for a few weeks or for an indefinite period. As it is the longest phase, it is the one most frequently observed. There are no division stages in the blood during this phase.
- **Phase IV:** There is a complete absence of flagellates in the peripheral blood and the fish appear to be free of trypanosomes. However, sometimes, the flagellates may be found in their preferred sites, the mesonephros, pseudobranchia or rete mirabile of the eye. The disappearance of flagellate is a manifestation of premunition, of a non-sterile rather than of a sterile immunity as these fish appearing apparently as infection–free may later suffer a sudden, severe relapse.

The ploidy of trypanosomes reproducing largely by binary fission still remains an open question. The main reason for uncertainty in this basic fact is the absence of the cell cycle phase with distinct easily counted chromosomes. For long, trypanosomes have been considered to be asexual (Hoare, 1972), however, Tait (1980) doubted on this concept by studying enzyme patterns of *T. brucei* and the existence of sexuality has been suspected. The assumptions of the existence of sexual processes in Trypanosomatids were predominately based on zymograms and genetic equilibrium studies and supported the idea of diploidy indirectly (Tait, 1980). However, there is as yet no indication of sexual reproduction in fish trypanosomes.

A comparison of the division pattern of *Trypanosoma trichogasteri* with those reported above indicate some deviations in the division pattern. The multiplication pattern of *T. trichogasteri* is different to that reported by Woo (1981) for *T. danilewskyi* who reported that the first division stage was the production of a new flagellum followed by division of the kinetoplast. However, in *T. trichogasteri,* division initiated with the kinetoplast and the new flagellum arose from the daughter kinetoplast either simultaneous or subsequent to cytokinesis.

T. danilweskyi divides by transverse binary fission whereas *T. trichogasteri* by longitudinal binary fission. However, both are symmetrogenic as characteristic of mastogophereans (Levine et al., 1980).

Reilly and Woo (1982) also observed flagellar origin prior to cytokinesis. Another conspicuous difference occurred in the sequential fission of the cytoplasm. The pattern of duplication during the present investigations is also dissimilar to that reported by Burreson (1982) who mentioned nuclear division prior to that in the kinetoplast in a related form, *Trypanoplasma bullocki.* Reilly and Woo (1982) observed cytoplasmic fission from the anterior flagellar end backwards.

The above microscopic and ultra-structural studies indicate that the pattern of division in piscine trypanosomes is apparently by binary fission, however, variations exist whether the division is longitudinal or transverse and the sequence of division of cell organelles. These variations maybe species specific, however further studies are in demand for other fish trypanosomes species in order to make some conclusive remarks.

Division in *T. trichogasteri* is almost similar to that reported by Wenyon (1926) and Lewis and Ball (1981) where division commences in the kinetoplast followed by nuclear division. Our findings support the latter in view of the fact that forms possessing two kinetoplasts and a single nucleus were abundant, whereas no forms with two nuclei and a single kinetoplast were visible. This was followed by cytokinesis and the separation into two daughter trypanosomes, nearly similar and equal to each other. Longitudinal as well as transverse binary fission took place in the kinetoplast as was also shown by Lewis and Ball (1981).

2.6. EVOLUTIONARY TREND IN TRYPANOSOMES

Trypanosomes are ubiquitous parasites and can be found in all classes of vertebrates, from fish to mammals, and are transmitted by a wide variety of arthropod and leech vectors. The application of molecular techniques has greatly increased our knowledge of the evolutionary history and population biology of trypanosomes, and for most trypanosome groups, the findings of early studies have been superseded by the results of more modern studies, which, in general, offer increased levels of resolution, robustness and accuracy. However, the application of new methodologies across the full range of trypanosome groupings has been by no means equal. For example, the evolution of some groups (notably those of medical importance) has been studied using DNA sequence data from a range of complementary sequence markers, e.g., 18S rRNA and GAPDH, which demonstrate trypanosomes to be a monophyletic assemblage; in turn, such a finding is key in offering insights into the evolutionary origins of trypanosomes. In contrast, our understanding of some other groups still rests, in some cases almost entirely on the results of earlier techniques, e.g., isoenzyme analyses. Recently, Stevens et al. (2011) has focused on the evolution of key trypanosome groupings (e.g., subgenera) in the light of literature published until 2005, and has evaluated the contribution of molecular evidence to our understanding of trypanosome nomenclature, taxonomy and evolution.

Trypanosomes are transmitted by a variety of ectoparasite vectors such as leeches and bloodsucking insects. They have thus adapted to life in a diverse range of hosts and evolved a number of different and complex developmental pathways.

2.6.1. Evolution of Trypanosoma

Reports on the occurrence of parasites, changes induced in the host body and host reaction to invasion are gradually accumulating. But when and how these organisms evolved is often a big question mark for the parasitologist.

Since the first attempts to classify the evolutionary history of trypanosomes, there have been conflicting reports regarding their true phylogenetic relationships and, in particular, their relationships with other vertebrate Trypanosomatids as well as with many insect parasitizing Trypanosomatids (Hughes and Piontkivska, 2003; Hamilton et al., 2004; Simpson et al., 2006). Perhaps the issue that has provided most debate is that concerning the monophyly (or otherwise) of genus *Trypanosoma* and, even with the advent of molecular methods, the findings of numerous studies have varied significantly. While of arguably limited applied importance, resolution of the question as to whether or not trypanosomes are monophyletic is critical to accurate evaluation of competing, mutually exclusive evolutionary scenarios for these parasites, namely the 'vertebrate-first' or 'insect- first' hypotheses (Leonard et al., 2011).

Current views relating to the nature of evolutionary mechanisms are based chiefly on studies of highly specialized organisms, and because parasites are specialized, these views are particularly appropriate. Many attempts have been made to trace the phylogenies of parasites but most of them are speculative and often make assumptions without evidence and without a critical analysis of opposing views. The evolution of the blood parasite, *Trypanosoma* the causative agent of Sleeping Sickness in man is an excellent example of heme-bioinvasion. When and how the parasite entered the blood has for long puzzled the protozoologist. To bridge this gap, the evolution of *Trypanosoma* was traced out based on some concrete facts during the visit to the Institute of Parasitology, Warsaw with Prof. Irena Wita under the Indian National Science Academy (INSA) and Polish Academy of Sciences (PAS) Senior Scientist Exchange Program.

Apparently, the mode of transmission as a means of understanding the evolutionary history of trypanosomes has assumed great importance. Blood sucking arthropods or leech vectors transmit most vertebrate trypanosomes from one host to the other. The parasites are taken up by a vector with a blood meal and undergo one or more cycles of development and multiplication in the alimentary tract of the invertebrate before infective forms are transmitted to a new vertebrate host. But how did the *Trypanosoma* reach the vector and subsequently the definitive host warrants a deeper critical examination.

The genus appears to have a common ancestor with the euglenoid Protozoa early in the evolution of eukaryotic cells. It has probably evolved from a *Phytomonas* like ancestor, a closely related genus of flagellates. *Phytomonas* is a parasite of euphorbiacean plants and *Trypanosoma* has evolved from this ecologically distant organism. Through pre-adaptation, this change has incurred. The genus *Phytomonas* may infect the fruit, latex, sap, phloem and flowers of many plant families. These plant feeding bugs accidentally transmit *Phytomonas* from one plant to the other. These insects feed with mouth parts adapted to pierce the plant cuticle and while sucking the nutritive plant juices, suck up *Phytomonas* (Plate XLIXA). In doing so, the original stage in the plant increasingly became adapted to its insect intermediate host till it became more of a parasite of the insect than the plant (Plate XLIXB). The common occurrence of *Herpetomonas* in insects today supports the above evolutionary trend and the adaptation of *Phytomonas* to the insect vector.

Due to further evolution, the piercing and sucking mouth parts of the bug, pre-adapted to suck plant tissue modified to pierce the skin and drink the blood of vertebrates (Plate LA). Biological evidence indicates that bugs belonging to the family Reduviidae (*Triatoma megista*) are the chief carriers of *Trypanosoma cruzi*, the causative agent of Chagas Disease. Flagellates in the gut of the bug gradually became physiologically and biochemically adapted to living in the blood (Plate LB). Thus, a number of small consecutive steps, made possible by morphological, ecological or physiological pre-adaptation and suitability made *Trypanosoma* evolve from plant latex to blood micro-niche.

The Trypanosome Research Group of the School of Biological Sciences, University of Bristol, U.K. is currently engaged in building phylogenetic trees based on gene sequences in order to study trypanosome evolution. Due to the absence of a fossil record and the limited number of distinctive morphological characters, this appears to be the only realistic approach and it is presumed that this study will provide some concrete basis for the evolution of *Trypanosoma.*

The school has so far constructed Trypanosomatid phylogenies based on the genes for the small subunit ribosomal RNA (ssu rRNA) and for glycosomal glyceraldehyde phosphate dehydrogenase (gGAPDH). Trees based on gGAPDH nucleotide and amino acid sequences robustly support monophyly of genus *Trypanosoma*, revealed to be a deeply branching lineage of the family Trypanosomatidae. Other Trypanosomatids, including *Leishmania*, branch paraphyletically at the base of the trypanosome clade. These relationships are reiterated in most ssu rRNA trees, depending on the sequence alignment used and exclusion of highly divergent taxa.

The phylogenetic position of trypanosomes from Australia has been influential in the construction of theories on trypanosome evolution. Three trypanosome isolates from a kangaroo, wombat and duck-billed platypus collected by Harry Noyes of the Liverpool School of Tropical Medicine were included in the first trees, and Patrick Hamilton undertook a further survey of trypanosomes from Australian vertebrates. This resulted in several new trypanosome isolates: five from platypus, two from the common wombat, one from a wallaby, one from a European Rabbit and one from a Currawong (bird). The trypanosomes from the three marsupials fell in different clades, and were more closely related to trypanosomes from non-Australian mammals than to each other, suggesting that Australian marsupial trypanosomes have divergent evolutionary origins. The Currawong trypanosome clustered with *T. bennetti*, an avian trypanosome from North America.

According to the school, current evolutionary hypothesis proposes that gGAPDH trees support the hypothesis that trypanosomes evolved from an ancestral insect parasite, which adapted to a vertebrate/insect transmission cycle as was concluded above at the Institute of Parasitology, Warsaw during the present observations. The reason could be because the trypanosome clade arises from within the wider group of Trypanosomatids, which are fundamentally insect parasites, some of which in addition have a vertebrate or plant host. The switch from insect to leech vectors for fish and some amphibian and reptile trypanosomes was probably secondary. The trees show these aquatic leech- transmitted trypanosomes in a single, robust clade, and it is clear that the majority of insect-transmitted trypanosomes did not evolve from this clade.

It has also been suggested that host parasite relationships coevolve. This rather obvious statement forms the basis for many current ideas about sexual selection (Hamilton and Zuk, 1982) and the maintenance of sexual reproduction (Hamilton et al., 1990; Ladle, 1992). However, in most other areas of research on host-parasite relationships, the possibilities of co evolution is generally neglected. Most studies, for example, on the evolution of parasite virulence consider evolutionary pressure on the parasite while holding the host constant (Frank, 1996). A reasonable justification for this approach is that parasites often have shorter generation times, higher population densities and higher mutation rates than their host; thus the parasite is expected to evolve much more rapidly (Hafner et al., 1994).

Other changes of the host life-history can be seen as an adaptation by the host to reduce the parasite's effect. Such changes can come in two varieties: phenotypic and genetic evolutionary changes.

If the life history theory traits and parasites resistance have a genetic basis, parasite pressure can lead to an evolutionary change in a host's life history. The theoretical background for this can be illustrated by considering the evolution of age at first reproduction: if hosts are subjected to severe parasite pressure and if they are unable to resist the parasite by other means, they will evolve to mature earlier (Hochberg et al., 1992). Examples documenting either a genetic correlation of resistance with a host's life history or an evolution response of the life-history are rare.

Although detailed studies on many systems are lacking, it is presumed that the host's life-history often constrains the evolutionary possibilities of the parasite. Macdonald (1957) considered the development of malarial parasite within a mosquito and has shown that the adult mosquito's life expectancy is similar to the parasite's developmental time. The mosquito may dic before the parasite has completed its development and is ready to transmit to the next host. Therefore, any increase of the host mortality, whether due to genetic or environmental factors, will decrease the parasites transmission success; the parasites will have to modify its development to compensate for changes of its host's life history in order to survive.

The evolutionary origins of host manipulation by parasites may go all the way back to the establishment of complex life cycles (Smith Trail, 1980; Moore, 1984). Any individual parasite manifesting the ability to alter the behavior of its host with a resulting increase in its transmission success is favored over its conspecifics by natural selection. Given enough genetic variation, what originally may have been an incidental side-effect of infection could then have been shaped by selection into a more refined manipulation mechanism. No doubt several living parasite species share an ability to manipulate host behavior that was inherited, perhaps with modification, from a common ancestor. On the other hand, it is also likely that host manipulation has evolved independently in different, more distant parasite taxa. Different evolutionary constraints and selective forces related to the ecology of the host-parasite system may have interacted to generate the observed variability in the ability of parasites to manipulate host behavior.

The parasite's life cycle appears to be well-suited for a host living in patchy and ephemeral habitats, combining vertical transmission to spread its infection among sub-populations and with horizontal infections to infect individuals with in sub-populations (Koella et al., 1998). The life cycle appears to be tuned to use the mode of transmission with the higher potential of transmission success. The parasites transmitted vertically in good growth conditions and its reproductive success can be expected to be high; it transmits horizontally the environmental conditions that would lead to low reproductive success of the host and thus would allow only low vertical transmission success of the parasite (Koella et al., 1998).

The parasite can not only exert considerable selection pressure on the host's life history, but it also affects the genetic correlation among life history traits, which in turn determines the host response to selection. These complex instructions may lead to two different pathways of evolution, depending upon the parasite's prevalence. The evolutionary trend, however, will be governed by the selection in the uninfested multiplication sites. The response of the host's life history to the parasite maybe be negligible, which can maintain virulent parasites with low potential of vertical transmission. Therefore, if the parasite's prevalence is low, selection is likely to lead to virulent parasites with horizontal transmission. The decreases of the genetic correlations to low and negative values enhance, rather than counteract the selection imposed by the parasite. Thus, high prevalence is likely to select predominantly vertical transmission of avirulent parasites (Koella et al., 1998).

Recent interest in host-parasite co-evolution has followed in the footsteps of progress in phylogenetic reconstructions of the evolutionary history of related groups of animals (Jackson, 1999). This emphasis on phylogenetic relationships should be reflected in any measure of host specificity because the host specificity of parasites is a simple concept encompassing their evolutionary history of associations with different host taxa (Caira et al., 2003). Poulin and Mouillot (2003) proposed a new index of host specificity providing a brief summary of a parasite's history of host switching and of the average taxonomic (or phylogenetic) distance between hosts during host switches.

Often interest in host specificity is motivated by studies of host-parasite co- evolution or by assessments of the risks of a parasite species colonizing a new host species for example, following a species introduction. If we consider species of parasites from an evolutionary perspective, it is not the prevalence or intensity of infection that matters the most, but the phylogenetic relationships among the host species (Poulin and Mouillot, 2003).

It thus appears that many factors determine the evolution of a host's life-history or a parasite's life cycle. However, the interactions between the life-history and life cycle form an important part of the host parasite relationship. Suitable details of the interactions can determine the outcome of selection; thus, without explicitly considering these interactions, it is difficult, if not impossible, to predict the direction of evolution.

Organic environment is equally important to the inorganic environment in influencing the adaptations of a given parasitic species. The organic or living environment includes all forms of life with which a given individual or species comes into ecologic relation. The evolution of *Trypanosoma* is an example of its adaptability in different organic environments to facilitate its journey from plant sap to human blood.

2.7. PLEOMORPHISM

2.7.1. Trypanosome Infectivity and Pleomorphism

The taxonomic studies conducted in the thesis would appear incomplete without mentioning pleomorphism of trypanosomes which is one of the most remarkable feature of trypanosomes. Research on these parasites conducted over the past 34 years in the investigator's laboratory revealed that the parasites discovered from fishes up to mammals could range from monomorphic to polymorphic forms. The facts given below are the compiled information of the entire work done by the research team from the commencement of the work till date.

24 species of trypanosomes from 12 fish hosts have been discovered. Out of these, 12 species were monomorphic, 6 dimorphic and 6 polymorphic. Out of these, from *Heteropneustes fossilis*, 2 monomorphic (*T. kargenensis* Gupta and Gupta, 1994; *T. karelensis* Gupta et al., 2000), 2 dimorphic (*T. artii* Gupta et al., 2002; *T. heteropneusti* Gupta et al., 2006) and one polymorphic (*T. singhii* Gupta and Jairajpuri, 1981) species were discovered. From *Channa punctatus*, one monomorphic (*T. rohilkhandae* Gupta and Saraswat, 1991), two dimorphic (*T. aligaricus* (Gupta and Jairajpuri, 1982); *T. saulii* Gupta et al., 2006) and one polymorphic (*T. bareilliana* Gupta et al., 1987) were discovered. From *Colisa* (*=Trichogaster*) *fasciata*, 2 monomorphic (*T. trichogasteri* Gupta and Jairajpuri, 1981; *T. piscidium* Gupta et al., 2003), one new variety (*T. trichogasteri* var. nov. *fasciatae*) and one polymorphic (*T. colisi* Gupta, 1986) species were discovered. From *Mystus vittatus*, one monomorphic (*T. bagroides* Gupta and Gupta, 1996) and one dimorphic (*T. tengari* n.sp., unpublished data) whereas from *Catla catla*, one monomorphic (*T. monomorpha* Gupta and Jairajpuri, 1985) and one new variety (*T. monomorpha* var. nov. *catlae* Gupta et al., 2000) were discovered. Amongst the remaining species, three of them were monomorphic (*T. notopteri* (Gupta and Jairajpuri, 1985) from *Notopterus notopterus; T. ritae* Gupta and Yadav, 1989 from *Rita rita*; *T. danilewskyi* (Laveran and Mesnil, 1904) from *Cyprinus carpio*), one dimorphic (*T. batrachi* (Gupta and Jairajpuri, 1983) from *Clarias batrachus*) and three polymorphic (*T. attii* Gupta and Jairajpuri, 1981 from *Wallago attu; T. barbi* (Gupta et al., 1987) from *Barbus barbus; T. ticti* (Gupta et al., 1998) from *Puntius ticto*) species have been put on record (Table 2.21).

In addition to the above, representative vertebrates from Amphibia (*Rana tigrina, Bufo melanostictus*), Reptilia (*Hemidactylus flaviviridis, Calotes versicolor*), Aves (*Columba livia*), and Mammals (*Rattus Rattus* from India and *Clethrionomys glareolus*, *Apodemus flavicollis,* and *Microtus oceanus* from Poland) were also examined for flagellate parasites. Amongst these, the infected hosts were *Bufo melanostictus* (*T. bisalpurensis* n. sp. unpublished data), *Rana tigrina* (*T. rotatorium*), *Rattus Rattus* [*Trypanosoma (Herpetosoma) lewisi* (Laveran and Mesnil, 1901)], *Clethrionomys glareolus* [*Trypanosoma* (*Herpetosoma*) *evotomys* (Molyneux, 1969)], *Apodemus flavicollis* [*Trypanosoma* (*Herpetosoma*) *grosi kosewiense* (Karbowiak and Wita, 2004)], and *Microtus oeconomus* [*Trypanosoma* (*Herpetosoma*) *microti* (Laveran and Pettitt, 1909)] (Table 2.21). It was interesting to conclude that amongst fishes, monomorphic, dimorphic as well as polymorphic forms were observed but from amphibians and mammals, only monomorphic forms were encountered. The reptiles (*Hemidactylus flaviviridis, Calotes versicolor*), however, were not prone to trypanosome infection. Host-wise analysis indicated that *Heteropneustes fossilis* was most susceptible to trypanosomes (Table 2.22).

2.7.2. Trypanoplasm Infectivity and Pleomorphism

Trypanoplasms are characteristic of fishes. Ten species of this parasite have been recorded during this entire period. These parasites too exhibited pleomorphism: 6 forms being monomorphic, 2 dimorphic and 2 polymorphic. *Mystus vittatus* was infected with *T. guptii* Gupta and Gupta, 1987 (polymorphic) and *T. tengari* Gupta et al., 1988 (monomorphic), *Cyprinus carpio* with *T. golysiana* Gupta and Pilarczyk, 1994 (monomorphic) and *T. polygolysiana* Gupta and Pilarczyk, 1994 (polymorphic); *Clarias batrachus* with *T. (Cryptobia) maguri* Gupta and Gupta, 1997 (dimorphic), *T. hematalis* Gupta and Gupta, 2010 (monomorphic) and *T. monomorpha.* n.sp (unpublished data) (monomorphic). *Puntius ticto* was infected with *T. cyprinoides* Gupta et al., 1998 (dimorphic) and a new variety, *T. sarnae* var. nov. *sophorae* and *Catla catla* with *T. catli* Gupta et al., 1999 (monomorphic) (Table 2.23). Host-wise analysis for trypanoplasm infectivity indicated that *Clarias batrachus* was the most prone species (Table 2.24).

2.7.3. Discussion

Pleomorphism is the sequential phenotypic manifestation of a single genotype as it appears in the trypomastigote stage in the vertebrate host. The phenomenon is triggered by changes in the host antibody spectrum. There are a few fish trypanosomes which seem to be virtually monomorphic, displaying only variability in size and/or in length/width ratio. The developmental stages in their vertebrate host, if any, seem to be limited to other than morphological manifestations.

Table 2.21. List of Trypanosoma Discovered by the Investigator

Fish Hosts				
Sl. No.	**Parasites**	**Hosts**	**Forms**	**Authors**
1.	*T. trichogasteri* n.sp.	*Trichogaster fasciata*	Monomorphic	Gupta & Jairaj-puri (1981)
2.	*T. singhii* n.sp.	*Heteropneus-tes fossilis*	Polymorphic	Gupta & Jairaj-puri, (1981)
3.	*T. attii* n.sp.	*Wallago attu*	Polymorphic	Gupta & Jairaj-puri, (1981)
4.	*T. aligaricus* n.sp.	*Ophiocepha-lus punctatus*	Dimorphic	Gupta & Jairaj-puri (1982)
5.	*T. batrachi* n.sp.	*Clarias batra-chus*	Dimorphic	Gupta & Jairaj-puri (1983)
6.	*T. monomorpha* n.sp.	*Catla catla*	Monomorphic	Gupta & Jairaj-puri (1985)
7.	*T. notopteri* n.sp.	*Notopterus notopterus*	Monomorphic	Gupta & Jairaj-puri (1985)
8.	*T. colisi* n.sp.	*Colisa fasciata*	Polymorphic	Gupta (1986)
9.	*T. bareilliana* n.sp.	*Channa punc-tatus*	Polymorphic	Gupta et al. (1987)
10.	*T. barbi* n.sp.	*Barbus barbus*	Polymorphic	Gupta et al. (1987)
11.	*T. ritae* n.sp.	*Rita rita*	Monomorphic	Gupta & Yadav (1989)
12.	*T. rohilkhandae* n.sp.	*Channa punc-tatus*	Monomorphic	Gupta & Saraswat (1991)
13.	*T. kargenensis* n.sp.	*Heteropneus-tes fossilis*	Monomorphic	Gupta & Gupta (1994)
14.	*T. danilewskyi*	*Cyprinus carpio*	Monomorphic	Laveran & Mesnil (1904)
15.	*T. bagroides* n.sp.	*Mystus vittatus*	Monomorphic	Gupta & Gupta (1996)
16.	*T. trichogasteri var. nov. fasciatae*	*Colisa fascia-tus*	Monomorphic	Gupta et al. (1998)

17.	*T. ticti* n.sp.	*Puntius ticto*	Polymorphic	Gupta et al. (1998)
18.	*T. monomorpha var. nov. catlae*	*Catla catla*	Monomorphic	Gupta et al. (2000)
19.	*T. karelensis* n.sp.	*Heteropneus-tes fossilis*	Monomorphic	Gupta et al. (2000)
20.	*T. artii* n.sp.	*Heteropneus-tes fossilis*	Dimorphic	Gupta et al. (2002)
21.	*T. piscidium* n.sp.	*Colisa fasciata*	Monomorphic	Gupta et al. (2003)
22.	*T. saulii* n.sp.	*Channa punc-tatus*	Dimorphic	Gupta et al. (2006)
23.	*T. heteropneusti* n.sp.	*Heteropneus-tes fossilis*	Dimorphic	Gupta et al. (2006)
24.	*T. tengari* n.sp.	*Mystus vittatus*	Dimorphic	Unpublished data
Amphibian Hosts				
1	*T. bisalpurensis* n.sp.	*Bufo mela-nostictus*	Monomorphic	Unpublished data
2	*T. rotatorium*	*Rana tigrina*	Monomorphic	Mayer (1843)
Mammalian Hosts				
1	*Trypanosoma (Herpetosoma) lewisi*	*Rattus rattus*	Monomorphic	Laveran and Mesnil (1901)
2	*Trypanosoma (Herpetosoma) evotomys*	*Clethrionomys glareolus*	Monomorphic	Molyneux (1969)
3	*Trypanosoma (Herpetosoma) grosi kosewiense*	*Apodemus flavicollis*	Monomorphic	Karbowiak and Wita (2004)
4	*Trypanosoma (Herpeto soma) microti*	*Microtus oeconomus*	Monomorphic	Laveran and Pet-titt (1909)

Table 2.22. Host-Wise List of Trypanosomes Discovered by the Investigator

Fish Hosts			
Hosts	**Parasites**	**Forms**	**Authors**
Heteropneustes fossilis	*T. singhii* n.sp.	Polymorphic	Gupta & Jairajpuri (1981)
	T. kargenensis n.sp.	Monomorphic	Gupta & Gupta (1994)
	T. karelensis n.sp.	Monomorphic	Gupta et al. (2000)
	T. artii n.sp.	Dimorphic	Gupta et al. (2002)
	T. heteropneusti n.sp.	Dimorphic	Gupta et al. (2006)
Channa punctatus	*T. aligaricus* n.sp.	Dimorphic	Gupta & Jairajpuri (1982)
	T. bareilliana n.sp.	Polymorphic	Gupta et al. (1987)
	T. rohilkhandae n.sp.	Monomorphic	Gupta & Saraswat (1991)
	T. saulii n.sp.	Dimorphic	Gupta et al. (2006)
Colisa fasciata	*T. trichogasteri* n.sp.	Monomorphic	Gupta & Jairajpuri (1981)
	T. colisi n.sp.	Polymorphic	Gupta (1986)
	T. trichogasteri var. Nov. fasciatae	Monomorphic	Gupta et al. (1998)
	T. piscidium n.sp.	Monomorphic	Gupta et al. (2003)
Mystus vittatus	*T. bagroides* n.sp.	Monomorphic	Gupta & Gupta (1996)
	T. tengari n.sp.	Dimorphic	Unpublished data
Catla catla	*T. monomorpha* n.sp	Monomorphic	Gupta & Jairajpuri (1985)
	T. monomorpha var. nov. catlae	Monomorphic	Gupta et al. (2000)
Wallago attu	*T. attii* n.sp.	Polymorphic	Gupta & Jairajpuri (1981)
Clarias batrachus	*T. batrachi* n.sp.	Dimorphic	Gupta & Jairajpuri (1983)
Notopterus notopterus	*T. notopteri* n.sp.	Monomorphic	Gupta & Jairajpuri (1985)

Barbus barbus	*T. barbi* n.sp.	Polymorphic	Gupta et al. (1987)
Rita rita	*T. ritae* n.sp.	Monomorphic	Gupta & Yadav (1989)
Cyprinus carpio	*T. danilewskyi*	Monomorphic	Laveran & Mesnil (1904)
Puntius ticto	*T. ticti* n.sp	Polymorphic	Gupta et al. (1998)
Amphibian Hosts			
Bufo melanostictus	*T. bisalpurensis* n.sp.	Monomorphic	Unpublished data
Rana tigrina	*T. rotatorium*	Monomorphic	Mayer (1843)
Mammalian Hosts			
Rattus rattus	*Trypanosoma (Herpetosoma) lewisi*	Monomorphic	Laveran and Mesnil (1901)
Clethrionomys glareolus	*Trypanosoma (Herpetosoma) evotomys*	Monomorphic	Molyneux (1969)
Apodemus flavicollis	*Trypanosoma (Herpetosoma) grosi kosewiense*	Monomorphic	Karbowiak and Wita (2004)
Microtus oeconomus	*Trypanosoma (Herpetosoma) microti*	Monomorphic	Laveran and Pettitt (1909)

Table 2.23. List of *Trypanoplasma* Discovered by the Investigator

Fish Hosts				
Sl. No.	**Parasistes**	**Hosts**	**Forms**	**Authors**
1.	*T. guptii* n.sp.	*Mystus vittatus*	Polymorphic	Gupta & Gupta (1987)
2.	*T. tengari* n.sp.	*Mystus vittatus*	Monomorphic	Gupta et al. (1988)
3.	*T. golysiana* n.sp.	*Cyprinus carpio*	Monomorphic	Gupta & Pilarczyk (1994)

4.	*T. polygolysiana* n.sp.	*Cyprinus carpio*	Polymorphic	Gupta & Pilarczyk (1994)
5.	*T. (Cryptobia) maguri* n.sp.	*Clarias batrachus*	Dimorphic	Gupta & Gupta (1997)
6.	*T. cyprinoides* n.sp.	*Puntius ticto*	Dimorphic	Gupta et al. (1998)
7.	*T. catli* n.sp.	*Catla catla*	Monomorphic	Gupta et al. (1999)
8.	*T. haematalis* n.sp.	*Clarias batrachus*	Monomorphic	Gupta & Gupta (2010)
9.	*T. monomorpha.* n.sp.	*Clarias batrachus*	Monomorphic	Unpublished data Gupta & Gupta (2014)
10.	*T. sarnae var. nov. sophorae*	*Puntius ticto*	Monomorphic	Gupta & Gupta (2014)

Table 2.24. Host-Wise List of Trypanoplasms from Fish Hosts Discovered by the Investigator

Sl. No.	Hosts	Parasistes	Forms	Authors
1.	*Mystus vittatus*	*T. guptii* n.sp.	Polymorphic	Gupta & Gupta (1987)
2.		*T. tengari* n.sp.	Monomorphic	Gupta et al. (1988)
3.	*Cyprinus carpio*	*T. golysiana* n.sp.	Monomorphic	Gupta & Pilarczyk (1994)
4.		*T. polygolysiana* n.sp.	Polymorphic	Gupta & Pilarczyk (1994)
5.	*Clarias batrachus*	*T. (Cryptobia) maguri* n.sp.	Dimorphic	Gupta & Gupta (1997)
6.		*T. haematalis* n.sp.	Monomorphic	Gupta & Gupta (2010)
7.		*T.monomorpha.*n.sp.	Monomorphic	Unpublished data

8.	*Puntius ticto*	*T. cyprinoides* n.sp.	Dimorphic	Gupta et al. (1998)
9.	*Catla catla*	*T. catli* n.sp.	Monomorphic	Gupta et al. (1999)
10.	*Puntius ticto*	*T. sarnae* var. Nov. *sophorae*	Monomorphic	Gupta & Gupta (2014)

It has been observed in the past that during the course of infection, most fish trypanosomes display forms successively differing conspicuously in their shape and size; they were designated by early authors (Laveran and Mesnil, 1912) as the pleomorphic forms, "*var. parva* "or "*var. magna* "and sometimes these forms were erroneously credited with separate taxonomic status (Fantham, 1919), as even reported in later times (Bauer, 1984). However, it soon became clear that these are but extremes in the sequence of forms occurring during the course of infection with one species (Minchin, 1909; Breindl, 1915; Laird, 1951; Khan, 1976). Out of the large number of piscine trypanosomes, only in a few has the full sequence been established.

In such pleomorphic species, the first trypanosomes to appear in the blood of the infected host are the "young" forms – rather small and slender with a few, shallow waves of the undulating membrane and nucleus revealing no distinct karyosome. After some time, intermediate forms appear until, eventually, "adult" forms prevail to persist during the chronic phase of infection. They are large, stout, sometimes very wide, and usually have a richly spiraling undulating membrane with numerous bends, a nucleus with a prominent karyosome and sub-surface striation. They often have more chromatic granules in their cytoplasm and free flagellum is shorter than in the young forms. Depending on the phase of infection at which the fish is examined, one finds young, slender or large adult forms only, or intermediate ones. A mixture of forms is probably the result of sequential leech feeds.

The existence of a sequence of different morphological types of trypanoplasms during the course of infection in carps was firmly established as early as in 1906 by Keysselitz. Although he misinterpreted numerous stages both in the blood and in the leech as 'gametes,' his reproduction of polymorphism is the most complete to be found in existing reports. The sequence of stages was confirmed by Roberstson (1911) in goldfish, Breindl (1915) in loaches and Kipp (1968) in tench trypanoplasms. Bower and Woo (1977) studied *Trypanoplasma catostomi* in parasite-free fingerlings of the host fish and noted a remarkable polymorphism including the size as well as differing body structure coupled with the progress of the infection originally started by the introduction of a single flagellate into the host. In view of the paucity of information, sequential pleomorphism in blood-stream stages of trypanoplasms may or may not be a generally distributed phenomenon (Lom, 1979).

Pleomorphism is expressed in the following features:

1. Size changes.
2. Changes in length/width ratio i.e., slender or broad forms.
3. The number, width and depth of the waves of the undulating membrane.
4. The presence or absence of a distinct karyosome in the nucleus by light microscopy.
5. Presence and number of stainable cytoplasmic granules.
6. Length of the free end of the flagellum.
7. Distance of the kinetoplast from the posterior end.
8. Shifts in the position of the nucleus in the body.

Presence of subsurface pellicular striation ("myonemes") in stained preparations by light microscopy (longitudinal or spirally oriented striations, probably sub-pellicularly located ribbons of mitochondrial systems.

Separate morphological stages may be defined by the combination of the above characters; marked changes of points 1 to 5 are proper to truly polymorphic species while changes in points 1, 2, 6, 7, 8, 9 may also be found in "monomorphic" species.

Pleomorphism in *Trypanosoma* is well marked in having a markedly variegated shape and five different stages in varying combinations at various periods of the digenetic life cycle (vertebrate and invertebrate hosts) may occur (trypomastigote, amastigote, promastigote, spheromastigote and metacyclic stages). However, trypomastigotes are the predominant stages reported from fish blood, but in the vertebrate blood too, pleomorphism may be visible as seen above.

Pleomorphism in fish trypanosomes has been reported by some authors without assigning any names (Becker and Overstreet, 1979; Joshi, 1979a) whereas others have given specific names to the various forms. Laveran and Mesnil (1907) distinguished the 'large' and 'small' forms of *T. remarki*; Tanabe (1925) Type I, II and III; Dutton et al. (1907) small, medium and large; Qadri (1955) small and large forms of *T. striati;* Qadri (1962a) large and stumpy forms of *T. batrachi*; Becker (1967) slender and broad forms of *T. occidentalis*; Joshi (1982b) short, elongated and stumpy forms of *T. aori.*

The form that predominates depends on the course of infection, which again in its turn is subject to many factors. Although this time sequence of morphs has been verified in some species only (e.g., *T. barbatulae* Breindl, 1915; *T. murmanensis* Khan, 1976), it most probably holds true for all of the polymorphic species. The most marked examples of pleomorphism can be drawn from among marine species, e.g., those from skates- *T. giganteum, T. rajae, T. gargantua.* The freshwater examples are *T. granulosum, T. percae and T. remarki.*

In fish blood flagellates, a condition similar to Trypanosomatids of other host groups exists. Pleomorphism is presumed to be sequential and is regarded to be a phenotypic expression of the genotype (Lumsden and Evans, 1976). However, as aptly stated by Bardsley and Harmsen (1973), only observation of the parasite in its vertebrate host when it is subjected to a complete variety of naturally- occurring conditions and factors can reveal whether it really is monomorphic. In experimental transfers among various hosts (carps, goldfish, gudgeon, pike, *Barbus conchonius*), varying temperature regimes or study of various stages of infection failed to reveal any true pleomorphism (Lom, 1979).

Bloodstream forms of the initial stages of infection are always rather small and of a regular shape (young stages have comma-like or crescent-like shapes). In more advanced infections, bigger, very irregular forms prevail, reflecting in their shape wave-like undulatory movements of the whole body. In the chronic, often long lasting, period of infection, the number of trypanoplasms was reduced; assuming a large size and an irregular shape, sometimes with numerous large bends of the undulating membrane; there was a nucleus with a conspicuous endosome not seen in young stages (Lom, 1979). The length of flagellum relative to the body length may also greatly vary during the cycle, as well as the number of stainable granules in the body.

Given the extreme metaboly of the cell shape, it is difficult to characterize the shape differences in young and advanced infection stages in more precise terms but they are quite conspicuous as are the differences in size. For example, in *Trypanoplasma* sp. from pikes (*T. guerneyorum*), the average size of a young form is 20×4 µm (excluding flagella), while in the chronic stage form it is 38×9 µm.

The large forms of trypanosomes and trypanoplasms found in late infection in the circulating blood were considered to be "adult" forms and were conveniently used as reference stages for taxonomic comparisons (Lom, 1979). While it seems certain that they are competent to initiate growth in the vector (as proven by Khan, 1978; in *Trypanosoma murmanensis*), there is no evidence as to whether it is this or some cryptic, tissue-inhabiting, persisting, stage which is responsible for relapses of parasitemia. The survey done so far indicates that pleomorphism amongst fish hosts appears to be a regular feature as noted above by other authors too. However, amongst amphibian and mammalian hosts, monomorphism was dominant suggestive of chronic trypanosomiasis. But even in fish hosts, it is safe to consider that monomorphism dominates (12 species) as compared to dimorphism (6 species) and polymorphism (6 species) as summarized by the researcher.

Chapter **3**

Ecoparasitism

Contents

3.1. ECOLOGICAL FACTORS AND PARASITISM

Owing to the in extricate and intrinsic difficulties involved in studying the ecology of intimate associations, most of the work on ecological factors and parasitism has lately been concerned with free living organisms. However, only relatively recently some reliable, quantitative investigations or theoretical insights have begun to exert from the findings on intimate associations, particularly parasitism.

The probability of parasitic infection of a host is influenced by both environmental and genetic factors. Traditionally, it was assumed that a host was either resistant or susceptible to a particular parasite and therefore the interaction between a parasite and potential host had only two possible outcomes: either the resistant host rebuffed the parasitic attack and remained uninfected or the parasite successfully invaded and significantly reduced the reproductive success of the susceptible host (Minchella, 1985).

The only much unique aspect of a parasite's ecology is the utilization or inhabitation of another species as its niche instead of occupying "non-living" niches as the free-living organisms do. The processes undertaken by parasites to find and infect their hosts, both of which are indispensable for them to continue their genre of life, however, do not show significant diversities from those used by free-living organisms to encounter and invade their niches in the non-living environment. Not only this, the analysis of the associating organisms has proved that their relationship is truly ecological.

The composition of the community of parasites of animals is considered to be a product of their present-day ecology, their physiological traits, and the phylogenetic history of the hosts (Vickery and Poulin, 1998). Many studies have mentioned the importance of host phylogeny, few studies have considered it in analyses of parasite communities (Poulin, 1995).

Host ecology and phylogeny can greatly influence parasite communities, but more importantly, they usually do so in different ways (Morand et al., 2002). Janovy et al. (1992) suggested that ecological variables should influence the overall parasite population structure (i.e., prevalence and abundance), whereas the evolution of host species, and thus phylogeny, should only affect the species of parasites.

Parasite communities are often complex because several hosts may be involved in the parasite's life-cycle, the transmission rates between hosts can vary, and also, host specificity varies among parasites. Parasite communities can be described quantitatively using a wide range of so-called descriptors including the abundance, intensity, richness, prevalence, incidence and diversity of parasites. All of these provide different information about the structure of the parasite community (Bush et al., 1997). Many parasitological studies that have used different databases, however, have found varied patterns in the relationships between descriptors and ecological variables of hosts, especially whether phylogeny is included or not (Poulin, 1995; Poulin and Rohde, 1997).

Ecologists need the ability to describe an organism's surroundings. Parasites are not fundamentally different from other living organisms in this regard, but, because parasites often have complex life cycles, various descriptive terms can be rather misleading (Bush et al., 1997).

Various terms have been used in this thesis to accurately describe the parasites, with terminology primarily based on Bush et al. (1997).

3.1.1. Prevalence

Prevalence is the number of hosts infected with 1 or more individuals of a particular parasite species for taxonomic group divided by the number of hosts examined for that parasite species. It is commonly expressed as a percentage when used descriptively and as a proportion when incorporated into mathematical models.

Prevalence is intended as a descriptive statistic for presence-absence data on parasites in a sample of hosts and is used when it is desirable to classify hosts into 2 categories, infected and uninfected. Prevalence is one of the most widely reported descriptors of parasitic infection because it requires detection of the presence of the parasite and not enumeration of the individuals present. It is one of the most commonly used ecological terms in parasitology.

3.1.2. Intensity of Infection

Intensity of infection is the number of individuals of a particular parasite species in a single infected host, i.e., the number of individuals in an infrapopulation.

Intensity is a form of density with the sampling unit specifically defined as an individual infected host. Therefore, intensity is a convenient measure for parasitologists because hosts are discrete and natural sampling units.

A survey carried on freshwater fishes of northern India belonging to 15 genera and 20 species showed variation in trypanosome infectivity under different environmental factors. The following three such factors were found to influence the relative prevalence and infectivity of this hematozoan parasite in fishes.

- Age, length and weight
- Season and annual cycles
- Locality of the host

3.1.3. Impact of Age, Length, and Weight on Parasite Prevalence

The presence of larger parasite infrapopulations and infracommunities in older, larger hosts is a common characteristic of fish host-parasite systems (Dogiel, 1970). Unfortunately, the size and age of the fish hosts are often tightly correlated, making it impossible to isolate the individual effects of either variable on parasite abundance.

3.1.4. Methodology

Twenty species of freshwater fishes were collected from different localities of Northern India as shown in Table 3.1.

3.1.4.1. Age

The age of the fishes were calculated according to the Length Frequency Method (Qayyum and Qasim, 1964).

Table 3.1. List of Fishes Sampled for Trypanosome Infectivity for Environmental Studies

Host Species	Taxonomic Status	Common Name	Length of Fish (cm)
Notopterus notopterus	Notopteridae	Patra	14–16
Catla catla	Cyprinidae	Bhakar	20–25
Cyprinus carpio	Cyprinidae	Common carp	15–20
Labeo calbasu	Cyprinidae	Karaunchar	18–25
L. rohita	Cyprinidae	Rohu	16–20
Puntius ticto	Cyprinidae	Sidhri	6–9
P. sophore	Cyprinidae	Sidhri	6–9
Wallago attu	Siluridae	Padhani	15–35
Mystus tengara	Bagridae	Tengara	6–9
Mystus vittatus	Bagridae	Tengara	7–11
Rita rita	Bagridae	Rita/ Hunna	5–8

Bagarius bagarius	Sisoridae	Gonch	8–15
Heteropneustes fossilis	Heteropneustidae	Singhi	10–80
Clarias batrachus	Claridae	Magur	10–80
Xenentodon cancila	Belonidae	Kauwa	17–21
Channa punctatus	Ophioceohalidae	Saul	5–20
C. striatus	Ophioceohalidae	Sauri	5–20
C. gachua	Ophioceohalidae	Chanaga	7–11
Anabas testudineus	Anabanditae	Kawai	12–13
Colisa fasciatus	Anabanditae	Khosti	5–8

3.1.4.2. Length Groups

The total length (T.L.) of *Channa punctatus* and *C. striatus* was measured from the tip of the snout to the tip of the tail in centimeters. The fishes were grouped arbitrarily into 3 length groups:

Group A	5–10 cms
Group B	11–15 cms
Group C	16–20 cms.

3.1.4.3. Weight Groups

Heteropneustes fossilis and *Clarias batrachus* were selected as the test models for making observations on weight-related trypanosome prevalence. Three weight groups for each of the above genus were made as follows:

- Group A: 10–20 gm
- Group B: 21–40 gm
- Group C: 41–80 gm

Blood was collected from the caudal vein or directly from the heart. Hanging drop preparations were made to detect the flagellates alive and thin blood smears were fixed in methanol and stained in Giemsa to study fixed parasites as given earlier in "Materials and Methods."

3.1.5. Results

3.1.5.1. Age Related Incidence

The effect of the age of the host on trypanosome incidence was investigated in *Channa striatus* and *Heteropneustes fossilis*. In *C. striatus*, fish of age groups 0+ to 2+ were infection-free, infection gradually rose from 3+ (4.4%) to 6+ (16.5%) aged fish. *H. fossilis* of 0+ age were infection-free, and those of 1+ year (3.5%) were poorly infected as compared to those of 5+ years (52.5%). In both the species of fishes examined, the ones belonging to the older age groups were more prone to infection than the younger ones. It was further observed that *H. fossilis* not only in general, had a higher prevalence than *C. striatus*, but amongst similar aged fish, the former fish succumbed to infection more easily (Table 3.2).

Table 3.2. Effect of age of *Channa striatus* and *Heteropneustes fossilis* on Trypanosome Infection

% infection		
Age	***C. striatus***	***H. fossilis***
0+	–	–
1+	–	3.5
2+	–	16.1
3+	4.4	20.0
4+	5.2	46.0
5+	14.2	52.5
6+	16.5	Not sampled

3.1.5.2. Length-Related Incidence

The prevalence and intensity of trypanosomes was observed in two species of fishes, *Channa punctatus* and *C. striatus* according to the length groups mentioned above. The samples of the fishes indicated that young fish ranging in length from 5–10 cms (Group A) were free from trypanosome infection. Group B fish showed a prevalence of 10.5% (*C. punctatus*) and 6.0 % (*C. striatus*), intensity being similar in both species (2–5 ts/100 RBC) (Table 3.3.). The maximum prevalence of 20.0% and 5.5% was recorded in *C. punctatus* and *C. striatus* respectively of Group C fish.

3.1.5.3. Weight-Related Incidence

Weight-related incidence of trypanosomes was observed in *Heteropneustes fossilis* and *Clarias batrachus*, according to the weight groups mentioned above. The results indicated that young fish having low weight were less infected than the heavier, older ones in both species of fish examined. The prevalence of 15.5% (Group A) rose to 45.0% in Group C and the intensity of 5–10 ts/100 RBC (Group A) to 8–20 ts/100 RBC in *H. fossilis*. Similarly, prevalence (9.6%) and intensity (4–8 ts/100 RBC) was lowest in Group A *Clarias* as compared to those in Group C (30.0% and 7–15 ts/100 RBC respectively) (Table 3.3).

Table 3.3. Trypanosome Prevalence and Intensity in Freshwater Fish in Different Length Groups

Fish	**Group**	**Length (cms)/ Weight (gm)**	**Prevalence (%)**	**Intensity (ts/100 RBC)**
Channa punctatus	A	5–10 cms	Nil	Nil
	B	11–15 cms	10.5	2–5
	C	16–20 cms	20.0	4–10
Channa striatus	A	5–10 cms	Nil	Nil
	B	11–15 cms	6.0	2–5
	C	16–20 cms	15.5	4–8
Heteropneustes fossilis	A	10–20 gms	15.5	5–10
	B	21–40 gms	20.2	6–12
	C	41–80 gms	45.0	8–20
Clarias batrachus	A	10–20 gms	9.6	4–8
	B	21–40 gm	10.5	4–10
	C	41–80 gm	30.0	7–15

3.1.5.4. Discussion

It has frequently been observed that one of the most important factors influencing the incidence of parasite fauna of any host animal is the stage of the animal's growth. Study of parasite fauna of man has clearly shown that the appearance of many parasites is associated with a certain age of the host, the newly born being usually parasite-free. It has been pointed out that older animals have a larger number of parasites than younger ones of the same species.

For fishes, numerous studies have examined the role of host body size (mostly body length) as a predicator of infra- and component communities of parasites (Grutter and Poulin, 1998). A higher incidence of trypanosomes in older *Channa punctatus*, *Clarias batrachus* and *Trichogaster fasciata* has been reported (Gupta and Jairajpuri, 1984) as compared to younger ones of the same species.

Parasite infrapopulations that increase as the result of simple accumulation (requiring that the life span of the parasite be comparable to that of the host) are expected to do so as a result of increasing host age. Such parasites should have either a direct life cycle or utilize intermediate hosts whose ingestion is not limited by fish size and should, therefore, be present in most host size and age classes (Dogiel, 1964).

The infrapopulation size of a parasite utilizing intermediate hosts whose proportion in the fish host's diet changes over time is expected to be related to host size rather than host age. Yellow perch do not undergo a strict diet shift during ontogeny but do display changes in the proportion of certain prey items as fish become larger (Cannon, 1973). Although there will be some accumulation of these parasites, it will not begin until the host is large enough to consume the intermediate hosts as prey, an event that will occur at a later age for male perch than for female perch. Parasites that locate fish hosts through chemotaxis (a phenomenon seen only in species infecting hosts with limited vagility (Haas, 1994) or phototaxis should have an infrapopulation size that is dependent on host size as larger hosts make better targets.

It is expected that the parasite infracommunity will be composed of both, infrapopulations accumulating with host age and infrapopulations increasing with host size. The colonization of age-dependent species should not be affected by host size, and they should therefore be present in most host age and size classes, maybe at different intensities (Zelmer and Arai, 1998).

The segregation of infections by host age, size classes and host weight would contribute to the aggregation that is often observed in parasite populations when host age, size classes and host weight are combined (Crofton, 1971a,b; Shaw and Dobson, 1995). Once the relationship between parasite number and host size or age is determined for a parasite species, the degree of aggregation within age or size classes can be compared to the overall aggregation within the host population to resolve its relationship to the distribution of the parasite's component population (Bush et al., 1997). The probability of exposure to a parasite using insects as intermediate hosts would be higher for larger fish, leading to a size dependent increase in parasite number. Indeed, the slopes of the increase in abundance with host size have been shown to be similar for male and female perch, indicating that this relationship is size dependent and is not affected by potential behavioral differences between male and female perch. The sexual size dimorphism inherent to yellow perch (*Perca flavescens*) populations in Garner Lake, Alberta forms two size classes within each age class, facilitating the separation of the independent effects of age and size on parasite recruitment by these hosts. All of the parasites that were analyzed showed an increase in abundance in older, larger perch (Zelmer and Arai, 1998).

On the other hand, in older fish, there are reports indicating either a decreasing rate of infection, or the infection is completely absent. Sometimes, however, probably due to ecological factors inhibiting ready transmission, the youngest age class may be the least infected [(e.g., *T. salmositica* in *Cottus rhotheus* (Becker and Katz, 1965a)].

Majority of the studies conducted above suggest that from fresh water fishes it is clearly possible to postulate the rule of general increase in the intensity and prevalence with the growth of the host. However, blood parasites are transmitted by leeches which do not form the food of fish. But similar results (blood parasites increasing with fish age/length/weight as observed in the present studies) maybe because with the increase in fish age/length/weight, the surface area available for fish-leech transmission increases resulting in greater occurrence of blood parasite infectivity in fish.

3.1.6. Impact of Season and Annual Cycles on Parasite Prevalence

The impact of season and annual cycles on parasite prevalence was observed during different phases of the study.

During the first phase, the prevalence of trypanosomes was categorized from the survey carried as above on freshwater fishes of northern India belonging to 15 genera and 20 species.

3.1.6.1. Trypanosome Infectivity in 20 Species of Freshwater Fishes in Relation to Seasonal Variations

The annual rhythm of trypanosome fluctuations was evidenced in four seasons of the year: summer (May to July); autumn (August to October); winter (November to January) and spring (February to April). The minimum infectivity of 2% during the summers reached its peak value of 31% during the winters. In autumn, the infectivity was 22.3% whereas in spring 8.6%.

3.1.6.2. Trypanosome Infectivity in 7 Species of Freshwater Fishes in Relation to Seasonal Variations

Fishes were collected by various methods between January 2006 to December 2009. During the first year of sampling, 7 species of freshwater fishes(*Cirrhinus mrigala, Puntius ticto, Mystus tengara, Heteropneustes fossilis, Clarias batrachus, Channa punctatus* and *Colisa fasciatus*) were sampled for trypanosome infectivity. However, as the first 3 species were infection–free during 2006, their data were therefore eliminated and the subsequent work was executed on the remaining 4 species. A total of 2,304 fishes (12 specimens of each fish species each month) were examined for the 4 consecutive years.

Leeches were also collected from the same ponds as the fishes and infectivity in them recorded. Trypanosome abundance was correlated with temperature and leech abundance and infectivity.

Trypanosomes could be detected by hanging drop preparations: blood films were also fixed and stained as given earlier. Low parasitaemias were further confirmed by hematocrit centrifugation. A careful scrutiny for 20 mins at X400 was performed before declaring a fish negative for trypanosomes.

Morphometric studies showed that *Heteropneustes fossilis, Clarias batrachus, Channa punctatus* and *Colisa fasciatus* were infected with *Trypanosoma singhii* (Gupta and Jairajpuri, 1981), *T. batrachi* (Qadri, 1962), *T. aligaricus* (Gupta and Jairajpuri, 1982) and *T. colisi* (Gupta, 1986), respectively.

Trypanosome infectivity in fishes varied in different months of the year. The data reflects that infectivity was least during the summers (May to July, temperature 25–30°C) and reached its peak during the winters (November to January, temperature 3–10°C) in all the four species and during each annual cycle (Plate LI, LII). A significant rise in *T. singhii* ($P < 0.01$) and *T. colisi* ($P < 0.01$), an insignificant rise in *T. batrachi* ($P > 0.01$) and a significant fall in *T. aligaricus* ($P< 0.001$) was recorded from 2006 to 2009 (Table 3.4). The total infectivity of trypanosomes in all the four species of fishes from year to year showed a general pattern of increase from 2006 to 2009 except for an insignificant fall in 2008.

The leeches (*Hemiclepsis marginata*) were collected in association with fish as well as astry. The leeches could not be collected corresponding to the number of fishes. However, infectivity in leeches also showed a continuous pattern of increase from 2006 to 2009 (Table 3.4). As the specific stages of the trypanosomes in leeches could not be traced, gross infection in leeches was plotted against gross infection in fishes. It is evident from the study that the leeches were more heavily infected as compared to the fishes. Moreover, in both cases, there was an increase in infectivity from 2006 to 2009.

Table 3.4. Yearly Percentage Infectivity of Trypanosomes from 2006 to 2009 in Fishes and Leeches

Parasite	**Host**	**Year**			
		2006	**2007**	**2008**	**2009**
Trypanosoma singhii	*Heteropneustes fossilis*	38.8	40.2	40.2	43.7
T. batrachi	*Clarias batrachus*	20.1	20.8	22.8	21.5
T. aligaricus	*Channa punctatus*	17.3	15.9	15.2	13.8
T. colisi	*Colisa fasciata*	24.0	35.4	31.2	38.8
Total infection (All 4 parasites)	All fish	27.9	28.1	26.7	29.5
Leech (*Hemiclepsis marginata*)		45.3	48.6	58.1	59.2

3.1.6.3. Discussion

Many species of piscine trypanosomes have been reported but erudite studies which deal with infection prevalence during different seasons (at different macrohabitat temperatures) and annual changes in infectivity correlated with infectivity and abundance of leech population remains a less resolved enigma.

Seasonal changes in the incidence of some species of parasites are not infrequent and have often been recorded. The first attempts to study the qualitative and quantitative seasonal fluctuations of the entire parasite fauna were made by Elton et al. (1931). Individual data were first summed up by Dogiel (1964).

Seasonal changes of the nature are very sharply reflected in organic life and it is natural to expect that trypanosomes should not remain indifferent to the annual cycle of climate changes. Seasonal variations of piscine trypanosomes have drawn attention of various workers. Khayboulajev (1970) reported that higher specificity of fish trypanosomes exists with respect to leech, rather than fish and that the trypanosomes are more common in fish exposed to leeches than in more mobile fish. Evidently, the waters during winters are not only comparatively more stagnant but also during winters (post-spawning period and resting phase) in India, the fishes are spent–up and exhausted due to the energy consumed during spawning and are lethargic in nature. These criteria may favor maximum fish-leech association leading to the transmission of parasitic trypanosomes from the intermediate host to the final host.

Cottrell (1977) recorded higher incidence in March which declined during the summer and he attributed the rise in incidence of trypanosome infections to a decline in the serum globulins. Infection was recorded only in December (Joshi, 1978) and in the colder months (November to February, 1970–73) by screening for trypanosomes round the year (Joshi, 1979a). Letch and Ball (1979) during 33 months of sampling in four fish categories- gudgeon, stone loach, bullhead and minnow recorded highest trypanosome infection in fishes from the river Lee in August which decreased till May, then again elevated in the gudgeon during 1976–77. This view finds support from other workers (Briendl, 1911; Mokhayer et al., 1976). However, Khan, Barrett et al. (1980) and Khan, Murphy et al. (1980) reported higher prevalence of trypanosomes in the Labrador area where the coldest bottom temperature was recorded. Woo (1981) observed division in *T. danilewskyi* from the goldfish, *Carassius auratus* at 20°C. Gupta and Jairajpuri (1984) also recorded highest infection during the winters. Karbowiak and Wita (2001b) recorded maximum prevalence of *T. evotomys* in the bank vole, *Clethronomys glareolus* in August (45%) and September (38%). The peak infection was observed in the winter months during the present course of studies when the temperature fell to about 10°C.

Zelmer and Arai (1998) reported that in carp of the 0+ and of the 1+ year classes in European polycultures, parasitemia often persists during the whole summer, whereas in 2+ and older carp, the parasitemic periods are usually limited to early spring periods.

The present findings have shown a higher prevalence of infection during the winters. Also, a higher incidence of trypanosomes during the colder months of each year, was recorded. The results of the present survey suggest that in this country, the examination of freshwater fishes, at least of the species examined, during the winter would give the greatest opportunity to detect infections and prove useful for other experimental purposes. The annual infectivity during the consecutive years also showed a variation, which might correspond with the climatic changes each year and the relative abundance and infectivity of the vector.

It was further observed that the annual changes in infectivity showed an increase in 2006, 2007 followed by a marginal fall in 2008 with a subsequent elevation during the last quarter of experimentation. The drop in infectivity in 2008 might have been correlated with the fact that during this year the northern part of India received less rains resulting in poor availability and infectivity of the leeches. The alterations in trypanosome infectivity might correspond with the climatic changes each year and the relative abundance and infectivity of the vector. The seasonal and annual changes in the incidence of parasite fauna are complex phenomena determined by many abiotic and biotic factors, the most important among them are the climatic factors, specially temperature. Season, as with most ecological parameters, will have varied effects or multiple effects depending on the species being considered and the subtle combination of the parameters in any given area. A very general statement on levels of parasite intensity in fish would be that the summer months usually show the greatest abundance in numbers and variety of parasites since this is the period when the fish host is most actively feeding (Dogiel, 1958). This would then be followed by a decline in the winter months; but this is not always the case. Thomas (1958), working with two species of *Crepidostomum* of *Salmo trutta* and *S. salar*, showed the degree of infection was at its maximum during the winter and at the minimum during the summer. Moreover, blood parasite infection does not appear to be related with fish food as these parasites are transmitted by blood sucking activity of the leeches.

Many biotic and abiotic factors may play a role in determining the seasonal and year to year changes (Dogiel, 1964). The role of temperature (abiotic) and vector (biotic) may play attributable roles in the infectivity of piscine trypanosomiasis.

The role of water temperature cannot be overruled. Ambient temperature is important to the infection course as too high or too low a temperature inhibits infection. During winters, the water in the ponds, lakes and rivers remains quite quiescent, highly oxygenated (0.182–0.365 ml/l of O_2) and least turbid (Joshi, 1979a). These conditions favor maximum fish–leech association. Trypanosomes are transmitted to fishes by aquatic leeches, an ectoparasitic annelid belonging to its smallest class, Hirudinea which sucks the blood of fishes and may store it in special pouches. They are predominantly found in fresh water habitats and are more abundant in tropical countries than in temperate zones (Hickman et al., 1979).

The variation in trypanosome infectivity during the different seasons of the annual cycle may be attributed to the fact that the fishes during summers are most active and the leeches have least chances to suck fish blood. In autumn, most fishes spawn, their activities slow down and during the post spawning and resting phase i.e., winter, the fishes are exhausted and leech-fish association is the greatest resulting in maximum infectivity during this part of the annual cycle.

Khan et al. (2004) stressed the role of season on helminth infestations although the effect was different on different parasitic groups which could be related to their life cycle patterns e.g., recruitment, intermediate hosts etc. Inspite of resemblances of the observations of these authors with some previous researchers, there is a noticeable variation with others. The variation maybe due to the different water bodies, hosts, feeding habits of the host, diversity of the climatic conditions and availability of intermediate hosts. This warrants the study of life cycle in these parasites and the detailed study under experimental conditions in order to find out the possible role of a particular factor in the recruitment and maturation of a specific parasite.

It has been proposed that water temperature affects fish trypanosomes directly causing increased division (Needham, 1969). Sypek and Burreson (1983) postulated that the temperature regulates the host's immune system against hemoflagellate infections and immunity of fish to trypanosomes may vary at different temperatures. A deeper study on inter-current infections, blood loss due to leech feeding, impact of stress and vector biology are necessary for studying the mechanism underlying changes in parasite incidence due to climate and season.

3.1.7. Impact of Locality of the Host on Parasite Prevalence

3.1.7.1. Site, Location and Habitat

The terms 'site' and 'location' have a long history of use among parasitologists, and, to them, the terms may carry specific connotations. To others, such connotations are unknown or vague. According to Bush et al. (1997), the site or location of a parasite is the topological or spatial location in a host from where a particular sample of parasites is collected. Site and location are thus anatomical parallels to geographic locality. Habitat refers to the typical local environment in which parasites occur.

Organs and tissues provide the local environment (including physical, chemical, and biological surroundings) for the parasites, therefore, they are appropriately called habitats. Bush et al. (1997) proposed that the term habitat is preferable because it has a long history of similar use in the literature on free-living forms and it can easily be compounded, e.g., microhabitat, habitat-specific. Further, recognizing the complexity of many parasite life cycles, it is often necessary to describe the surroundings of free-living phases, particularly in autecological studies.

3.1.7.2. Locality

The term 'locality' refers to a geographic locale of the external environment where the parasite is found. Locality is used widely for geographic position, and the term is restricted to an identification of where, geographically, the individual, population, or community is obtained. Locality is the spatial region where a host (or hosts) is Locality is an important factor which plays a contributory role in the prevalence of parasitic species. The prevalence of parasites is affected by the environmental factors as they have a controlling system on the availability of parasites in their host due to ecological and biological factors. The characteristics of a typical locality have a high degree of impact on all aspects of animal life history. According to this hypothesis, a relationship between habitat structure and the main components of fitness is predicted and consequently also predicts a relationship between habitat structure.

In different localities, blood parasite occurrence can differ due to non-availability and non-infectivity of invertebrate and vertebrate hosts. Hematozoa in *Agelaius phoeniceus* in Eastern Ontario were sampled by Weatherhead and Bennett (1991) who found that parasite prevalence was significantly higher in hosts sampled in lake Opinicon marshes than in Beaver marshes. It has also been observed that mature forests produced nestlings in better condition but with a higher prevalence of hemoparasites than young forest, maybe because mature forests serve as a good site for both the host and the parasite vector (Sanchas et al., 2007).

The co-relation between type of habitat and species of parasite occurrence has been studied over the years (Ingles, 1936; Rankin, 1945). It has been shown by Tella et al. (1999) that a positive correlation exists between the number of parasite species per host and host geographic range in some host-parasite associations.

The habitat effects on physiological stress response in nestling blue tits were investigated by Arriero et al. (2008) who showed that recently acquired ectoparasite and hemoparasite infections were linked with forest habitat structural characteristics, higher prevalence of fleas and blood parasites in more mature forests and higher prevalence of blow flies in degraded forests. Leech parasitism was reported to be highest on bottom walker and adults and differed throughout the year. Moreover, leech intensity was highest in turbid ponds causing greater infectivity in these ponds.

The present studies infer that locality of the hosts may be important factors for assessing the prevalence and intensity of blood parasites in the fish hosts. Factors responsible for the variance in infectivity may be varied as each locality has its typical fauna and environmental characteristics.

The prevalence of trypanosome parasites was categorized locality-wise based on collection of twenty host samples from each of the six localities, namely:

1.	Moradabad	2. Aligarh	3. Bareilly fish market
4.	Shahjahanpur	5. Nakatia pond	6. Faridpur pond
7.	Budaun		

Table 3.5 gives us evidence to infer that infectivity ranges from 0% (Nakatia pond) to 28% (local fish market, Bareilly) showing variability in infection due to different localities. It is apparent that infectivity was highest at Bareilly during this piece of work. Restriction and variation in trypanosome infectivity in different localities has been shown by investigators (Misra et al., 1973; Bower and Woo, 1979). The seasonal ponds of Nakatia were free from trypanosome infection of fishes. It is interesting to note that this locality failed to show the presence of leeches which act as intermediate hosts of parasites. The complete absence of the latter might be correlated with the lack of leech population in that area. The maximum infection was noted in the local fishes of Bareilly (28%). Each locality is typified by its own chemical composition, invertebrate fauna and environmental conditions constituting the "macroenvironment" of the parasite. Moreover, variations also exist in the population and infectivity of the intermediate host of each locality.

The studies conducted above on the effect of growth of the host, seasonal variations and locality of different fish species are suggestive of the fact that ecological factors play an all round influence on trypanosome incidence. The studies emphasize that the ecological approach in the study of trypanosome problems carries utmost weightage as in the case of other parasitological problems. It may be stressed that although the effect of each factor has been studied separately, factor interaction opens up a new line for further research on this problem.

It can be concluded that the physical condition from where the hosts are collected, physical condition of the host, the location of parasite within the host, season of parasite collection, life cycle of the parasite as well as the morphological features of the parasites may play a contributory role in parasite prevalence. Each geographical locus is basically an ecological complex composed of parasite, vector, the host and macro-environmental characteristics constituting an ecological network. Therefore, the locality from where the host is collected may play a targeted role in causing variance in prevalence and intensity of any parasite population in fish fauna.

Moreover, it cannot be said that infection with a particular parasite species is static for all times with respect to any ecological parameter as the environment is not a simple summation of the parts of the environment. "New properties emerge because the components interact, not because the basic nature of the components is changed." It would appear that the hierarchical systems of the environment, its composition, distribution, amount, number, and changing status of organisms including vectors create an ever-changing environmental web and therefore, cause differential susceptibility in various host species during different time periods.

Table 3.5. Effect of Locality of the Host on Trypanosome Infection

Locality		**Prevalence (%)**
A.	Moradabad	18
B.	Aligarh	22
C.	Bareilly fish market	28
D.	Shahjahanpur	9
E.	Nakatia pond	0
F.	Faridpur pond	6
G.	Budaun	12

3.2. CONCURRENT INFECTIVITY: INTERACTION AMONGST BLOOD PARASITES

Parasites in the same host organism (idiohostal units; Kisielewska, 1970) and living in the same environment may, interact negatively with each other. These investigations cover a broad taxonomic range of both parasites and hosts, but are only a few among all the records which are available. This indicates that negative interactions are significant phenomenon in the ecology of parasites.

The mechanisms involved in these interactions and their ecological results are often complex, but the different relationships are usually categorized as amensalism, predation, parasitism or competition (Odum, 1971). It has proved difficult to define parasitism and to draw the line between parasitism and predation. Competition has also proved difficult to define strictly as is illustrated by Birch (1957), Milne (1961), DeBach (1966), Miller (1967), Odum (1971), and Krebs (1972) who all define competition differently.

Understanding community structure has been the topic of numerous studies, in which the roles of interactive and/or non-interactive processes have been addressed. Parasitic organisms provide interesting models for studying the organization and structure of communities (Souza, 1994). One important point that has been highlighted is the role of interactions, mostly competition, between species in determining their relative abundances (Cornell and Lawton, 1992). Dobson and Roberts (1995) emphasized the role of interactions between parasites and their hosts (regulatory effects) in determining community structure rather than direct interactions between parasite species.

In order to observe the phenomenon of interaction amongst concurrent hemoflagellates, an experiment was designed to observe the impact of dual infectivity of *Trypanosoma* and *Trypanoplasma* on the population density of the two hemoflagellates. The test fish was *Mystus vittatus* and the studies were executed as follows:

3.2.1. Methodology

One hundred *Mystus vittatus* were procured form different freshwaters of the region. Hanging drop and routine hematological preparations were made to observe the flagellates (Gupta and Jairajpuri, 1983). Fishes were sorted out into uninfected, infected with the monoflagellates, *Trypanosoma*, infected with the biflagellate, *Trypanoplasma* and concomitant having both species of parasites. Three aquaria containing 10 trypanosome-infected *Mystus* each were maintained in the laboratory. Fishes of aquarium A were inoculated with 1 ml of *Trypanoplasma tengari* and those of aquarium B were inoculated with *T. maguri* (from *Clarias batrachus*) infected blood. Aquarium C acted as the control. Infection of both the flagellates was recorded on every alternate day for 30 days.

3.2.2. Results

3.2.2.1. Prevalence and Intensity of Hemoflagellates

Mystus vittatus were either found to be infected with the monoflagellate, *Trypanosoma tengari* Gupta et al., 1988 (prevalence 18%; intensity 9–21 forms/ 15 mins scan) or the biflagellate, *Trypanoplasma tengari* (prevalence 8%; intensity 3–7 forms/15 mins scan) or even 2% of the fishes had concomitant infection with *Trypanosoma* (intensity 10–17 forms/ 15 mins scan) and *Trypanoplasma* (intensity 2–4 forms/ 15 mins scan) (Table 3.6). As under natural concomitant conditions, trypanosomes were more intense (10–17 forms/15 mins scan) as compared to the trypanoplasms (2–4 forms/ 15 mins scan), an experiment was designed by inoculating lone *Trypanoplasma*-infected blood (donor fish) into lone *Trypanosoma*- infected recipient fish in order to observe the effect of co-occurrence on the population density of the two parasites.

Table 3.6. Prevalence and Intensity of Hemoflagellates in *Mystus vittatus*

	Lone Infection		**Concomitant**	
	Trypanosoma	***Trypanoplasma***	***Trypanosoma***	***Trypanoplasma***
Prevalence (%)	18	8	2	
Intensity (parasites/15 mins scan)	9–21	3–7	10–17	2–4

Table 3.7. Survival of Hemoflagellates Upon Inoculation of Trypanoplasms into Trypanosome-Infected Hosts (Recipient *Mystus vittatus* Parasitized with *Trypanosoma Vittati*)

Aquarium	**Inoculum (1 ml blood)**	**Parasite**	**Duration (days)** 2 4	6	8 10	12	14 16 18 20 22 24 26 28 30
A.	*Try-*	*Try-*	++ ++	++	++	+	+ + + + + +
	pano-	*pano-*	- +	+	++	++	+ + +
	plasma	*soma*			++		++ ++ +++ +++ +++
	tengari	*Try-*			++		+++ +++ +++ +++
		pano-					
		plasma					
B.	*T. ma-*	*Try-*	++ ++	++	++	++	++ ++ ++ ++ ++ ++
	guri	*pano-*	- +	++	++	-	++ ++ ++
		soma			+ -		- - - - - -
		Try-					- - -
		pano-					
		plasma					
C.	Nil	*Try-*	++ ++	++	++	++	++ ++ ++ ++ ++ ++
		pano-	- -	-	++	-	++ ++ ++
		soma			- -		- - - - -
		Try-					- - - -
		pano-					
		plasma					

+++ : heavy intensity

++ : moderate intensity

+ : mild intensity

- : no infection

3.2.2.2. Inoculation Experiments

Inoculation of 1 ml of *Trypanoplasma tengari* infected blood in trypanosome infected *M. vittatus* (aquarium A) could induce infection from day 4 post infection (p.i.). However, trypanosome intensity began to decline from day 10 being a meagre 5–8 forms/15 mins scan by day 16. The biflagellates appeared in the circulation on day 4 p.i. and survived till the end of experimentation.

The hemoflagellates in fishes of aquarium B met with a different fate. Here trypanosome infection was unabated whereas trypanoplasm infection reduced in intensity continuously and by day 8 p.i., the trypanoplasm infection subsided.

Trypanosomes appeared normally in the control group (Table 3.7).

3.2.2.3. Discussion

Hematozoans of fishes have been aptly explored from different waters of the globe (Becker and Overstreet, 1979; Mandal, 1979; Joshi, 1982a; Khan and Newman, 1982). However, most published reports are based on hematological surveys and deal with descriptions of new species or new host and locality records and some have even worked on the biochemical impact of hematozoans (Tandon and Joshi, 1974; Gupta and Jairapuri, 1983; Gupta et al., 2000). However, a deeper probe into parasite co–existence, interspecific interactions, survival, competition, inhibition and host specificity are aspects which are not yet sufficiently elucidated and warrant a deeper probe.

Kinetoplastids are renowned for their ability to use surface glycoproteins in their interactions with the host (Adamson and Caira, 1994). Competition clearly affects realized niches in parasites (Holmes, 1986). This feature maybe difficult to demonstrate specially with comparative abundance data even where strong effects occur (Dobson and Pacala, 1992). Parasite competition may occur within the host population by the intermediary of pathogenicity and parasites remote from one another in microhabitat can have very strong interspecific effects. The parasite niche includes transmission requirements and therefore availability of susceptible hosts. When one parasite species compromises this resource for another, interspecific competition occurs and it is conceivable that such competition restricts host range. It is often difficult to demonstrate competition, therefore, its potential role in restricting niches in evolution becomes doubly difficult to demonstrate (Adamson and Caira, 1994). The present studies have shown that trypanosomes have a higher peak of prevalence as well as intensity as compared to the biflagellated forms under natural infectivity. Concomitant infection, though observed is rare. Mandal (1979) could record isolated hemoflagellates in different specimens though a concomitant condition with *Trypasosoma vittati* and *Dactylosoma* sp. was also observed and reported.

3.2.2.4. Classification of Interactions

Examples of negative interaction may be categorized as competition where the host and the resources it supplies or the site in the host is the thing (Milne, 1961) for which the parasites compete.

Historical reviews of the meaning of competition in biological literature and discussion of its definition have been given by Birch (1957), Milne (1961), DeBach (1966) and Miller (1967). Competition is sometimes used in a very broad sense and may include predation. Most authors, however, attempt to restrict the meaning of competition to interaction among organisms for obtaining a common limited resource (Birch, 1957; Milne, 1961; Miller, 1967; Ayala, 1972). According to Milne (1961), most of the things for which animals compete can be classed as either food or space.

Several studies of fish have shown an increase of parasite abundances with host size or host age (Dogiel, 1964). Comparative analyses have also shown positive relationships between fish size and parasite species richness (Poulin, 1995). However, intra-specific studies taking into account fish size (and age), parasite species diversity and parasite abundances are still very scarce.

Older and larger hosts harbor rich and diversified parasite communities. Polyansky and Bychowsky (1963) and Lo et al. (1998) considered diet, body mass, and school formation as factors responsible for the number of parasite species harbored by a host.

As with free-living animals, one has in most cases to accept certain observed numerical, developmental, and distributional patterns of individuals and species as indications of competition having taken place, since knowledge about the mechanisms and processes which have led to the observed situation is lacking. It is important to realize, however, that similar results, i.e., negative correlation in the occurrence of parasite species, may be caused also by other factors such as food preference and habitat selection by the host. The patterns of parasites in idiohostal units where competition is believed to have taken place are very similar to those situations which are normally interpreted as resulting from competition by free-living animals. Most of the examples given are examples of exclusion or reduction in numbers of one species by another, which are the most often quoted results of competition. The examples given by Chappell (1969), MacKenzie and Gibson (1970) and Holmes (1973) of changes in site of parasites in multispecies, as compared to single species, infections, fall within the interactive segregation of Brian (1956) and Nilsson (1967). Interactive segregation is believed to be replaced in the course of evolution by genetically regulated segregation, called selective segregation by Brian (1956). Segregation of related species of parasites into different organs in the host is relatively common and may be interpreted as evidence for selective site segregation (Holmes, 1973).

3.2.2.5. Negative Interactions amongst Parasites

Parasites of one or different populations may interact with each other, mostly either for food or space, and thus can affect each other negatively. The ecological consequences of such an interaction are often complicated and difficult to define. The negative interactions may operate between different species (Interspecific interactions) or within species (Intraspecific interactions).

3.2.2.5.1. Interspecific Interactions

1. **Exclusion:** In innumerable studies it has been observed that if an organism is infected with more than one parasite species, one of them will normally show exclusion or great reduction in numbers; or, in other words, the presence of one parasite species in the host may exclude or reduce infection with that of another species. Such a predominance of single species infection has been seen among the ectoparasites as well as in the parasites of alimentary canal and closed systems from where they are released only for the continuation of their life cycles (Kennedy, 1975).

Wilson (1916) found that out of 100 crappies concurrently exposed to copepod *Ergasilus coeruleus* and glochidia of *Lampsilis recta*, only 25 were infected with the species. The occurrence of even 10 copepods upon the gills reduced the number of glochidia that could infect the fish. Conversely, on a fish that naturally harboured more than 300 glochidia, the copepods were found to be totally absent. Cross (1934) reported that the fishes infected with more than 14 acanthocephala of the genus *Neoechinorhynchus* harboured very few cestodes of the genus *Proteocephalus exiguys*, whereas those having more than 25 individuals of the latter species had few or no acanthocephala. Similarly, Paperna (1964) reported that in case of intense infection with *Dactylogyrus vastator* on gills of carps, the infection with *D. extensus* was relatively very low and the carps were much susceptible to infestation with the latter species only in the absence of the former species.

Several experimental studies have been performed to illustrate the consequences of concurrent infections with various pairs of parasitic species. Larsh and Donaldson (1944) experimentally observed the effects of concurrent infections in mice with the larvae of nematode *Nippostrongylus muris* and eggs of the tapeworm *Hymenolepis nana* var. *fraterna*. Their studies showed that the number of cysticercoids of *H. nana* were significantly lowered when *N. muris* was established first. However, although such negative correlations have been reported to occur for several other pairs of parasites, but most clearly defined cases of this genre are found between species that inhabit the same microhabitat.

Examples of exclusion or reduction in numbers of one or more blood parasite species in the presence of another have also attracted attention of authors (Halvorsen, 1976). Schilling (1936) observed that *Trypanosoma brucei* and *T. congolense* could supersede each other in mixed infections. In most cases, *T. congolense* appeared to be "stronger" of the two species. When an unaltered strain of trypanosome was mixed with a strain that either had lost its kinetoplast or that had been rendered arsenic-resistant, the strain with changed characteristics invariably, and more or less rapidly, disappeared from a population maintained by serial transfer in rats or mice. The observed rate of disappearance closely followed the rate of disappearance calculated from the multiplication rates of the parent strains. It was also shown by Adler (1954) that hamsters previously inoculated with virulent strains of *Leishmania infantum* were strikingly less susceptible to *Plasmodium berghei* than normal hamsters. The infection with *L. infantum* inhibited the multiplication of the malarial parasite but the reciprocal situation was not confirmed. Concurrent infections of certain fishes with intestinal helminths of two or more different species caused competitive exclusion resulting in relative spatial segregation in the host's intestine. In cases of exceptionally heavy infections, the presence of one species in significant numbers was associated with a lowered density of the other species, compared to its regular density in single infections (Amin, 1975).

Concurrent infections with *Trypanosoma* and *Trypanoplasma* have indicated that in most weak hemoflagellate infections of carp, both *Trypanosoma carassii* and *Trypanoplasma borreli* are present. In massive infections, however, one or the other species always prevails.

2. **Segregation:** The concurrent infections with two species of parasites, in addition to the above mentioned effects, may influence the localization of one or both species within the host. This has experimentally been observed by Holmes (1961, 1962) in concurrent infections of rats with the cestode *Hymenolepis diminuta* and the acanthocephalan *Moniliformis dubius*. It was found that the anteriad ontogenic migration of the former species did not occur in rats which were already infected with that of the latter. Mackenzie and Gibson (1970) observed that the nematode distribution of *Contracaecum heterochorus* in the gut of flounder was largely even only when *C. minutes* was absent. Similar change was also found in the distribution of the digenean *Crepidostomum metoecus* in the intestine of brown trout in the presence of the cestode *Cyathocephalus truncatus* (Halvorsen and Macdonald, 1972). Holmes (1973) examined some of the basic features of microhabitat specificity of parasites and explored its relationships with interspecific interactions and the development of parasitic communities.

3.2.2.6. Intraspecific Interactions

An inverse relationship between the size of the worms and their numbers in the intestine of the vertebrate host has been reported for numerous tapeworms (Read, 1959; Roberts and Mong, 1968; Andersen, 1972). In addition, the "crowding effect" also results in the development of the so-called primary strobila with the delaying of strobilization (Halvorsen and Andersen, 1974). Although both of these effects are common in crowding of single tapeworm species, but similar effects in concurrent tapeworm infections are not infrequent (Read and Phifer, 1959; Guttowa, 1967).

Tandon (1973) has shown the effects of crowding on the development of reproductive organs and the size of the digenetic trematodes. But, unlike other parasites, the crowding has a little effect (if any) on the size and rate of development in nematodes. However, their survival like most other parasites may be low in case they had an intense worm burden in initial stages. For example, it has been observed that populations of *Nippostrogylus muris* with a high density in rats have low longevity, whereas those with low density survive to the longest which maybe owing to competitive stress amongst densely populated individuals.

3.2.2.7. Mechanisms of Competition

The mechanisms and processes involved in competition are rarely known. Those who have observed results indicative of competition among parasites have often suggested that competition has involved mechanical interference, acquired immunity, and/or the production of materials by one species which are deleterious to the other. In most cases, there is no evidence that these are the mechanisms behind the observed situations. In some cases, however, the investigations have penetrated deeper into the causal mechanisms.

- **Food Uptake:** Utilization of dietary resources has been shown to be a mechanism involved in competition amongst parasites. Read and Phifer (1959) demonstrated that host dietary carbohydrate quantity and quality might be the limiting factor in interaction between the tapeworms *Hymenolepis citelli* and *H. diminuta,* and among individuals of *H. diminuta*, leading to stunted specimens in crowded situations.
- **Reproduction:** The outnumbering of one species by another as a result of differences in reproductive potential has been regarded as a mechanism of competition by Beament (1961) and DeBach (1966) among others. It has been shown that the rate of disappearance of the strain of trypanosomes with changed characteristics in mixed

infections with unaltered strains closely followed the rate of disappearance calculated from the multiplication rates of the parent strains.

Damian (1964) suggested that antigen sharing by two parasite species may represent convergent evolution by the parasites toward greater antigenic compatibility with a common host, while Schad (1966) alternatively proposed that when co-occurring parasites are likely competitors, non-reciprocal cross immunity may be a device evolved to limit the abundance of a competing species.

Even if the mechanisms involved in host-mediated competition among parasites are necessarily different from those operating in "habitat-mediated" competition among free-living animals, of which Rose and Rose (1961), Milne (1961) and Nilsson (1967) give examples, the principal similarity between the two situations appears strong. It is, however, necessary to consider that host-mediated reactions are also directed toward the species provoking them. The statement by Wilson and Bossert (1964) that there is no reason to doubt that the essential characteristics of intraspecific chemical communication are also true of interspecific chemical communication is relevant in this context. Miller (1967) expressed doubt as to whether mechanisms involved in interspecific territoriality would have been evolved unless they had first arisen among members of the same species. This is also most likely to be the case for the mechanisms involved in host-mediated competition among parasites.

- **Interference and Exploitation:** The process of competition may, according to Miller (1967) involve interference and/or exploitation. Interference refers to any activity which either directly or indirectly limits a competitor's access to a necessary resource or requirement. It usually operates in a spatial context. Exploitation is the utilization of a resource once access to it has been achieved.

Competition is believed to be of special importance among closely related species (Williams, 1951). It is therefore interesting that many examples of competition among parasites involve species that are not closely related to each other. They therefore support Levin (1970), who used the hypervolume niche concept of Hutchinson (1957) to create a model which showed that certain dimensions of the hypervolumes were of paramount importance in the competition between two species, and therefore species which appear to fill different niches may be serious competitors.

- **The Ecological Implications of Negative Interaction amongst Parasites:** Investigations of negative interactions amongst parasites have been restricted to idiohostal units, but there have been speculations that the results may have applications at other levels of ecological complexity.

Interspecific competition, and other forms of interspecific negative interaction in idiohostal units leading to reduced survival or multiplication of one or more parasite species in these units, theoretically lead to fewer parasites of these species within the host population (synhostal units, Kisielewska, 1970). The distribution pattern and the occurrence of the parasites are, together with the separate and combined effects of the parasite species on the growth rate of the host population, important factors in determining the ecological effect of the negative interaction among the parasites. If the combined influence of the parasites is strong and the probability of concurrent infections high, negative interaction among the parasites may be important in reducing the stress on the host population.

Negative interactions among parasites leading to reduction of idiohostal unit size, and thus in reduction of the demands of the parasites on the hosts, may have population dynamics effects similar to those of reduction of idiohostal unit size caused by host immune reactions. Crofton (1971a) discussed the difficulties in assessing the role of immunological reactions on populations. For the purpose of model-making, Crofton (1971b) assumed that a minimal stimulation, expressed in numbers of parasites, was needed to evoke an immunological response which would either remove the parasites or render them sterile, and that the immunological response could be swamped by high infection pressures and host death ensuing before an immunological reaction was manifested. By making these assumptions, Crofton (1971b) found that the introduction of an immune factor into a mathematical model of host-parasite relationships produced an oscillating system, and not invariably higher population levels of host. The inoculation experiments have shown some interesting results and throw light on interaction of the forms concerned. When trypanoplasms from the same host species are inoculated into trypanosome infected hosts (Group A), the former establish themselves from day 4 p.i. and exert an inhibitory effect on trypanosomes. Dykova and Lom (1979) found that histopathological changes associated with trypanoplasm infection in goldfish were more extensive than those imparted by a trypanosome. The decline in trypanosome population in group A maybe correlated with the fact that due to interaction of the two species, the trypanosome population declines to a lower intensity. It has been shown that the presence of one parasite species in the host may exclude or reduce infection with that of another species (Kennedy, 1975). Pontin (1982) has referred to interspecific competition as an interaction between two species which results in reduced population size of both the competing species. The consequences of concurrent infections with various pairs of parasitic species have shown a negative correlation and are more pronounced between species which inhabit the same microhabitat (Sharma and Ratnu, 1984). Gupta et al. (1998) conducted cross transmission tests by inoculating *Trypanoplasma cyprinoides* infected blood from *Puntius ticto* to *Trypanosoma ticti* infected fish. On day 21 p.i., the trypanoplasm population suppressed the trypanosome population indicating that trypanoplasms are the dominant species maintaining their density under concomitant conditions whereas the monoflagelate population fell considerably.

The present studies have shown that trypanoplasms exert a negative interaction on trypanosomes compelling it to decline and also a reduction in intensity of both the parasites under experimental concurrent conditions (Table 3.7). The present experiment is an example of negative interaction manifested as "exclusion" where *Trypanoplasma* excludes or reduces the *Trypanosoma* population and thus appears to be the "stronger" species.

In group B, *Trypanoplasma maguri* when inoculated could barely survive for a couple of days, rapidly declining in intensity and completely subsiding by day 8 p.i. Under natural conditions, however, *T. maguri* may not survive in *M. vittatus* even up to day 8 p.i. as potential specificity is usually wider than that actually observed (Kirshenblat, 1941). The dying out of the parasite is probably due to its specificity due to which it is unable to survive in another host. The donor parasite is from the host belonging to the family Clariidae and the recipient host belongs to the family Bagridae which may account for the unadaptability of the parasite in a new host environment.

Mukherjee and Haldar (1982) have shown that fish trypanosomes are host specific and failed to initiate infection when they were inoculated artificially into another host. Arrea et al. (1998) inoculated white rats with 10**6** trypomastigotes of *Trypanosome lewisi*, simultaneously or two days before and after inoculation with 10**5** oocysts of *Toxoplasma gondii*. A greater number of cysts were found in the brain of animals having concomitant inoculations as compared to rats inoculated with either one of the two parasites. Since both organisms can be found in rats, it is possible that infections with *T. lewisi* could make this rodent another intermediate host for *Toxoplasma* infections. It was postulated by the authors that *T. lewisi* infections could induce an increase of *Toxoplasma* multiplication in the rats after oocyst ingestion. Immunosuppression in trypanosome models has been reported. Darji et al. (1992) explained the mechanisms for that effect in African trypanosomes. Channon and Kasper (1996) have found that *T. gondii* induces immunosuppressive factor such as gamma interferon. Unfortunately, such experiments on trypanoplasms have not gained due weightage perhaps because they are comparatively rare and do not occur in warm blooded animals due to which scientists have focused lesser attention on their biological and behavioral patterns. It is evident from the study that trypanoplasm from *Mystus vittatus* is host specific and incidence is the outward manifestation of specificity (Bykhovski, 1957). Specificity and interaction amongst co-existing species is a complex phenomenon and erudite studies on the geographic distribution, intraspecific variations, vector, developmental cycle, transmission, pathological and physiological affects and ecological factors affecting epizootiology have to be properly elucidated. This may also be true for trypanoplasms which may further throw some light and justify for the mechanism underlying co-existence of

3.3. HOST SPECIFICITY AMONGST BLOOD PARASITES

The erroneous, outdated rule "different host = different parasite" was in the past responsible for many dubious descriptions and, unfortunately, it still survives in works of some authors. Experimentally verified evidence for strict host specificity is rare, while a broad range of hosts has been demonstrated in many species.

One of the most fundamental properties of any parasite species is its host specificity (Adamson and Caira, 1994). The specificity of a parasite is determined by the number of host species it can exploit successfully ('host range' *sensu* Lymbery, 1989). It is accepted that its measurement is not as straightforward as it might seem, this single feature of parasites is important for many reasons. It reflects the breadth of their ecological niche and their exact position in the biosphere better than any other parameter. In addition, host specificity determines the likelihood that a parasite will successfully invade a new habitat, or adjust to new hosts after it enters a new geographical area. Host specificity is therefore an important life history trait of parasites, just as important as body size and fecundity of the parasite. It has however, proven extremely difficult to quantify in any way that would allow meaningful comparison among species.

Host specificity of a parasite is not merely a function of how many host species it can exploit, but also of how closely related these host species are to each other. Caira et al. (2003) proposed a new index of host specificity that incorporated basic phylogenetic information on the host species used by a parasite. This index considered not only how many host species were used by the parasite but also how many genera, families and orders these hosts belonged to.

Poulin and Mouillot (2003) added another new index of host specificity taking into account the average taxonomic or phylogenetic distance between pairs of host species used by a parasite.

Parasites are specialists, not just of hosts, but of microhabitat or tissue site and different developmental stages often possess entirely different sets of specificities. Parasite specificity is not fundamentally different from specificity, or niche restriction in free living organisms except that the niche of parasites is part of another organism. The degree of specificity a parasite exhibits is the result of an interplay of factors that act on various temporal scales.

Circumstances surrounding the origins of parasitism can be strong determinants of the subsequent course of evolution in the group. Thus, specificity of a parasite lineage often reflects that of its free living ancestors. This is due to the fact that parasitism arises from ecological associations between potential hosts(s) and parasite that already involve some degree of specificity (Adamson and Caira, 1994).

Parasites, because of the nature of their life cycles, come into contact with a restricted array of hosts. Parasites are first and foremost, site or microhabitat specific. They live and feed in particular regions of the host. The flagellate parasites of fish are specific to the blood microniche where they live and feed. In order to elucidate the phenomenon of host specificity in blood parasites and their adaptability to blood microniche, certain experiments were conducted.

The viability of blood parasites to establish themselves in potential/alternate hosts was worked out by conducting cross inoculation experiments to demonstrate host specificity in fish and amphibian hosts.

For this purpose, the experiments were conducted in different series as per the following protocol.

1. Class, family, genus and species of recipient identical to donor.
2. Class, family and genus identical but species of recipient different from the donor.
3. Class and family identical but genus and species of recipient different from the donor.
4. Class identical but family, genus and species of recipient different from the donor.
5. Class, family, genus and species of recipient all different from the donor.

3.3.1. SET I

Experimental Protocol: The recipient group belonged to the following categories:

Category 1: Class, family, genus and species of recipient identical to donor.

Category 2: Class, family and genus identical but species of recipient different from the donor.

Category 3: Class identical but family, genus and species of recipient different from the donor.

Controls were run throughout the experiment for comparing the results.

3.3.1.1. Methodology

A total of 1,190 fishes belonging to the species *Notopterus notopterus*, *Catla catla, Cirrhinus mrigala, Cyprinus carpio, Labeo rohita, Puntius ticto, Wallago attu, Mystus vittatus, M. tengara, Heteropneustes fossilis, Clarias batrachus, Xenentodon cancila, Channa punctatus, C. striatus and Colisa fasciatus* were examined during the survey. For parasite study, live flagellates were observed in hanging drop preparations and methanol fixed forms were stained as mentioned in "Materials and Methods."

The blood of *Channa punctatus* was infected with *Trypanosoma aligaricus* (Gupta and Jairajpuri, 1982) and this parasite was used as the test inoculum. The host specificity of piscine trypanosomes was investigated by inoculating infected blood from *Channa punctatus* to three groups of fishes: Group A (*C. punctatus*), Group B (*C. striatus*) and Group C (*H. fossilis*). It was ensured that the experimental fishes, destined to act as recipients were free from flagellate infection by examining them twice a week for 30 days before inoculation. Three ml of cardiac blood was collected in heparinized syringes from the infected donors, pooled and 0.6 ml was inoculated intraperitoneally into the fishes of group A, B and C. Recipient fish were maintained in glass aquaria under favorable conditions of temperature, food and oxygen. Controls for each group were maintained throughout the duration of the experiment. After inoculation, the blood of recipient fishes was examined after every 24 hr for 21 days to detect trypanosomes.

3.3.2. SET II

Category 1: Class, family, genus and species of recipient identical to donor.

Category 2: Class and family identical but genus and species of recipient different from the donor.

Category 3: Class, family, genus and species of recipient all different from the donor.

Controls were run throughout the experiment for comparing the results.

3.3.2.1. Methodology

The blood of *Catla catla* was infected with *Trypanosoma monomorpha* (Gupta and Jairajpuri, 1985) as indicated from the above survey and this parasite was used as the test inoculum in Set II experiment. The procedure for live parasite study was similar as mentioned above for Set I.

In this set of experiment, host specificity of trypanosomes was investigated by inoculating infected blood from *Catla catla* to three groups of fishes: Group A (*C. catla*) (Category I), Group B (*Labeo rohita*) (Category II) and Group C (*Bufo melanostictus*) (Category III). It was ensured that the recipients were free from flagellate infection by examining them as mentioned for Set I experiment.

T. monomorpha-infected blood was collected and 0.6 ml was inoculated intraperitoneally into the animals of Group A, B and C. Recipient fish (*Catla catla* and *Labeo rohita*) were maintained in glass aquaria under favorable conditions of temperature, food and oxygen whereas *B. melanostictus* were maintained in an amphibious environment in plastic cages. The blood of recipient fishes/toads was examined after every 24 hr for 30 days to detect trypanosomes.

3.3.3. Results

Trypanosoma batrachi (Qadri, 1962), *T. batai* (Joshi, 1978), *T. trichogasteri* (Gupta and Jairajpuri, 1981), *T. singhii* (Gupta and Jairajpuri, 1981), *T. aligaricus* (Gupta and Jairajpuri, 1982), *T. attii* (Gupta and Jairajpuri, 1982), *T. notopteri* (Gupta and Jairajpuri, 1985) and *T. monomorpha* (Gupta and Jairajpuri, 1985) were encountered from the blood of 15 species of fresh water fishes (Table 3.8).

Table 3.8. Prevalence of Trypanosomes in Some Fresh Water Fishes of Northern India

Host Species	**Taxonomic Status**	**Common Name**	**Length of Fish (cm)**	**No. Sampled**	**No. Infected**	**% Infection**	**Infection Intensity (no. of ts/film)**	**Locality**	**Trypanosome Species**	**Authors**
Notopterus	Notopteridae	Patara	14–16	82	5	6.0	10–15	Bareilly	*T. notopteri*	Gupta and Jairajpuri (1985)
Catla catla	Cyprinidae	Bhakur	20–25	90	4	4.4	15–20	Bareilly	*T. monomorpha*	Gupta and Jairajpuri (1985)
Cirrhinus mrighala	Cyprinidae	Nain	15–20	40	-	-	-	Nakhatia, Bareilly	-	-
Cyprinus carpio	Cyprinidae	Common carp	15–20	60	-	-	-	Nakhatia, Bareilly	-	-
Labeo rohita	Cyprinidae	Rohu	16–20	64	4	6.2	6–10	Aligarh	*T. batai*	Joshi (1978)
Puntius ticto	Cyprinidae	Sidhri	6–9	98	-	-	-	Badaun	-	-
Wallago attu	Siluridae	Padhani	15–35	45	14	31.1	15–24	Faridpur, Bareilly	*T. attii*	Gupta and (1982) Jairajpuri
Mystus tengara	Bagridae	Tengara	6–9	95	-	-	-	Shahjahanpur	-	-
M. vittatus	Bagridae	Tengara	7–11	51	-	-	-	Shahjahanpur	-	-

Heterop-neustes fossilis	Heteropne- us-tidae	Singhi	23–28	120	56	46.6	40–60	Bareilly	*T. sing-hii*	Gupta and Jaira-jpuri (1981)
Clarias batrachus	Claridae	Magur	20–25	75	18	24.0	20–30	Kwarsi, Aligarh	*T. ba-trachi*	Qadri (1962) Gupta and Jaira-jpuri (1983)
Xenentodon cancila	Belonidae	Kauwa	17–21	100	-	-	-	Moradabad	-	-
Channa punctatus	Ophiocepha-	Saul	8–18	50	3	6.0	15–21	Fish mar-ket, Aligarh	*T. alig-aricus*	
C. striatus	Ophiocepha-lidae	Sauri	14–19	122	-	-	-	Fish mar-ket, Aligarh	-	-
Colisa fasciatus	Anaban-tidae	Khosti	5–8	98	15	15.3	60–100	Drains of Aligarh	*T. tricho-gasteri*	

Table 3.9 Effect of Cross Transmission of *Trypanosoma aligaricus* (Gupta and Jairajpuri, 1982) (0.60 ml Inoculum/Fish) Into Fish Hosts, *Channa punctatus*, *C. striatus* and *Heteropneustes fossilis*

		Duration (in days)																					**Intensity (No. of Try-panosomes/ Blood Film After 21 days p.i.) *Donor Experimental***	
Fish (Group)	**No. of fishes**	1	2	3	4	5	6	7	8	9	10	11	12	13	14	15	16	17	18	19	20	21		
***Channa punctatus* (A)**	**Control (4)**	-	-	-	-	-	-	-	-	-	-	-	-	-	-	-	-	-	-	-	-	-	-	-
	Experi-mental (4)	-	-	+	+	+	+	+	+	+	+	+	+	+	+	+	+	+	+	+	+	+	15–25	10–16
***C. striatus* (B)**	**Control (4)**	-	-	-	-	-	-	-	-	-	-	-	-	-	-	-	-	-	-	-	-	-	-	-
	Experi-mental (4)	-	-	-	+	+	+	+	+	+	+	+	+	+	+	+	+	+	+	+	+	+	8–14	
***Heteropneus-tes fossilis* (C)**	**Control (4)**	-	-	-	-	-	-	-	-	-	-	-	-	-	-	-	-	-	-	-	-	-	-	-
	Experi-mental (4)	-	-	-	-	-	-	-	-	-	-	-	-	-	-	-	-	-	-	-	-	-	-	-

Fresh inoculum **Fresh inoculum**

Table 3.10. Effect of Cross Transmission of *Trypanosoma monomorpha* (Gupta and Jairajpuri, 1985) (0.60 ml Inoculum/Fish) into Fish Hosts, *Catla catla*, *Labeo rohita* and Amphibian Host, *Bufo melanostictus*

		Duration (in days)																													**Intensity (No. of trypanosomes/ blood film after 21 days p.i.)**	
																															Donor	***Experimental***
Fish (Group)	**No. of fishes**	1	2	3	4	5	6	7	8	9	10	11	12	13	14	15	16	17	18	19	20	21	22	23	24	25	26	27	29	30		
***Catla catla* (A)**	**Control (5)**	-	-	-	-	-	-	-	-	-	-	-	-	-	-	-	-	-	-	-	-	-	-	-	-	-	-	-	-	-	-	-
	Experi-mental (5)	-	-	-	+	+	+	+	+	+	+	+	+	+	+	+	+	+	+	+	+	+	+	+	+	+	+	+	+	+	8–15	6–12
***Labeo rohita* (B)**	**Control (5)**	-	-	-	-	-	-	-	-	-	-	-	-	-	-	-	-	-	-	-	-	-	-	-	-	-	-	-	-	-	-	-
	Experi-mental (5)	-	-	-	-	-	+	+	+	+	+	+	+	+	+	+	+	+	+	+	+	-	-	-	-	-	-	-	-	-	8–14	
***Bufo melanostictus* (C)**	**Control (5)**	-	-	-	-	-	-	-	-	-	-	-	-	-	-	-	-	-	-	-	-	-	-	-	-	-	-	-	-	-	-	-
	Experi-mental (5)	-	-	-	-	-	-	-	-	-	-	-	-	-	-	-	-	-	-	-	-	-	-	-	-	-	-	-	-	-	-	-

↑ **Fresh inoculum** (day 15) ↑ **Fresh inoculum** (day 21)

3.3.4. SET I

The blood of group A and B gave a positive response of trypanosomes on the 3rd and 4th day respectively which persisted till day 21 p.i. The intensity of trypanosomes in experimental fish was 10–16 forms/20 mins scan and 8–14 forms/20 mins scan in Group A and B fish respectively on day 21 p.i. However, group C failed to show any flagellate up to a week. Hence a fresh inoculum from infected *Channa punctatus* was again injected to group C on day 8 p.i. and day 15 p.i. Despite critical examination of group C films they failed to show any response (Table 3.9). The infection prevalence was approximately 75% in the experimental fishes as compared to the donors (Table 3.9).

3.3.5. SET II

Five fishes/toads of each group were inoculated intraperitoneally with *T. monomorpha* inoculum. The blood sample of Group A responded favorably to the inoculum and trypanosomes appeared in the peripheral blood on day 4 p.i. and persisted till the termination of the experiment (day 30) when the intensity was 6–12 forms/20 mins scan as compared to the donor fish where the intensity was 8–15 forms/20 mins scan.

Group B fish also responded to the inoculum and trypanosomes were visible in the blood of *Labeo rohita* on day 6 p.i. These parasites persisted in the blood till day 19 p.i. but on day 20, there was a sharp decline of trypanosomes. The experiment was therefore prolonged up to 30 days to observe the fate of the parasites. Interestingly, the infection subsided and the trypanosomes disappeared from the blood from day 21 p.i.

Group C toads were devoid of any flagellate infection throughout the experiment. A fresh inoculum was again repeated on day 10 p.i. and day 20 p.i. without any positive results (Table 3.10).

3.3.6. Discussion

A plethora of flagellated species from fresh water, estuarine and marine fishes are listed in the literature. Most species are named only on the basis of the host they infect or on minor morphologic and morphometric differences. Hoare (1955) suggested that only constant, non-intergrading morphological or physiological differences should be used as interspecific criteria among parasitic protozoa. The blood milieu of a particular species, temperature, state of parasite development, host specificity, vector specificity and geographic area have to be carefully scrutinized in order to establish a new species (Becker, 1970). Grogl et al. (1980) regarded geographic isolation as a sounder criterion for hemoparasites of freshwater fishes than for blood parasites from mammals or birds.

The present survey records eight species of trypanosomes. No two parasites inhabited the blood of the same fish.

Host specificity refers to the peculiar mutual adaptation which restricts a parasite to its host species (Noble and Noble, 1982). Most early studies on fish trypanosomes assumed a high degree of host specificity by assigning species status of populations of trypanosomes found in different species of fish host. A different host was often the only distinction for newly erected species (Henry, 1913a,b); unfortunately, this practice survives in many present papers. Some of thc early cross-infection experiments yielded results which fortuitously seemed to corroborate the strict specificity concept (Breindl, 1915; Ioff et al., 1926).

Other records of experiments indicate that the hemoflagellates of fish are not species specific, although the results are sometimes ambiguous. Kipp (1968), using inoculation of Tench blood infected with *T. borreli*, failed to infect pikes, perches and chubs, but succeeded with carps, breams and crucian carps. Khan (1977b) using leech infected with *Trypanosoma murmanensis* successfully transmitted the parasite to 13 species of marine fish belonging to 4 different orders. He fed infection-free leeches on four species of infected perciform and pleuronectiform fish and thus transmitted the trypanosome to a total of six other taxonomically related and unrelated fish. He assumed that all the trypanosomes were referable to the single species, *T. murmanensis*.

More dependable evidence for strict host specificity is rare; such data were presented by Lebailly (1905) and Cottrell (1977) for *Trypanosoma platessae*. On the contrary, there is evidence that many fish trypanosomes, e.g., *T. carassii, T. cobitis, T. murmanense, T. rajae* can inhabit a variety of hosts, which may sometimes be taxonomically remote. However, there are always some host species in which cross infection fails.

The problem of host specificity among fish trypanosomes has gained impetus during the recent years (Becker, 1967; Khan, 1972; Letch, 1979; Mandal, 1979, 1984; Mukherjee and Haldar, 1982). *T. mukasai* (Baker, 1960), *T. occidentalis* (Becker, 1967) and *T. murmanensis* (Khan, 1977b) are euryhostpitalic. On the other hand, sufficient records exist where not only one parasite selects only one host for its existence, but different species have been reported from the same host:

T. punctati Hasan and Qasim, 1962; *T. elongatus* Raychaudhari and Misra, 1973; *T. channai* Narasimhamurti and Saratchandra, 1980 and *T. aligaricus* (Gupta and Jairajpuri, 1982) from *Channa punctatus*; *T. sacchobranchi* Castellani and Willey, 1905; *T. danilewskyi* var. nov. *saccobranchi* Qadri, 1962; *T. mukundi* Raychaudhuri and Misra, 1973 and *T. singhii* Gupta and Jairajpuri,

1981 from *Heteropneustes fossilis* and *T. batrachi* Qadri, 1962; *T. maguri* Tandon and Joshi, 1973 and *T. qadrii* Narasimhamurti and Saratchandra, 1980 from *Clarias batrachas*. These findings support the specificity of the parasites under speculation.

Barrow (1954), using parasite-free leeches, failed to transmit trypanosomes from eels to goldfish, but succeeded in infecting goldfish with undetermined trypanosomes from nine-spined and blue-spined stickly back, minnow, tench, perch and bullhead; he did not say whether the goldfish used were parasite-free; unless the inoculated fish had a parasite-free anamnesis it proved insusceptible because of super-infection immunity, or the introduction of parasites may have caused relapse of a dormant infection. However, Robertson (1911) used trypanosome-free goldfish; she infected them with leeches carrying trypanosomes from perches and breams. Lom (1973), using syringe inoculation to parasite-free reared goldfish, was able to repeatedly, and without failure, infect the goldfish with *T. danilewskyi* from carps, and with populations of trypanosomes from tenches assignable to the same species. He also succeeded with populations of trypanosomes, morphologically identical to *T. danilewskyi* isolated from eel and gudgeon, but failed with *T. anguillae* and *T. elegans* from the same hosts. Inoculations of *T. percae* from perches, *T. aceriane* from *Acerian cernus*, *T. premaki* from pikes, and trypanosomes from breams, barbs and roaches gave repeatedly negative results. *Trypanoplasma guerneyorum* from pikes also failed to establish itself in the goldfish. On the other hand, *T. danilewskyi* isolated from carps and passaged in goldfish was easily established in parasite-free reared pikes and in crucian carps from the wild. *Trypanoplasma borreli*, isolated from carps and also passaged in goldfish, failed to infect parasite-free reared pikes and *Puntius conchonius*, but did establish itself in minnows, crucian carps, roaches and gudgeons. Sometimes, a susceptible host is protected only by an ecological barrier against its potential parasite (e.g., the susceptibility of *Puntius conchonius* (Lom, 1973) or of *Lebistes reticulatus* (Khaybulyayev, 1976) to *T. danilewskyi*. Interestingly, and similar to salivarian trypanosomes, fish trypanosome species can thrive in hosts which occupy very different taxonomic positions.

Discrepancies between the transfer experiments of Robertson (1911), Barrow (1954) and Lom (1973) (e.g., in the infection of goldfish with trypanosomes from perches and eels) can be explained, at least in part, by the fact that one fish host may, even if not simultaneously, act as the host of two (or more) di- or poly- xenuous trypanosome species having differing preferences for other hosts. In favor of this assumption, are observations on North American trypanoplasms. *T. cataractae* (Putz, 1972a,b) also infects, apart from its principal host, *Rhimichtys cataractae* (prevalence of infection 97%)

and four more cyprinid species; intraperitoneal injection can extend the list to two more hosts, one of which belongs to a different family. *T. salmositica* in addition to its principal host, *Oncorhynchus tschawytscha* which is affected by the heaviest infections (Becker and Katz, 1965a-c), infects in the wild a total of 9 species of salmonid fishes, 5 species of cottids, 3 of cyprinids and the three-spined sticklebacks; the only common feature of the hosts is their cold-water habitat. The marine species *T. bullocki* infects four species from two fish families. The case of *T. salmositica* was used by Khaybulyayev (1976) as a part of the evidence for his claim that blood flagellates of fish while having a low specificity for fishes are strictly specific for leeches – which certainly does not hold true for European leeches. Bower and Woo (1977) on the evidence of unsuccessful transmissions of *T. commersoni* to 16 species of taxonomically related or alien hosts claimed a strict host specificity for this trypanoplasm.

Lom (1979) was able to infect, by syringe-inoculation of parasitaemic blood, parasite-free reared goldfish with *T. carassii* from carps and Tenches, with morphologically similar populations from eel and gudgeon (but not with *T. granulosum* and *T. elegans* from these two hosts). Inoculation of *T. percae, T. acerinae and T. remaki* also gave repeatedly negative results. On the other hand,

T. carassii isolated from carps and passaged in goldfish was easily established in parasite-free reared pike (= host to *T. remarki*) and in seemingly uninfected perch from the wild. These and other data, although fragmentary and inconsistent, seem to indicate that not only one trypanosome species may infect various hosts but also that one fish species may act as a host for two or more trypanosome species, perhaps with differing preferences for their hosts. It is obvious – although experimental data are fragmentary and inconsistent that strict host specificity seems to be an exception rather than a rule in fish hemoflagellates, which may have – in addition to one or more principal hosts (e.g., common carp, crucian carp and Tench in *T. carassii*) – additional "supplementary" hosts, which may also serve as source of infection for their principal hosts. The latter may have – in addition to one or more principal hosts (e.g., carp and Tench in *T. danilewskyi)* which best meet their growth requirements – more hosts which at the same time may serve as principal hosts for other hemoflagellates (e.g., eel – such as *T. granulosum* and *T. danilewskyi*). Thus, important commercial fishes such as salmon or carp may contract their leech-vectored infections from sources in "trash" fishes which act in this sense as reservoir hosts. Gupta et al. (1998) determined the host range of *Trypanosoma kargenensis*, a natural parasite of *Heteropneustes fossilis* by experimentally inoculating the parasite in *Clarias batrachus* and *Channa punctatus*. Although the parasite was capable of infecting its natural host and *Clarias batrachus*, it failed to establish itself in the circulation of *Channa punctatus* despite repeated inoculations on day 7 and day 14 p.i. The authors also discussed the significance of ecological and physiological factors in determining specificity. The results concluded that trypanosomes were capable of infecting closely related species (*H. fossilis* and *Clarias batrachus* are both siluroid fishes) more easily as compared to distantly related ones (*Channa punctatus*: Ophiocephaliformes).

Specificity is a characteristic feature of parasitism. To elucidate the concept of host specificity of piscine trypanosomes, transmission experiments were conducted. Two sets of experiments were conducted. In Set I, infected blood from *C. punctatus* was inoculated into the three groups of fishes and the effect noted: Group A which consisted of recipient *C. punctatus* (class, family, genus and species of recipient identical to donor) showed a positive response and comparable intensity as in the normal course; group B consisted of an intra-generic variant of the donor, *C. striatus* (class, family and genus identical but species of recipient different to donor), also responded favorably to the inoculation which otherwise was infection-free. But group C (*H. fossilis*) (class identical but family, genus and species of recipient different from donor) failed to reveal any trypanosome inspite of two inoculation passages. It was inferred that the monoflagellates of fish may infect closely related host species more readily as compared to distantly related ones.

In Set II experiment, *T. monomorpha* infected blood from *Catla catla* (family Cyprinindae) was inoculated into the same host (*Catla catla*) in Group A (Class, family, genus and species of recipient identical to donor), into *Labeo rohita* (Group B) which belonged to the same family Cyprinindae (Class and family identical but genus and species of recipient different from the donor) and group C, *Bufo melanostictus* (Class, family, genus and species of recipient all different from the donor) where the recipient belonged to the class Amphibia and the donor to Pisces.

Here again, as Group A fish (*Catla catla*) were the natural hosts for *T. monomorpha*, the parasite could survive satisfactorily, Group B fish (*Labeo rohita*) belonged to different genus and species: the parasite tried to establish itself, but perhaps due to the immune response of the fish, the parasite population gradually declined and by day 21 p.i., the parasite disappeared from the blood circulation. The experiment was therefore prolonged for 30 days, but till the termination of the experiment, the parasite was not observed in any blood film despite rigourous examination in live, fixed and centrifuged preparations. Group C recipients (*Bufo melanostictus*) failed to host the parasite despite repeated inoculations on day 10 and 20 p.i. It appears that hemoparasites do not adapt to divergent conditions of micro-environment. It maybe added that other species of trypanosomes have been reported from amphibian blood (Miyata, 1978; Ray, 1992) but the latter failed to host piscine trypanosomes: this may be due to host specificity of the parasite concerned.

Set II experiment would therefore infer that *T. monomorpha* can successfully establish itself in its natural host, *Catla catla* (Group A); when the recipient is changed to a host with a different genus and species (*Labeo rohita*), but both belonging to the same family (Cyprinidae), the parasite tries to survive but it ultimately is not able to establish itself as shown by its disappearance on day 21 p.i. Bower and Woo (1977b) showed that the narrow host range of *Cryptobia catostomi* was maintained by the fact that parasites in abnormal hosts stimulated a complement cascade that resulted in their lysis. Wehnert and Woo (1980) demonstrated similar mechanism mediating specificity in *C. salmositica*. When the class, family, genus and species of the recipient are all changed (Group C), the parasite does not survive in this new host at all despite the fact that both fish and toad are cold blooded animals. It is difficult to predict how parasites are able to survive in their natural hosts. It is questionable whether they do not elicit complement cascade or are they resistant to complement attack by the normal host.

Host specificity amongst piscine trypanosomes is thus apparent from the experiments conducted above. In either case, parasites from the same host face no difficulty in establishing themselves in the same host, when the genus is changed, the parasite can survive too under experimental conditions, but genus onwards, at the inter-family level, the experimental hosts are unable to act as natural hosts and at the inter-class level, the question of trypanosomes adapting themselves to survive does not arise as indicated by the experiment.

It has been postulated that the mutual adaptation of host and parasite manifest as a strict specificity and usually develops during the course of prolonged period of close association between phylogeny of parasite and hosts (Dogiel, 1964). The capacity for changing hosts might depend on the range of reactions of the parasite to the hosts and if the physiological and morphological mechanisms of the parasite are flexible enough to permit the parasite to overcome the resistance of the new host, then the parasite might be successful in adapting itself in the new host environment. New host transfer is also determined by the degree of natural immunity of the parasite and the factors which help to overcome it. Shulman (1958) emphasized the importance of duration of association between the parasite and hosts suggesting that host change becomes difficult in older host-parasite associations. This would implicate that *T. aligaricus* and *T. monomorpha* have an old association with *Channa punctatus* and *Catla catla* respectively rendering them difficult to survive in other newer hosts.

Specificity as such, has its complexity which may be resolved by studying the geographical distribution, intra-specific variations, vector, developmental cycle, transmission, pathological and physiological effects as well as ecological factors affecting epizootiology. On elucidation of these facts, specificity amongst piscine trypanosomes may be regarded as a lone criterion for species differentiation.

The strongest evidence for host specificity of mammalian trypanosomes comes from the parasites of the subgenus *Trypanosoma* (*Herpetosoma*). Noyes et al. (2002) have shown that *T.* (*Herpetosoma*) species does not infect an alternative host. Intra peritoneal injections of *T.* (*H.*) *musculi, T.* (*H.*) *microti, T.* (*H.*) *grosi and T.* (*H.*) *evotomys* are only infective to their natural hosts (Molyneux, 1969; Maraghi and Molyneux, 1989; Santos-Gomes et al., 1993). German strains of *T. evotomys* isolated from *Clethrionomys glareolus* (Arvicolinae) are not infective to either Asian or North American species of *Clethrionomys* (Krampitz, 1959). Ionizing radiation and splenectomy did not modify the resistance of *Mus musculus* to *T. lewisi* of rats (Albright and Albright, 1991). *T. rangeli* was believed to be an exceptional *T.* (*Herpetosoma*) species as it is infective to a wide range of hosts. It has been shown that *T. rangeli* is a member of the subgenus *T.* (*Schizotrypanum*) which includes parasites like *T. cruzi* with a wide vertebrate host range (Stevens et al., 1999).

Host defense considerations should be more important in generating specificity in vertebrates than invertebrates because the defense system is more elaborate in vertebrates. This would depend on the nature of the interaction (Adamson and Caira, 1994).

The studies on specificity of hemoparasites of piscine hosts indicates that the parasites are specific in nature. It is not known whether host specificity is brought about by the recognition of special secretions of a suitable host or by other means (Chappell, 1980). Several reasons have been postulated for variation in host specificity and parasitic Protozoa with a change of hosts in their life cycles also frequently show great differences in their specificity toward their final and intermediate hosts. Whereas two strains of a given parasite may infect a final host with equal facility, they may not be equally infective to a given strain of intermediate host. Khan (1991) studied the distribution, abundance, host selection and prevalence of trypanosomes in the marine leech, *Johanssonia arctica* in the Northwest Atlantic. Jones and Woo (1992) observed that *T. catostomi* could be infected only by one leech, *Actinobdella inequiannulata* despite the fact that two other species of leeches also occurred in the same environment (*Hemiclepsis marginata* and *Desserobdella phalera*) which illustrates specificity of the parasites towards the intermediate host as well. The existence of contact between the parasites and the host is one of the indispensable factors for the development of specificity. Genetic, physiologic, biochemical, ecologic or physical factors have generally been postulated for determining host specificity (Noble and Noble, 1982) but according to Cheng

(1986), the governing factor could be ecological or structural. Ecological specificity may have a food selection basis. Schiller (1959) and Smith and Haslewood (1963) suggested that some native biochemical components of the host govern specificity. Williams (1968) reported structural basis of compatibility as a viable cause. The manifestation of a minimal amount of host response in the form of cellular response (encapsulation, phagocytosis) and/or production of antibodies is an important mechanism governing host specificity.

Valid criteria for determining host specificity in parasitic Protozoa are still partly obscure. The genetic complex of a parasite and its host is obviously the basis of biochemical and physiological patterns in determining the character of parasite host specificity. The biochemical pattern embraces the production of metabolites that play a decisive role in parasite host adjustments.

Many parasites require two or more hosts to complete their life cycles. Specificity to these hosts may differ, presumably depending on their functional role. Parasites that use intermediate hosts as trophic channels may exhibit reduced specificity in these hosts to increase their chances of being channeled to the final host. Intermediate hosts may also act as vectors which actively introduce the parasite to the final host. The parasite must interact more extensively with vectors and this may favor specificity. It appears in the present case that trypanosomes too interact with the vectors, the leeches resulting in specificity of the trypanosome parasite concerned. Finally, the intermediate hosts for amphibian trypanosomes are mosquitoes and not leeches as for fish trypanosomes. It has been shown that experimental infection with frog trypanosomes could be established in laboratory reared mosquitoes but frogs could not be infected with flagellates (Perez-Reyes, 1967). Therefore, it is but obvious that in this case (Set II, Group C) interplay of specificity between intermediate and final hosts maybe crucial for generating host specificity.

Chapter **4**

Pathobiological Studies

Contents

4.1. BLOOD PARASITES VIS-A-VIS HOST HEMATOLOGY

Hematological investigations have been ever increasing in importance not only in basic ichthyology but also in practical fish pathology. Hematology today encompasses many aspects of biochemistry, molecular and cellular biology, immunology, pharmacology and oncology and diagnosis by hematological techniques have proved to be an excellent aid.

Hematological techniques as an ideal tool for the diagnosis of fish disease has been widely accepted. Information regarding estimation of total erythrocyte counts, total leukocyte counts, differential leukocyte counts, PCV, ESR values and hemoglobin contents, is very limited specially in context to the changes in their values under flagellate infected conditions. The studies assume greater importance as they are being utilized as valuable clinical tools for the piscine flagellates diagnosis. The application of these techniques will assist in acting as important diagnostic tools for fishery scientists in assessing the fish health.

The hematology in different fish hosts was investigated under different conditions of parasite infectivity. Six different hematological parameters were investigated:

1. Total Erythrocyte Count (TEC)
2. Total Leukocyte Count (TLC)
3. Differential Leukocyte Count (DLC)
4. Packed Cell Volume (PCV) /Hematocrit (Hct)
5. Erythrocyte Sedimentation Rate (ESR)
6. Hemoglobin percentage (Hb%)

All the above hematological parameters were assayed during different sets of experiments. For the sake of convenience and comparison, the findings are being categorized into different groups. The different host-parasite combinations and corresponding hematological parameters investigated are as follows:

4.1.1. Host-Parasite Systems

Group	**Host**	**Parasite Parameters**	**Hematological**
I. *Clarias batrachus*	1.	*Trypanosoma batrachi* Qadri, 1962	TEC, TLC, DLC, Hb
II. *Channa punctatus*	1.	*Trypanosoma alig-aricus* (Gupta and Jairajpuri, 1982)	TEC, TLC, DLC, Hb

III. *Wallago attu*	1.	*Trypanosoma attii* (Gupta and Jairajpuri, 1982)	TEC, TLC, Hb
IV. *Cyprinus carpio*	1.	*Trypanosoma danilewskyi* (Laveran and Mesnil, 1904)	Hb, PCV
	2	*Trypanoplasma golysiana* Gupta and Pilarczyk, 1994	
	3.	*Trypanoplasma poly-golysiana* Gupta and Pilarczyk, 1994	
V. *Mystus vittatus*	1.	*Trypanosoma bagroi-des* Gupta and Gupta, 1986	Hb, ESR, PCV
	2.	*Trypanoplasma ten-gari* Gupta et al., 1988	
VI. *Clarias batrachus*	1.	*Trypanosoma batrachi* Qadri, 1962	Hb, ESR, PCV
	2.	*Trypanoplasma he-matalis* Gupta and Gupta, 2010	
VII. *Het-eropneustes fossilis*	1.	*Trypanosoma kar-genensis* Gupta and Gupta, 1994	DLC
VIII. *Colisa fasciata*	1.	*Trypanosoma colisi* Gupta, 1986	DLC

4.1.2. Hematological Studies: An Approach

Diseases are the most serious limiting factors in aquaculture, because of increased density of fish in restricted water where the fish pathogens can easily transmit from one fish to another. Much of economic loss is however, preventable with proper fish health management (Kabata, 1985).

Being poikilothermic aquatic animals, the physiology of fishes changes with environment, season, sex, maturity, temperature, pollutants and other intrinsic and extrinsic factors. Therefore, the physiology of fish also alters due to the parasitization by pathogens especially by blood flagellates which are known to be metabolically dependent on the blood of their hosts (von Brand, 1973). The scope of fish hematology and its use is gradually gaining importance in order to understand the basic principles of fish physiology for their better maintenance. Specific biological characters of each fish species make the comparison of their hematological indices exceptionally difficult. Therefore, it becomes essential to investigate the hemogram of each species separately if these data are to be utilized as diagnostic tools for fish diseases. These views are supported by Hickey (1976), Joshi et al. (1980), and Kaur (1996).

Fish parasites may cause fish mortality in culture fishes where the entire fish population of pond may be killed, resulting in loss of potential food and economic loss to the culturists. The success of the implementation of various fishery development programs depends to a certain extent on the intensification of fish parasitological research, as the improvement of fish yield can mainly be achieved from healthy fish stock (Srivastava, 1975).

Fish hematology has occupied a prominent place in fishery management (Lucky, 1980). Fishery management requires establishment of healthy fish. The role of hematological values as a pre-requisite for diagnostic values has been stressed (Blaxhall, 1972; Sandness et al., 1988).

Blood is the most accessible component of the vertebrate body fluid system. Accordingly, it is the tissue most frequently examined as it attempts to assess the physiological status of an animal. The value of the classical blood variables (total erythrocyte count, total leukocyte count, differential leukocyte count, hemoglobin, erythrocyte sedimentation rate, hematocrit) and their derivatives are regarded as valid indicators of systemic response to external stimulus and parasitic invasion and any changes are therefore reflected in their morphology and distribution in the blood (Srivastava and Choudhary, 2010). The latter authors also assessed the effect of artificial photoperiod on the hematological values of adult *Clarias batrachus*. These are to be assessed from the view point of measurement precision and prevention of distortion of information content by sampling procedures and storage condition. Metabolic disorders and dysfunction are physiological manifestations of the pathological state of the fish.

When and in what manner blood was first examined is unknown. However, in the past 50 years, hematology has become a broad science, which, in seeking to understand the normal and pathologic physiology of hematopoietic system employs all the methodologies of diverse scientific disciplines.

The reasons for the growing interest in hematological studies of fish blood is due to its diagnostic potential in fish toxicology and fish pathology. It is thus essential to generate a library of information by collecting basic data related to hematological variation in different fish species in relation to ecological, environmental and physiological conditions to promote future comparison of hematological data which is necessary for the proper health management of fish biota.

The investigations of blood parameters, physiological profiles and morphological/cytochemical characterization of blood cells in fish plays a significant role in fish health assessment and therefore, any exo and/or endogenous effect can sharply reflect the fish heme-indices. They can therefore be crucial in providing vital information on the general well-being of fish (Barton and Iwama, 1991; Wendelaar Bonga, 1997; Tavares-Dias and Morares, 2007). Stoskopf (1993) considered evaluation of blood cells, blood biochemistry to be useful for the diagnosis of diseases and for evaluating the physiological status of fish. Hematological studies on Indian freshwater fishes have been conducted by Bhatt and Singh (1981), Murad et al. (1990), among others. Houston (1997) reviewed the role of hematological variables in the assessment of fish health.

4.1.2.1. Fish Blood

Fish blood consists of cellular components suspended in plasma. The cellular components of fish blood comprise of red blood cells (RBC), white blood cells (WBC) and thrombocytes, which are designated as formed elements (Ross and Reith, 1985). Erythrocytes are elliptical, and as such have two different diameters: CLD ("cell long diameter") and CSD ("cell short diameter"). NLD ("nucleus long diameter") and NSD ("nucleus short diameter") are similarly used for nuclear dimensions. Unlike humans, fish red cells are nucleated having small black dots at the middle of each cell. Hemoglobin is predominantly present in the cytoplasm of the erythrocytes (Dube and Munshi, 1973), an iron-containing biomolecule that can bind oxygen and the blood owes its typical red color due to this pigment. It takes a pink color with Giemsa stain. The nucleus occupies a central position under normal conditions and may be round or oblong in shape. The primary function of the erythrocyte is to serve as a carrier of the respiratory pigment, hemoglobin. Functionally, these cells take up oxygen from the gills and release it while squeezing through the body's capillaries.

4.1.2.2. Fish Hemogram

Hemograms of man are routinely utilized for adjudging the pathological state. But if the fish population is to be saved in order to achieve the goals of "Blue revolution," the hemograms of fishes have to be designed properly to assess the health status of the fish.

Realizing the above, hemograms are increasingly being used by fishery biologists as a basis for the evaluation of dietary adequacy for the demonstration of pathological status, the assessment of physiological response to stress conditions and for the establishment of hematological norms for various species (Khan, 1974).

In fishes, the pathogens are being commonly studied but hematological responses of fishes on natural and experimental infection and pathogenecity are comparatively rare (Cottrell, 1977; Dubey and Pandey, 1985; Islam and Woo, 1991a; Gupta et al., 2003). Siddique and Naz (2009) investigated the hematological changes in fishes infected with helminth parasites from the fresh waters near Jhansi and observed conspicuous fall in the TEC, Hb%, PCV and MCHC and a rise in the TLC, ESR and MCV values of infected fish.

The pathogenic potential of fish trypanosomes was not appreciated until relatively recently. Depending on the intensity, the infection induces a series of changes in the organism:

1. Changes in blood values, such as a decrease in serum protein levels, increase in leukocyte but decrease in erythrocyte number and hemoglobin content, increase in serum globulins. Anemia is induced by hemolysins secreted by live trypanosomes which lyse the RBC, and by hemodilution, which is a result of generalized oedema.
2. Changes in somatic indices and condition factors.
3. Histopathological changes, which may be transient or irreparable.
4. Eventual mortality.
5. Looking into the above necessity for estimating the pathogenic potential of blood parasites in fish, the present venture was undertaken to observe the hematology and biochemistry of fish blood vis a vis blood parasite invasion.

4.1.3. Blood Parasites Vis-a-Vis Total Erythrocyte Count (TEC)

Total Erythrocyte Count (TEC) is the number of red cells per volume of blood, measured in microliters (µL) or cubic millimeters (mm^3). Red blood cells, also known as RBC's, red blood corpuscles, hematids, erythroid cells or erythrocytes (from Greek *erythros* for "red" and *kytos* for "hollow," with *cyte* translated as "cell" in modern usage) are the most common type of blood cell and are the principal means of delivering oxygen (O_2) to the body tissues via the blood flow through the circulatory system in vertebrates.

Swammerdam was the first to observe erythrocytes in 1654 who called them as "ruddy globules" but they were first accurately described in human blood in 1674 by the Dutch microscopist, Leeuwenhoek. However, the functional significance of red cells was appreciated only when Hoppe-Seyler demonstrated that hemoglobin has the property of readily taking up and discharging oxygen (Wintrobe, 1981).

For evaluating this important parameter in fish due to trypanosome invasion, the following host-parasite systems were taken into consideration.

4.1.3.1. Host-Parasite Systems

Parasite	Host	Host Site	Group
Trypanosoma batrachi Qadri, 1962	*Clarias batrachus*	Aligarh	IA Control IB Infected
Trypanosoma aligaricus Gupta and Jairajpuri, 1982	*Channa punctatus*	Aligarh	IIA Control IB Infected
Trypanosoma attii Gupta and Jairajpuri, 1982	*Wallago attu*	Aligarh	IIIA Control IIIB Infected

Group I

- **Parasite:** *Trypanosoma batrachi* (Qadri, 1962)
- **Host:** *Clarias batrachus*

Group II

- **Parasite:** *Trypanosoma aligaricus* (Gupta and Jairajpuri, 1982)
- **Host:** *Channa punctatus*

Clarias batrachus and *Channa punctatus* collected from Chautal pond, Aligarh were transported alive to the laboratory. They were allowed to rest in order to recover from the stress caused by transportation. Care was taken that the experimental fish were of the same size and age. After careful observations, blood films of the uninfected and infected fish were separated and comparative evaluation on blood indices of the two groups performed.

The blood of *Clarias batrachus* revealed the presence of *Trypanosoma batrachi* (Qadri, 1962) and that of *Channa punctatus*, *Trypanosoma aligaricus* (Gupta and Jairajpuri, 1982).

TEC was lower in both the species of infected fishes as compared to the values in uninfected fish. The TEC value of $2.81 + 0.16 \times 10^6$/c mm in uninfected *Clarias* declined to $2.01 + 0.20 \times 10^6$/c mm in trypanosome-infected fish indicating a 28.46% fall whereas that of *Channa*, it fell from the value of 2.41 + 0.31 × 10^6/c mm in uninfected fish to 1.61 + 0.23 × 10^6/c mm in trypanosome-infected fish indicating a 33.19% fall (Table 4.1).

Table 4.1. Changes in Total Erythrocyte Count (TEC) and Total Leukocyte Count (TLC) in Fresh Water Fishes Infected with Flagellates

Parasite	Host	Status of Host/ Groups	TEC (×106/c mm)	TLC (/c mm)
Trypanosoma batrachi Qadri, 1962	*Clarias batrachus*	Control (10) IA Infected (10) IB	2.81 + 0.16 2.01 + 0.20	20115 + 1102 27110 + 2005
		% Change	**(-) 28.46**	**(+) 34.70**
Trypanosoma aligaricus Gupta & Jairaj-puri, 1982	*Channa punctatus*	Control (10) IIA Infected (10) IIB	2.41 + 0.31 1.61 + 0.23	17756 + 2116 22120 + 1180
		% Change	**(-) 33.19**	**(+) 24.50**
Trypanosoma attii Gupta & Jairajpuri, 1982	*Wallago attu*	Control IIIA Infected IIIB	2.48 + 1.10 1.69 + 0.90	19552 + 958 24055 + 753
		% Change	**(-) 31.85**	**(+) 23.03**

Group III

- **Parasite:** *Trypanosoma attii* (Gupta and Jairajpuri, 1982)
- **Host:** *Wallago attu*

Wallago attu (25–36 cms) were collected from the freshwater ponds in and around Bareilly and from the river Ramganga. They were subsequently maintained in tanks (100 × 80 × 50 cms) and the water was kept well aerated. The fish were held for one month prior to experimental infection and during this period, their blood was examined for trypanosome prevalence. The lethargic and apparently diseased fish (ts-infected) were separated and the physically damaged or abnormal fish were discarded. The fish were fed natural food supplemented with artificial fish food.

For the confirmation of etiological agent, an experiment was set up consisting of three tanks containing 10 fish each. Aquarium 1 and 2 contained uninfected fish, while aquarium 3 had heavily infected (15–24 forms/20 minutes scan) fish. Aquarium 1 fish (Group I) served as the control and fish of aquarium 2 (Group II) were inoculated with 1 ml of trypanosome infected blood taken from fish of aquarium 3 (Group III) under sterile and aseptic conditions. Blood from fish of Group I (10 samples) and Group II (10 samples) was examined at an interval of five days each to determine Hb concentration and cell counts.

The first microscopic scan on the blood of 50 fish revealed active, wriggling trypanosomes identified as *Trypanosoma attii* (Gupta and Jairajpuri, 1982) in 15 apparently lethargic and diseased fish at intensities ranging from 3–15 forms/20 minutes scan (low) to 15–24 forms/20 minutes scan (high).

TEC values fell from $2.48 + 1.10 \times 10^{6}$/c mm to $1.69 + 0.90 \times 10^{6}$/c mm in Group IIIB at day 30 p.i. The decline in TEC was 31.85% (Table 4.1). The flagellate- infected films also showed various abnormal erythrocyte cells [vacuolated erythrocytes (Plate LIIIA) macrocytes, punctuated erythrocytes and enucleated erythrocytes showing karyorrhexis (Plate LIIIB,C)].

The maximum fall in TEC was recorded in *Channa punctatus* infected with *T. aligaricus* (33.19%) (Plate LIVA) amongst the three species of fishes examined followed by *W. attu* (31.85%) and *Clarias batrachus* (28.46%) (Plate LIVB).

4.1.3.2. Discussion

Fish infected with blood parasites indicate decreased values of TEC. Smirnova (1970) reported low TEC values in infected *Lota lota.* Tandon and Joshi (1973) and Joshi (1989b) observed a decline in TEC content of *Clarias batrachus* infected with *Trypanosoma*. Joshi and Tandon (1980) reported the mean values of TEC in 10 species of ts-infected and uninfected fish. Kumar et al. (1984) observed that TEC, Hb and PCV values declined in the diseased fish, *Schizothorax plaglostomus* infected with Black Spot Disease whereas ESR and TLC values increased. TEC values fell by 35.0% in *Nemacheilus rupicola* (Joshi, 1985). Lowered blood values after infection of trypanosomes in hill-stream fish *Garra gotyla* has also been reported (Joshi, 1989a). The prevailing anemia, symptomatic of metabolic dysfunction among higher vertebrates is due to the metabolic strain consequential to parasite infection. Sharma and Joshi (1991) reported 21.96% fall in TEC in *Garra gotyla*. Rauthan et al. (1995) studied TEC in a hillstream fish, *Barilius bendelisis* recording a low value of $2.0 + 0.15 \times 10^{6}$/c mm in trypanosome infected fish as compared to $3.0 + 0.40 \times 10^{6}$/c mm in uninfected ones. Saha et al. (1997) also reported a decline in TEC values of *Heteropneustes fossilis* infected with *T. mukundi*. Gupta et al. (2001) recorded the TEC content in uninfected (Group A), *Trypanosoma*-infected (Group B) and a condition of *Trypanosoma* + *Trypanoplasma* infection (Group C) in *Clarias batrachus*. They recorded a maximum change (24.05% fall) in TEC of Group C as compared to other indices examined (Hb-14.48% fall; TLC 14.22% rise; PCV-8.85% fall) indicating that TEC was the most vulnerable hematological component to flagellate infection. Gupta et al. (2003) observed that the control values of $2.76 \pm 0.04 \times 10^6$ /c mm of blood (uninfected) decreased to $2.44 \pm 0.03 \times 10^6$ /c mm of blood (*T. piscidium*- infected) in *Colisa fasciatus* and again regarded the parasite to cause anemia in its host.

A controversy exists regarding the mechanism underlying the phenomenon. A reduction in TEC values indicates that the parasites are responsible for the destruction of erythrocytes and/or inadequate erythropoiesis by which the bone marrow is affected (Richardson and Kendall, 1957). The most arguable ones are the views of George et al. (1966) who assume phagocytosis of erythrocytes in the spleen of mammals to be sufficiently increased to account for the ensuing anemia; others contend that immune factors (specially autoimmune) are contributory (McGhee, 1964). Phagocytosis of erythrocytes in the spleen may be insufficiently increased or the toxic factors released by the parasites (Maegraith, 1969) may play a decisive role for the ensuing anemia. Evidence also suggests that at least some trypanosomes generate soluble material which is capable of causing very slow lysis and sometimes, due to rapid and lasting type of mobility of the flagellates in the circulating blood, direct trauma to the erythrocytes may be considered as a possible cause of hemolysis (von Brand, 1973). Trypanosomes probably lower the oxygen carrying capacity of the blood, hamper the normal red cell production and produce distorted cells.

The distortion of erythrocytes may occur due to the alteration in the hemoglobin orientation and crystallization of hemoglobin. Micro and macrocytic anemia was conspicuous in infected fish suggesting that these pathological manifestations are under the influence of various physio-pathological conditions as reported in higher animals too (Wintrobe, 1981). Poikilocytosis, common in infected fishes depends on factors like physiological adaptation, activity of fish, cell size and micro and macro environment directly influencing the osmotic fragility of the cell. Factors for red cell destruction in fish infected with trypanosomes may lie within the red cell as well as in the plasma of infected animals (Yadav, 1990).

Islam and Woo (1991) postulated two factors to be responsible for causing anemia in gold fish which were correlated with trypanosome infection:

1. Secretion/excretion of hemolysin associated with acidosis; and
2. Hemodilution.

Firdaus et al. (1994) were of the opinion that when TEC decreases, ESR rate increases and the volume of plasma also increases which indicates low plasma viscosity allowing the cells to sediment more acutely. The significant decrease in TEC even at low parasitemia is symbolic of immediate pathological changes.

Marked reduction of TEC and Hb values in high infection further accentuates the possibility of fish mortality due to acute anemic conditions. It has been shown that TEC values also decline in *Panthera tigris* infected with trypanosomiasis (Singh et al., 2000). Such observations are in conformity with the findings of Tandon and Joshi (1973), Khan (1977) and Lom et al. (1986).

The predominant or sole cause continues to be a debatable issue, but it can be almost confirmed that trypanosomes probably lower the oxygen carrying capacity of the blood. The flagellates may hamper the normal process of erythropoiesis thereby producing distorted cells. In the view of the above facts, the presence of the trypanosomes is found to be concurrent with anemia.

4.1.4. Blood Parasites Vis-A-Vis Total Leukocyte Count (TLC)

The role of white blood cells in the defense mechanism of organisms has been established since long and it is evident that each of the several leukocyte cell lines has special properties and a unique role in the body defense. The details and mechanism by which the functions of leukocytes are accomplished are still incompletely understood. "Leukocyte Kinetics" deals with cell multiplication, maturation, storage and delivery via blood to the tissue and site of infection or cell damage and is different not only for each leukocyte type but varies from species to species and with the nature of infectivity. Therefore, a single standard for interpreting the leukocyte pattern during disease in all animals is not possible because the normal differential leukocyte count is not the same in all animals (Jain, 1986). The leukocyte pattern also varies from one fish species to another and parasitization further complicates the situation. TLC was estimated in similar sets of experiments as mentioned for TEC.

4.1.4.1. Host-Parasite Systems

Parasite	Host	Host Site	Group
Trypanosoma batrachi Qadri, 1962	*Clarias batrachus*	Aligarh	IA Control IB Infected

Trypanosoma aligaricus Gupta and Jairajpuri, 1982	*Channa punctatus*	Aligarh	IIA Control IIB Infected
Trypanosoma attii Gupta and Jairajpuri, 1982	*Wallago attu*	Bareilly	IIIA Control IIIB Infected

Group I

- **Parasite: *Trypanosoma batrachi* Qadri, 1962**
- **Host: *Clarias batrachus***

Group II

- **Parasite: *Trypanosoma aligaricus* (Gupta and Jairajpuri, 1982)**
- **Host: *Channa punctatus***

The method of collection, transportation, maintenance of fish and film preparation was similar to that mentioned for TEC and the procedure for TLC estimation was as given in "Materials and Methods."

The blood of *Clarias batrachus* revealed the presence of *Trypanosoma batrachi* Qadri, 1962 and that of *Channa punctatus*, *Trypanosoma aligaricus* (Gupta and Jairajpuri, 1982).

TLC was higher in both the species of infected fishes as compared to the values in uninfected fish. The TLC value of 20,115 + 1,102/c mm in uninfected *Clarias* rose to 27,110 + 2,005/c mm in trypanosome-infected fish indicating a 34.7% rise whereas that of *Channa*, it rose from the value of 17,756 + 2,116/c mm in uninfected fish to 22,120 + 1,180/c mm in trypanosome-infected fish indicating a 24.5 % rise (Table 4.1).

Group III.

- **Parasite: *Trypanosoma attii* (Gupta and Jairajpuri, 1982)**
- **Host: *Wallago attu***

The experimental protocol was similar to that mentioned for TEC. TLC values rose from 19,552 + 958/C mm in Group IIIA to 24,055 + 753/ C mm in Group IIIB at day 30 p.i. in *Wallago attu* infected with *T. attii* (Plate LIVB). The rise in TLC was 23.03% (Table 4.1).

Amongst the three species of fishes examined, the maximum rise occurred in *Clarias batrachus* infected with *T. maguri* (34.7%). The elevation in TLC in *Channa punctatus* (24.5%) and *W. attu* (23.03%) was not significantly different from each other (Plate LIVB).

Joshi (1973) observed that the TLC values increased corresponding to the decrease in TEC and Hb concentration due to trypanosomiasis in fish. Abnormal cell types have also been observed in fishes harboring trypanosomes (Tandon and Joshi, 1973). The total leukocyte count was 206% and 170% higher in *Channa gachua* and *Notopterus notopterus* respectively infected with trypanosomes as compared to the normal ones (Joshi and Tandon, 1980). Joshi (1985) recorded an elevation of 74.8% ($P > 0.001$) in trypanosome infected *Nemacheilus rupicola* and 43% rise in *Puntius ticto* (Joshi, 1989a). Sharma and Joshi (1991) recorded an increase in TLC in parasitized fish. Contrarily, Dabral (1980) observed an increase in TLC values under mild infection which decreased under heavy infection in two species of catfishes. However, these findings were not supported by subsequent workers. Often cellular immunity is consequential to parasitization expressed in the form of increased lymphopoiesis as well as an overall increase in leukocytes culminating in leukocytosis (Islam and Woo, 1991c). The TLC values were again found to increase in *Barilius bendelisis* from 10,601.30/c mm to 11,780.4/c mm in infected fish (Rauthan et al., 1995). Saha et al. (1997) reported that the TLC values increased abnormally in high infections but remained within the normal ranges in low infectivity. The TLC values in *Clarias batrachus* infected with hemoflagellates increased from 15,250/c mm to 17,156/c mm under concomitant infection of *Trypanosoma* and *Trypanoplasma* (Gupta et al., 2001). In *H. fossilis* infected with *T. artii*, the values increased from 11,631/c mm to 14,364/c mm. The data reveals that except for neutrophil concentration, all the leukocyte components elevated during parasitization: thus the increase in TLC is self- evident.

Incidentally, concrete reasons for these changes in the blood picture have not been provided so far. Trypanosome infection may produce some toxicants, not fully characterized chemically which may induce these pathogenic changes (von Brand, 1973). The present studies reveal significant ($P < 0.001$) differences in the cellular components of the blood which may be due to the interference of some by products or end products of trypanosome metabolism.

4.1.5. Blood Parasites Vis-a-Vis Differential Leukocyte Count (DLC)

Differential Leukocyte Count (DLC) also known as Leukocyte Differential Count, Peripheral Differential, White Blood Cell Morphology, WBC Differential is formally known as White Blood Cell Count Differential. It is a common hematological test useful to identify changes in the distribution of white cells which may be related to specific types of disorders. The test measures the percentage of each type of white blood cell (WBC) present in the blood and also reveals if there are any abnormal or immature cells. The leukocytes, are classified into two main groups: granulocytes and agranulocytes. The granulocytes, which include neutrophils, eosinophils, and basophils, have granules in their cell cytoplasm. Neutrophils, eosinophils, and basophils also have a multilobed nucleus. As a result they are also called polymorphonuclear leukocytes or "polys." The nuclei of neutrophils also appear to be segmented; therefore, they are also called as segmented neutrophils or "segs." The nongranuloctye white blood cells, lymphocytes and monocytes, do not have granules and have nonlobular nuclei. They are sometimes referred to as mononuclear leukocytes.

DLC is a popular clinical test for disease diagnosis in humans. It is a useful clinical examination to detect morbid situation in an animal. However, it is not very popular but equally demanding for the diagnosis of fish diseases. Therefore, the present studies were undertaken to make a clinical diagnosis of fish based on the results of this test. The Differential Leukocyte Count (DLC) in four species of freshwater fishes, *Clarias batrachus, Channa punctatus, Colisa fasciatus,* and *Heteropneustes fossilis* was evaluated as per the following host-parasite association.

4.1.5.1. Host-Parasite Systems

Parasite	**Host**	**Host Site**	**Group**
Trypanosoma batrachi Qadri, 1962	*Clarias batrachus*	Aligarh	IA Control IB Infected
Trypanosoma aligaricus (Gupta and Jairajpuri, 1982)	*Channa punctatus*	Aligarh	IIA Control IIB Infected
Trypanosoma colisi Gupta 1986	*Colisa fasciatus*	Bareilly	IIIA Control IIIB Infected
Trypanosoma kargenensis Gupta and Gupta, 1994	*Heteropneustes fossilis*	Bareilly	IVA Control IVB Infected

The method of collection, transportation, maintenance of fish and film preparation was similar to that mentioned for TEC and the procedure for TLC estimation was as given in "Materials and Methods."

The blood of *Clarias batrachus* was infected with *Trypanosoma batrachi* (Qadri, 1962), *Channa punctatus* with *Trypanosoma aligaricus* (Gupta and Jairajpuri, 1982), *Colisa fasciatus* with *Trypanosoma colisi* (Gupta, 1986) and *Heteropneustes fossilis* with *Trypanosoma kargenensis* (Gupta and Gupta, 1994).

The trend of change in DLC values followed the same pattern in all the species of fishes sampled (Plate LVA-D).

4.1.5.2. Lymphocytes

Lymphocytes of *Clarias batrachus* increased by 17.1% (uninfected: 46.1 + 4.9; infected 54.0 + 3.1), in *Channa punctatus* by 18.3% (uninfected: 46.2 + 2.1; infected 54.7 + 3.7), in *Colisa fasciatus* by 17.5% and in *Heteropneustes fossilis* by 23.1% in infected fishes as compared to the values in uninfected fish. The maximum rise was recorded in *H. fossilis* (23.1%) (Table 4.2). In ts-infected blood films of *H. fossilis*, lymphocytosis was clearly visible (Plate LVIA).

4.1.5.3. Monocytes

These agranular leukocytes showed a conspicuous rise in infected fishes, 34.3%, 31.1%, 64.8% and 25.1% in *Clarias, Channa*, *Colisa* and *Heteropneustes* respectively. Maximum change was recorded in *Colisa* (64.8%).

4.1.5.4. Neutrophils

These granulocytes showed a reverse pattern to that of agranulocytes. The number of neutrophils declined sharply indicating neutropenia in infected fish being 13.1 + 4.0 in *Clarias* and 14.1 + 3.2 in *Channa* as compared to the uninfected values of 30.0 + 4.3 and 32.0 + 4.1 in *Clarias* and *Channa* respectively. Similarly, the control values of 34.3 + 5.1 and 30.5 + 3.5 declined to 12.6 + 2.1 and 9.7 + 2.2 in *Colisa and Heteropneustes* respectively. Maximum change was recorded in *H. fossilis* (68.1% fall) and minimum in *Channa punctatus* (55.9% fall). Although the number of neutrophils decreased, it was interesting to observe the presence of neutrophilic myelocytes in ts-infected blood (Plate LVIB). These myelocytes are characterized by an eccentric, round or oval nucleus with one side usually flattened. The nuclear chromatin is coarse and nucleoli are small and often not visible.

4.1.5.5. Eosinophils

The eosinophils showed a marked elevation in the infected fishes. The mean values indicate that eosinophils elevated most sharply in infected *Colisa* (154.5% rise) followed by *Heteropneustes* (142.8% rise), *Channa* (39.1% rise) and *Clarias* (35.2% rise). As a result, eosinophilia was a common feature (Plate LVIC).

4.1.5.6. Basophils

Basophils were usually the least frequent white blood cells observed in the blood films of infected and uninfected fishes. But in all cases, there was a rise in these leukocytes in infected blood. Maximum rise of 325% was recorded in *Channa* and minimum in *Colisa* (100%) (Table 4.2).

These observations show that in general, the percentage of agranulocytes and granulocytes is higher in infected fishes except for the neutrophils which decline significantly (P< 0.001) in both the species of infected fishes.

Table 4.2. Differential Leukocyte Counts (DLC) in Fresh Water Fishes Infected with Trypanosomes

Fish	**Parasite**	**Differential Leukocyte Counts**						
		Status of host/Group		**Lympho-cytes**	**Mono-cytes**	**Neutro-phils**	**Eosin-ophils**	**Basophils**
Clarias batrachus	*Trypanosoma batrachi* Qadri 1962	Control (10) Infected (6)	1A 1B	46.1 ± 4.9 54.0 ± 3.1	19.5± 2.1 26.2 ± 2.3	30.0 ± 4.3 13.1 ± 4.0	3.4 ± 0.3 4.6 ± 0.8	1.0±0.2 2.1 ± 1.2
		% Change		(+) 17.1	(+) 34.3	(-) 56.3	(+) 35.2	(+) 110
Channa punctatus	*T. aligaricus* Gupta & Jairajpuri 1982	Control (10) Infected (8)	2A 2B	46.2 + 2.1 54.7 ± 3.7	18.6 ± 3.2 24.4 ± 3.7	32.0 ± 4.1 14.1 ± 3.2	2.3 ± 0.5 3.2 ± 1.1	0.8 ± 0.08 3.4 ± 1.01
		% Change		(+) 18.3	(+) 31.1	(-) 55.9	(+) 39.1	(+) 325

Colisa fasciatus	*T. colisi* Gupta, 1986	Control (10) Infected (8)	3A 3B	47.8 ± 4.8 56.2 ± 5.1	14.8 ± 2.8 24.4 ± 3.3	34.3 ± 5.1 12.6 ± 2.1	1.1 ± 0.5 2.8 ± 0.3	2.0 ± 0.3 4.0 ± 0.8
		% Change		(+) 17.5	(+) 64.8	(-) 63.2	(+) 154.5	(+) 100
Hetero – pneustes fossilis	*T. kargenensis* Gupta & Gupta 1994	Control (10) Infected (8)	4A 4B	44.9 ± 3.1 55.3 ± 2.8	21.1 ± 3.1 26.4 ± 2.1	30.5 ± 3.5 9.7 ± 2.2	2.1 ± 0.4 5.1 ± 0.3	1.4 ± 0.3 3.5 ± 0.7
		% Change		(+) 23.1	(+) 25.1	(-) 68.1	(+) 142.8	(+) 150

4.1.5.7. Discussion

The science of immunology is expanding by leaps and bounds. The lymphocytes play a pivotal role in the initiation and execution of the immune response. The increasing recognition of immunological disorders has prompted in-depth studies of lymphocyte physiopathology.

Lymphocytes are a heterogeneous group of cells both morphologically and functionally. Apparently, the lymphocytes divide spontaneously upon parasitic invasion due to their immunogenic properties, thus accounting for an increase in concentration.

Lymphocyte numbers are known to fluctuate according to the physiological condition of the fish (Klontz, 1972). Decreased lymphocyte numbers have been observed under stressed conditions – hypoxia, cortisol induced or during handling and transport (Ellsaesser and Clem, 1986, 1987). Lymphocyte numbers were also found to be diminished in trouts exposed to constant illumination (Valenzuela et al., 2008). These results infer that the number of lymphocytes decline under stressful condition of the fish. However, ts-infected fish had higher small lymphocyte counts in *Garra gotyla* (Sharma and Joshi, 1991) and *Barilius bendelisis* (Rauthan et al., 1995) indicating that immunological response against trypanosomes may account for the rise in lymphocytes and the present results fall on the same lines.

Monocytes are considered to be components of Mononuclear Phagocyte System (MPS). Contributions on the origin, function and fate of the blood monocyte have expanded considerably during the recent years. Monocytes are derived from hematopoietic stem cell in the bone marrow and shortly after entering the circulation, they migrate in to various tissues and body cavities to become either "fixed" or "free" macrophages.

Neutrophils form the first line of cellular defense against any infection. The functions of neutrophils have become better defined during the recent years and more has become known about their pathophysiology. Neutrophil kinetics in infection varies with the degree, type and duration of the disease process. It has been postulated that the reduction in neutrophil numbers (neutropenia) is due to the increased marginalization and migration in to tissues (Walker and Willemze, 1980). As noticed above, neutrophic myelocytes were quite frequent in the blood films. It is quite possible that the development of the neutrophils upto the myelocytes is normal in such fish but the presence of trypanosomes somehow prevents the maturation of these myelocytes into neutrophils causing the prevailing neutropenia.

The last decade has witnessed extensive research on eosinophils to elucidate mechanisms of eosinophilia commonly associated with allergic diseases, controls of eosinophil production and eosinophil functions. This research has revealed an unifying concept of the genesis of eosinophilia seen in diverse diseases and the role of eosinophils in controlling infection.

Eosinophil-mediated destruction of parasites is an important mechanism of protective parasitic immunity. Observations have been supported from *in vitro* and *in vivo* studies on several parasites including blood stream form of *T. cruzi* and *T. theileri* (Sullivan, 1979; Van Dallen, 1981).

Eosinophils have been assigned the function of detoxification and it has been suggested that they inactivate the histamine or histamine-like toxic substances (Vaughn, 1953). Hemoflagellates produce some toxic substance in blood and these eosinophils serve the purpose of detoxification in infected fish. The basophilic granulocytes are also said to be heparinocyte, which in higher mammals release the anticoagulant material at the time of requirement (Sherburne, 1973).

Basophils are numerically insignificant but functionally important. Due to its scarcity in blood, it has received less attention. Newer approaches are enhancing our understanding of basophils and their physiological importance. These cells do not functionally represent an end-stage; they are capable of resynthesizing their granules and perhaps have the potential to undergo blast transformation under appropriate conditions (Dvorak, 1982).

NLI and MLI are employed for evaluating the activity and intensity of infection. When these indices are considered individually or in continuation with each other, they provide a fairly good idea about the severity and progress of infection. Unfortunately, these indices are not well documented even from normal fish hosts. The Neutrophil Lymphocyte Index (NLI) is significant for DLC values. NLI has been estimated for higher warm-blooded animals (dog, cat, horse, cattle) (Schalm, 1962) and even for man (Jain, 1986).

Sharma and Joshi (1991) reported conspicuous changes in trypanosome infected blood in the hill stream fish, *Garra gotyla*. The number of large and small hemoblasts, small lymphocytes, basophils and eosinophils rose but the number of neutrophils and thrombocytes fell in the diseased fishes as compared to the uninfected ones. The number of large lymphocytes like those of neutrophils also fell by a little margin in infected fish. Rauthan et al. (1995) also observed noticeable changes in DLC in *Barilius bendilisis* infected with trypanosomes. The number of hemoblasts, small lymphocytes, basophils and monocytes rose while the number of neutrophils, eosinophils and thrombocytes fell.

It is now agreed that in all the five vertebrate taxa including fish, natural stressors or exogenous stressors in the form of parasites elicit a stress response in their leukocyte profile (Davis et al., 2008). From an ecological standpoint, change in ratio between lymphocytes and neutrophils reliably suggest stressful conditions of the environment, and therefore serve to assess stress. It is well known that fish are under stress due to parasitic invasion and it is therefore natural to expect that white blood cell counts undergo deviation from their normal numbers as shown in the present study.

4.1.6. Blood Parasites vis-a-Vis Packed Cell Volume (PCV)/Hematocrit (HCT)

The hematocrit (Ht or Hct) or packed cell volume (PCV) or erythrocyte volume fraction (EVF) is the percentage of blood volume that is occupied by red blood cells. In other words, the erythrocyte mass at the bottom obtained by the centrifugation of blood is the Packed Cell Volume (PCV) and through the years, the word hematocrit (Gk: *haima*, blood; *krinein*, to separate) has become synonymous with PCV. The term "hematocrit" comes from the German/ Swedish "hematokrit" coined by Blix at Upsala in 1891 (Hedin, 1891). The "Hematokrit" was modeled after the "Lactokrit" which was used in dairy farming. It is the most readily determined, commonly measured and widely employed hematobiological variable. The hematocrit test provides information regarding the presence and intensity of morbid processes within the body or the presence of anemia, polycythemia, leukopenia, leukocytosis or alterations in the quantity of blood platelets and the appearance of blood plasma (Wintrobe, 1981).

Hematocrit, originally described by Wintrobe (1981) was the name applied to the tube which he designed to perform the experiment and VPRC (Volume of Packed Red Cells) was the unit of measurement. However, the term hematocrit has been commonly used to describe the measurement, rather than the instrument and refers to the hematocrit as the measurement (percentage of blood sample occupied by the erythrocytes) and is a valuable tool for diagnostic purposes. Adams et al. (1993) included Hct with other variables in their health assessment index and Allen (1993) commented hematocrit to be one of the primary diagnostic parameters for fish health. However, according to Houston (1997), Hct in fish comparable in information content to their mammalian counterparts was questionable.

During the present course of studies, PCV was calculated in Groups IV, V and VI as per the following protocol:

4.1.6.1. Host-Parasite Systems

Parasite	**Host**	**Host Site**	**Group**	
Trypanosoma danilewskyi Laveran and Mesnil, 1904		Golysz Wyszni II Baginiec III	Control	IVA(i) IVA(ii) IVA(iii)
Trypanoplasma golysiana Gupta and Pilarczyk, 1994	*Cyprinus carpio* (Group IV)	Golysz Wyszni II Baginiec III	Infected	IVB(i) IVB(ii) IVB(iii)
Trypanoplasma polygolysiana Gupta and Pilarczyk, 1994				
Trypanosoma bagroides Gupta and Gupta, 1986 *Trypanoplasma tengari* Gupta et al., 1988	*Mystus vittatus* (Group V)	Kargena, Bareilly	Control Ts-infected Tpl-infected Ts+Tpl-infected	VA VB VC VD
Trypanosoma batrachi			Control	VIA

Qadri, 1962			Mild Ts	VIB(i)
Trypanoplasma haema-	*Clarias*	Natural	Heavy Ts	VIB(ii)
talis	*batrachus*			
Gupta and Gupta, 2010	(Group VI)	infection	Ts+Tpl)	VIB(iii
		Exptl (30	Ts inoc.	VIC(i)
		days p.i.)	Tpl inoc.	VIC(ii)

Group IV

- **Parasites:**
  - o *Trypanosoma danilewskyi* (Laveran and Mesnil, 1904)
  - o *Trypanoplasma golysiana* Gupta and Pilarczyk, 1994
  - o *Trypanoplasma polygolysiana* Gupta and Pilarczyk, 1994
- **Host: *Cyprinus carpio***

Investigations were carried out on three year old carps from four ponds: Golysz, Wyszni II, Wyszni III and Baginiec III as shown in Table 9. From each pond, fish showing poor growth (s – small fish) weighing less than 600 gm as well as normal carp (b – big fish) weighing 760–940 gm were examined for parasites. Hematocrit was investigated in carps of both groups from each pond. Fish from Wyszni III pond were uninfected, therefore, investigations were conducted in fish from the remaining three ponds. PCV was investigated in the control group (Group A) and flagellate-infected (Group B) *Cyprinus carpio* collected from three different ponds: Golysz [Groups IVA(i), IVB(i)], Wyszni II Groups IVA(ii), IVB(ii)], and Baginiec III [Groups IVA(iii), IVB(iii)] of the Fish Culture Experimental Station, Golysz, Chybie, Poland. The results indicated that the values of 41.0 + 4.3, 35.5 + 3.8 and 40.5 + 4.9 fell to 32.0 + 3.1, 30.3 + 4.3 and 33.0 + 3.0 indicating 21.95%, 14.64% and 18.51% fall in the three ponds respectively (Plate LVII). Moreover, the fall was maximum in carps of Golysz pond and the average decline was 21.95% (Table 4.3). It was further observed that the hematocrit values in 's' fish were lower than the 'b' fish thereby confirming the pathological status of 's' fish.

Group V

- **Parasites:**
  - o ***Trypanosoma bagroides* Gupta and Gupta, 1986**
  - o ***Trypanoplasma tengari* Gupta et al., 1988**
- **Host: *Mystus vittatus***

Fishes from Kargena pond, Bareilly were transported alive and acclimatized to the laboratory conditions. Hanging drop preparations and hematocrit were employed for parasite detection and permanent smears prepared as given in "Techniques."

The fishes were divided into four sub-groups for the investigation of hematological indices which were calculated according to Wintrobe (1981) as follows:

Group VA: Control

Group VB: Trypanosome- infected

Group VC: Trypanoplasm- infected

Group VD: Concomitant *Mystus vittatus* ranging in size from 7–9 cms were prone to hemoflagellate infection revealing *Trypanosoma*, *Trypanoplasma* and 11% concomitant infectivity in the sampled fishes. The infections were moderate but the trypanoplasms outnumbered the trypanosomes in the same sample under concomitant conditions. The intensity of trypanosomes was lower in Group VD as compared to Group VB.

The PCV values furnish valuable criterion of the degree of anemia or polycythemia. These values suffered maximum depletion in Group VD fishes making them more prone to anemia. Accelerated ESR and lower PCV values are characteristic of anemia in man also (Oser, 1965).

The PCV value in Group VA (control) was 25.0 + 6.6 under ts-infected conditions (Group VB), the fall was not of much significance (8.0%) where the value declined to 23.0 + 3.8 (Plate LVII). However, with a further fall, the values declined to 13.4 + 3.7 in Group VC indicating 46.4% fall. Maximum decline of 59.2% (Table 4.3) was recorded in *Mystus* having mixed infection of *Trypanosoma* and *Trypanoplasma* (Group VD) (Plate LVIIIA).

Group VI

- **Parasites**
  - ***Trypanosoma batrachi (*Qadri, 1962);**
  - ***Trypanoplasma hematalis (*Gupta and Gupta, 2010);**
- **Host: *Clarias batrachus***

Live *C. batrachus* (n = 110) (20–26 cms. in length) were obtained from the local fishing grounds, acclimatized to laboratory conditions and hanging drop preparations and hematocrit techniques were employed for parasite scan. Thin blood smears were prepared and stained as given earlier. The photomicrographs were taken under oil immersion in LEICA-DMLB photoautomat microscope.

***Groups*: For the assessment of hematological parameters, fishes were divided into the following groups and sub-groups: Group VIA – control Group VIB – Natural Infection B(i) – Mild Ts – infected B(ii) – Heavy Ts – infected B(iii) – Ts + Tpl – infected Group VIC – Experimental infection C(i) – Ts – inoculated C(ii) – Ts + Tpl inoculated *Inoculum*. Recipient fish were maintained in glass aquaria under favorable laboratory conditions of food, oxygen and temperature. It was ensured that Group VIC fishes were free from flagellate infection by examining them twice a week for 30 days prior to inoculation. Three ml of cardiac blood was collected in a heparinized syringe from the infected donors (Group VIBii for Group VIC (i); and Group VIBiii for Group VIC (ii)), pooled for each group separately and inoculated intraperitoneally into the fishes of Group VIC(i) and VIC(ii) respectively. Controls for each group were maintained throughout the experimentation duration. After inoculation, the blood of recipient fishes was examined after every 24 hours for 21 days to detect flagellates.

PCV of *C. batrachus* showed a significant fall ($P < 0.01$) under concomitant (Group VIB(iii)) and under heavy infection of monoflagellates (Group VIB(ii)) as compared to Group VIA. PCV showed no significant change in fish of Group VIB(i) being 40.8 ± 0.05 as compared to 44.0 ± 2.51 of Group A fish. The maximum decline (65.45%) was evident in Group VIB(iii) (Plate LIXA) whereas that in the experimental groups was 12.50% (Group VIC(i)) and 26.59% (Group VIC(ii)) (Table 4.3) (Plate LX).

From the present experiment, trypanosomes may be responsible for the depletion of PCV index finds justification from the fact that the value of this parameter correspondingly decreased in Group VIB(i) (mild trypanosome infection), Group VIB(ii) (heavy trypanosome infection) and Group VIB(iii) (concomitant infection). Natural infectivity values are supported by the experimental values where the trend was similar, though not in competence with the changes under natural infectivity.

4.1.6.2. Discussion

Studies on PCV values of fishes under flagellate infectivity are comparatively scant (Pandey and Pandey, 1976; Dabral, 1980; Raizada and Singh, 1982; Yadav, 1990) as compared to other hematological parameters. Khan (1985) investigated the pathogenesis of *Trypanosoma murmanensis* in four species of marine fish of the north western Atlantic. In sculpins (control 29.0 + 2.4; infected 16.0 + 2.0) and flounder (control 32.0 + 4.4; infected 18.0 + 2.3) hematocrit was significantly lower in infected fish as compared to the control. Attempts to observe Hct changes under varying degrees of infectivity have also been made. The PCV values have not only been recorded to decline in fishes (*Clarias batrachus*) due to trypanosomes (Joshi, 1989b) but also in fishes infested with leeches (Joshi, 1989a), the transmitive agents for trypanosomes in fishes thereby indicating that a fall in PCV values is indicative of parasitemia. Rauthan et al. (1995) detected changes in PCV in *Barilius bendelisis* infected with trypanosomes (control 38.0 + 3.6; infected 29.0 + 2.2). Saha et al. (1997) observed a sharp, highly significant reduction in Hct values reflecting high parasitaemia. Gupta et al. (1998) recorded a significant fall ($P<0.001$) under trypanosome-infected condition (uninfected 37.20 + 0.30; mild infection 31.66 + 0.55; heavy infection 23.77 + 0.66) in *Colisa fasciatus*.

PCV changes under concomitant infectivity with *Trypanosoma* and *Trypanoplasma* have also indicated variations. Gupta et al. (2001) reported decline in PCV values in *Clarias batrachus* (uninfected 42.25; Ts-infected 41.76; Ts + Tpl 38.51). In *H. fossilis*, the PCV values of 38.25 declined to 34.05 in infected fish (Gupta et al., 2002).

The findings of various authors as indicated above are suggestive of the fact that: trypanosomiasis in fishes causes a reduction in Hct/PCV values; the fall in Hct is greater due to trypanoplamosis as compared to trypanosomiasis and under concurrent conditions, the fall is again greater as compared to lone infectivity.

PCV values below the normal range are indicative of anemia, the PCV values declined in parasitized hosts inferring that the parasites maybe regarded as the etiological agents of anemia.

Further, a direct relationship is normally found to exist between Hb concentration and PCV. It has also been indicated that the ratio of PCV to Hb concentration is 3:1 (Jain, 1986). In the present findings also, the ratio is almost similar or higher supporting the hematological relationship cited above.

4.1.7. Blood Parasites Vis-a-Vis Erythrocyte Sedimentation Rate (ESR)

The origin of Erythrocyte Sedimentation Rate (ESR) dates back to 1918 when Fahraeus while seeking an early test for pregnancy, discovered that the suspension stability of the blood is altered during pregnancy. He also found that the speed of sedimentation of red cells is accelerated in many diseases (Fabrikant, 1972). A test was thus discovered which, because of its simplicity and wide applicability soon became popular. ESR still continues to be a useful test. Although the sedimentation rate is increased in a wide variety of diseases and is thus quite non-specific, nevertheless, it constitutes a valuable screening procedure for detecting the presence or absence of active disease.

With an objective to infer whether similar conditions of ESR change ensues during flagellate infection in fish, observations on ESR were made in Groups V and VI as follows.

4.1.7.1. Host-Parasite Systems

Parasite	Host	Host Site	Group	
Trypanosoma bagroides Gupta and Gupta, 1986 *Trypanoplasma tengari* Gupta et al., 1988 *Trypanosoma batrachi* Qadri, 1962	*Mystus vittatus*	Kargena, Bareilly	Control Ts-infected Tpl-infected Ts+Tpl-infected	VA VB VC VD
Trypanoplasma haematalis Gupta and Gupta, 2010	*Clarias batrachus*	Control Natural infection Exptl (30 days p.i.)	Mild Ts Heavy Ts Ts+Tpl Ts inoc. Tpl inoc.	VIA VIB(i) VIB(ii) VIB(iii) VIC(i) VIC(ii)

Group V

- **Parasites**
  - ***Trypanosoma bagroides* (Gupta and Gupta, 1986)**
  - ***Trypanoplasma tengari* (Gupta et al., 1988)**
- **Host: *Mystus vittatus***

The ESR value in Group VA (control) was 0.5 + 0.15 mm/hr. This value rose to a maximum of 3.2 + 0.5 mm/hr in Group VC (540% rise). The rise under mixed infection (Ts + Tpl, Group VD) was 220% (Table 4.3) (Plate LVIIIB).

Table 4.3. Changes in Erythrocyte Sedimentation Rate (ESR) and Packed Cell Volume (PCV) in Fresh Water Fishes Infected with Flagellates

Parasite	Host	Status of Host Groups		ESR (mm/ hr)	PCV (Hemtocrit)
T. danilewskyi Laveran & Mesnil, 1904	*Cyprinus carpio*	**Control**	**IVA**		
		(i) Golysz	IVA(i)	-	41.0 + 4.3
		(ii) Wyszni II	IVA(ii)	-	35.5 + 3.8
		(iii) Baginiec III	IVA(iii)	-	40.5 + 4.9

Trypanoplasma golysiana Gupta & Pilarczyk, 1994 *Trypanoplasma polygolysiana* Gupta & Pilarczyk, 1994		**Infected** (i) Golysz (ii) Wyszni II (iii) Baginiec III	**IVB** IVB(i) IVB(ii) IVB(iii)	 - - - - - -	 32.0 + 3.1 30.3 + 4.3 33.0 + 3.0 **-21.95%** **-14.64%** **-18.51%**
		% Change **(i) Golysz** **(ii) Wyszni II** **(iii) Baginiec III**	 **IVB(i)** **IVB(ii)** **IVB(iii)**		
		Control	(VA)	0.5 + 0.15	25.0 + 6.6
Trypanosoma tengari Gupta et al., 1996	*Mystus vittatus*	Ts-infected Tpl-infected Concomitant	(VB) (VC) (VD)	1.12 + 0.19 3.2 + 0.5 1.6 + 0.20	23.0 + 3.8 13.4 + 3.7 10.2 + 2.8
		% Change			
Trypanoplasma tengari Gupta et al., 1988		**Ts-infected** **Tpl-infected** **Concomitant**	**(VB)** **(VC)** **(VD)**	**120%** **540%** **220%**	**-8.0%** **-46.4%** **-59.2%**
		Control	VI A	0.96 + 0.05	44.0 + 2.51
	Clarias	**Natural infection**	VIB	1.0 + 0.12	40.8 + 0.05

Trypanosoma *batrachi* 1962 Redescription Gupta & Jairaj- puri 1983	*batra-* *chus*	Mild Ts-infected VIB (i) Heavy Ts- VIB (ii) infection VIB (iii) Ts+Tpl-infected VIC **Ex-** **perimental** VIC (i) **inoculation** VIC (ii)	2.30 + 0.80 2.41 + 0.21 1.31 + 0.21 1.62 + 0.32	26.0 + 1.02 15.2 + 0.12 38.5 + 1.22 32.3 + 2.21
		Ts-inoculated		
		Tpl-inoculated		
		% Change **VI B (i)**	**+ 4.1 %**	**-7.2 %**
Trypanoplasma *haematalis* Gupta & Gupta 2010		**VI B (ii) VI B** **(iii)** **VI C (i)**	**+ 139.5** **%** **+ 151.0** **%** **+ 36.45** **%** **+ 68.7 %**	**-40.9%** **-65.4 %** **-12.5 %** **-26.5 %**
		VI C (ii)		

Group VI

- **Parasites**
  - ***Trypanosoma batrachi* Qadri, 1962**
  - ***Trypanoplasma hematalis* Gupta and Gupta, 2010**
- **Host: *Clarias batrachus***

Sedimentation rate of red blood cells increased by 151.0% in the fish infected with both the flagellates (Group VIB(iii)) and 139.5% (Plate LIXB) in fish heavily infected with monoflagellates only (Group VIB(ii)) as compared to Group A fish (value- 0.96±0.05 mm/hr). The increase in Group VIC (i) was 36.45% and that in VIC(ii), 68.75% (Table 4.3) (Plate LXB).

4.1.7.2. Discussion

In the first instance, a significant increase in ESR ($P < 0.01$) was evident in parasitized *Mystus vittatus* and the findings are concurrent with those of Schumacher et al. (1956) in the brook trout suffering from furunculosis and Preston (1960) noted the same in *Pleuronectes* showing necrosis of the caudal fin. However, Khan (1977b) observed insignificant changes in ESR in *Gadus morhua* infected with *T. murmanesis*.

In the second instance, ESR increased by 151.0% in Group B(iii) and the increased viscocity was accomplished by a decreased blood flow. Rauthan et al. (1995) recorded a rise in the ESR level of ts-infected *Barilius bendilisis* where the values of 1.65 + 0.18 mm/hr rose to 2.4 + 0.25 mm/hr.

The rate of erythrocyte sedimentation depends upon the interaction between opposing physical forces. Settling occurs because the density of the red corpuscles is greater than the density of the medium. The fall of the red corpuscles causes an upward displacement of the medium, thus producing an upward current and a retarding force. Under normal conditions, the downward and upward forces are nearly equal, and consequently little settling occurs. When the concentration of red cells is less than normal, there is less retardation of sedimentation caused by the red cells themselves (Wintrobe, 1981). ESR is significantly influenced by the concentration of certain plasma proteins and due to hemoflagellates, certain proteins increase which might account for a significant increase in the ESR of *Clarias*. Increase in ESR is probably related to physiology, hemodilution and erythropenia. It has been proposed that ESR itself can be trusted to ascertain the intensity and duration of disease in fishes (Joshi, 1989a). Gupta et al. (1998) recorded an appreciable rise in ESR in *Colisa fasciatus* infected with *T. trichogasteri* var. *fasciatae* (uninfected: 1.10 + 0.22; mildly infected: 1.62 + 0.24; heavily infected 1.95 + 0.20).

4.2. BLOOD PARASITES VIS-A-VIS HEMOGLOBIN

Hemoglobin (hem + protein), is a respiratory pigment that carries and delivers oxygen necessary for cell survival and vastly increases the binding power of the blood for oxygen. Hemoglobin is not just a single type of molecule but really a class of structurally similar molecules that vary in their structure and in their affinity for oxygen under different conditions. This conjugated protein enjoys an important position in physiology because it carries and delivers oxygen necessary for cell survival. Under physiological conditions, hemoglobin production and destruction are balanced under physiological conditions and any deviation in either one is bound to cause a severe hematological disorder. As hemoglobin has immense physiological importance as a transporter of oxygen, much research has been focused in this direction.

An erythrocyte contains 400 million molecules of hemoglobin constituting 95% of its dry weight. Faulty hemoglobin molecule production is linked to altered function and instability of the hemoglobin leading to multifarous hematological disorders. The enhancement in the knowledge of hemoglobinopathy provides an answer to many unsolved problems. More than hundred structurally distinct hemoglobin variants have been reported from humans (Harris and Kellermeyer, 1970). Fish hemoglobin is of two basic types: monomeric and tetrameric. Monomer hemoglobins consists of single-heme polypeptide molecules, each with a molecular weight of about 17,000. They are characteristic of lampreys and hagfishes (Agatha). Tetramer hemoglobins are characteristic of all "higher" fishes. They are composed of four chains of amino acids much like mammalian hemoglobins, and have molecular weights of approximately 65,000. There are many different kinds of tetrameric hemoglobins, and several kinds may be found in one fish. The significance of synthesizing more than one hemoglobin type appears to be related to the different functional properties of each, therefore, different combinations of hemoglobin types reflect adoptions to different environments or ways of life. The hemoglobin concentration was evaluated in all the six groups as follows:

4.2.1. Host-Parasite Systems

Parasite	**Host Host**	**Locality**	**Group**	
Trypanosoma batrachi Qadri, 1962	*Clarias batra-chus*	Aligarh	Control In-fected	IA
Trypanosoma aligaricus Gupta and Jairajpuri, 1982	*Channa punc-tatus*	Aligarh	Control In-fected	IIA
Trypanosoma attii (Gupta and Jairajpuri, 1982)	*Wallago attu*	Bareilly	Control In-fected	IIIA IIIB
Trypanosoma danilewskyi Laveran and Mesnil, 1904 *Trypanoplasma golysiana* Gupta and Pilarczyk, 1994 *Trypanoplasma polygolysi-ana* Gupta and Pilarczyk, 1994	*Cyprinus carpio*	Golysz Wyszni II Baginiec III Golysz Wyszni II Baginiec III	Control A Infected B	IVA(i) IVA(ii) IVA(iii) IVB(i) IVB(ii) IVB(iii)

Trypanosoma bagroides Gupta and Gupta, 1986 *Trypanoplasma tengari* Gupta et al., 1988	*Mystus vittatus*	Kargena Bareilly	Control Ts-infected Tpl-infected Ts+Tpl-infected	VA VB VC VD
Trypanosoma batrachi Qadri, 1962 *Trypanoplasma hematalis* Gupta and Gupta, 2010	*Clarias batrachus*	Bareilly	Control (A) Natural Mild Ts Infection Heavy Ts (B) Ts+Tpl Exptl Ts inoc. (30 days Tpl inoc. p.i.) (C)	VIA VIB(i) VIB(ii) VIB(iii) VIC(i) VIC(ii)

Group I

- **Parasite: *Trypanosoma batrachi* Qadri, 1962**
- **Host: *Clarias batrachus***

Group II

- **Parasite: *Trypanosoma aligaricus* (Gupta and Jairajpuri, 1982)**
- **Host: *Channa punctatus***

The experimental protocol was similar to that described for TEC for the same groups. Hb was lower in both the species of infected fishes as compared to the values in uninfected fish. The Hb values of 14.30 + 1.6 gm/100 ml in uninfected *Clarias* declined to 9.12 + 0.6 gm/100 ml in trypanosome-infected fish indicating a 36.2% fall whereas that of *Channa*, it fell from the value of 12.24 + 1.1 gm/100 ml in uninfected fish to 9.12 + 0.8 gm/100 ml in trypanosome-infected fish indicating a 25.4% fall (Table 4.4) (Plate LXI).

Group III

- **Parasite: *Trypanosoma attii* (Gupta and Jairajpuri, 1982)**
- **Host: *Wallago Attu***

A frequent cause of anemia is a reduction in the amount of Hb per unit volume of blood which may be either due to faulty production of erythrocytes or to a shortening of lifespan. The present study was designed to ascertain the etiological basis of anemia by experimentally observing the effect of *T. attii* on the Hb level, TEC and type and concentration of cells in normal and infected *Wallago Attu.* The experimental protocol and groups were similar to those mentioned for TEC.

In Group III, Hb % decreased from 11.4 + 0.48 g/100 ml (Group IIIA) to 8.0 + 0.42 g/100 ml (Group IIIB) showing a 29.8% decline at day 30 p.i. as compared to the control (Plate LXIIA). It was further observed that there was a gradual downfall in the Hb% being 11.0 + 0.45 g/100 ml at day 10 p.i., 10.3 + 0.48 g/100 ml at day 15 p.i., 9.5 + 0.52 g/100 ml at day 20 p.i., 8.3 + 0.44 g/100 ml at day 25 p.i. and finally the lowest value of 8.0 + 0.42 g/100 ml at the completion of the designed experiment (day 30 p.i.) (Table 4.5). When these values were compared to those of TEC (Plate LXIIB), the trend was found to be similar where the values of 2.32×10^6 /c mm declined to 1.6×10^6 /c mm at day 30 p.i. The observations as revealed in the experimental study intended for finding out the probability of trypanosomes in the causation of anemia are shown in Table 4.5.

Group IV

- **Parasites:**
  - *Trypanosoma danilewskyi* (Laveran and Mesnil, 1904)
  - *Trypanoplasma golysiana* Gupta and Pilarczyk, 1994
  - *Trypanoplasma polygolysiana* Gupta and Pilarczyk, 1994
- **Host: *Cyprinus carpio***

Hemoglobin was estimated in carps collected from three ponds, Golysz, Wyszni II and Baginiec III in the uninfected (Group A) and infected (Group B) fish according to the experimental set-up as mentioned for PCV.

The hemoglobin level of 10.22 + 1.1 gm/100 ml [Group IVA(i)] declined to 8.33 + 0.5 in Group IVB(i) indicating 18.49% fall; that in Group IVA(ii) from 7.63 + 0.5 to 6.55 + 0.6 gm/100 ml in Group IVB(ii) showing 14.47% fall whereas in Group IVA(iii) the values of 7.18 + 0.7 gm/100 ml fell to 6.48 + 0.4 gm/100 ml in Group IVB(iii) indicating a 9.74% fall (Plate LXIIIA). Maximum fall thus occurred in Group IVB(i) which coincided with the finding that maximum 's' (small) fish were harvested from this pond although all ponds were stocked initially under similar conditions (Table 4.4).

Group V

- **Parasites**
  - ***Trypanosoma bagroides* (Gupta and Gupta, 1996)**
  - ***Trypanoplasma tengari* (Gupta et al., 1988)**
- **Host: *Mystus vittatus***

Mystus vittatus ranging in size from 7–9 cms were prone to hemoflagellate infection revealing *Trypanosoma*, *Trypanoplasma* and 11% concomitant infectivity in the sampled fishes. The infections were moderate but the trypanoplasms outnumbered the trypanosomes in the same sample under concomitant conditions.

Table 4.4. Hemoglobin Changes in Fresh Water Fishes Infected with Flagellates

Parasite	**Host**	**Status of Host Groups**		**Hb (gm/100 ml)**
	Clarias	Control	IA	14.30 + 1.6
	batrachus	(10)	IB	9.12 + 0.6
Trypanosoma batrachi		Infected (10)		
Qadri, 1962				
		% Change		(-)36.20
T. aligaricus Gupta & Jairajpuri, 1982				
	Channa punctatus	Control (10) Infected (10)	IIA IIB	12.24 + 1.1 9.12 + 0.8
		% Change		(-)25.49
T. attii Gupta & Jairajpuri, 1982	*Wallago attu*	Control Infected	IIIA IIIB	11.4 + 0.48 8.0 + 0.42
		% Change		(-)29.80
T. danilewskyi Laveran & Mesnil, 1904 *Trypanoplasma golysiana* Gupta & Pilarczyk, 1994 *Trypanoplasma polygolysiana* Gupta & Pilarczyk, 1994	*Cyprinus carpio*	Control (i) Golysz (ii) Wyszni II (iii) Baginiec III	 IVA(I) IVA(II) IVA(III)	 10.22 + 1.1 7.63 + 0.5 7.18 + 0.7

		Infected (i) Golysz (ii) Wyszni II (iii) Baginiec III	IVB (I) IVB (II) IVB (III)	8.33 + 0.5 6.55 + 0.6 6.48 + 0.4
		% Change (i) Golysz (ii) Wyszni II (iii) Baginiec III	II III	(-)18.49% (-)14.47% (-) 9.74%
		Control		
Trypanosoma bagroides Gupta & Gupta, 1996	*Mystus vittatus*	Ts- infected Tpl-infected	(VA) (VB) (VC)	8.2 + 2.2 5.0 + 1.5 4.8 + 1.4
		Concominant	(VD)	4.2 + 1.0
Trypanoplasma tengari Gupta et al., 1988		% Change (i) Ts-infected (ii) Tpl-infected (iii) Concomi	(VB) (VC) (VD)	(-)39.02% (-)41.46% (-)48.78%
		Nant		
Trypanosoma batrachi Qadri, 1962 Redescription Gupta & Jairajpuri, 1983	*Clarias batrachus*	Control Natural infection Mild Ts-infected Heavy Ts- infection Ts+Tpl-infected Experimental inoculation	VI A VIB VIB (i) VIB (ii) VIB (iii) VIC	13.8 + 0.28 10.8 + 0.23 8.5 + 0.50 6.1 + 0.12

		Ts-inoculated Tpl-inoculated	VIC (i) VIC (ii)	11.3 + 0.47 10.6 + 0.53
Trypanoplasma hematalis Gupta & Gupta, 2010		% Change	VI B (i) VI B (ii) VI B (iii) VI C (i) VI C (ii)	(-)21.70% (-)38.40% (-)55.79% (-)18.10% (-)23.18%

Table 4.5. Hemoglobin Values (g/100 ml blood) in Normal (Group I) and Trypanosome Infected (Group II) *Wallago attu* (All values are ± SE)

Group	**Days (p.i.)**					
	5	**10**	**15**	**20**	**25**	**30**
I (Control)	11.8±0.62	11.4±0.34	11.7±0.41	11.2±0.61	11.3±.0.12	11.4±0.48
II (Ts-infected)	11.5±0.50	11.0±0.45	10.3±0.48	9.5±0.52	8.3±0.44	8.0±0.42

The intensity of trypanosomes was lower in Group VD under concomitant conditions as compared to Group VB (ts-infected only) while Hb% was lower being 4.2 ± 1.0 gm/100 ml in Group VD as compared to 5.0 ± 1.5 gm/100 ml in Group VB (ts-infected). Both the values however had depleted from the control group (Group IVA: 8.2 ± 2.2 gm/100 ml). The present results indicate that due to flagellate infection, the Hb% falls in infected fish and also the fall is greater under tpl-infected conditions (41.46% fall) as compared to ts-infection (39.02% fall). However, the maximum fall (48.78%) occurred under concomitant conditions (Group VD) even though the intensity of parasites was lower in this group indicating that trypanoplasms cause greater depletion in hemoglobin as compared to trypanosomes (plate LXIIIB).

Group VI

- **Parasites**
    - o *Trypanosoma batrachi* **(Qadri, 1962)**
    - o *Trypanoplasma hematalis* **(Gupta and Gupta, 2010)**
- **Host:** ***Clarias batrachus***

The findings of this set of experiment were most interesting because an experiment was designed to demonstrate that blood flagellates induce Hb loss and cause anemia in *Clarias batrachus* by inoculating parasites into uninfected fish. There were six sub-groups as mentioned in the beginning of this chapter consisting of one control group, three naturally infected groups and two experimental groups. The inoculum was prepared as mentioned for PCV.

Hb% declined in all the five groups of infected fish as compared to the control (Group A) and a maximum fall of 55.79 % was observed in fish of Group VIB(iii) (Group VIA – 13.8 ± 0.28 gm/100 ml; Group VIB(iii) – 6.1 ± 0.12 gm/100 ml). The percent fall in Group VIC(i) was 18.10% and in VIC(ii) 23.18% (Table 4.4) (Plate LXIVA,B).

Though Hb% declined in all the infected groups, the fall was greater under natural infectivity as compared to experimental inoculation.

4.2.1.1. Discussion

The capacity of transporting the respiratory gas is reflected by the hemoglobin content in unit volume of blood of different species of fishes as well as in specimens of different body weight of the same species. The hemoglobin content in fishes is known to be influenced by age (Aider, 1970), intrinsic activity of the fish (Love, 1980) or time and period of the year (Chanchal et al., 1979). Concentration of hemoglobin in most teleosts is said to be higher than the elasmobranchs. Active fishes with higher rate of body metabolism are reported to contain more hemoglobin in their blood than the sluggish ones (Yadav, 1990).

ATP is the major organic phosphate in the red blood cells of a wide variety of fish species. Its role as an allosteric modifier of fish hemoglobins is similar to that of 2,3-diphosphoglycerate (2,3-DPG) in mammalian red blood cells, in that it decreases the affinity of hemoglobin for oxygen. After exposure to the anoxia of high altitude or in certain conditions of anemia, mammals increase red blood cell 2,3-DPG to facilitate oxygen unloading at the tissues. By contrast, eels have been shown to decrease erythrocyte organic phosphate (that is, ATP) and increase hemoglobin–oxygen ($Hb{-}O_2$) affinity when acclimated to low environmental oxygen. Greaney et al. (1977) acclimated *Fundulus heteroclitus*, a euryhaline minnow, to hypoxic conditions and confirmed that this fish also lowered red blood cell ATP by as much as 40% suggesting that this is a general response to hypoxia among water-breathing vertebrates.

Studies on the normal hemoglobin content of fishes has attracted attention of several workers (Pradhan, 1961; Bhatt and Singh,1981; Raizada and Singh, 1982; Sharma and Joshi, 1984; Daniela et al., 2007) but those concerned with hemoglobin concentration variation due to trypanosomes still demand considerable attention. Joshi (1973) observed higher Hb content in more active fish as compared to the inactive or lethargic ones and it is well evidenced that infected fish are not active in nature.

Trypanosome-infected fish have shown various degrees of decline in hemoglobin concentration. Hb% has been reported to decline in *Clarias batrachus* and *Mystus vitatus* (Tandon and Joshi, 1973); *Nemacheilus rupicola* (Joshi, 1985) and *Tor putitora* (Joshi, 1989a). Khan (1977b) recorded a 40.42% fall in the parameter being discussed in young (16–24 cm) *Gadus morhua* infected with T. *murmanensis* whereas Khan (1985) observed 25.0–46.9% fall in cod, shorthorn sculpin and winter flounder harboring the same parasite. Joshi and Tandon (1980) reported comparatively low Hb values in infected fish recording the maximum fall in *Heteropneustus fossilis* besides other fishes examined (*Clarias batrachus, Channa punctatus, Labeo bata* and *Mystus vittatus*. Joshi and Dabral (1981) made similar observations on *H. fossilis* infected with *T maguri* and Joshi (1985) in *Nemacheilus rupicola*. In *Wallago attu*, the fall was 11.4 % in parasitized samples (Tandon, 1986). Joshi (1989a, b) made similar observations on *Nemacheilus rupicola* and *Clarias batrachus.* In *Garra gotyla,* the fall was 12.90% (Sharma and Joshi, 1991)) Rauthan et al. (1995) recorded a decline from 9.90 to 8.00 gm% in a hill stream fish *Barilius bendelisis.*

Hb concentration showed a significant fall in *Colisa fasciata* in mildly infected and heavily infected fish infected with *Trypanosoma trichogasteri* var. *fasciatae* showing a maximum fall of 40% (Gupta et al., 1998) and again, Gupta et al. (2003) recorded almost similar fall of 44% in the same fish infected with *T. piscidium*. In *Clarias batrachus*, *T. batrachi* reduced the Hb% from 13.19% to 11.28% (Gupta, Gupta, and Gupta, 2001) whereas *T. qadrii* reduced the Hb% from 13.80 + 0.48% to 7.90 + 0.60% in the same fish (Gupta, Gupta, and Gangwar, 2001). Gupta et al. (2002a) further observed that younger saccobranchid fish showed a greater depletion in Hb% as compared to the larger fish. In *H. fossilis* infected with *T. artii*, the fall in Hb% was lower (16.18%).

Hemoglobin concentration in fish of the same size but with different degrees of intensity showed a fall which was more marked in fish having heavy infection (Gangwar, 2001) which are in tune with the earlier findings of Tandon and Joshi (1973) in *C. batrachus* infected with *T. maguri* and in *Mystus vittatus* parasitized by *T. vittati*.

It has been shown for parasitic infection in fish hosts that the secretion/ excretion of hemolysin associated with acidosis and hemodilution maybe regarded as important criteria in inducing anemia (Islam and Woo, 1991a). The depletion in hemoglobin concentration appears to be concurrent to the fall in the TEC value indicating anemia due to trypanosomiasis. Investigations on the hemoglobin concentration of parasitized fish provide ample indication of anemia and such fishes suffer from asphyxiation and suffocation, the metabolic activities decrease due to the fact that such fishes are lethargic and cause consequential lowering of Hb%. Several etiological bases of anemia have been proposed. Herbert and Inglis (1973) suggested that some type of immunological reaction may cause anemia whereas Khan et al. (1980) attributed lytic factors released directly or indirectly by the parasite to be the major cause. It has also been proposed (Valli and Mills, 1980) that anemia is mainly due to hemolysis and the bone marrow is generally hyper proliferative due to parasitic burden. The rate of metabolism decreases at low temperaturc and fish became motionless with occasional slow movement, resulting in a decrease in Hb concentration (Joshi, 1985). The flagellates probably inhibit erythrocyte and hemoglobin synthesis in the host fish. It is obvious that anemic condition of the fish may tend to cause the use of some anaerobic mode of metabolism by the existing parasite within the fish.

Of the protozoal parasites which neither penetrate the red cells nor attach themselves to the erythrocyte surface, the trypanosomes are of greatest importance (Soulsby, 1976). The anemia they produce is preceded by a hemo dilution effect and plasma volumes may be increased by as much as 50% (Clarkson, 1968; Naylor, 1971; Mamo and Holmes, 1975). The characteristics of anemia are macrocytosis, reticulocytosis, normoblastic hyperplasia of the bone marrow and spleen and increased hemosiderin deposits and erythrophagocytosis in the spleen. These, together with the shortened circulating half-life of 51 Cr-labelled red cells the accelerated urinary excretion of 51 Cr from the labelled red cells and the retention of injected 59 Fe in the body are all indicative of an anemia of hemolytic origin.

The mechanism whereby red cells are "processed" for increased erythrophagocytosis in all the protozoal diseases is still a matter for speculation and there are a number of hypotheses to account for this phenomena.

1. Trypanosome antigen may become attached to the surface of the red cells and thus make it susceptible to erythrophagocytosis. The first part of the process, i.e., the attachment to the red cell has been shown to occur (Herbert and Inglis, 1973; Woo and Kobayashi, 1975).
2. Trypanosome antigen once attached to the surface of the red cell may combine with anti- trypanosome antibody from the circulation. This may make the red cells susceptible to phagocytosis or alternatively, to lysis if complement is fixed to the immune complex (Woo and Kobayashi, 1975).
3. Once complement has become activated and bound to erythrocyte

surfaces, antibodies to complement (mainly the C_3 fragment), immunoconglutinins may be produced and exacerbate the hemolytic process.

4. Another possibility which is receiving attention at the present time is that trypanosomes produce a hemolytic factor which causes direct hemolysis of the circulating red cells. This was first noted by Fiennes (1954) who found that if cells from normal bovines were incubated with infected plasma then the cells were hemolyzed. Workers in Geneva (Murray et al., 1975) have found that trypanosomes produced a small molecular weight protein which if incubated with normal red cells is capable of causing lysis. This is an attractive hypothesis because of its simplicity and can explain the rapid onset of the aemia, i.e., in 2 to 3 days in many experimental infections. It can be visualized in the case of very high parasitaemias that sufficient of this hemolysin would be produced to cause anemia; however, it is more difficult to envisage this happening where parasitemias are very low.
5. Trypanosomes are highly active and it is conceivable that they compete with the red blood cells for essential metabolites; there have been suggestions of a hypoglycemia especially in the terminal stages of the disease. This is true for the present findings as well as mentioned in the next section of the thesis. If the red cells are unable to regenerate ATP their surfaces will be altered and they may become susceptible to phagocytosis (Maruta and Mizuno, 1971).
6. The etiology of the anemia in some cases of trypanosomiasis at least has been attributed to thrombocytopenia with disseminated intravascular coagulation.

It is natural to expect that due to the presence of these parasites, the demand for more oxygen carriage within the blood stream should automatically increase because of the demand of the parasites themselves. But, the flagellates probably inhibit erythrocyte (and thus also hemoglobin) synthesis in the host fish consequently lowering the erythrocyte/hemoglobin content in the infected host. Apparently, the metabolic activity of the infected fish decreases, the anemic condition of the fish makes use of some anaerobic mode of metabolism, the fishes suffer from asphyxiation and suffocation, and these factors, maybe responsible for the ensuing anemia in infected fishes. Based on the present studies, TEC, TLC, DLC, PCV, ESR and Hb can be utilized as hemogram indices for the diagnosis of trypanosomiasis and trypanoplasmosis in fish.

4.3. BLOOD PARASITES VIS-A-VIS HOST SERUM BIOCHEMISTRY

A knowledge of chemical composition of fish is of paramount importance to evaluate its nutritive value. It is well documented that fluctuations in the biochemical constituents and metabolic activity of the host may occur as a result of parasitic infection. In higher vertebrates, protozoan parasites are known to cause immunosuppression (Kretttli and Pereira, 1981). However, studies pertaining to immunosuppression caused by protozoan pathogens in fishes are not so common. Reports on changes in blood biochemistry due to parasitic invasion are sporadic. The blood biochemistry in fish hosts was investigated in various fish hosts under different conditions of infectivity. Three different biochemical parameters were investigated:

1. Glucose (Chapter: Blood parasites and serum glucose)
2. Total Serum Protein (TSP) (Chapter: Blood parasites and TSP)
3. Cholesterol (Chapter: Blood parasites and serum cholesterol)

All the above parameters were assayed during different sets of experiments. For the sake of convenience and comparison, the findings are being categorized into different groups. The different host-parasite combinations and major groups are as follows:

Host-Parasite Systems		
Group	**Host**	**Parasite**
I	*Wallago attu*	*Trypanosoma attii* (Gupta and Jairajpuri, 1982)
II.	*Hetero-pneustes fossilis*	*Trypanosoma singhii* (Gupta and Jairajpuri, 1981)
III.	*Channa punctatus*	*Trypanosoma aligaricus* (Gupta and Jairajpuri, 1982)
IV.	*Cyprinus carpio*	1. *Trypanosoma danilewskyi* (Laveran and Mesnil, 1904) 2. *Trypanoplasma golysiana* (Gupta and Pilarczyk, 1994) 3. *Trypanoplasma polygolysian*a (Gupta and Pilarczyk, 1994)
V.	*Clarias batrachus*	1. *Trypanosoma batrachi* (Qadri, 1962) 2. *Trypanoplasma maguri* (Gupta and Gupta, 1987)

4.3.1. Blood Parasites Vis-à-Vis Serum Glucose

Hemato-indices are valuable indicators of the pathogenic state of the host. Alterations in these indices are known to occur due to trypanosomes (Sharma and Joshi, 1991; Gupta et al., 2000) as well as trypanoplasms (Lom and Dykova, 1984; Gupta et al., 1990). Trypanosome infectivity may alter and interfere with the normal biochemical make-up and physiology of the blood of various small and large warm blooded animals (Ikeziani, 1946; Hoppe and Chapman, 1947; Marciacq and Seed, 1970). Piscine blood is subjected to variable degrees of changes due to inhabitation by trypanosomes (Smirnova, 1970; Tandon and Joshi, 1974; Gupta and Jairajpuri, 1983; Tandon and Masih, 1983). This part of the research work reports the effect of *Trypanosoma attii* (host *Wallago attu*), *T. singhii,* (host *Heteropneustes fossilis*), *T. aligaricus* (host *Channa punctatus*), *T. batrachi, Trypanoplasma maguri* (host *Clarias batrachus*), and *T. danilewskyi, Trypanoplasma golysiana, T. polygolysiana,* (host *Cyprinus carpio*) on the blood glucose level of their respective host.

4.3.2. Methodology

Fishes of the same size range (*W. attu* – 25 to 30 cms, *H. fossilis* – 18 to 22 cms, *Channa punctatus* – 16 to 20 cms, *Cyprinus carpio* – 18.5 to 24.5 cms) *Clarias batrachus* – 20 to 25 cms) were collected from the fresh water ponds of Bareilly, India. At Poland, investigations were carried out on three year old carps from four ponds (Table 9). From each pond, fish showing poor growth (s – small fish) weighing less than 600 g as well as normal carp (b – big fish) weighing 760 to 940 g were examined for pathogens and parasitological infection.

Blood was collected from the caudal vein and glucose content was determined as given in "Materials and Methods."

In order to estimate the effect of biflagellates on the glucose concentration of its host, it was ensured that the experimental animals were free from other protozoans and microbial infection.

4.3.2.1. Results

For the sake of convenience, the observations are being arbitrary divided into the following groups:

Experimental Groups			
Group	**Parasite**	**Host**	**Infectivity**
IA	Nil	*Wallago attu*	Uninfected

IB	*Ts. attii*	Ts-infected	
IIA	Nil	*Heteropneustes fossilis*	Uninfected
IIB	*Ts. singhii*	Ts-infected	
IIIA	Nil	*Channa punctatus*	Uninfected
IIIB	*Ts. aligari-cus*	Ts-infected	
IVA (i) Golysz	Nil	Uninfected	
(ii) Wyszni II	Nil	Uninfected	
(iii) Baginiec III	Nil	*Cyprinus carpio*	Uninfected
IVB (i) Golysz	Ts +/ Tpl	Ts +/ Tpl	
(ii) Wyszni II	Ts +/ Tpl	Ts +/ Tpl	
(iii) Baginiec III	Ts +/ Tpl	Ts +/ Tpl	
VA	Nil	*Clarias batrachus*	Uninfected
VB	*Ts. batrachi*	*Clarias batrachus*	Ts-infected
VC	*Tpl. maguri*	*Clarias batrachus*	Tpl-infect-ed

Trypanosoma attii (Gupta and Jairajpuri, 1982) (host *Wallago attu*), *Trypanosoma singhii* Gupta and Jairajpuri, 1981 (host *Heteropneustes fossilis*), *T. aligaricus* (Gupta and Jairajpuri, 1982) (host *Channa punctatus*), *T. batrachi* Qadri, 1962 and *Trypanoplasma maguri* Gupta and Gupta, 1987 (host *Clarias batrachus*) (India), *Trypanosoma danilewskyi*, (Laveran and Mesnil, 1904), *Trypanoplasma golysiana* and *T. polygolysiana* Gupta and Pilarczyk, 1994) (Poland) were discovered from the blood of infected fish. It was further proposed to observe the effect of these parasites on the blood glucose level of their respective hosts under different conditions of infectivity.

A hypoglycemic condition was evident in the infected hosts. The maximum intensity of trypanosomes was observed in Group IIB whereas, *Trypanoplasma*-infected fish were most poorly infected. It was interesting to observe the hypoglycemic condition in the infected fish (Table 4.6).

In *Wallago attu,* the normal values of 80.2 + 7.8 mg/100 ml in Group IA declined to 60.3 + 5.7 mg/100 ml in Group IB indicating 24.7% fall in the glucose values. In *Heteropneustes fossilis* (par. *Trypanosoma singhii*) glucose concentration fell to 38.7 + 2.8 mg/100 ml in Group IIB from the normal value of 62.3 + 3.2 mg/100 ml in Group IIA. In *Channa punctatus* infected with *T. aligaricus*, the values scrolled down by 23.9% in Group IIIB fish. In Group IV, glucose concentration was determined in uninfected (Group IVA) vs infected (Group IVB) hosts in carps from three different ponds of the Fish Culture Experimental Station, Chybie, Poland: Golysz [Group IVA(i), IVB(i)]; Wyszni II [Groups IVA(ii), IVB(ii)] and Baginiec III [Groups IVA(iii), IVB(iii)]. Fish of all three ponds indicated a fall in glucose value in the flagellate infected (Ts+/Tpl) infected fish. The maximum fall (28.8%) was recorded in carps of Baginiec III pond [Group IVB(iii)] as compared to that in Group IVB(i) (14.0%) and Group IVB(ii) (22.5%).

In *Clarias batrachus* (Group V), the values were determined under trypanosome- infected and *Trypanoplasma*-infected conditions and compared with the control. The values in *Clarias* infected by *Trypanoplasma* fell more sharply (Group VA uninfected: 65.3 + 2.2 mg/100 ml; Group VC Tpl-infected: 29.6 + 7.8 mg/100 ml) as compared to fish infected with trypanosomes (Group VB Ts-infected: 41.2 + 3.2 mg/100 ml) showing a significant ($P < 0.001$) fall in infected fish (Table 4.6).

Amongst all the groups, the lowest values (29.6 + 7.8 mg/100 ml) were observed in group VC (Tpl-infected) fish indicating maximum consumption of glucose by *Trypanoplasma* where glucose concentration depleted by 54.6% as compared to that of *Trypanosoma* (36.9% decline) in the same fish, *Clarias batrachus*. The glucose values of the different groups are shown in Table 4.6 (Plates LXV-LXIX).

Table 4.6. Biochemical Changes in Fresh Water Fishes Infected with Flagellates

Parasite	**Host**	**Status of Host Groups**	**Protein (gm/ 100 ml)**	**Glucose (mg/ 100 ml)**	**Cho-lesterol (mg/ dl)**
Trypanosoma attii Gupta & Jairajpuri, 1982	*Wallago attu*	Control (10) IA	5.1 ± 0.5	80.2 ± 7.8	340.8 ± 12.4
		Infected (10) IB	4.8 ± 0.35	60.3 ± 5.7	290.2 ± 5.7
		% Change	**(-)5.8%**	**(-) 24.7%**	**(-) 14.8 %**

Trypanosoma singhii Gupta & Jairajpuri, 1981	*Hetero-pneustes fossilis*	Control (10) IIA Infected (10) IIB	4.21 ± 0.31 4.92 ± 0.55	62.3 ± 3.2 38.7 ± 2.8	344.3 ± 9.3 291.7 ± 7.4
		% Change	**(+) 16.6%**	**(-) 37.8%**	**(-) 15.2 %**
T. aligaricus Gupta & Jairajpuri, 1982	*Channa punctatus*	Control IIIA Infected IIIB	3.8 ± 0.42 3.3 ± 0.21	53.5 ± 2.9 40.7 ± 4.2	280.7 ± 11.3 257.3 ± 8.4
		% Change	**(-)13.1%**	**(-)23.9%**	**(-) 8.3 %**
T. danilewskyi Laveran & Mesnil, 1904 *Trypanoplasma golysiana* Gupta & Pilarczyk 1994 *Trypanoplasma polygolysiana* Gupta &Pilarczyk, 1994	*Cyprinus carpio*	**Control** IV A (i) Golysz IVA (i) (ii) Wyszni II IVA (ii) (iii) Baginiec III IVA (iii) **Infected** IV B (i) Golysz IVB (i) (ii) Wyszni II IVB (ii) (iii) Baginiec III IVB(iii)	3.60 + 0.34 3.51 + 0.42 3.71 + 0.21 2.81 + 0.28 3.28 + 0.31 3.22 + 0.29	70.5 ± 7.8 75.3 ± 4.8 68.3 ± 3.9 60.5 ± 3.8 58.3 ± 2.9 48.5 ± 7.1	340.2 ± 25.2 401.4 ± 32.3 480.3 ± 20.4 280.5 ± 10.7 302.7 ± 8.9 390.2 ± 17.2
		% Change Golysz IVB(i) Wyszni II IVB(ii) Baginiec III IVB(iii)	**(-) 22.0% (-) 6.5% (-)13.2%**	**(-) 14.0% (-) 22.5% (-) 28.8%**	**(-) 17.5 % (-) 24.8 % (-) 22.5 %**

Trypansoma batrachi Qadri, 1962 *Trypano-plasma ma-guri* Gupta &Gupta, 1987	*Clarias batrachus*	Control (10) VA Ts-infected (8) VB Tpl-infected (7) VC	4.35 + 0.38 4.95 + 0.51 3.81 + 0.32	65.3 + 2.2 41.2 + 3.2 29.6 + 7.8	401.8 ± 14.1 349.2 ± 10.7 320.4 ± 6.5
		% Change Ts-infected VB Tpl-in-fected VC	**(+) 13.7 % (-) 11.6 %**	**(-) 36.9 % (-) 54.6 %**	**(-) 12.0 % (-) 20.0 %**

Table 4.7. Carbohydrates and Related Compounds Utilized by Parasitic Flagellates*

	Pen-toses	**Hexoses**	**Di-acch-arides**	**Trisa-ccha-ride**	**Polysacc-harides**	**Alcohols**	**Gluco-side**
Species	Arbinose Rham-nose Xylose	Fructose Galac-tose Glucose Man-nose	Lactose Maltose Saccharose	Raffinose	Dextrin Inulin Soluble starch	Glycerol Mannite Sorbite	Amygdalin
Trypano-soma equiper-dum	°	• • • •	•	-	-	• °	-
Trypano-soma brucei brucei	-	• ° • •	•	-	-	-	-
Trypano-soma lewisi	-	• • •	•	-	-	-	-

Trypanosoma cruzi	-	•	•	-	-	-	-

*Cheng, 1986

° Very slightly used

•Used by flagellate

- Not used

Table 4.8. End Products of Aerobic Glucose Metabolism in Trypanosomes of Vertebrate Hosts Product[a]

Morphological Type	**Pyruvate**	**Glycerol**	**Acetate**	**Lactate**	**Succinate**	**CO_2**
Long-slender trypomastigote	++++	±b -	-	-	-	±
Short-stumpy trypomastigote	++++	±b	++	-	±	+

[a]The amount of carbon recovered in each produced is expressed as a percentage of the carbon of glucose utilized, where ++++ represents 51%; ++++, 26–50%; ++, 6–25%; +, 1–5%; ± 1%; 0, no available data.

[b] *Variable amounts of glycerol have been reported by various authors, which is probably due to partial anaerobiosis (Fairlamb and Opperdoes, 1986).*

4.3.2.2. Discussion

Glucose is one of the important nutrients found in blood and its concentration maybe influenced by the presence of parasites (Smyth, 1996). The negative influence innately associated with many parasites aggravates the severity of morbidity of such parasitic infestations. In view of this, several studies have evaluated the occurrences, causes and consequences of the severe morbidity (Agbolade et al., 2009).

Due to the above mechanism of glucose utilization by the trypanosomes, the glucose content of the blood is subjected to alterations when infected by trypanosomes indicating terminal hypoglycemia. Tandon and Joshi (1974) observed a decrease in the blood glucose levels of certain freshwater fishes

– *Clarias batrachus*, *Heteropneustes fossilis*, *Channa punctatus,* and *Mystus seenghala* infected with *Trypanosoma*. A significant ($P < 0.001$) decline occurred in the glucose concentration (uninfected: 65.3 + 2.2 mg/100 gm; infected 41.2 + 3.21 mg/100 gm) of *Clarias batrachus* infected with trypanosomes due to the exhaustion of carbohydrate reserves of the host (Gupta and Jairajpuri, 1983). Glucose content showed the maximum depletion of 40.0% under Ts-infected conditions in *Nemacheilus rupicola* (Joshi, 1985).

Earlier studies have indicated that concurrent infectivity due to *Trypanosoma* and *Myxosoma* in *H. fossilis* culminated in terminal hypoglycemia showing a significant ($P < 0.01$) fall in the glucose concentration of the infected fish. The normal glucose values of 62.2 + 3.0 mg/100 ml declined to 50.4 + 4.2 mg/100 ml in Ts-infected fish and to 32.6 + 3.2 mg/100 ml under concomitant condition. The maximum decline (47.5%) was observed under heavy and concomitant parasitemic condition (Gupta et al., 2001).

The physiology of carbohydrate metabolism in trypanosomes has been a matter of concern over the recent years. Carbohydrates are not stored in significant quantities within Trypanosomatids. Therefore, most blood forms depend on exogenous glucose for their main energy source. Although glucose has been recognized as the most important exogenous substrate, other hexoses and glycerol can also be utilized as shown for some other species of trypanosomes (Table 4.7, Cheng, 1986).

Blood forms of certain trypanosome species do not possess a functional TCA cycle or the cytochrome type of terminal respiratory pathway. Energy (ATP) is produced when glucose is metabolized to pyruvate through glycolysis. In trypanosomes, the cytochrome system has apparently been replaced by a particular terminal L-α-glycerophosphate oxidase system in which NADH, reduced during glycolysis, is deoxidized to NAD+ in the presence of dihydroxyacetate phosphate. Glucose is preferred as a respiratory substrate. The absence of a functional TCA cytochrome system in blood trypanosomes is correlated with the presence of mitochondrion that is thin and tubular with few cristae (Noble and Noble, 1982). In culture forms, glucose is metabolized through both anaerobic and aerobic pathways to produce CO_2, succinate, acetate and pyruvate as end products.

The bloodstream trypanosomes have ample access to glucose where they derive ATP predominantly from anaerobic glycolysis through the Embden-Myerhoff pathway. Trypanosomes have a high glycolytic rate amongst the eukaryotes. Most of the enzymes of the glycolytic pathway are confined to an organelle unique to trypanosome, the glycosome.

Kinetoplastida possess two unique features: (a) the presence of a peculiar organelle, the *glycosome;* and (b) the presence of *trypanothione*, a novel thiol-containing substance, first identified by Fairlamb and Cerami (1985), which plays an essential role in protection against oxidative stress.

The morphology and biochemical activities of the glycosome have been reviewed by Opperdoes (1987, 1991), Early reports indicated that the glycolytic enzymes in the Trypanosomatidae might be particle-bound. This was confirmed by the association of specific enzymes with a microbody organelle which was eventually named the glycosome. This organelle plays a dominant role in trypanosome metabolism and, to date, nine enzymes have been identified in it.

The glycosome contains the first seven enzymes of the glycolytic pathway and two additional enzymes involved in glycerol metabolism: hexokinase, glucose phosphate isomerase, phosphofructokinase, fructose 1,6-bis-phosphate aldolase, triosephosphate isomerase, D-glyceraldehyde-3-phosphate dehydrogenase, phophoglycerate kinase, glycerol 3-phosphate dehydrogenase and glycerol kinase. Glucose is channeled through a chain of enzyme-active sites to provide impetus for the observed rapid rates of glycolysis as shown recently by Sarkar (2010). Three of these enzymes may have isoenzymes in the cytosol. About 240 glycosomes appear to be present in a blood stream form.

Kinetoplastids have a peculiar metabolism in that the first seven steps of their glycolytic pathway are separated into the glycosome, a property unique to this group of eukaryotes. These organelles seem to be related to the peroxisomes and glyoxyosomes of other eukaryotes (Clayton and Michels, 1996; Michels et al., 2006). Trypanosomes in the vertebrate bloodstream, unlike the forms in their invertebrate hosts, are dependent on glucose from their host since glycolysis is their sole source of ATP. As they neither possess a functional citric acid cycle nor display oxidative phosphorylation, this makes glycolysis a very attractive target for the design of drugs against trypanosomes (Verlinde et al., 2001).

Glucose + 2 ADP + 2Pi + 2NAD+ = 2 Pyruvate + 2 ATP + 2NADH + 2H+

Helfert et al. (2001) have elucidated the role and importance of triose phosphate isomerase (TPI) in the glycolysis of *T. brucei* in the mammalian bloodstream form. Most of the glycolytic reactions in *T. brucei* are located in the glycosome, whose membrane is assumed to be impermeable to both ATP/ADP and NAD+/NADH. The essential role of the glycosome has been indicated in prior computational studies by some of the authors (Bakker et al., 2000) to prevent hexose kinase (HK) and phospho-fructose kinase (PFK) from being exposed to the high ATP concentrations of the cytosol, as these enzymes seem to be hardly regulated by allosteric effectors as commonly is the case in other eukaryotes. In their vertebrate hosts, under anaerobic conditions, the trypanosomes would have a yield of only one ATP per glucose consumed, as they need to regenerate NAD+, which they manage by reducing dihydroxy acetone phosphate (DHAP) to glycerol-3-phosphate (Gly3P) and – after regeneration of ATP – produce glycerol.

Under aerobic conditions, reduction equivalents are exported from the glycosome in form of Gly3P, which then is oxidized to DHAP under consumption of O_2 by a mitochondrial glycerol phosphate oxidase (GPO). The DHAP produced is then reimported into the glycosome and transformed into glyceraldehyde 3-phosphate (GA-3-P) by triose phosphate isomerase (TPI). Subsequent transformation via the main branch of glycolysis to pyruvate (Pyr) doubles the ATP yield per equivalent glucose compared to anaerobic conditions. To explain why TPI depletion had that harsh an effect on growth rate, a computational model of glycolysis was developed by Helfert et al. (2001).

A model was proposed by Bakker et al. (1997) in which a detailed computational reconstruction of the entire glycolytic pathway of trypanosomes in mammals had been created. It was enhanced to incorporate more detailed kinetics with measured parameters for glucose-6-phosphate isomerase (PGI) and TPI. Variation of the maximal velocity of TPI indicated that, as was to be expected for this enzyme, TPI had a large overcapacity, with little changes to the ATP levels or glucose fluxes over a wide range of activities. The authors reported that at low relative activities-below 15% of the measured maximal velocity of the wild type--the flux through pyruvate started to decrease and that through the glycerol branch increased as expected. In contrast to the anaerobic scheme, however, no regime with equimolar glycerol and pyruvate production was reached, but all glycolytic steady state fluxes decreased to zero. This was found to stem from the activity of the mitochondrial GPO, which unlike under anaerobic conditions, still oxidized Gly-3-P exported from the glycosome to DHAP. The export of reduction equivalents lead to an increase of the NAD+/NADH ratio and therefore to a possible accumulation of DHAP, as it cannot be reduced to Gly-3-P. This accumulation in turn could lead to inhibition of ALD and a reduction of glycolytic flux. To confirm these predictions, measurements of Pyr production, intracellular ATP and DHAP levels were performed (Helfert et al., 2001). Qualitatively, these measured values fitted the predictions quite well, while there were still some quantitative disparities. Use of an inhibitor of GPO, salicylhydroxamic acid, indicated that GPO activity might be essential.

Trypanosoma brucei can take up glucose equal to its own dry weight in one to two hours (Cheng, 1986). It has been shown for the blood stream trypomastigote forms of *T. brucei* that they possess no energy stores, are unable to oxidize amino acids or fatty acids and are thus entirely dependent on an exogenous supply of carbohydrate (glucose, fructose, mannose or glycerol) for their energy production (Fairlamb and Opperdoes, 1986). Glucose is said to be utilized by trypanosomes and its transport in the bloodstream form of *T. brucei* was characterized by enzymatically measuring the d-glucose uptake (Seyfang and Duszenko, 1991).

Short-stumpy trypomastigotes can also respire d-oxoglutarate at a rate about half that of glucose. Aerobic glycolysis proceeds through the Embden-Meyerhof pathway, all the enzymes of which have been detected in the glycosomes, the mitochondrion and the cytosol. The glycolysis process differs from the normal, vertebrate pattern in that the end product is almost exclusively (98 per cent) pyruvate, with traces of CO_2 and glycerol (Table 4.8), the process supplying all the ATP synthesized by the organisms, and proceeds at a rate faster than observed in any other organism (Opperdoes et al., 1990).

Pyruvate is not metabolized further by long slender forms as they lack the enzymes pyruvate decarboxylase and lactate dehydrogenase (Fairlamb and Opperdoes, 1986). However, some pyruvate is further metabolized to acetate, CO_2 and succinate in the short stumpy forms.

Molecular oxygen serves as the terminal electron acceptor, resulting in the formation of water. A glycerol 3-phosphate shuttle transports the reducing equivalents from NADH in the cytosol into the mitochondrion, where, via a glycerophosphate oxidase system, it passes to molecular oxygen. Under anaerobic conditions, long-slender blood stream forms continue to utilize glucose, as under aerobic conditions, but, because the glycerophosphate is inoperative, the glucose is metabolized into equimolar amounts of pyruvate and glycerol.

Due to the above mechanisms, the pathogenic trypanosomes may consume so much sugar in the bloodstream that they may cause exhaustion of the carbohydrate reserves of the host. As a result of this strain on the liver, its function breaks down leading to the death of the infected animal. It is an established fact that among higher vertebrates, a considerable fall in the blood sugar level occurs and may in cases, cause glucopyruvic intoxication leading to the death of infected animal (von Brand, 1973). Terminal hypoglycemia and lowered glycogen reserves occur in small laboratory rodents with 2–3 billion trypanosomes per milliliter blood and in rabbits with so few parasites that it is often difficult to find them in peripheral blood. Feeding sugar to infected animals prolongs their lives for short periods but does not prevent the development of terminal hypoglycemia (von Brand, 1973). It has also been reported that infected guinea pigs had very low level of glucose-6-phosphatase activity and it was suggested that due to this, they were unable to release glucose in to the blood stream and to carry out gluconeogenesis efficiently. Trypanosomes not only consume the blood sugar but also inhibit the enzymatic activity of liver causing a reduction in the release rate of glucose from glycogen (Marciaq and Seed, 1970). It is probable that due to parasitic infection, a significant (P

< 0.01) increase in the serum potassium level coupled with the production of specific toxin by the parasite maybe responsible for the ensuing hypoglycemic condition (Gupta et al., 2000).

Miletti et al. (2006) characterized the D-glucose transport activity present in 2 life-stage forms of *T. rangeli* (epimastigotes and trypomastigotes) using D-[U-14C] glucose as substrate. The authors indicated that *T. rangeli* transports D-glucose with high affinity in both epimastigote (Km 30 μM) and trypomastigotes (Km 80 μM) life-forms. Both transport activities were inhibited by Cytochalasin B and Phloretin, indicating that probably D-glucose uptake in *T. rangeli* is mediated by facilitated diffusion of the sugar. Significant differences were observed between epimastigotes and trypomastigotes in relation to their affinity for D-glucose analogues, and the predicted amino acid sequence of a putative D-glucose transporter from *T. rangeli* (TrHT1) showed a larger identity with the *T. cruzi* D-glucose transporter encoded by the TcrHT1 gene than with other transporters already characterized in Trypanosomatids.

The biflagellate, *Trypanoplasma* are also known to cause depletion in glucose content in the infected host. *T. salmositica* causes glycemia in *Salmo gairdneri* (Woo, 1979). It is interesting to note that the trypanosomes reported by Gupta and Jairajpuri (1983) were much smaller in size (10.2–33.2 × 1.0–3.8 μm) and had a higher intensity as compared to *Trypanoplasma maguri* (22.2–48.7 × 2.6–6.5 μm), nevertheless, the decline in glucose concentration due to *Trypanoplasma* was 57% as compared to the depletion by trypanosomes of the same fish (37%) revealing that the fall is significantly higher ($P < 0.01$) due to the former in contrast to the latter. These findings infer that as trypanoplasms are more voluminous, they require and consume greater quantities of glucose which is in agreement to the views of Cheng (1986) who made similar observations for *Trypanosoma brucei* which could take up glucose equal to its own dry weight. It has also been observed that heavier the intensity (4–5 forms/per 20 mins scan), greater is the consumption leading to higher depletion in glucose concentration.

It is noteworthy to put on record the experimental findings of Evans and Brown (1972) who showed experimentally that the bloodstream forms of *T. brucei* transferred to liquid medium rapidly utilized glucose over the first 12 hr of culture and this was accomplished by a lowering of free pyruvate in the medium and a rise in the proline oxidase activity of the trypanosomes thereby indicating that glucose was the food of choice for trypanosomes.

The present observations also show that the glucose concentration declines significantly ($P < 0.001$) in the five species of fishes inhabited with trypanosomes/ trypanoplasms suggesting that these parasites invariably consume the glucose content of their respective hosts. It is just possible that the trypanosomes consume so much sugar in the blood stream that they cause exhaustion of the carbohydrate reserves of the host. Further, these parasites may hamper the normal glycolytic pathway of the host itself and therefore, may also play a contributory role in the hypoglycemic condition of the host. Moreover, as was evident from the present findings, during trypanosomiasis, the greatest fall in the glucose content of the host was recorded in *H. fossilis* (Group IIB) and during overall infection with flagellates highest fall was recorded in Group VC (54.6%) having dual infection of *Trypanosoma* and *Trypanoplasma*. Trypanosomiasis and trypanoplasmosis compared, the latter caused more severe hypoglycemia as compared to the former. The studies would infer that the trypanoplasms too have a high glycolytic rate. As the trypanoplasms are confined only to fish blood, erudite studies on glycolysis in trypanoplasms are not as exhaustive as those of trypanosomes and are still wanting. However, Opperdoes et al. (1988) concluded that in *Trypanoplasma borelli,* the glycolytic pathway is compartmentalized in a way similar to that found in Trypanosomatidae, where seven glycolytic enzymes and two enzymes of 4-glycerol metabolism are associated with glycosomes as shown above. Apparently the presence of glycosomes is a characteristic of all Kinetoplastida. Therefore, it appears that the biflagellates have a stronger glycolytic activity as compared to the monoflagellates resulting in the observed low values of glucose in fishes infected with these parasites.

4.3.3. Blood Parasites Vis-a-Vis Total Serum Protein (TSP)

In order to observe the effect of flagellates on the protein content in the blood, investigations were conducted on similar host-parasite systems as mentioned for glucose.

4.3.3.1. Host-Parasite Systems

Group Host	Parasite
I. *Wallago attu*	*Trypanosoma attii;* Gupta and Jairajpuri (1982)
II. *Hetero-pneustes fos-silis*	*Trypanosoma singhii;* Gupta and Jairajpuri (1981)
III. *Channa punctatus*	*Trypanosoma aligaricus;* Gupta and Jairajpuri (1982)
IV. *Cyprinus carpio*	*Trypanosoma danilewskyi;* Laveran and Mesnil (1904)

	Trypanoplasma golysiana; Gupta and Pilarczyk (1994)
	Trypanoplasma polygolysiana; Gupta and Pilarczyk (1994)
V. *Clarias batrachus*	*Trypanosoma batrachi;* Qadri (1962)
	Trypanoplasma maguri; Gupta and Gupta (1987)

The observations on protein content of parasitized fish showed a trend of decline in most of the groups [Groups IB (5.8%), IIIB (13.1%), IVB(i) (22.0%), IVB(ii) 6.5%, IVB(iii)(13.2%) and VC (11.6%), but in two groups, IIB (16.6%) and Group VB (13.7%), there was a rise in the values of the protein content (Table 4.6). However, in most of the cases, the percent change was insignificant (P > 01) (Plates LXV – LXIX).

There is little conclusive evidence to indicate that protein utilization takes place during parasitic infection or there is significant impairment of protein absorption. Moon et al. (1968) observed that only GOT and GPT increase in *Trypanosoma*-infected animals whereas Mc Dougald and Hansen (1970) observed that serum enzymes like asparatate, amino- transferase GOT, alanine aminotransferase, GPT, lactate dehydrogenase, LDH, glutamate dehydrogenase GLDH, malate dehydrogenase MDH, sorbitol dehydrogenase SDH and creatine kinase CPK increase when hosts are infected with *Histomonas meleagridis*. It has been shown by Freeman et al. (1970) that more than 95% increase in IgG occurred during infection with *Trypanosoma* and *Schistosoma* in their respective hosts.

The requirement for amino acids by parasites has not been thoroughly examined, as a consequence of which, it is not possible to list the essential aminoacids for any parasite species. Probably the greatest body of information on requirement for, and biosynthesis of aminoacids concerns the parasitic Protozoa, particularly the trypanosomes and malarial parasites. Despite this, even for Protozoa, there is only limited information available at present and the available ones are mainly centered around mammalian protozoans. It has been estimated that the total protein content of parasite tissues varies from around 20% of the dry weight to as much as 80% (Chappell, 1980).

Studies on protein synthesis have proceeded on two fronts: one involving the study of incorporation of radioactively-labelled precursors into the protein of intact organisms, the other utilizing cell-free ribosomal preparations. Here again, most studies are based on parasitic protozoa. The hemoflagellates possess both mitochondrial and cytoplasmic protein synthesis. The mitochondrial system is sensitive to the potent inhibitor chloramphenicol, whereas the cytoplasmic system is inhibited by cycloheximide. This is true for most protozoans (Chappell, 1980). It has been shown that protozoan parasites, including *T. rhodesiense* can also utilize aminoacids for energy metabolism.

Parasitic protozoa require the same 20 amino acids for their proteins. These can be obtained either from the environment or by synthesis, similar to mammals. From the fragmentary knowledge, it appears that some are required preformed. Table 4.9 lists the amino acids taken up by some parasitic mammalian trypanosomes (Gutteridge and Coombs, 1977). Unfortunately, similar data on fish trypanosomes is yet to be formulated.

Trypanosomes possess the ability to hydrolyze proteins within the cells by inactivation of enzymes, as this is an essential requirement for the control of growth and development. This autolysis yields aminoacids which are available for protein synthesis. The parasites may also use exogenous protein as a source of aminoacid. In most cases, these proteins are ingested by the protozoan, either through a cytosome, or by phagocytosis or pinocytosis into food vacuoles. The protein is then hydrolyzed in the vacuole by the lysosomal enzymes (cathepsins and other peptidases) which are presumably released into the vacuole when lysosomes fuse with it. It has also been suggested that protozoa may excrete proteases which hydrolyze host proteins outside the parasite and the aminoacids produced can then be absorbed. However, careful investigations are required to determine the localization of any enzyme, and in most cases these have not been carried out on the proteases of parasitic protozoa (Gutteridge and Coombs, 1977).

Table 4.9. Amino Acid Biosynthesis in Some Trypanosome Species*

Parasite	**Source Molecule**	**Amino Acid Synthesized**
Trypanosoma rhodesiense	Carbohydrate serine glucose	Alanine, glycine, serine, aspartic acid, glutamic acid. Alanine, glutamic acid, aspartic acid,
T. cruzi	alanine aspartate glu-	glycine, cystine, threonine. Alanine, cystine, taurine, serine.
T. gambiense	tamic acid arginine	Cystine, taurine, threonine, taurine, glutamic acid, proline.
T. brucei	Proline	Alanine, aspartic acid.

*Gutteridge and Coombs (1977).

It is presumed that parasitic protozoa utilize free aminoacids from their environment although their mechanisms of entry into the cells have been investigated rarely. Aminoacids enter the trypanosomes both by simple diffusion and by mediated processes involving several sites, the specificities of which are unusual and different from those of mammalian cells. In *T. brucei*, N_1 site specific for serine, threonine and alanine and N_2 site for neutral amino acids have been characterized. Blood stream stages of *T. rhodesiense* can synthesize alanine, glycine, serine, aspartic acid and glutamic acid from carbohydrate, *T. cruzi* culture forms can produce alanine, aspartate, glutamate, glycine, systeine and threonine from serine and culture forms of *T. brucei* are capable of converting proline to alanine and aspartate. It has further been reported that in Trypanosomatids, both cytoplasmic and mitochondrial protein synthesis occurs, possibly in the blood stream trypanosomes which possess acristate mitochondria. However, mitochondrial protein synthesis accounts for only about 3% of the total synthesis. Cytoplasmic protein synthesis of trypanosomes is inhibited by cycloheximide, but not chloramphenicol (Gutteridge and Coombs, 1977). According to Cazzulo (1995), epimastigotes of *T. cruzi*, catabolize proteins and amino acids with production of NH_3. Guimarães et al. (2008) reconstructed the amino acid metabolic pathways of *T. cruzi in silico*. Again, these studies are mainly confined to trypanosomes of higher vertebrates and man, similar findings on fish hemoflagellates are still in demand.

Changes in protein similar in principle to those found in man's serum also occur in the cerebrospinal fluid and specially pronounced are increases in γ globulin (von Brand, 1973). It may be added that like in all parasitic infections, the formation of antibodies may add up in causing quantitative changes in serum proteins.

According to Cornille and Hornung (1968), IgM and IgG levels increase in humans infected with *T. rhodesiense* and have approximately normal γ globulin level. Gupta and Jairajpuri (1983) recorded a rise in protein level from 4.63 + 0.12 gm/100 gm to 5.31 + 0.21 gm/100 gm in *Clarias batrachus* infected with *T. batrachi*. Gupta et al. (2001) again recorded a rise in protein in the same fish infected with trypanosomes in another set of experiment under mild and heavy trypanosome infection and the rise was higher in the latter group. However, Joshi (1985) reported that the protein content in infected *Nemacheilus rupicola* declined marginally from 3.00 + 0.12 mg/dl to 2.70 + 0.11 mg/dl in ts-infected fish.

It is well established that immune responses are generated after experiencing parasitic infection and they provide the animal with acquired immunity. For this purpose, antibodies are produced which are chemically fractions of proteins (immunoglobulins). Smyth (1996) opined that trypanosomes are good examples of extracellular protozoans against which antibody responses can provide effective immunity.

It has often been observed that a progressive decrease in albumin and an increase in globulins, most often γ globulin occurs which leads to more or less marked changes in albumin/ globulin ratio. It is due to this change that both instances may occur: either a rise or fall in protein concentration of parasitized hosts. The production of antibodies implicate the production of immunoglobins that is fractional proteins. The quantitative relationship between antibodies and immunoglobulins may be variable.

It is postulated that due to the fractional increase or decrease in the protein content of parasitized hosts, both types of changes maybe witnessed.

It may be added that while considering the variations in plasma protein values, the stage or severity of infection, the host species examined, the presence or absence of previous infections, hormonal balance and similar factors must also be taken into account.

4.3.4. Blood Parasites Vis-À-Vis Serum Cholesterol

In order to observe the effect of flagellates on the cholesterol content in the blood, investigations were conducted on the following freshwater edible fishes which were infected with different species of flagellate parasites:

4.3.4.1. Host-Parasite Systems

Group Host	**Parasite**
I. *Wallago attu*	1. *Trypanosoma attii* (Gupta and Jairajpuri, 1982)
II. *Heteropneustes fossilis*	1. *Trypanosoma singhii* (Gupta and Jairajpuri, 1981)
III. *Channa punctatus*	1. *Trypanosoma aligaricus* (Gupta and Jairajpuri, 1982)
IV. *Cyprinus carpio*	1. *Trypanosoma danilewskyi* (Laveran and Mesnil, 1904) 2. *Trypanoplasma golysiana* (Gupta and Pilarczyk, 1994) 3. *Trypanoplasma polygolysiana* (Gupta and Pilarc-zyk, 1994)
V. *Clarias batrachus*	1. *Trypanosoma batrachi* (Qadri, 1962)
	2. *Trypanoplasma maguri* (Gupta and Gupta, 1987)

Lipids play an important role in the interaction process between pathogens and host cells with lipid rafts being highly specialized regions of the plasma membrane that are enriched in cholesterol and sphingolipids (Priotto et al., 2009). The important biochemical constituent, the common sterol, cholesterol is abundantly encountered in animal tissues. It has been well recognized in various clinical disorders.

The name "cholesterol" originates from the Greek *chole-* (bile) and *stereos* (solid), and the chemical suffix *-ol* for an alcohol. François Poulletier de la Salle first identified cholesterol in solid form in gallstones, in 1769. However, it was only in 1815 that chemist Eugène Chevreul named the compound "cholesterine." (Olson, 1998). It is the most abundant sterol and serves as an important precursor of many other steroids.

Cholesterol is slightly soluble in water; it can dissolve and travel in the water-based bloodstream at exceedingly small concentrations. Since cholesterol is insoluble in blood, it is transported in the circulatory system within lipoproteins, complex discoidal particles which have an exterior composed of amphiphilic proteins and lipids whose outward-facing surfaces are water-soluble and inward- facing surfaces are lipid-soluble; triglycerides and cholesterol esters are carried internally. Phospholipids and cholesterol, being amphipathic, are transported in the surface monolayer of the lipoprotein particle.

Cholesterol is susceptible to oxidation and easily forms oxygenated derivatives known as oxysterols. These can be formed by three different mechanisms; autoxidation, secondary oxidation to lipid peroxidation, and cholesterol metabolizing enzyme oxidation. A great interest in oxysterols arose when it was shown they exert inhibitory actions on cholesterol biosynthesis, termed as the "oxysterol hypothesis." Additional roles for oxysterols in human physiology include their participation in bile acid biosynthesis, function as transport forms of cholesterol, and regulation of gene transcription (Russell, 2000).

The method of collection of fish and the experimental groups were as described in the section of "Glucose" and these studies are also based on the same five groups as mentioned earlier. The analysis of serum for cholesterol estimation involves the isolation of cholesterol from other blood constituents and the separation of unesterified cholesterol from the more soluble esters. This sterol was estimated by the procedure as mentioned in "Materials and Methods."

4.3.5. Results

Hypocholesterolemia was consequential in all five infected groups. Amongst ts- infected fish, the fall was 14.8% in Group IB, 15.2% in Group IIB, 8.3% in Group IIIB. In Group IVB (i) (Golysz pond), the percent fall was 17.5, in Group

IVB; (ii) (Wyszni II pond), 24.8% and that in Group IVB; and (iii) (Baginiec III pond) it was 22.5% in carps having either/mixed infection of trypanosomes and trypanoplasms. The minimum fall (8.3%) was recorded in Group IIIB and the maximum in Group IVB (ii) (24.8%). However, the fall was lower as compared to that of glucose in similar groups (Table 4.6; Plates LXV-LXIX).

4.3.6. Discussion

Cholesterol changes in various clinical disorders are well recognized (Seshadri et al., 1981). Abnormally low levels of cholesterol are termed hypocholesterolemia. Research into the causes of this state is relatively limited, but some studies suggest it to be linked with depression, cancer, and cerebral hemorrhage in humans. In general, the low cholesterol levels seem to be a consequence of an underlying illness, rather than a cause. (Lewington et al., 2007).

In fish, attempts have been made during the recent years for cholesterol estimation under different physiological stages of fish. Exposure of *Channa* to four concentrations of DAP up to 96 hrs revealed marked changes in its cholesterol level (Singh and Singh, 1997). With regard to cholesterol level in trypanosome-infected fish, it was found that serum cholesterol levels of six species of fresh water fishes were lowered due to trypanosome infection. The percentage of infection in the different species of hosts ranged from 3.27% in *Clarias batrachus* to 16.66% in *Wallago attu*. Loss in serum cholesterol level was highest (42.39%) in the spiny eel *Mastacembelus armatus* and lowest in the carp *Cirrhina mrigala* (4.56%) (Tandon and Chandra, 1977) thereby stressing the role of trypanosomes in inducing hypocholesterolemia. Gupta and Jairajpuri (1983) investigated the cholesterol content in fish where the values of 402.1 + 14.31 mg/100 gm fell to 354.7 + 11.80 mg/100 gm in *Clarias batrachus* infected with *T. batrachi*. The cholesterol level in *Nemacheilus rupicola* also declined from 248.50 + 30.40 to 214.30 + 70.80 in ts-infected fish (Joshi, 1985). A similar pattern of fall in trypanosome infected fish was reported by Tandon (1986).

The blood levels of certain lipids are quite variable in protozoan diseases. Blood cholesterol level has been found to be normal (Linton, 1930), elevated (Ada and Fulton, 1948) or decreased (Lippi and Sebastiani, 1958) during trypanosomiasis. Corresponding contradictions exist in respect to the phospholipid and fatty acid levels during trypanosomiasis (Randall, 1934). Dixon and Williamson (1970) showed that when trypanosomes grow in lipid rich medium, they maintain control over their lipid composition. Such control might be exerted by specific absorption mechanisms or at the biosynthetic level. Little information is available about the biosynthesis of lipids by trypanosomes. The blood form of *T. rhodesiense* shows a slight incorporation of acetate into lipid *in vitro* (Grant and Fulton, 1957), *T. cruzi* can synthesize sterol and complex lipids from acetate (Halevy, 1962) and the culture form of *T. lewisi* incorporates acetate into long chain fatty acids (Korn et al., 1965). Dixon et al. (1971) studied lipid biosynthesis, uptake and oxidation of fatty acids by blood and culture forms of *T. lewisi* and *T. rhodesiense.* Nutritional requirements for lipids by trypanosomes are difficult to ascertain because of the requirement for blood components in their culture media, although *T. cruzi* growing on a partially defined medium shows requirement for stearic acid (Bone and Parent, 1963) and monogenetic Trypanosomatids require exogenous lipid when growing at supranormal temperatures (Guttman, 1963).

Cholesterol is the major sterol of blood stream trypanosomes and has been reported to accumulate in the stumpy forms of *Trypanosoma rhodesiense* (Gutteridge and Coombs, 1977). Parasites in general appear to possess only a limited ability to synthesize long chain fatty *de novo* and therefore rely to a great extent upon the host to provide their essential requirement (Chappell, 1980). Both the blood and culture forms of *T. rhodesiense* contain cholesterol as the main sterol constituent (Love, 1980). It has also been reported that exogenous cholesterol could be taken up by these trypanosomes and to a limited extent, esterified. Sterols have two functions in trypanosomes, one structural for which exogenous sterol is used and the other, metabolic which is fulfilled by biosynthesis. *T. cruzi* contains no cholesterol when grown on a cholesterol-free medium but becomes quantitatively the most important sterol when it was a component of the growth medium which is suggestive that the composition of the microenvironment (blood) may deeply influence the nature and relative abundance of cholesterol in trypanosomes.

However, there is a paucity of data on the effect of concomitant infection (*Trypanosoma* and *Trypanoplasma*) on the quantitative distribution of blood cholesterol in the fresh water fishes.

During the present course of investigations, under concomitant conditions, the fall was higher than 20% which was probably due to greater cholesterol requirement of the trypanoplasms for their metabolic activities. The physiological basis for such changes is unknown, but they definitely indicate a metabolic disturbance. Moreover, the fact finds support from the present findings that the metabolic requirements of the trypanoplasms warrant greater demands of cholesterol thereby being responsible for causing a more severe hypocholesterolemia in *Trypanoplasma* /concomitant infected fish as compared to those of lone trypanosome infection.

Plate LXIX provides a comparison of the changes incurred in the biochemical parameters investigated in the different groups. Out of all the parameters, the figure depicts a maximum change in glucose concentration (37.8%) in *Heteropneustes fossilis* infected with *Trypanosoma singhii.*

Chapter 5

Trypanosome Culture

Contents

5.1. CULTIVATION OF FISH TRYPANOSOMES IN MONOPHASIC AND DIPHASIC MEDIA

The development of an *in vitro* system which allows a continuous propagation of hematozoic trypomastigotes, retaining their infectivity to the host has been a long-standing requisite for biochemical and immunological studies. The diversity of organisms parasitized by kinetoplastids is vast and therefore, the nutritional requirements of the parasitic kinetoplastid flagellates can differ greatly from one another. These differences are mirrored in a variety of culture media employed for their *in vitro* culture (Taylor and Baker, 1978).

A number of methods for preparing such media have been developed during the past years (Simmons et al., 1964; Taylor and Baker, 1978) and *Trypanosoma congolense* from cattle (Brun, 1982), *T. acomys* from the spiny mouse, *Acomys cahirinus* (Abdallah et al., 1990) and *T. freitasi* from the opossum (Thomaz and Deane, 1990) have been cultivated from different mammals. However, attempts to culture piscine trypanosomes are less popular This is the first endeavor to culture any hemoflagellate (*T. kargenensis*) discovered so far from *H. fossilis* for which various monophasic and diphasic media were tried and their relative effectiveness is discussed.

5.1.1. Inoculum

Culture studies were conducted on 100 specimens which were initially screened for trypanosome infection. The infected fish were separated and 0.1 to 0.3 ml of blood was drawn into a sterile, heparinized capillary tube.

The blood was blown out into 1 ml of buffered phosphate saline glucose (PSG) solution (Lanham and Godfrey, 1970) and the process repeated to get a final suppression of 0.5 ml infected blood in 1 ml of PSG. The diluted blood was centrifuged at 1000 rpm for 5 mins and the buffy coat layers were pooled to get 0.05 buffy layer diluted to 1 ml PSG (ratio 4:6). The suspension contained 1×10^4 ml^{-1} of viable trypanosomes which was used as inoculum.

5.1.2. Culture Media

The suspension prepared as above was seeded in the following monophasic and diphasic media and incubated at 25°C BOD incubator and 37°C in a thermostatically controlled incubator.

1. RPMI 1640 without serum (pH 7.2) buffered by HEPES (SIGMA).
2. RPMI 1640 as above with 10% FCS.
3. Medium 199 without serum (pH 7.2) buffered by carbonate bicarbonate (GIPCO).

4. Medium 199 as above supplemented with 10% FCS (SIGMA).
5. Medium TYM.
6. Medium TYM as above with 50% distilled water.
7. NNN blood agar according to Baker (1966). The cultures were monitored at 4 hourly intervals.

*5.1.2.1. In Vitro Culture **of** Trypanosoma kargenensis*

In NNN blood agar overlaid with Ringer Locke's saline and containing penicillin at 100 IU/ml, streptomycin at 10 microgram/ml, the trypanosomes were actively motile for up to 48 hrs. After this, the parasites appeared fewer in number and invariably sluggish. The first sub-passage was attempted at 14 hrs. Here again, the trypanosomes were actively motile but were lesser in number. The second sub-passage was done at 48 hrs after which there was a decline in trypanosome population. Amongst the monophasic media, the flagellates were active in medium 199 up to 24 hrs, thereafter, their growth was suppressed. At the end of 48 hrs, there were hardly any motile forms when the culture tube was examined under an inverted microscope. The results are tabulated in Table 5.1.

Table 5.1. *In vitro* Culture of *Trypanosoma kargenensis* Gupta and Gupta, 1994 in Monophasic and Diphasic Media

Time (Hours)														
Medium	0.5	2	6	10	14	16	20	24	28	32	36	40	44	48
Monophasic RPMI 1640	+	±	-	-	-	-	-	-	-	-	-	-	-	-
RPMI 1640+serum No movement	-	-	-	-	-	-	-	-	-	-	-	-	-	
MEM 199	+++	+++	+++	++	++	+	++	++	+	+	+	±	-	-
MEM 199+serum	+++	+++	+++	++	++	+	++	++	+	+	+	±	-	-
TYM No movement	±	-	-	-	-	-	-	-	-	-	-	-	-	
TYM+50% D.W.	+	+	-	-	-	-	-	-	-	-	-	-	-	-
Diphasic NNN+Locke's soln	+++	+++	+++	+++	+++	+++	+++	+++	+++	+++	+++	+++	+++	+++

+antibiotics ↑ ↑

1st sub passage; 2nd Sub passage.

The culture stages were more difficult to stain than flagellates in blood smears. Smears were fixed for 5 min in a mixture of 5 parts of methylalcohol and 1 part of tincture of iodine, rinsed with methylalcohol. After drying, they were hydrolyzed in 1 N HCI at 60°C for 2 min, rinsed with neutral buffered water and stained with Giemsa stain.

5.1.3. Discussion

In the beginning of the twentieth century, McNeal and Novy first recorded that they had obtained cultures of *T. lewisi* and since then, attempts were made to culture mammalian trypanosomes, specially the more pathogenic ones like *T. brucei* and *T. evansi*. These successful attempts were subsequently followed by cultivation of blood parasites of birds (Baker, 1966), amphibians and those found in the stomachs of mosquitoes and other insects in artificial media.

In fish, Brumpt (1906) was unable to obtain trypanosome cultures in the medium of McNeal and Novy. Lebailly (1906) reported that he had found profound changes in fish trypanosomes *in vitro*, but he did not use any medium, simply he kept films of blood containing trypanosomes between coverslip and slide and sealed the edges of the coverslip with paraffin wax. By the end of eight to ten days, he reported "small granular spherical masses each with a flagellum moving."

However, these cannot be cited as true examples of fish trypanosome culture, but indeed they were the first such attempts to be made. It was Thomson (1908) who first successfully cultivated *T. danilewskyi* from the blood of goldfish in McNeal and Novy medium, slightly modified by Nocht and Mayer. This was subsequently followed by other workers. Qadri (1962c) cultured *T. striati* in Shortt's modification of NNN for 16 months by maintaining suitable sub-cultures. Later on, Letch (1979) cultured fish trypanosomes in 4N blood agar (as used by Baker, 1966) overlaid with Locke's solution (as used by Tobie et al., 1950) containing the antibiotics, penicillin and streptomycin and sub-cultures were supplemented with Basal Minimum Essential (BME) medium containing 10% (v/v) FCS.

Smolikova (1977) and Suchankova and Lom (1977) tried to cultivate *T. danilewskyi* in MEM (Minimal Essential Medium) tissue culture medium but obtained growth in initial passages only. Letch (1979) mentioned growth of *T. cobitis* in BME (Basal Minimum Essential) Medium containing foetal calf serum. Rouskova and Kulda (1983) found complex liquid medium L4NHS suitable for obtaining trypomastigote-rich suspensions of *T. danilewskyi* and this was the first report of a successful cultivation of fish trypanosomes in a monophasic liquid medium. Jones and Woo (1991) compared the culture characteristics of *T. catostomi* and *T. phaleri* from north American freshwater fishes by growing them in undiluted and 50% dilutions of Eagle's basal medium (BME, Hank's salts and L-glutamine) each supplemented with heat-inactivated Foetal Bovine Serum (FBS) to 2.5, 5 or 10% (v/v) and buffered with 25 mM HEPES (pH 7.3).

Overath et al. (1998) studied one lower vertebrate parasite/host pair and the carp trypanosome *T. carassii* [syn. *T. danilewskyi* (Lom and Dykova, 1992)] as a model.

T. carassii infects a variety of cyprinid fish such as carp, goldfish, crucian carp, and Tench as well as some members of noncyprinid families, other members of which, e.g., rainbow trout, are refractive (Lom, 1979; Lom and Dykova, 1992; Woo, 1994). The prevalence in nature is very high and may approach 100% in densely populated fish cultures. Natural infections tend to be long-lasting and patent parasitaemia may alternate with latent states. Immunity has been suggested to be mediated by antibodies (Barrow, 1955; Woo, 1981; Islam and Woo, 1991; Ahmed, 1994). In fish blood, the parasites have a trypomastigote morphology. Upon transfer to the vector, blood-sucking leeches (i.e., *Hemiclepsis marginata*), the parasites transform into proliferating epimastigotes in the crop. Thereafter, non-dividing metacyclic trypomastigotes move to the proboscis of the leech and can be transferred to another fish during a second blood meal (Lom and Dykova, 1992; Woo, 1994).

A number of reports on the cultivation of *T. carassii in vitro* have been published (cf. reviews in Lom, 1979; Lom and Dykova, 1992; Woo, 1994;). Lom and Dykova (1992) cultured fish trypanosomes and trypanoplasms. The parasites were grown in the biphasic medium SNB 9. The authors isolated the blood parasites and maintained them in the above medium consisting of a solid phase of Neopepton Difco 2 g, Bacto agar Difco 2 g, NaCl 0.6 g, distilled water 100 ml and a liquid overlay. The above were dissolved and sterilized by autoclaving at 120°C for 20 min. When cooled to about 45–50°C, 25 ml sterile, fresh defibrinated rabbit blood or sterile human blood was added aseptically for transfusion and mixed thoroughly by swirling. While still liquid, 2 ml aliquots were dispensed, using a pre-warmed pipetting unit and working in a hood, into sterile culture tubes and were allowed to set in a slant at an angle of about 12°C. Culture tubes, 13 mm in diameter with autoclavable screw caps, were conveniently used and stored in a refrigerator before adding the liquid overlay composed of N e o p e p t o n Difco 2 g, NaCl 0.6 g and distilled water 100 ml. After autoclaving, the tubes were cooled, and 100 ml of overlay was supplemented with 0.25 ml of a vitamin solution composed of Cholinechloride 0.001 g, I-inositol 0.002 g, Folic acid 0.001 g, Nicotinamide 0.001 g, D-calcium panthotenate 0.001 g, Pyridoxolchloride 0.001 g, Riboflavine, 0.0001 g, Thiamine chloride 0.001 g, NaCl 0.1 g dissolved in 2.5 ml of distilled water. To this overlay, 0.2 ml of penicillin (potassium salt) solution (200 units/ml), and 0.15 ml streptomycin-sulphate solution (150 µg/ml) were added. In the medium used for isolation, a higher antibiotic concentration: 3000 units/ml of penicillin, 1000 mg/ml of kanamycin and 500 mg/ml of streptomycin was used. For *Trypanosoma,* before use, 2.0 ml of liquid overlay was dispensed into the culture tubes. The inoculum was about 0.05–0.1 ml. For *Trypanoplasma,* before use, 0.5 ml of liquid overlay was dispensed into the culture tubes. The inoculum was about 0.1–0.2 ml. These were then sub-cultured at 7–10 day intervals. By this procedure, the authors achieved maximum growth of 6×10^7 flagellates per ml of medium. For *Trypanosoma*, maximum density obtained was 2.5×10^7 flagellates per ml medium. Cultures were kept at temperatures of 18 to 25°C, depending on the species and strain of the parasite and the temperature requirements of the host.

In this culture system, the bloodstream stages transformed to dividing epimastigotes, which differentiated to non-dividing, metacyclic trypomastigotes in the stationary phase. Upon dilution, the latter forms again reverted back to epimastigotes. Short-term cultures retained their infectivity for fish, cultivation lead to loss of infectivity. Some strains were incapable of transforming from bloodstream forms to epimastigotes *in vitro* (Lom and Dykova, 1992). There are only a few brief reports on the cultivation of trypomastigote bloodstream stages. An exceptional *T. carassii* isolate designed Ma grew only as trypomastigotes in either biphasic blood-agar media or in fish-tissue cultures (cf. review by Lom and Dykova, 1992). The requirement of fish serum for growth of *T. carassii*, presumably as bloodstream forms, was also reported (Bienek and Belosevic, 1997).

A pre-requisite for the biochemical characterization of *T. carassii* is an adequate supply of bloodstream-stage parasites. Bienek and Belosevic (1997) also described the cultivation of *T. carassii* [isolate TrCa from a crucian carp (*Carassius auratus gibelio*)] taken directly from fish blood. Growth was dependent of fish serum, whereas mammalian serum did not support growth. The generation time was longer and the final cell density was lower in these primary cultures than in the established culture system. Details on the morphological features of the parasites and their infectivity for fish were not reported. The culture system described by the authors proved useful for the biochemical, cell biological, and molecular genetic characterization of *T. carassii* and its interaction with the fish immune system. Overath et al. (1998) have shown that adaptation of methods in use for *T. brucei* allows purification of *T. carassii* from fish blood and its long- term cultivation as infective trypomastigotes *in vitro*. The authors opined that the culture system should be useful for the biochemical characterization of trypanosomes and its interaction with the fish immune system. Amongst the other fish trypanosomes cultures, Zintl et al. (2000) cultured *T. granulosum* by injecting trypanosome-infected blood from wild eels caught from Mayo, west Ireland into blood agar slopes (Davies et al., 1992). The trypanosomes were transferred into modified SDM79 (Davies et al., 1995). The authors used this culture for inoculating trypanosomes to eels in order to observe the infectivity of different stages in the life cycle.

The mechanisms leading to the non-sterile immunity typical of these trypanosome-vertebrate systems are unknown; in particular, it has not been excluded that antigenic variation of parasite surface structure contributes to parasite persistence.

Some attempts to culture fish trypanoplasms have also been made. Ponselle (1913) cultured *Trypanoplasma varium* from the European loach, *Cobitus barbatula* using a simple blood agar medium. The culture forms resembled the blood forms, however, their infectivity for fish blood was not tested by the author. Putz (1972b) cultured *T. cataractae* from cyprinids in the southeastern states in three diphasic media, originally developed for mammalian trypanosomes and the flagellates could be maintained only for 7–9 days. Woo (1979) successfully cultured and sub-passaged *T. samositica*, the culture forms were morphologically similar to blood forms from the rainbow trout. It was also observed that the culture forms were infective to rainbow trout and the dividing forms seen in the cultures were similar to those found in the blood of trout (Woo, 1978). However, Peckova and Lom (1990) grew different strains of trypanoplasms in axenic culture. Cultures of *T. borreli* (from *Blicca bjoerkna, Cyprinus carpio, Scardinius erythrophthalmus* and *Tinca tinca*), *T. guerneyorum* (from *Esox lucius* and *Trypanoplasma* spp.) were not infective for susceptible fish. These authors grew the trypanoplasms in biphasic blood agar medium, SNB-9 supplemented with vitamins and antibiotics. In cuture, the flagellates transformed into smaller, elongated stages with a poorly developed undulating membrane and into short forms that were morphologically similar to stages in the leech vector. The authors passaged the cultures weekly at 17–20°C but they also grew at 4°C. During the present culture experiment, it was found that medium 199 could maintain the parasites for 36–40 hrs when they began to dwindle in numbers. The diphasic method adopted from Letch (1979) proved to be better in which the trypanosomes were actively motile for 48 hrs and continued to survive after the first sub-passage and were alive and motile at 48 hrs when the second sub- passage was attempted.

The parasite could thrive in MEM 199 for 36–40 hrs but in diphasic medium (NNN + Lockes soln.+ antibiotics) the parasites were actively motile even after the second sub-passage (48 hrs). It thus appears that the latter medium is more appropriate for maintaining trypanosome culture as compared to the former.

Chapter 6

Control Strategy Program

Contents

6.1. LYMPHOCYTE CULTURE

Interest in the *in vitro* reactivity of lymphocytes has greatly intensified since the discovery of their involvement in immunological function. The past few decades have witnessed a rapid advance in the culture techniques of lymphocytes. Mitogens, are chemical substances (usually proteins) that encourage cells to commence cell division, triggering mitosis. They trigger signal transduction pathways in which mitogen-activated protein kinase is involved, leading to mitosis. Duly recognizing the importance of mitogens for triggering lymphocyte division, and mitogen-stimulated transformation of blood lymphocytes in culture, the technique has been extensively used for cytogenetic, immunological, and metabolic studies. As a result, many clinical and research laboratories are now routinely doing short-term lymphocyte cultures. However, the techniques are not as popular with fish lymphocytes, and therefore, the present experiment was designed keeping in view the role of lymphocytes in generating immunity against parasites.

Fish are the oldest animal group with an immune system showing clear similarities with the defense systems of mammals and birds. The cellular and humoral responses have the expected characteristics of specificity and memory. Moreover, the appearance of immunoglobulin (Ig) as well as the other members of the 'Ig superfamily' is observed for the first time in this animal group.

Parasitic diseases sometimes enact tremendous tolls on fish production specially when fish are under intensive culture. In recent years, there have been concerted efforts made to understand the piscine immune system better specially as it relates to protection against infectious agents. Some of the recent advances in the fish diseases include the prospects of using immunological strategies against a limited number of parasitic infections and the adaptation of immunological techniques to study the disease process.

In many parasitic groups, the immunological responses of fish to the parasites are little understood. However, these techniques can help us better understand the host-parasite relationship and to resolve phylogenetic relationships between species or groups of parasites. In addition, they are of use in serodiagnostics of infections and for evaluating protection in recovered and immunized fish.

The immune responses of fish are influenced by many of the factors which also affect immunity in homoeothermic vertebrates. These include stress, hormonal changes, seasonal effects, intercurrent infections, drugs and environmental pollution. In addition, as aquatic poikilotherms, fish are particularly susceptible to changes in water quality and to the effects of ambient temperature. Their gills provide a large surface area which can become a route for antigen uptake. Also, as amniotes, many of which have free living larvae, young fish maybe vulnerable to antigens encountered at a stage when their immune system is relatively immature.

Over the recent years, the lymphocyte has become the center of interest for immunological and cytogenetic studies as it can be easily obtained from peripheral blood, can be induced to transform into a precursor blast cell which can elaborate various substances such as antibodies and lymphokines and can also be made to mitose, making chromosomes readily available for karyotype studies. In mammalian immunology, separation procedures are now established and have been well reviewed (Denman, 1973). However, this work is still in infancy with fish blood.

The cultivation of fish lymphocytes appears to have similar factors affecting it, as in mammals. Although culture conditions can be reasonably controlled, there are a number of variables which make consistency impossible. Many of the factors can have significant effects on the cyclic AMP responses of culture cells and during the past years, the importance of cyclic nucleotides in lymphocyte proliferation and differentiation has become apparent (Hadden, 1977). It has been suggested that cAMP is associated with inhibition of lymphocyte proliferation and that cGMP is involved in induction of lymphocyte proliferation (Hadden, 1978). Calcium has also been reported to be important in lymphocyte proliferation (Parker and Lukes, 1971). Therefore, changes in intracellular cAMP, cGMP and calcium in the cell have been suggested as the regulatory signals for cell growth and replication (Schonhofer and Peters, 1977).

Fish lymphocyte population itself is a source of variability. Very little is known about their types (T and B), responses to mitogens, functions and possible sub-populations that it is difficult to work with a population of known types. It was therefore proposed to culture fish lymphocytes in order to observe blastogenesis in these WBC's and to assess their role in inducing cellular immunity.

During this part of the study, lymphocytes of *Heteropneustes fossilis* and *Clarias batrachus* were isolated by centrifugation of diluted blood and Plasmagel method (Monoculture) and MLC (Mixed Lymphocyte Culture) directly from the spleen and were cultured in order to elucidate the phenomenon of cellular immunity against parasites. For this purpose, Growth Medium A (mitogen, Con A; antibiotic, penicillin) and Growth Medium B (mitogen PHA, antibiotic streptomycin) in medium RPMI 1640 with 10% FCS and HEPES buffer were used as the culture media. Blastogenesis and proliferation of lymphocytes were observed for assessing the viability of lymphocytes as tools for immunization techniques.

6.1.1. Methodology

6.1.1.1. Lymphocyte Isolation

Live specimens of *Heteropneustes fossilis* (size 15–20 cms) and *Clarias batrachus* (size 20–25 cms) were collected from ponds of Rohilkhand region, India and acclimatized to laboratory conditions for 10 days. Blood was collected aseptically from the caudal vein using a sterile syringe containing heparin (500 units heparin in 0.2 ml). 1 ml of blood was diluted four times in Hanks balanced salt solution, mixed and the lymphocytes separated by centrifugation of diluted blood and plasmagel method according to Blaxhall (1985). Observations were made on 10 samples for each method. All glassware used was sterilized in hot air and autoclaved prior to experimentation.

i. Centrifugation of Diluted Blood: The diluted blood was placed in sterile, plastic tube (10 cm × 1.3 cm) and centrifuged at 300–350 rpm at room temperature. The supernatant, rich in leukocytes was removed at 5 mins interval for 3 times.

ii. Plasma Gel: Plasmagel (1–2 ml) was mixed with diluted blood and centrifuged at 40–50 g for 10 min. The supernatant leukocyte rich plasma was removed and the original tube was centrifuged again.

6.1.1.2. Lymphocyte Culture

The lymphocytes were cultured either after separation from the blood as described above or directly from the spleen (Mixed Lymphocyte Culture-MLC). In the latter case, the spleen was collected from sacrificed fish and minced carefully in phosphate buffer at pH 7.2. The composition of growth media (A and B) are given in Table 6.1.

Table 6.1. Composition of Growth Media (A and B) for Culture of Lymphocytes of *Heteropneustes fossilis* and *Clarias batrachus*

Composition	**Growth Medium A**	**Growth Medium B**
RPMI 1640	100 ml	100 ml
10% FCS	10 ml	10 ml
HEPES buffer	10 ml	10 ml
Antibiotic Penicillin Streptomycin	100 IU/ml	100 µg/ml
Mitogen Con A PHA	10 mg/ml	1 mg/ml

The media were membrane filtered (0.34 μ) prior to the addition of the mitogens. Lymphocytes were cultured at 20°C. Observations on blastogenesis were made every 24 hr for 96 hr. The lymphocytes were smeared on glass slides, fixed in methanol, stained in Giemsa, washed 2–3 times in acetone, cleared in Xylene and mounted in Canada balsam. All investigations were conducted aseptically under a laminar flow hood.

6.1.2. Results and Discussion

6.1.2.1. Lymphocyte Isolation

The mean recovery of lymphocytes was 45% by centrifugation of diluted blood and was significantly higher ($P < 0.01$) than the plasmagel method (30%).

In mammalian immunology, lymphocyte separation procedure have been established and reviewed by Denman (1973) who stressed the role of pH osmolarity of the gradient, cell content, volume and dilution of the sample to affect the separation of cells. However, similar studies on fish blood are far from adequate.

The aim of the separation technique is to obtain maximum recovery of viable lymphocytes for subsequent use. During, experimentation, the low yield of lymphocytes by Plasmagel method is probably due to the reason that this procedure yields a mixed suspension of lymphocytes, granulocytes, thrombocytes and a variable number of erythrocytes. Careful controlling of the centrifugation speed is required to avoid undue sedimentation of lymphocytes and gross contamination with erythrocytes. Erythrocyte contamination is generally 5–10% after careful handling. The method may be more effective using larger volume of blood.

On the other hand, the cells or their receptors are not damaged due to low centrifugation speed and the absence of additives in the centrifugation method (Blaxhall, 1985). The method proved to be simple and yielded a better recovery rate of lymphocytes than the plasma gel method (as mentioned above). Nevertheless, accurate dilution of the blood, appropriate adjustment of the speed (300–350 rpm) and time (5 min interval) are required to prevent the loss of lymphocytes into the erythrocyte layer.

6.1.2.2. Lymphocyte Culture

Results indicate that in both species of catfish, the lymphocytes in growth medium A started to develop in 24 hr but the lymphocyte concentration was comparatively very low (Table 6.2). Apparently, Con A binds to glycoprotein containing α-mannosyl moieties. On the other hand, the growth of lymphocyte in medium B peaked at 24 hr and persisted up to the culmination of experiment (96 hr). Phytohemagglutinin is known to induce lymphokine production in cultures (Bellanti, 1978).

Table 6.2. Cultivation of Fish Lymphocytes in RPMI 1640 with Con A and PHA as Mitogens (pH 7.2)

Cultivation Method	**Growth Medium**	**Lymphocyte Culture (hrs)**					
		24	**48**	**60**	**72**	**84**	**96**
Monoculture							
Hetero-pneustes fossilis	A	+	+	+	+	+	+
	B	++	++	++	++	++	++
Clarias batrachus	A	+	+	+	+	+	+
	B	++	++	++	++	++	++
MLC*	A	++	++	++	++	++	++
	B	+++	+++	+++	+++	+++	+++

* Mixed Lymphocyte Culture.

The best results were obtained in MLC where maximum lymphocytes could be harvested throughout the test period in both the growth media. Mixed Lymphocyte Culture is an ideal *in vitro* test for cell mediated immune reaction towards cell surface alloantigens encoded by Major Histocompatibility Complex (MHC) and can be used for the assay of blastogenesis. During this cultivation, genetically dissimilar lymphocytes are cultured together, blast transformation and mitosis occur and each population of lymphocyte reacts to the foreign histocompatibility antigens of the other population which may account for the enhanced division.

Mitogens are plant proteins which can be added to cultures of non-sensitized lymphocytes to initiate blast transformation (Bellanti, 1978). The most important of these compounds is a family of proteins called lectins for example Phytohemagglutinin (PHA, source-kidney bean, *Phasoleus vulgaris*) and Concanavalin A (Con A, source-jackbean *Canavalia insiformis*) which have been used as mitogens to initiate lymphocyte blastogenesis during the present course of investigations. Both mitogens are known to be specific for T cell sub- set (Chakravarty, 1996). The lymphocytes are stimulated to divide by membrane perturbations induced by cross linking of surface molecules. Mitogens have the potential to bind to and activate large sub-populations of lymphocytes. Con A and PHA are regarded as non-specific stimulants which selectively stimulate the lymphocytes to synthesize DNA and divide thereby inducing blastogenesis in the lymphocyte sub-populations (Oppenhein and Rosenstreich, 1976). These lectins can bind sugars specifically, for example, PHA binds N-acetyl-D-galactosamine and Con A binds α–D-mannose and α–D-glucose. The binding of lectin to lymphocyte membranes stimulates nucleoside incorporation, phospholipid synthesis, DNA synthesis and cell division. All lymphocytes do not respond equally to all lectins: PHA primarily stimulates T-cell division, slightly affecting B-cells and Con A is also a T-cell mitogen (Tizard, 1995).

The dosage of mitogen, especially PHA, required to obtain optimum mitosis is about twice that required for mammals when using reagent grade PHA. When using purified PHA, the optimum is about 0.001–0.01 units/ml, the response showing a plateau effect, not seen with human lymphocytes. The reason for the higher dose required in fish may be a function of the greater number of lymphocytes contained in a unit number of leukocytes than in mammals, thus requiring more PHA to be present for reaction with the increased number of lymphocytes. When using reagent grade PHA with fish lymphocytes it is impossible to obtain mitosis consistently, although the lymphocytes show morphological changes indicating commencement of stimulation.

Many morphologic and metabolic changes accompany blast transformation. It has been proposed that in the initial stages, lectin binding to specific receptors or lymphocytes leads to changes in membrane permeability and transport of various nutrients and cations, particularly calcium (Segel, 1981).

Lymphocyte stimulation has lately been defined in terms of biochemical changes accompanying the process of transformation (Wintrobe, 1981). The mitogens interact with a variety of specific membrane receptors (Douglas, 1971). In PHA stimulated cells, this step is followed by an increased turnover of membrane phosphatidilinositide and is followed in the next 24 hr by increased accumulation of phospholipids (Fischer and Mueller, 1968). An increased acetylation of histones appears to precede the increase in nuclear RNA synthesis indicating changes in the fine structure of chromatin and the ability of DNA to serve as a template for RNA systhesis (Cooper, 1971). This is followed by an increased protein synthesis. Morphological changes begin within 24 hr, followed by endocytosis and redistribution of acid hydrolases within lysosomes. Within 24 hr to 34 hr, there is a gradual increase in the incorporation of 3H-thymidine (Bloom, 1971) which has been used as an indicator for the measure of DNA synthesis and the degree of lymphocyte stimulation (Wintrobe, 1981). It has been reported (Thomas and Woo, 1995) that T-cell mitogens (PHA and Con A) and B cell mitogens (e.g., lipopolysaccharides) induce *in vitro* proliferation of respective cell lines. This resulted in increased DNA synthesis which was measured by the incorporation of radio-labelled DNA precursors (e.g., tritiated thymidine) in the cells. The cells were cultured in a suitable culture medium with the mitogen for 3–5 days. Radiolabelled DNA precursor was added to the cultures 24 hrs before harvesting. The DNA of the harvested cells was precipitated and the radioactivity was measured in a liquid scintillation counter. This technique was employed to study the responsiveness of lymphocytes in *Atheresthes stomias* infected with *Phrixocephalus cincinnatus* (Faisal et al., 1990).

A number of variables may play a role in the cultivation of fish lymphocytes viz. cyclic nucleotides (Hadden, 1978), atmosphere of fish lymphocytes, dose of the mitogen, serum added to the culture medium, stress, hormones, diet, age, disease and drugs (Barta, 1983).

Results of the present investigations indicate that PHA is a superior mitogen specially in the species investigated which can trigger non-specific blastogenesis and proliferation of non-sensitized lymphocytes and MLC, a better technique for lymphocyte culture as compared to monoculture. The present work is an attempt to stimulate studies on antibody formation and lymphokines under adequate sensitization as well as the probability to utilize lymphocytes as biotechnological tools in order to determine their immunizing potency. In the long term, the relationship between lymphocyte blastogenesis and immune response due to sensitization by parasites may prove to be an exciting new area of research in order unravel many unsolved questions on fish immunity due to an array of fish pathogens.

The relationship between fish genetics and immune reactivity will emerge as an exciting new area of research. All the data on fish cell populations (e.g., characterization by monoclonal antibodies), regulation of the response by factors and the major histocompatibility complex will be needed to solve the question of differential disease resistance in different fish strains.

The establishment of a fish lymphocyte culture technique giving as consistent results as with mammalian lymphocytes appears to a difficult. The areas still requiring further investigation are as follows:

1. The development of a suitable alternative to serum.
2. An investigation of the action of cell nucleotides in mitogen stimulated fish lymphocytes, together with their subsequent loss as a result of possible depletion of membrane lipids.
3. The application of lymphocyte culture techniques in monitoring the effects of environmental pollutants and in fish cytogenetics.

6.2. CELLULAR IMMUNITY

Cell-mediated cytotoxicity has been regarded as one of the protective mechanism in fish in addition to complement fixing antibodies and phagocytosis by macrophages (Woo, 2003).

In the immunologically mature individual, contact with an antigen leads not only to the production of circulating antibody, but also to the development of a separate cell-mediated form of response. Cell-mediated immunity occurs particularly in infections by agents which enter the cells.

The thymus-dependent lymphocytes can be stimulated to differentiate and divide by plant extracts (PHA). The mitogens act on the small lymphocytes of all vertebrates and induce enlargement of the cells, increased synthesis of RNA followed later by DNA synthesis. The plant extracts are specific for sugar determinants present in the lymphocyte cell membrane, and interaction leads to 'triggering' of the cell activity. This involves biochemical changes and include transport of ions (e.g., Ca^{++}), amino acid and nucleotides through the cell membrane leading to purine synthesis and finally cell division. This process known as lymphocyte transformation can be brought about in sensitized lymphocytes on subsequent exposure to the sensitizing antigen. The non-specific plant mitogens are used as a test of the competence of the cell-mediated immune system (Weir, 1983).

Steps in lymphocyte transformation can be summarized as follows:

1. Reaction of inducer at plasma membrane-PHA or antigen
2. Increased synthesis of RNA in 30 min to 2 hours
3. Morphological changes – enlargement of the cell, changes in nucleus, increases in (transformation) lysosome granules and endocytosis (commences at about 20 hours)
4. Increase in thymidine incorporation into DNA about 36 hours for PHA and 40 hours for antigens.
5. Mitosis

Cell mediated immune responses in fish against parasites is little explored despite the well documented occurrence of functional T lymphocytes and a number of lymphokines including the macrophage migration inhibition factor, mitogenic factors, chemotactic factors and interleukin 1 and 2 (Pasquier, 1989; Thomas and Woo, 1995). The present study is thus an initial attempt to stimulate blastogenesis in lymphocytes in the presence of mitogens and this would prove helpful for the control of infection in fish by enhancing cell mediated immunity.

6.3. INDUCTION OF IMMUNOLOGICAL MEMORY AND DEVELOPMENT OF IMMUNE SERA AGAINST FISH TRYPANOSOMES

Fish, as the first vertebrate group appearing in evolution after adaptive radiation during the Devonic period, still represents the most successful and diverse group of vertebrates. This heterogenous group of organisms occupies an apparent crossroads between the innate immune response and the appearance of the adaptive immune response. Importantly, immune organs homologous to those of the mammalian immune system are present in fish. Some traits of the fish immune system anticipate mammalian sophistication. However, their structural complexity is less, potentially limiting the capability to generate fully functional adaptive immune responses against pathogen invasion (Tort et al., 2003).

The immune system must recognize a large number of cells and molecules (antigens) that it has never seen before and it must decide how to respond to them. Some antigens that make up the individual's body are not attacked. Other antigens, such as viruses, bacteria, parasites and toxins are responded to by mixtures of the T cell response and the different types of antibody responses. The immune system remembers antigens it has seen before and on its next encounter, it is often capable of eliminating them before disease occurs. This 'memory' is the basis of vaccination. The immune response and immunological memory are complex and this exposition is necessarily simplified and restricted.

The main characteristics of the mammalian immune response can be observed in fish: recognition, specificity and memory. Fish can reject allografts (Botham et al., 1980) and synthesize specific antibodies after antigenic challenge (Carton, 1973). Specific immunoglobulins can be produced by the teleost fish in response to antigenic stimulation. In teleostean fish, only one Ig class (Desvaux and Charlemagne, 1981) was said to be synthesized (IgM-like, tetrameric) but it is now recognized that the fish immune system contains five classes of immunoglobulins: IgM (Warr, 1995), IgD (Hirono et al., 2003), IgZ (Danilova et al., 2005), IgT (Hansen et al., 2005) and IgH (Savan et al., 2005). However, out of these, the predominant immunoglobulin class in fish blood serum is still tetrameric IgM-like molecule (Koumansvandiepen et al., 1995) consisting of 8 light (L) and eight heavy (H) chains and has a molecular weight (MW) between 600,000 and 800,000 depending on the fish species. Molecular weight for most teleost H and L chains range between 70–81 and 23–28 kDa respectively. Desvaux and Charlemagne (1981) characterized the humoral response in goldfish to particulate antigens and discussed the kinetics and regulation of antibody synthesis. Ellis (1982) demonstrated immunoglobulins in the plasma, gut, skin, bile and mucous of fish and differentiated between the immune mechanisms of fish and higher vertebrates.

The ability of the body to defend itself against specific invading agents such as bacteria, toxins, viruses and foreign tissues is called *immunity*. Immunity is said to have a 'memory' for most invading agents encountered before, because a second encounter with the same agent prompts a rapid and vigorous response. This is called *immunological memory* which leads to a perception that an individual is immune to a particular agent.

This important feature of the immune system, the capacity to develop immunological memory is the ability of the body's immune system to remember an encounter with an antigen due to the activation of B cells or T cells having specificity for the antigen and to react more swiftly to the antigen by means of these activated cells in a later encounter. It derives its associative and robust nature from a sparse sampling of a huge input space by recognition units (B cells and T cells) and a distribution of the memory among many independent units (B cells and T cells in the memory population of the immune system).

Immunological memory, created from a primary response to a specific pathogen, provides an enhanced response to secondary encounters with that same, specific pathogen. This process of acquired immunity is the basis of vaccination. Primary response can take 2 days and up to 2 weeks to develop. After the body gains immunity towards a certain pathogen, when infection by that pathogen occurs again, the immune response is called the secondary response.

A first contact with an antigen usually induces relatively short-lived effector cells (activated Th, plasma cells or cytotoxic T cells). There are also long-lived memory cells among the progeny of the original non-primed lymphocytes. These memory cells retain the capacity to be stimulated by the antigen. The development of immunological memory is often examined indirectly by monitoring the secondary response. In the case of positive memory, this response is faster and more vigorous than the primary response. The height of the secondary response is dependent on the amount of priming antigen. A relatively low priming dose is usually optimal for memory development in carp (Rijkers et al., 1980: Lamers et al., 1985). In carp, the ratio between secondary and primary antibody responses never reached the high level found in mammals (10–20-fold in carp and up to 100-fold in mammals) (Woo, 1995). Elegant *in vitro* studies in rainbow trout also showed that the B precursor cell frequency in fish immunized with the hapten- carrier TNP-KLH increased about 15-fold (Arkoosh and Kaattari, 1991). The same authors also showed that there was no evidence for antibody affinity maturation during the primary or secondary response against this T- dependent antigen. This would indicate that memory in fish is probably due to an expansion of the antigen-specific precursor cell pool. However, a qualitative change in affinity maturation, as seen in mammals, was not observed by Woo (1995).

6.3.1. Immune Serum

The persistent population of memory cells is the mechanism by which the immune system remembers. If the body is exposed to the same antigen, the memory population quickly produces large quantities of antibodies and many a times the antigen is cleared before it causes diseases. The repeated recognition of the same antigen by the immune system (B cells and T cells) and the distribution of the memory to produce immunity results in the formation of the immune serum. This immune serum is refractive to identical antigen challenges and is ideal for immunization purposes.

Keeping the above facts in view, the experimental work executed in this part of the thesis was conducted to observe if repeated challenges of the same antigen (*Trypanosoma kargenensis*) in *Heteropneustes fossilis* is capable of generating immunological memory in the host, and if so, after how many passages can immune serum be obtained which can be recommended for use for immunization purposes in order to control the manifestations of the said parasite.

The experiment was conducted as follows:

6.3.2. Methodology

Heteropneustes fossilis (20–25 cm) were maintained in the laboratory for the experimental assays.

- **Parasite Inoculum**: The parasite inoculum was prepared from the blood of *H. fossilis* infected with *Trypanosoma kargenensis*. Blood samples from ts- infected fish (0.1 ml fish-1) were obtained from the caudal vein containing 25,000 trypanosomes and inoculated intraperitoneally with 0.3 ml of cold-blooded vertebrate Ringer's solution (CBVR).

The experiment was executed in 6 sets of experiments in different groups as follows:

Set I:	Group I (Control): Group II:	Not inoculated with trypanosomes Inoculated with ts-infected blood
Set II:	Group III (Control): Group IV:	Inoculated with ts-infected blood Recovered fish of Group II challenged with ts-infected blood
Set III:	Group V (Control): Group VI:	Inoculated with ts-infected blood Recovered fish of Group IV challenged with ts-infected blood

Set IV:	Group VII (Control): Group VIII:	Inoculated with ts-infected blood Recovered fish of Group VI challenged with ts-infected blood
Set V:	Group IX (Control): Group X:	Inoculated with ts-infected blood Recovered fish of Group VIII challenged with ts-infected blood
Set VI:	Group XI (Control): Group XII: Group XIII:	Inoculated with ts-infected blood Naïve fish inoculated with immune serum of Group X and challenged with ts-infected blood Naïve fish inoculated with heated immune serum of Group X and challenged with ts-infected blood

Through this experiment, the *in vivo* immunity of *T. kargenensis* was determined in *H. fossilis* through serial passage of the parasite. For this purpose, the experiment was executed in 6 sets of experiments as mentioned above. Prior to experimentation, the blood of naïve fish was examined for parasites and categorized into uninfected fish and infected fish. Out of these, uninfected fishes of Group I (n=10) were not inoculated with parasites and served as the control group whereas Group II (n=20) fish were inoculated intra-peritoneally with approximately 25,000 parasites obtained from infected fish. Out of the recovered fish of Group II (n=10), fishes were selected for Group IV and were again inoculated with parasites. Similarly, the recovered fish of Group IV were selected as Group VI fish (n=6); recovered fish of Group VI as Group VIII fish (n=6), recovered fish of Group VIII as Group X fish (n=5). Group XII and XIII comprised of naïve fish, the former being inoculated with immune serum of Group X and ts-inoculum (as above) and the latter, inoculated with heat-inactivated immune serum of Group X and ts-inoculum. The odd Groups III, V, VII, IX and XI were inoculated with the parasites and acted as the control groups (n=10 each) for each even group.

In order to detect infection in the samples, parasites were counted in a hemocytometer (Archer, 1965). The slides were examined every week up to 15 weeks. Examination of 40 microscopic fields (objective X 10, ocular X 40) with no detectable infection were considered to have recovered from the infection and blood from these fish were used for the next set of experiment as mentioned above.

6.3.4. Results

The mean parasitemia was relatively high for the first 9 weeks and then declined to a low level for the duration of the experiment. The cumulative mortality was 7 (4 in Group II, 2 in Group IV and 1 in Group VI) after 15 weeks. Neither the parasites were detected in Group I (control group) nor did any of the fish of this group die.

Parasites were readily detected in Group II fish challenged with *T. kargenensis* up to 12 weeks but were not detected after that. Such recovered fish with no detectable parasitemia following challenge were termed as immune fish and the plasma from these termed as immune plasma. Plasma from naïve fish were termed normal plasma.

Group IV fish (recovered fish of Group II) were again inoculated with 0.1 ml of ts-infected blood and in this group, infection was detected up to 8 weeks. Similarly, Group VI and VIII fish remained infected up to 6 weeks and 2 weeks respectively and then the infection subsided. Interestingly, when Group X fish were inoculated with the trypanosomes, no infection was detected right from 1 week post infection (w.p.i.) up to 15 w.p.i. indicating that these fish were now completely immune against *T. kargenensis* (Plate LXX).

Now in the last set of experiment (Set VI), naive fish were inoculated with immune serum obtained from Group X (Group XII) and were challenged with 0.1 ml of ts-inoculum (Group XII) whereas naïve fish of Group XIII were inoculated with heat-inactivated immune serum and also challenged with 0.1 ml of ts-infected inoculum. Out of the 5 fish of Group XII, low grade infection was detected in 1 fish up to 5 w.p.i. and then the infection subsided and in the remaining 4 fish, no infection was detected from 1 to 15 w.p.i. In the last group (Group XIII) parasites were readily detected throughout the experimentation (Table 6.3).

6.3.5. Discussion

The final disappearance of trypanosomes from the blood stream is the outcome of an immune response. There are a few records of trypanocidal antibodies in fish. Barrow (1954) observed lytic action of the serum of fish that had recovered from a *Trypanosoma* infection at 20°C. This lytic capacity soon decreased and was lost by three weeks after the disappearance of the parasite from the blood. The parasite antigens stimulate the secretion of specific antibody by the host's B-lymphocytes into the blood plasma (Roitt, 1974) and may attack the parasites in various ways including complement mediated lysis (lysins), agglutination (agglutinins) and facilitate phagocytosis (opsonins). Cottrell (1976) found that plaice infected with *T. platesae* had elevated levels of serum immunoglobulins possessing the beta electrophoretic mobility which might be associated with antibody. In goldfish experimentally infected with *T. carassii*, the immunity to super infection could last up to 350 days (Lom and Dykova, 1992). Probably this was due to non- sterile immunity as evidenced by outbreaks of parasitemia in some of the fish apparently free from infection previously. This appears to be temperature dependent as when fish are kept at a low sub-optimal temperature, the infection does not reach the last phase and a certain level of parasitemia is maintained. This may be due to other factors as well- stress, nutritional deficiency and fishes showing low parasitemia may be super infected with the same trypanosome species by a leech or by a syringe passage which may in some cases result in a high level of parasitemia (Lom and Dykova, 1992).

Table 6.3. Serial Passage of *Trypanosoma kargenesis* in *Heteropneustes fossilis* for the Development of Immune Sera

Trypanosome Infection (Weeks)																
Set of Experiment	**Groups**	**1**	**2**	**3**	**4**	**5**	**6**	**7**	**8**	**9**	**10**	**11**	**12**	**13**	**14**	**15**
Set I	Group I (control)	-	-	-	-	-	-	-	-	-	-	-	-	-	-	-
	Group II	+	+	+	+	+	+	+	+	+	+	+	+	-	-	-
Set II	Group III (control)	+	+	+	+	+	+	+	+	+	+	+	+	+	-	-
	Group IV	+	+	+	+	+	+	+	+	-	-	-	-	-	-	-
Set III	Group V (control)	+	+	+	+	+	+	+	+	+	+	+	+	-	-	-
	Group VI	+	+	+	+	+	+	-	-	-	-	-	-	-	-	-
Set IV	Group VII (control)	+	+	+	+	+	+	+	+	+	+	+	+	+	+	-
	Group VIII	+	+	-	-	-	-	-	-	-	-	-	-	-	-	-
Set V	Group IX (control)	+	+	+	+	+	+	+	+	+	+	+	+	+	+	+
	Group X	-	-	-	-	-	-	-	-	-	-	-	-	-	-	-

Set VI	Group XI (control)	+	+	+	+	+	+	+	+	+	+	+	+	+	+	-
	Group XII Naïve fish (Untreated immune serum+ts-inoculum	+(1 fish)	+	+	+	+	-	-	-	-	-	-	-	-	-	-
		-(4 fish)	-	-	-	-	-	-	-	-	-	-	-	-	-	-
	Group XIII Naïve fish (Δ immune serum+ts-inoculum	+	+	+	+	+	+	+	+	+	+	+	+	+	+	+

The aim of the parasite is to persist in its host for as long as possible at an optimum level in order to ensure a favorable chance of onwards transmission but low enough to avoid undue risk of killing the host. An unusual antibody, "ablastin" maybe directed against certain *Trypanosoma* species which interferes with parasite division but has no other harmful effect (D' Alesandro, 1975). The antibody "ablastin" is thus ideal from the parasite's view point by preventing multiplication, it reduces parasitemia to a level which does not seriously endanger the host and probably does not furnish too great an antigenic stimulus to the latter. The adult parasites may thus be able to persist for long periods in the blood of the vertebrates host, waiting to continue its life cycle in the vector. "Ablastic" antibodies have been found to occur in amphibians (Lom, 1969), however, their presence in fish is uncertain. It may be possible that ablastin or ablastin-like antibodies are stimulated in fish hosts as well and may account for the same phenomenon. Under certain conditions, fish may develop protective, nonsterile immunity to homologous challenge (Woo et al., 1983).

The piscine immune system is well developed, and in many ways it is similar to that in mammals (Van Muiswinkel and Vervoorn-Van Der Wal, 2006) which include a comparable set of immunocompetent cells (Ardelli and Woo, 2006). In general, the adaptive immune response is slower to develop in fish than in mammals, and this is in part due to its lower body temperature. However, the innate immunity in fish is as well developed and is as responsive as that in mammals.

Heterologous sera may exhibit a toxic effect upon trypanosomes and trypanoplasms; for example, trout serum kills carp-derived *Trypanosoma danilewskyi catostomi* from *Catostomus commersonii* was sensitive to fresh blood plasma of five heterologous fish hosts, being killed by a 3 hours exposure to plasma diluted to the titer 1: 16 (in rainbow trout plasma) (Bower and Woo, 1977). On the evidence of their experiments, the authors concluded that rainbow trout had a natural properdin system activated by the parasite. This is suggested to be one of the mechanisms of natural immunity of fish.

Fish that recover from cryptobiosis are protected from challenges (Jones and Woo, 1987; Li and Woo, 1995) and this obviously is an impetus for the development of an acquired protective strategy. As the immunity in recovered fish is non-sterile, Woo (2003) assumed that the protection was for the life of the fish.

The role of antibody production by passive immunization has been examined in the present experiment. The initial inoculation of *T. kargenensis* could induce infection in *H. fossilis* which survived up to 12 weeks in Group II and then the parasites were probably lysed by the immune system of the fish. The recovered fish of this group, when challenged again with the same parasite could survive up to 8 weeks p.i. (Group IV) and on repeated serial passages, the time of survival decreased to 6 weeks (Group VI), 2 weeks (Group VIII) and ultimately, in Group X, the fish had probably become immune so that the group was refractive to challenge and no parasitemia was detected. The plasma of these immune fish is the immune plasma.

Naive fish when inoculated with immune serum and challenged with ts-inoculum (Group XII) was refractive to *Trypanosoma*, secondly, the immune serum was heated (Group XIII) so that the complement was inactivated and, in this case, after challenging with ts-inoculum, infection was again restored indicating that injection of heated immune serum did not have a protective effect, probably complement activation also plays an important role. The complement system appears to be one of the central immune responses in fish. The classic, alternative and possibly also the lectin pathways have been described in several teleost groups (Nonaka and Smith, 2000). Ahmed (1994) was able to protect carp from a lethal infection with *T. danilewskyi* by a single injection with plasma from carp that had recovered from an infection with the same parasite. The protective plasma had a greater lytic activity (probably much more than that observed in the present case) as detected *in vitro*, compared with control plasma. However, since complement activity was not inactivated, and no specific antibody levels were measured, the mechanisms responsible for protection remain undefined.

Humoral immune response is induced in fish by piscine hemoflagellates has been demonstrated by several authors (Barrow, 1954, 1955; Lom, 1973; Sypek and Burreson, 1983; Burreson and Frizzell, 1986). *Cryptobia bullocki* and other species of hemoflagellates are apparently controlled by lytic antibody (Barrow, 1955; Sypek and Burreson, 1983). The protective role of antibody in goldfish, *Carassius auratus*, recovered from infection with *Trypanosoma danilewskyi* was shown by Lom (1973) and Woo (1981). The latter author found that goldfish were refractive to challenge for as long as 190 days following recovery from *T. danilewskyi*. However, antibody alone may not be protective against some bacterial pathogens of fish (Cossarini-Dunier, 1986). In addition, *C. salmositica* and *Cryptobia* (=*Trypanoplasma*) *bullocki* in summer flounder, *Paralichthys dentatus* have been observed with in peritoneal macrophages suggesting the importance of cellular immunity (Woo, 1979; Sypek and Burreson, 1983).

Woo (1979) reported that during experimental infection of *Trypanoplasma salmositica* in *Salmo giardneri*, an acute stage characterized by high, fluctuating parasitemia is followed by a chronic stage which persists for the duration of infection. Wehnert and Woo (1981) reported that rainbow trout are refractive to homologous challenge following recovery. Non-sterile immunity occurs in mice apparently recovered from infection with *Trypanosoma musculi*. The renal *vasa recta* harbors trypanosomes that maintain host immunity to homologous challenge (Oliver and Viens, 1985). Lom (1979) suggested that immunity to piscine hemoflagellates may also be non-sterile because of the long duration of protection. Burreson and Frizzell (1986) attempted to immunize summer flounder with formalized *C. bullocki*. The antibody titer increased following inoculation of antigen; however, there was no protection against homologous challenge. The authors used a modified ELISA (immunoblot EIA) technique to demonstrate *C. bullocki*-specific immunoglobulin in summer flounder and suggested that immunity to *C. bullocki* was sterile because no parasite-specific antibody was detected prior to challenge. This was based on the immunological response of a single flounder on challenge. The authors demonstrated that four summer flounders that had recovered from infection with *Cryptobia* (=*Trypanoplasma*) *bullocki* were refractive to challenge. However, failure to infect challenged flounders may have been due to a non-viable inoculum because naïve controls were apparently not inoculated concurrently. This was contrary to Lamers et al. (1985) who showed that immunological memory was maintained for at least 12 months in *Cyprinus carpio* immunized with bacterial antigen.

Jones and Woo (1987) examined blood from 30 naïve rainbow trouts and inoculated them intraperitoneally with approximately 70,000 parasites. The authors did not detect any parasite in immune fish while the fish inoculated with blood from the challenged control was infected (>30 parasites in 50 microscopic fields). In the present observations also, the immune *Heteropneustes fossilis* were refractive to challenge with *T. kargenensis*.

Agglutinating antibody specific for *C. salmositica* was detected 1-month post-infection using indirect hemagglutination. In addition, the infectivity of cultured *C. salmositica* was neutralized and parasites were lysed by incubation in immune plasma and complement, suggesting the importance of humoral immunity (Jones and Woo, 1987). The authors also opined that protection was conferred by transferring leukocytes and plasma from immune to naïve trout. It is possible that antibody was required as opsonin since specific immunoglobulin increases the rate of phagocytosis by salmonid macrophages (Griffin, 1983; Sakai, 1984). In addition, *Cryptobia* spp. are phagocytosed by peritoneal macrophages in the infected host (Woo, 1979; Sypek and Burreson, 1983). Immune serum is necessary for attachment and subsequent ingestion of rodent trypanosomes by activated macrophages (Takayanagi and Nakatake, 1977; Stevens and Moulton, 1978). In infected or in recovered and challenged hosts, *C. salmositica* binds with complement fixing antibody and is lysed or phagocytosed. The failure of neutralizing or complement-fixing antibody to react *in vitro* with *C. salmositica* from infected fish suggests that antigenic sites on these parasites may be blocked or disguised as occurs in some mammalian trypanosome infections (Cohen, 1982).

Antigenic sites on cultured parasites are presumably not blocked or disguised. If blockage of antigenic sites prevents lysis of parasites *in vivo* as it appears to *in vitro*, then complement fixation and lysis may only be important during the acute stage of infection. In the acute stage, parasites multiply rapidly and new antigenic sites are available making the parasites more susceptible to complement mediated lysis. Thus, infection in immune fish was not established following challenge. The infection of *T. kargenensis* in *H. fossilis* in the present experiment also appears to be in the acute phase as the role of complement in parasite lysis is established in the last set of experiment (Set VI) where complement inactivation could not kill the parasites and the infection survived. Phagocytosis may be more important during the chronic stage of infection. Viens et al. (1974) demonstrated a similar response in *T. musculi* infection in mice. A trypanocidal, thymus-independent antibody is responsible for eliminating newly divided parasites while a thymus dependent factor (involving sensitized lymphocytes and macrophages) is important in eliminating the remaining, fully developed parasites from the blood.

Although *C. salmositica* is transmited by the leech, *Piscicola salmositica* (Becker and Katz, 1965b), it is also transmitted directly between rainbow trout (Woo and Wehnert, 1983) and between various *Oncorhynchus* spp. causing a rapid increase in prevalence in some hatcheries (Bower and Margolis, 1983). Jones et al. (1986) concluded that experimental infection of rainbow trout with *C. salmositica* suppresses the immune response to a secondary antigen (*Yersinia ruckeri)* and that this suppression is non-specific. This may further indicate the importance of a cellular immune response in protection. Successful immunization will require an understanding of the immune response to attenuated (for example, irradiated) strains or to 'killed' parasite antigen (for example, heat or sonicated). Jones and Woo (1987) indicated that there is protective immunity (*in vivo* and *in vitro*) to the parasite and suggested the possibility of control by vaccination. Jones et al. (1993) developed a laboratory model to monitor antibody levels in outbred carp, following infections with *T. borreli*, and found the significance of difference between various infective doses to be low because of the high variability in parasitemia. The peak antibody response in these carps coincided with a decline and eventual absence of parasitemia, resulting in the survival of the majority of fish. However, the exact nature of the immune mechanism responsible for eliminating the parasite still remains unclear.

Wiegertjes et al. (1995) examined the antibody response of carp (*Cyprinus carpio*) to infection with *Trypanoplasma borreli.* Outbred carp responded with a production of specific antibodies, but highly susceptible isogenic hybrid carp did not. This suggests an apparent relationship between susceptibility and the lack of specific antibody production. This relation was partially confirmed by the passive transfer of immunity with immune plasma. In addition, two isogenic homozygous carp lines were highly susceptible to the trypanoplasm (100% mortality), in contrast to outbred carp, the majority of which survived infection. None of the carp in either homozygous carp line produced an antibody response to parasite-unrelated antigen (DNP-KLH). This suggests that the low antibody response was not entirely due to a poor state of helminth, but that these carp have a genetically pre-determined low antibody response.

The above authors passively immunized susceptible hybrid carp (E20 × E6, n = 4 per group) i.m. with 1,000 *T. borelli* and gave regular i.v. injections of 0.1 ml serum, the first injection being given 3 days p.i. Fish were injected with (non) immune serum, either heated for 30 min at 50°C to inactivate complement activity, or left untreated. The serum injections were carried out on the first, third and fifth days of each week, for 4 w.p.i. Immune sera were obtained from outbred carp that had survived an infection with *T. borreli,* and the non-immune control sera were from trypanoplasm-free outbred carp. Serum pool antibody titres, determined by serial dilution with enzyme linked immunosorbent assay (ELISA), were 10 240 in immune and 40 in non-immune carp. Parasitemia was measured at the time of passive transfer, and mortalities were recorded daily for 5 w.p.i.

The apparent relation between susceptibility and the absence of an antibody response initiated an investigation into the role of specific antibody in the defense against the trypanoplasm. The passive transfer of immune serum that was left untreated, reduced the parasitemia approximately five-fold (between 2 and 4 w.p.i.), compared with the parasitemia in fish receiving non immune control serum. Although the fish that received untreated immune serum generally did not survive infection (1/4 only), the survival time of this group was somewhat prolonged (MST 33 days). The group that received untreated non-immune control serum, and also the fish that received heated (non) immune sera, all developed comparable parasitemia and subsequently died at a MST of 27 days.

All homozygous carp died when infected with a dose of only 100 *T. borreli*, in clear contrast to the situation in the outbred group, in which the majority of fish (71%) survived infection. This confirmed the observations of Jones et al. (1993) on outbred carp, in which an acute early phase of high parasitemia was followed by elimination of the parasite. Genetic variation among outbred individuals in the ability to mount an immune response against the parasite was suggested by extreme variability in parasitemia. Similarly, Bower and Margolis (1984) observed genetic differences in innate susceptibility of Pacific salmon (*Oncorhynchus* spp.) to the hemoflagellate *C. salmositica*. Some salmon stocks suffered high mortalities at low exposure (100 flagellates), whereas others were resistant to high exposure of 106 flagellates.

6.3.6. Adaptive (Acquired) Immunity/Vaccination

Adaptive immunity has been exploited to protect the susceptible *Oncorhynchus* spp. from cryptobiosis (Woo, 2010). Two distinctly different experimental vaccines (a live attenuated vaccine and a metalloprotease-DNA vaccine) were developed by the author. Fish inoculated with the live attenuated *Cryptobia* vaccine were protected from infection when challenged with the parasite. However, the metalloprotease-DNA vaccine did not prevent an infection in vaccinated fish but antibodies produced in the vaccinated fish neutralized the disease-causing factor secreted by the pathogen. Although the DNA-vaccinated fish was infected, it did not suffer from cryptobiosis, and it essentially turned the pathogenic *Cryptobia* into a nonpathogenic flagellate as in the case of the *Cryptobia*-tolerant brook charr (Woo, 2010).

Many pathogens change their surface epitopes to evade the immune response in hosts, consequently a surface epitope-based vaccine may not be protective against all isolates of a pathogen. An enzyme-based vaccine (e.g., cysteine) has a couple of obvious advantages in that it probably protects against all isolates because it is unlikely that metabolic enzymes are dissimilar in different isolates. It may also be protective against pathogens which change their surface epitopes to evade the host immune system and antibodies against an enzyme- based vaccine may also inhibit enzymatic activities in unrelated pathogens. Thus, an enzyme-based vaccine would be a specific but broad spectrum vaccine (Woo, 2003).

C. salmositica was attenuated by prolonged *in vitro* culture and the strain was cloned. The attenuated parasite was maintained in tissue culture medium and it has remained avirulent since 1990. It produces a low infection in rainbow trout, does not cause disease, circulates in the blood for at least 6 months, and is protective when the fish is challenged with the pathogen (Woo and Li,

1990). The vaccine strain was reported to be smaller in size and had lost a few polypeptides. It adapted to *in vitro* culture and hence multiplied much more readily than the pathogenic strain in tissue culture medium (Woo and Thomas, 1991, 1992). A single injection of the strain protected 100% of vaccinated fish and consequently it is being used routinely as an experimental vaccine to study the development and mechanism of protective immunity in salmonids and the pathobiology of the disease. The vaccine has no detectable bioenergetic cost to juvenile rainbow trout (Beamish et al., 1996) and it protects various species of juvenile and adult salmonids from the pathogen (Feng and Woo, 1997, 1998). Rainbow trout vaccinated in fresh water and transferred to sea water were also protected on parasite challenge (Li and Woo, 1997). A single dose of the vaccine protected rainbow trout for at least 24 months (Li and Woo, 1995).

Vaccinated fish are partially protected if they are challenged at 2 weeks post vaccination (wpv) while all vaccinated fish are protected (e.g., no drop in packed cell volume and virtually no detectable infection after parasite challenge) at 4 wpv. Protection is via the production of complement fixing antibodies and under *in vitro* conditions, activated macrophages from head kidneys of vaccinated fish show antibody-independent and antibody-dependent cytotoxicities. Also, in the presence of antiserum macrophages are very efficient in engulfing living parasites.

The complement fixing antibody titers (Li and Woo, 1995) and cell-mediated response (Mehta and Woo, 2002) in vaccinated fish rose significantly soon after parasite challenge (classical secondary responses), and the responses in vaccinated and challenged fish were similar to those in naïve fish at 6 weeks after infection. Humoral and cell-mediated immunity are reported to be important components of the protection against *C. salmositica* in both vaccinated and recovered fish (Mehta and Woo, 2002).

Many protozoa that are of medical and economic importance (e.g., *Trypanosoma* spp., *Leishmania* spp.) have metalloprotease and cysteine protease (Rosenthal, 1999). Some of these pathogens also modify their surface coats to evade the host immune response (e.g., antigenic variations as in the *Glossina*- transmitted mammalian trypanosomes in Africa); consequently vaccines based on surface membrane epitopes are not effective. However, an enzyme-based vaccine may be worth serious considerations as enzymes are quite conserve. Also, an enzyme-based vaccine most likely protects against all isolates of the pathogen including those from different geographical locations which may have different surface membrane antigens. This is true even among different pathogens; for example, sequences of PCR-derived fragments of the metalloprotease gene of *C. salmositica* (Jesudhasan et al., 2007) are similar to those in other kinetoplastids, such as *Leishmania chagasi*, *L. donovani*

(Roberts et al., 1993), *T. brucei* (LaCount et al., 2003), and *Trypanosoma cruzi* (El-Sayed et al., 2005). Major surface protease (MSP) also known as GP63 or leishmanolysin is a highly abundant zinc metalloprotease present on the cell surface of *Leishmania* spp. The NCBI-conserved domain search shows that the alignment for the metalloprotease of *C. salmositica* has 78.3% similarity with peptidase-M8, leishmanolysin domain (Jesudhasan et al., 2007).

In order to understand the immunological mechanisms behind susceptibility to *T. kargenensis* infection, the immune characteristics have to be selected which may have an influence on host resistance, as was done with the Biozzi mouse lines selected for high and low antibody response (Biozzi et al., 1979) and *Cyprinus carpio* infected with *Trypanoplasma borreli* infection in common carp (Wiegertjes et al., 1995). The mouse lines were shown to be resistant or susceptible to *Trypanosoma cruzi* infection, respectively, and resistance could be passed on to the susceptible line by passive immunization with immune serum (Kierszenbaum and Howard, 1976). The present study exemplifies protective immunity *in vitro* to *T. kargenensis* and suggests the possibility of control against the parasite by vaccination.

6.4. DRUG ACTION

Parasitic diseases have become a scourge for humans and animals alike. Early diagnosis and effective treatment are regarded as the key to successful intervention and the coupling of improved, early diagnosis and effective chemotherapy is a sacred grail for an effective control program.

The control of parasitic diseases requires an intricate interplay of activities in the area of animal health, education and quality research. However, this interplay is not common for all diseases as for every parasitic disease, the mechanism is different because the essential factors differ greatly. The main factors can be listed as:

1. The presence and absence of a vector or animal reservoir.
2. The effectiveness of the drugs, their acceptability and their extent of cost effectiveness.
3. Whether the politico-economic climate supports or retards animal health measures.

The search for better drugs is a perpetual process for the control of the broad spectrum of parasitic infection. The advances in parasitic genomic research and its related technology during the last decade have opened up avenues for the implementation of novel drugs which were unthinkable during the past. However, the sweeping cost of drug discovery program make research and development specially difficult to promote and sustain the global economy of the world. Today, the pharmacologist is engaged in discovering new chemotherapeutic agents commonly termed as the "*Rational drug design*" which encompasses structure-based drug design that relies heavily upon computer modeling to modify an existing drug or design a new drug which can interact specifically with a selected molecular target important in disease progression.

With regard to fish parasitic diseases, very little pharmacological data is available. The studies have mainly been restricted to limited species at a particular temperature. It is well known that dose rates vary with temperature, renal function, hydration and ambient humidity (Crawshaw,1983) therefore, there is much scope to enhance our knowledge on pharmacological data.

The applicability of anti-protozoan drugs have seldom been tested. Establishment of biochemical mode of action of a drug provides an explanation as to why protozoans are killed but the host remains unharmed, commonly known as "*Selective toxicity*," the ability of a chemical or drug to kill a microorganism without harming its host.

Chemotherapy is essentially differential toxicity; that is, the drug is more toxic to the target organism than it is to the host. Severity of the side effects of chemotherapy is dependent partly on tissue damage and adverse reactions by the host to the drug. However, the drug can be directed more specifically to the pathogen if it is conjugated to an antibody specific for the target organism. Immunochemotherapy obviously increases costs and is generally not meant for routine use; it may however be a useful tool under certain circumstances as it reduces the drug dosage and its side effects. For example, it can be used to treat infected brood fish as about 50% of brood fish annually die from cryptobiosis in some hatcheries on the west coast of North America (Woo, 2006). It is expected that the side effects and accumulation of drug residues in host tissues will be reduced in immunochemotherapy, and this may also lower the risk of the development of drug-resistance by the pathogen. Reduction in drug residue in host tissues is also an important consideration if the fish are to be used for human consumption (Woo, 2010).

Chemotherapy and chemoprophylaxis are the only practical methods available at present for the control of trypanosomiasis (Chaka and Abebe, 2003) but emergence of drug resistant strains limits the use of currently available anti-*Trypanosoma*l drugs (Tuntasuvan et al., 2003). Hence, sustainability of chemotherapy and chemoprophylaxis of this infection in India would depend on the introduction of drugs which are not hitherto in common practice.

Looking into the paucity of data on therapeutic effects of drugs on piscine hematozoans, the present venture was undertaken. The effect of two drugs,

Berenil and Samorin was tested on blood flagellates and the relative effectiveness of the drugs compared to infer on the applicability of these drugs for the control of piscine flagellates by means of chemotherapy.

6.4.1. Berenil (Diminazene Aceturate)

Remarkable therapeutic success has been obtained during the last years in the treatment of protozoal diseases in domestic animals using a novel drug developed in the research laboratories of Fabwerke Hoechst. In 1955, Fussganger wrote about his work on diminazene aceturate (Berenil), an aromatic diamidine compound discovered in 1944 as published by Fussganger (1995). This aromatic diamidine was developed from a drug called "Congasin" and other aminoquinaldines that were found to be active against trypanosomes and other babesias (Fussganger and Bauer, 1962). The authors found that for the treatment of *Trypanosoma congolense* in dogs and cattle, a single i.m. dose of 2.5 mg/kg resulted in complete cure. Diminazene is sold as Berenil (Intervet), Dimisol (Virbac), Babezene (Milborrow-Bayer AH), Berenil RTU (Intervet), Crede-Bab-Minazene (Experto Vet), Dizene Cattle (Virbac) and Veriben (Sanofi AH) and has historically also been marketed as Azidine and Ganasang.

6.4.2. Chemical Structure

Chemically, diminazene is an aromatic diamidine. Despite the mode of action being similar to that of other diamidines, it has significant quantitative differences from these preparations, which is presumed to lie in the nature of the bridge which joins the two benzamidine rings in the molecule. The diaceturate salt is found to favor the above dilactate because it was found to have a better solubility.

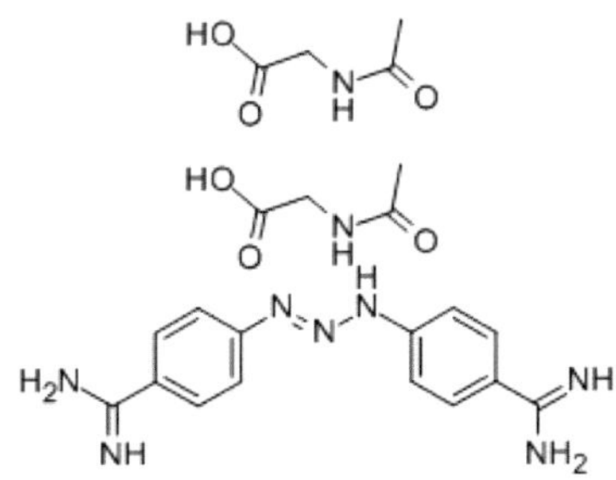

6.4.3. Berenil

Diminazine aceturate is an N-acetyl glycine compound chemically described as 4-4 (diazoamino) dibenamidine diaceturare; 1,3-bis(p-amidinophenyl) triazine bis (N-acetylglycenate) diaceturate; 1,3 bis (4-guanylphenyl) triazene diaceturate; and 4-4'-diamidinodiazoaminobenzene diaceturate.

6.4.3.1. Physio-Chemical Characteristics

The drug has a molecular weight of 515.5 and decomposes at 217°C. It is soluble in 14 parts of water at 20°C, is slightly soluble in alcohol and only slightly soluble in ether or chloroform. Due to the fact that diminazine aceturate consists of an organic base and an organic acid, once it is dissolved in water, it dissociates and each component has its own characteristics. Antipyrine is also an organic base.

6.4.3.2. Formulations

The innovator drug, Berenil, is composed of 45% m/m diminazene diaceturate and 555 m/m antipyrin. Aqueous solutions (pH 7) of this preparation are prepared by adding 25 ml sterile water to 2.6 g of Berenil granules containing 1.05 g of the active component diminazene diaceturate. This results in a 42 mg/ml of diminazene or a 4.2% m/v solution. The solution was dosed at 1 ml/10 kg representing a dose rate of 4.2 mg/kg. Ready-made or ready-to-use solutions [Dimisol (Virbac), Berenil RTU (Intervet), Dizene cattle (Virbac)] have been launched in the market.

6.4.3.3. Specifications of Berenil Used for the Experiment

Ready to use injectable solution of Diminazene Aceturate Berenil VET 7% RTU marketed by Intervet India Pvt. Ltd., Briahnagar, Off Pune-Nagar Road, Wagholi, Pune and manufactured in technical collaboration with Intervet India Pvt. Ltd. by Somrudh Pharmaceuticals Pvt. Ltd., J-174 MIDC, Tarapur, Thane was used for the experiment.

- **Composition (per ml):**

Diminazene Aceturate	70 mg
Phenazone BP	375 mg
Water for injection IP	q.s.

6.4.3.4. Properties of Berenil as a Therapeutic Drug

- Berenil RTU injectable solution is convenient and easy to use.
- Berenil RTU injectable solution is efficacious and ideally suited for commonly noticed mixed infections of the hemoprotozoans parasites.
- Berenil RTU injectable solution is well tolerated locally as well as systemically by young, feeble and chronically infected animals.

- Berenil RTU injectable solution is effective against protozoan strains resistant to other drugs.
- Experiments in the laboratory and in the field have shown that Berenil RTU injectable solution does not produce drug resistant strains.

6.4.3.5. Action on Resistant Trypanosomes

Berenil is widely accepted to overcome Quinapyramine resistance. Quinapyramine is slowly eliminated (metabolized and excreted) from the body over the month's period. Slow elimination of the drug leads to the development of resistance with Quinapyramine, if the treatment is repeated frequently.

Berenil is rapidly eliminated (within 24 hrs), therefore, changes of resistance are remote in the field conditions. Berenil has been found effective in the treatment of the resistant trypanosomes to Quinapyramine.

6.4.3.6. Samorin (Isometamidium Chloride), Synonym: Trypamidium

- **Additional Names:** 8-[3-(*m*-amidinophenyl)-2-triazeno]-3-amino-5-ethyl-6- phenylphenanthridinium chloride; 7-*m*-amidinophenyldiazoamino-2-amino-10- ethyl-9-phenylphenanthridinium chloride.
- Chemical Structure and Properties:
  - o **Molecular Formula:** $C_{28}H_{26}ClN_7$ Molecular Weight: 496.04
  - o **Percent Composition:** C: 67.80%, H: 5.28%, Cl: 7.15%, N: 19.77%

The commercially available products Samorin and Trypamidium have isometamidium chloride as the principle component with the remaining fraction comprising of two isomers, two analogs of a bis-species and homidium. Isometamidium is presented as a dark reddish-brown powder with a solubility in water of 6% (w/v) at 20°C.

6.4.3.7. Specifications of Samorin USED for the Experiment

6.4.3.7.1. Composition (per g) and Dosage Preparations

- Isometamidium chloride hydrochloride 1,000 mg:

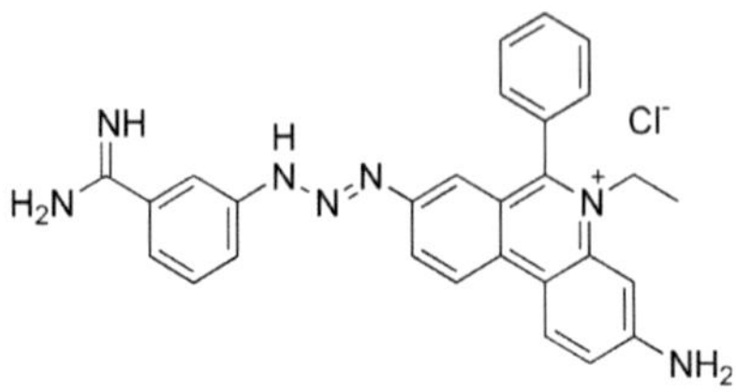

6.4.3.7.2. Samorin (Isometamidium Chloride)

The drug was used parenterally. The content of one sachet of the drug was dissolved in 50 ml or 100 ml of sterile water for injection to obtain a 2% or a 1% solution respectively and dilutions were made accordingly as per the scheduled dose. Sachet of 1 gram were available in the market.

6.4.3.8. Properties of Samorin as a Therapeutic Drug

Isometamidium is a phenanthridine aromatic amidine with a narrow therapeutic index which has been marketed for over 30 years as both, a prophylactic and a therapeutic trypanocidal agent in the field.

Different species of trypanosomes may have variable responsive to a specific drug. It thus becomes imperative to evaluate the efficacy of Isometamidium chloride against the Indian piscine trypanosomes before the drug can be adopted to control fish trypanosomosis and fish cryptobiosis as these parasites are known to cause hypocholerstrolemia (Tandon and Chandra, 1977b), hypoglycemia (Gupta et al., 2000), anemia (Gupta et al., 2003), splenomegaly, exophthalmia, ascites and abdominal distension in fish (Woo and Poynton, 1995).

6.4.3.9. Chemotherapy of Piscine Hemoflagellates by Trypanocidal Drugs

- **Test Fish**: *Clarias batrachus* Linn.
- **Parasites**:
  - o *Trypanosoma batrachi* Qadri, 1962; redescription Gupta and Jairajpuri, 1983
  - o *Trypanoplasma hematalis* Gupta and Gupta, 2010
- **Drugs**:
  - o Berenil (Diminazene aceturate)
  - o Samorin (Isometamidium chloride)
- **Parameters Investigated**: PCV, ESR, Hb% and parasitemia *Clarias batrachus* were selected and randomly divided into the following groups (n = 6 each):

Experiment A	**Drug: Berenil Experimental groups:**		
Group I:	**Control (Untreated)**		
	IA	Parasite	Nil
	IB	Parasite	*Trypanosoma batrachi*
	IC	Parasite	*Trypanoplasma haematalis*
Group II:	**Treated with 5.0 mg kg^{-1} Berenil**		
IIA:	Parasite	*Trypanosoma batrachi*	
IIB:	Parasite	*Trypanoplasma haematalis*	
Group III	**Treated with 10.0 mg kg^{-1} Berenil**		
	IIIA	Parasite	*Trypanosoma batrachi*
	IIIB	Parasite	*Trypanoplasma haematalis*
Group IV	**Treated with 12.5 mg kg^{-1} Berenil**		
	IVA	Parasite	*Trypanosoma batrachi*
	IVB	Parasite:	*Trypanoplasma haematalis*
Group V	**Treated with 15.0 mg kg^{-1} Berenil**		
	VA	Parasite	*Trypanosoma batrachi*
	VB	Parasite	*Trypanoplasma haematalis*
Experiment B	**Drug: Samorin**		
Group VI:	**Control (Untreated)**		
	VIA	Parasite	Nil
	VIB	Parasite:	*Trypanosoma batrachi*
	VIC	Parasite	*Trypanoplasma haematalis*
Group VII	**Treated with 0.1 mg kg^{-1} Samorin**		
	VIIA	Parasite	*Trypanosoma batrachi*
	VIIB	Parasite	*Trypanoplasma haematalis*

Group VIII	**Treated with 0.5 mg kg^{-1} Samorin**		
	VIIIA	Parasite	*Trypanosoma batrachi*
	VIIIB:	Parasite:	*Trypanoplasma haematalis*
Group IX	**Treated with 1.0 mg kg^{-1} Samorin**		
	IXA	Parasite	*Trypanosoma batrachi*
	IXB	Parasite	*Trypanoplasma haematalis*
Group X	**Treated with 2.0 mg kg^{-1} Samorin**		
	XA	Parasite	*Trypanosoma batrachi*
	XB	Parasite	*Trypanoplasma haematalis*

6.4.4. Treatment Regime

Blood samples (0.1 ml $fish^{-1}$) were obtained from the caudal vein prior to treatment and PCV, ESR, Hb% and parasitemia were recorded. Groups IB, IIA, IIIA, IVA, VA, VIB, VIIA, VIIIA, IXA, XA and Groups IC, IIB, IIIB, IVB, VB, VIA, VIIB, VIIIB, IXB, XB were inoculated intraperitoneally (i.p.) with 0.3 ml cold blooded vertebrate Ringer's solution (CBVR) containing 1,00,000 *Trypanosoma batrachi* and 10,000 *Trypanoplasma haematalis* respectively in fish. At 2 wk postinfection, Group I and VI (uninfected controls) fish were inoculated intramuscularly (i.m.) with 0.2 ml of CBVR. In the first set of experiment (Experiment A), fish were treated with Berenil at concentrations of 5.0 mg kg^{-1} (Group II), 10.0 mg kg^{-1} (Group III), 12.5 mg kg^{-1} (Group IV) and 15.0 mg kg^{-1} (Group V). In the second set of experiment (Experiment B), *Clarias* were treated with Samorin: Group VII (0.1 mg kg^{-1}) Group VIII (0.5 mg kg^{-1}), Group IX (1.0 mg kg^{-1}) and Group X (2.0 mg kg^{-1}). Blood samples were obtained prior to treatment (PT) and weekly for 5 wk post treatment (WPT) and each group was monitored for PCV, ESR, Hb% and intensity. The aim of drug action studies was to investigate the effect of the drugs on the altered hematology (PCV, ESR and Hb%) as also on the parasitemia.

The summary of the treatment regimes and the results are shown in Tables 6.4–6.7 (Berenil) and Tables 6.7–6.9 (Samorin). For each experimental group on different weeks post treatment (WPT), all the parameters were compared (Analyze Compare Means) for obtaining the mean value along with the standard error (SE). They were also analyzed separately (i.e., between groups and post infection treatment) by one way ANOVA and the significance (p value) was recorded at 5% ($p < 0.05$) level and 1% ($p < 0.01$) level. The statistical package for Social Scientists (SPSS) version 16 was employed for complete statistical analysis as mentioned in "Materials and Methods."

6.4.5. Results

6.4.5.1. Packed Cell Volume (PCV)

6.4.5.1.1. Berenil Treated

The PCV in the control group was 42.8 + 0.23 which fell to 28.3 + 0.56 and 21.5 + 0.62 in groups IB and IC respectively. There was no significant change in these values up to the 3rd week but during the 4th and 5th week, the values of 27.9 + 3.17 (3rd week) started increasing to 29.2 + 3.13 (4th week) and 32.4 + 4.13 (5th week) and from 22.5 + 2.54 (3rd week) to 23.9 + 3.31 (4th week) and 24.1 + 3.14 (5th week) in ts-inoculated and tpl-inoculated fish respectively. Under flagellate infectivity, the PCV values declined as seen earlier in the section of hematological changes. When these infected fishes were treated with the drug, they gradually started recovering clinically. The ts-infected fish recovered their PCV values up to 30.3 + 0.63, 38.1 + 0.51 and 38.2 + 0.49 at 5.0 mg kg^{-1}, 10.0 mg kg^{-1} and 12.5 mg kg^{-1} Berenil dose levels respectively at 5 WPT. In Groups III (dose 10.0 mg kg^{-1}) and IV (dose 12.5 mg kg^{-1}) the parasites were completely eradicated and there were no parasites at 5 WPT. However, at 12.5 mg kg^{-1} Berenil, out of 6 experimental fish, 3 died at 4th WPT and out of the remaining 3, 1 died at 5th WPT. The recovery of tpl- infected fish was slightly lower being 29.4 + 0.46, 33.4 + 0.42 and 33.4 + 0.06 at 5.0 mg kg^{-1}, 10.0 mg kg^{-1} and 12.5 mg kg^{-1} Berenil dose levels respectively at 5 WPT (Table 6.4). Here again, 2 fishes each out of the tpl-infected fishes died at 4th and 5th WPT. However, in both cases, the recovery in Berenil- treated fish was evident when compared to fishes of groups IB and IC. The insignificant increase in PCV values observed at 4th and 5th WPT in Groups IB and IC maybe due to the natural immunity acquired by the fish in reducing the infection and thus increasing the PCV values even though they were not treated with the drug. Although the parasitemia declined to 1.3×10^3 (Group IIA) and 1.1×10^2 (Group IIB) as compared to the untreated fish infected with trypanosomes [1.5×10^5 (Group IB)] and trypanoplasms [(1.6×10^4 (Group IC)], but the parasites were not completely eradicated. However, Group V fish treated with 15.0 mg kg^{-1} Berenil did not survive till the completion of the experiment. Three fish of Group VA and 4 fish of Group VB died on 3rd day and 4th day post treatment and even the remaining fish could not survive up to the 2nd observation date (2nd WPT). The gills of these fishes showed hemorrhage, signs of asphyxiation, convulsion and ultimately died.

6.4.5.1.2. Samorin Treated

The PCV in the control group was 44.0 + 0.66 (Group VIA) which fell to 27.2 + 0.12 and 22.0 + 0.40 in groups VIB and VIC respectively as is obvious in flagellate infected fish. There was no significant change in these values up to the 4th WPT but during the 5th WPT, the values of 27.4 + 1.98 (4th WPT) increased to 30.3 + 4.14 and from 23.5 + 2.26 (4th WPT) to 27.5 + 3.86 in ts-inoculated and tpl-inoculated fish respectively (Table 6.7). These changes might have been similar to those observed for Berenil. The changes in the Samorin-treated fishes at different dose levels were more significant. When the infected fishes were treated with Samorin, they gradually started showing clinical recovery. The ts-infected fish recovered their PCV values up to 31.7 + 0.16, 38.5 + 0.17 and 38.6 + 0.11 at 0.1 mg kg^{-1} Samorin, 0.5 mg kg^{-1} Samorin and 1.0 mg kg^{-1} Samorin dose levels respectively. The recovery rate of tpl-infected fish was again slightly lower being 25.6 + 0.14, 34.1 + 0.16 and 38.2 + 0.11 at 0.1 mg kg^{-1} Samorin, 0.5 mg kg^{-1} Samorin and 1.0 mg kg^{-1} Samorin dose levels respectively. In Group VII, the recovery occurred up to 2nd WPT and then the values started reverting back in the decreasing order. Moreover, there was a decline in parasitemia from 2.6×10^5 to 1.2×10^3 and from 1.8×10^4 to 0.9×10^2 in ts- and tpl-infected fishes respectively but the parasites were not completely eradicated. In Groups VIII (dose 0.5 mg kg^{-1}) and IX (dose 1.0 mg kg^{-1}) the parasites were completely eradicated and there were no parasites at 5 WPT. However, in both cases, the recovery in samorin-treated fish was evident up to 1.0 mg kg^{-1} dose level. All fish of the last group (Group V) treated with 2.0 mg kg^{-1} Samorin died within a few hours of injection. Their clinical signs were similar to those noted for Berenil- treated fish as above.

Table 6.4. Packed Cell Volume (PCV) in *Clarias batrachus* Treated with Different Doses of Berenil Under Flagellate – Infectivity

Parasite	Group*	Dose	PCV (Hematocrit) (Weeks Post Treatment)						Parasitemia (no./ml)	
			Pre-treatment (PT)	1	2	3	4	5	Pre. treatment (PT)	5 WPT
Nil	I A	-	42.8 + 0.23	43.3 + 2.51	44.5 + 3.72	43.8 + 2.63	42.9 + 2.72	43.7 + 3.82	Nil	Nil
Trypanosoma batrachi	I B	-	28.3 + 0.56	28.9 + 2.32	28.7 + 2.65	27.9 + 3.17	29.2 + 3.13	32.4 + 4.13	1,00,000	1.5×10^5

Trypanoplasma	I C	-	21.5 + 0.62	22.6 + 3.61	22.9 + 2.93	22.5 + 2.54	23.9 + 3.31	24.1 + 3.14	10,000	1.6×10^4
haematalis										
Ts. batrachi	II A	5.0 mg		29.5 + 2.72	35.3 + 1.94	27.8 + 2.45	28.1 + 3.15	30.3 + 0.63	-	1.3×10^3
Tpl. haematalis	II B	kg^{-1}		30.2 + 2.35	31.4 + 2.44	30.1 + 3.18	29.2 + 2.46	29.4 + 0.46	-	1.1×10^2
Ts. batrachi	III A	10.0		30.6 + 2.11	35.3 + 1.75	36.4 + 3.14	36.8 + 2.14	38.1 + 0.51	-	Nil
Tpl. haematalis	III B	mg kg^{-1}		30.2 + 3.22	33.4 + 2.16	33.3 + 2.66	33.4 + 3.16	33.4 + 0.42	-	Nil
Ts. batrachi	IV A			32.1 + 3.36	36.4 + 1.93	37.1 + 2.35	38.2 + 1.97	38.2 + 0.49	-	Nil
Tpl. haematalis	IV B	12.5		32.1 + 1.92	33.1 + 2.17	33.4 + 2.62	(3 died)	(1 died)	-	Nil
		mg					33.3 + 2.42	33.4 + 0.06		
		kg^{-1}					(2 died)	(2 died)		

Ts. batrachi	V A	15.0		30.8 + 3.32	100% mortality
Tpl. haematalis	V B	mg kg^{-1}		(3 died on 3rd day)	
				30.4 + 2.13	
				(4 died on	
				4th day)	

*n = 6

6.4.5.2. Erythrocyte Sedimentation Rate (ESR)

6.4.5.2.1. Berenil Treated

In untreated fish, the ESR values (0.89 + 0.03 mm/hr) increased to 1.61 + 0.11 mm/hr and 1.78 + 0.11 mm/hr in ts-infected and tpl-infected *Clarias* (Group I) respectively. Because these fish were not treated with the drug, there was no significant change in each sub-group up to 3rd week but during the 4th and 5th week, the values in Group IB and IC automatically started decreasing. This may perhaps be due to the production of protective antibodies which started controlling the parasites and therefore, the values started their return journey towards normalcy.

In the treated groups, the recovery rate was evident but differed with the dose-rate and parasite. In ts-infected fish, the recovery rate was 13.04% at 5.0 mg kg^{-1} and 32.90% at 10.0 mg kg 1. In tpl-infected fish, the recovery rate was 21.91%, 32.50% and 40.44% in Groups IIB, IIIB and IVB respectively. In Group IV, the situation differed. Up to 3rd WPT, the recovery was evident but at 4th WPT, 3 fish of IVA and 2 fish of Group IVB died. At the last observation date (5th WPT), 1 out of the 3-remaining fish of Group IVA and 2 out of the 4-remaining fish of Group IVB died. At 15.0 mg kg^{-1}, 3 fish of Group VA and

4 fish of Group VB died on days 3 and 4 post treatment respectively and the remaining by 2 WPT indicating that this dose is lethal for the fish (Table 6.5).

6.4.5.2.2. Samorin Treated

The ESR values in pre-treatment control fish (0.90 + 0.01 mm/hr) increased to 1.58 + 0.02 mm/hr and 1.81 + 0.03 mm/hr in ts-infected and tpl-infected *Clarias* (Group VI) respectively as is apparent under infected conditions. There was no significant change in each sub-group up to 4th week but during the 5th week, the values in Group VIB and VIC started decreasing even though they were not treated with the drug.

Recovery from infection was noted in the treated groups. In ts-infected fish, the values recovered up to 1.48 + 0.02 mm/hr, 1.06 + 0.02 mm/hr, 1.03 + 0.02 mm/hr at 0.1 mg kg^{-1}, 0.5 mg kg^{-1} and at 1.0 mg kg^{-1} dose levels of Samorin respectively. In tpl-infected fish, the values observed were 1.59 + 0.03 mm/hr, 1.16 + 0.02 mm/hr and 1.06 + 0.01 mm/hr in Groups VIIB, VIIIB and IXB respectively. At 2.0 mg kg^{-1}, there was 100% mortality of the fish, obviously it was the lethal dose (Table 6.8).

6.4.5.3. Hemoglobin (Hb)

6.4.5.3.1. Berenil Treated

As observed for PCV and ESR, the Hb% also declined in the untreated control group inoculated with trypanosomes (9.5 + 0.22 mg/100 gm) and trypanoplasms (6.6 + 0.19 mg/100 gm) as compared to the uninfected ones (13.4 + 0.42 mg/100 gm).

Fish treated with 5.0 mg kg^{-1} Berenil, the Hb values increased up to 2nd WPT (11.5 + 0.31 mg/100 gm) indicating recovery but thereafter, it started declining and the values of 9.9 + 0.27 (Group IIA) and 9.7 + 0.27 (Group IIB) were recorded indicating that the dose was too low for full recovery and even the parasites were not eliminated completely. In Groups III and IV, there was a consistent recovery of the hemoglobin level suggesting that these dose rates (10.0 and 12.5 mg kg^{-1}) were optimum for the treatment of hemoflagellate infection in fish. But when the dose level was further increased to 15.0 mg kg-1, at 1st WPT, 3 fish showed the value of 11.4 + 0.48 mg/100 and the remaining 3 died in group VA and in Group VB, 4 fish died, the remaining 2 showing the value of 10.2 + 0.31 mg/100 (Table 6.6). All the remaining fish also died up to 2nd WPT indicating the lethal nature of the dose.

Table 6.5. Erythrocyte Sedimentation Rate (ESR) in *Clarias batrachus* Treated with Different Doses of Berenil Under Flagellate-Infectivity

Parasite	Group*	Dose	ESR (mm/hr) (Weeks post treatment)						Parasitemia (No./ml)	
			Pre-treatment (PT)	1	2	3	4	5	Pre-treatment (PT)	5 WPT
Nil *Trypanosoma batrachi Trypanoplasma haematalis*	I A I B I C	- - -	0.89 +0.03 1.61 +0.11 1.78 +0.11	0.87 + 0.06 1.62 + 0.09 1.83 + 0.10	0.88 + 0.09 1.81 + 0.11 1.75 + 0.07	0.93 + 0.10 1.83 + 0.07 1.79 + 0.09	0.91 + 0.07 1.54 + 0.03 1.61 + 0.07	1.0 + 0.10 1.41 + 0.07 1.47 + 0.12	Nil 1,00,000 10,000	Nil 1.5×10^5 1.6×10^4
Ts. batrachi Tpl. haematalis	II A II B	5.0 mg kg^{-1}		1.51 + 0.05 1.57 + 0.07	1.40 + 0.08 1.39+ 0.07	1.73+ 0.10 1.75+ 0.05	1.71 + 0.06 1.74+ 0.04	1.74 + 0.11 1.76+ 0.11	- -	1.3×10^3 1.1×10^2
Ts. batrachi Tpl. haematalis	III A III B	10.0 mg kg^{-1}		1.40 + 0.06 1.42 + 0.09	1.12 + 0.07 1.31+ 0.05	1.07+ 0.09 1.21+ 0.12	1.08 + 0.06 1.20+ 0.09	1.08 + 0.04 1.21+ 0.09	- -	Nil Nil

Ts. batrachi *Tpl. haema-talis*	IV A IV B	12.5 mg kg^{-1}		1.31 + 0.12 1.36 + 0.07	1.11 + 0.05 1.17+ 0.08	1.08+ 0.10 1.06+ 0.09	1.09 + 0.07 (3 died) 1.07+ 0.09 (2 died)	1.07 + 0.02 (1 died) 1.06+ 0.04 (2 died)	- -	Nil Nil
Ts. batrachi *Tpl. haema-talis*	V A V B	15.0 mg kg^{-1}		1.17+0.07 (3 died on 3rd day) 1.20+ 0.12 (4 died on 4th day)	100% mor-tality	100% morality				

*n=6.

Table 6.6. Hemoglobin (Hb) Values in *Clarias Batrachus* Treated with Different Doses of Berenil Under Flagellate-Infectivity

Parasite	**Group***	**Dose**	**Hemoglobin (Hb) gm/100 gm (Weeks Post Treatment) (WPT)**						**Parasitemia (No./ml)**	
			Pre-treat-ment (PT)	**1**	**2**	**3**	**4**	**5**	**Pre-treat-ment(PT)**	**5 WPT**
Nil *Trypano-soma batrachi* *Trypano-plasma haematalis*	I A I B I C	- - -	13.4 + 0.42 9.5 + 0.22 6.6 + 0.19	13.2 + 0.41 9.7 + 0.43 7.1 + 0.31	13.7+ 0.71 9.8 + 0.51 7.4 + 0.41	13.1 + 0.32 8.6 + 0.51 7.4 + 0.31	13.8 + 0.67 10.4 + 0.21 8.1 + 0.61	13.3 + 0.42 11.1 + 0.61 8.8 + 0.52	Nil 1,00,000 10,000	Nil 1.5×10^5 1.6×10^4
Ts. batra-chi Tpl. haematalis	II A II B	5.0 mg kg^{-1}		11.0 + 052 9.8 + 0.51	11.5 + 0.31 10.1 + 0.37	10.6 + 0.61 10.8 + 0.52	10.1 + 0.42 10.2 + 0.64	9.9 + 0.27 9.7+ 0.27	- -	1.3×10^3 1.1×10^2
Ts. batra-chi Tpl. haematalis	III A III B	10.0 mg kg^{-1}		11.1 + 0.34 10.3 + 0.41	11.8 + 0.32 10.7 + 0.44	12.0+ 0.41 11.1+ 0.38	12.2 + 0.37 11.4 + 0.42	12.3+ 0.16 11.5+ 0.15	- -	Nil Nil

Ts. batrachi Tpl. haematalis	IV A IV B	12.5 mg kg^{-1}		11.6 + 0.62 10.5 + 0.38	11.8 + 0.51 11.7 + 0.47	12.3+ 0.35 11.4+ 0.42	12.5 + 0.42 (3 died) 11.6+ 0.91 (2 died)	12.6+ 0.11 (1 died) 11.5+ 0.12 (2 died)	- -	Nil Nil
Ts. batrachi Tpl. haematalis	V A V B	15.0 mg kg^{-1}		11.4 + 0.48 (3 died 3rd day) 10.2+0.31 (4 died on 4th day)	100 % mortality					

*n=6.

Table 6.7. Packed Cell Volume (PCV) in *Clarias batrachus* Treated with Different Doses of Samorin Under Flagellate-Infectivity

Parasite	**Group***	**Dose**	**PCV (Hematocrit) (Weeks post treatment) (WPT)**						**Parasitemia (No./ml)**	
			Pre-treat-ment (PT)	**1**	**2**	**3**	**4**	**5**	**Pre-Treat-ment (PT)**	**5 WPT**
Nil *Trypanosoma batrachi Try-panoplasma haematalis*	VI A VI B VI C	- - -	44.0 + 0.66 27.2 + 0.12 22.0 + 0.40	43.1 + 2.72 28.1 + 2.18 23.1 + 2.44	44.1 + 1.93 28.2 + 2.24 22.9 + 2.13	43.8 + 2.30 28.4 + 2.76 23.2 + 1.75	44.3 + 3.14 27.4 + 1.98 23.5 + 2.26	43.8 + 2.73 30.3 + 4.14 27.5 + 3.86	Nil 1,00,000 10,000	Nil 2.6×10^5 1.8×10^4
Ts. batrachi *Tpl. haema-talis*	VII A VII B	0.1 mg kg^{-1}		28.1 + 1.82 23.5 + 2.16	32.5 + 2.15 28.3 + 3.16	32.0 + 3.18 27.6 + 3.15	30.6 + 1.94 26.5 + 2.35	31.7 + 0.16 25.6 + 0.14	- -	1.2×10^3 0.9×10^2
Ts. batrachi *Tpl. haema-talis*	VIII A VIII B	0.5 mg kg^{-1}		30.2 + 2.22 24.7 + 2.32	34.5 + 1.92 28.9+ 2.71	35.2+ 3.76 31.2+ 2.18	36.1 + 2.17 33.2+ 1.92	38.5 + 0.17 34.1+ 0.16	- -	Nil Nil

Ts. batrachi Tpl. haematalis	IX A IX B	1.0 mg kg^{-1}		33.3+ 3.14 29.6 + 2.75	35.1 + 2.35 31.7 + 2.26	36.28+ 2.17 34.3 + 3.15	37.3 + 1.17 36.1 + 2.73	38.6 + 0.11 38.2 + 0.11	- -	Nil Nil
Ts. batrachi Tpl. haematalis	X A X B	2.0 mg kg^{-1}		100 % mortality						

*n=6.

Table 6.8. Erythrocyte Sedimentation Rate (ESR) in *Clarias batrachus* Treated with DIFFERENT doses of Samorin Under Flagellate-Infectivity

Parasite	**Group***	**Dose**	**ESR (mm/hr) (Weeks Post Treatment)**						**Parasitemia (No./ml)**	
			Pre-treatment (PT)	**1**	**2**	**3**		**5**	**Pre-treatment (PT)**	**5 WPT**
Nil *Trypanosoma batrachi Trypanoplasma haematalis*	VI A VI B VI C	- - -	0.90 + 0.01 1.58 + 0.02 1.81 + 0.03	0.86 + 0.06 1.59 + 0.08 1.92 + 0.09	0.94 + 0.12 1.63 + 0.12 1.83 + 0.07	0.88 + 0.17 1.93 + 0.07 1.97 + 0.09	1.0 + 0.07 1.87 + 0.09 1.99 + 0.13	0.99 + 0.12 1.38 + 0.42 1.57 + 0.31	Nil 1,00,000 10,000	Nil 2.6×10^5 1.8×10^4

Ts. batrachi *Tpl. haema-talis*	VII A VII B	0.1 mg kg^{-1}		1.52 + 0.08 1.72 + 0.09	1.37 + 0.17 1.41+ 0.09	1.42+ 0.23 1.47+ 0.07	1.51 + 0.04 1.57+ 0.06	1.48 + 0.02 1.59+ 0.03	- -	1.2 × 10^3 0.9 × 10^2
Ts. batrachi *Tpl. haema-talis*	VIII A VIII B	0.5 mg kg^{-1}		1.38 + 0.12 1.42 + 0.05	1.29 + 0.07 1.35+ 0.05	1.10+ 0.04 1.25+ 0.08	1.09 + 0.03 1.21+ 0.09	1.06 + 0.02 1.16+ 0.02	- -	Nil Nil
Ts. batrachi *Tpl. haema-talis*	IX A IX B	1.0 mg kg^{-1}		1.28 + 0.14 1.38 + 0.07	1.19 + 0.03 1.27+ 0.07	1.12+ 0.07 1.17+ 0.06	1.07 + 0.12 1.08+ 0.17	1.03 + 0.02 1.06+ 0.09	- -	Nil Nil
Ts. batrachi *Tpl. haema-talis*	X A X B	2.0 mg kg^{-1}		100% mortality						

*n=6.

Table 6.9. Hemoglobin (Hb) VALUES in *Clarias batrachus* treated with Different Doses of Samorin Under Flagellate-Infectivity

Parasite	Group*	Dose	Hemoglobin (Hb) gm/100 gm (Weeks post treatment) (WPT)						Parasitemia (No./ml)	
			Pre-treatment (PT)	1	2	3	4	5	Pre-treatment (PT)	5 WPT
Nil *Trypanosoma batrachi* *Trypanoplasma haematalis*	VI A VI B VI C	- - -	13.6 + 0.37 9.1 + 0.39 6.8 + 0.38	13.9 + 0.20 9.7 + 0.24 6.9 + 0.18	13.5+ 0.53 8.9 + 0.71 7.1+ 0.57	13.8 +0.57 8.6+ 0.53 7.4 + 0.64	13.3 + 0.42 9.8+ 0.72 7.9 + 0.65	13.4 +0.68 9.9+ 0.62 8.1 + 0.52	Nil 1,00,000 10,000	Nil 2.6 × 10^5 1.8 × 10^4
Ts. batrachi *Tpl. haematalis*	VII A VII B	0.1 mg kg^{-1}		10.8 + 0.57 9.0 + 0.47	11.5+ 0.41 10.7+ 0.71	10.4 + 0.71 10.0 + 0.68	9.4 + 0.73 9.3 + 0.45	9.2 + 0.12 9.4+ 0.13	- -	1.2 × 10^3 0.9 × 10^2

Ts. batrachi *Tpl. haema-talis*	VIII A VIII B	0.5 mg kg^{-1}		10.3 + 0.62 9.5 + 0.72	11.2+ 0.72 11.3+ 0.64	11.4+ 0.42 11.5+ 0.68	11.7 + 0.63 11.6+ 0.43	11.8+ 0.31 12.0+ 0.14	- -	Nil Nil
Ts. batrachi *Tpl. haema-talis*	IX A IX B	1.0 mg kg^{-1}		11.1 + 0.73 10.3 + 0.45	11.8+ 0.31 11.8+ 0.61	11.6+ 0.61 11.9+ 0.43	12.0+ 0.41 12.3+ 0.82	12.1+ 0.17 12.5+ 0.16	- -	Nil Nil
Ts. batrachi *Tpl. haema-talis*	X A X B	2.0 mg kg^{-1}		100 % mortality						

*n=6.

6.4.5.3.2. Samorin Treated

The Hb% also declined (9.1 + 0.24 mg/100 gm) in the untreated control groups (VIB) and VIC (6.8 + 0.18 mg/100 gm) infected with trypanosomes and trypanoplasms respectively as compared to the uninfected ones (13.6 + 0.20 mg/100 gm) (Table 6.9).

Clarias treated with 0.1 mg kg^{-1} Samorin, the Hb values increased upto 2nd WPT (11.5 + 0.41 mg/100 gm) showing signs of recovery but soon after, it started declining and the values of 9.2 + 0.12 mg/100 gm (Group VIIA) and 9.4 + 0.13 mg/100 gm (Group VIIB) were recorded indicating that the dose was not optimum for full recovery. The parasitemia in these fish reduced but was not eliminated. The consistent recovery of the hemoglobin level in Groups VIII and IX, suggested that these dose rates (0.5 and 1.0 mg kg^{-1}) were the optimum dose. However, 100% mortality at 2.0 mg kg^{-1} indicated its lethal nature.

The results indicate that both the drugs were able to reduce the trend of abnormalicy in hematological parameters as well as a bring a decline in the intensity of the parasites.

6.4.5.4. Discussion

The pathogenesis of trypanosomiasis is complex and the mechanism obscure. Damage to the tissues is brought about through the metabolic activities of trypanosomes, more certainly through the repeated insults offered by the emergence of successive trypanosome variants and the attempts made to suppress them by the host's defense mechanism.

Studies on the physiology and biochemistry of trypanosomes have tended to center on carbohydrate metabolism and specially on changes in oxidative pathways associated with the development cycle of trypanosomes. The drugs for the control of trypanosomes, therefore have to be accordingly devised and implemented.

Trypanocides, the drugs used to kill trypanosomes are classified into 2 broad categories based on the speed with which they lyse trypanosomes and their ability to affect the infectivity of trypanosomes. The trivalent arsenicals, acriflavine and most diamidines, such as Berenil, are rapidly lethal and do not affect infectivity of trypanosomes. However, suramin, antrycide, and phenanthridinium compounds, such as isometamidium, take longer to kill trypanosomes but reduce infectivity (Hawking, 1963).

Berenil is used extensively to combat trypanosomiasis and babesiosis in cattle, buffalo, sheep and several other domestic animals (Arowolo and Ikede, 1980). It has both, curative as well as prophylactic properties. Being extracellular parasites, the efficacy of the drug on trypanosomes has been found to be positive. Its efficacy was 100% against surra in buffaloes (Joshi and Singh, 2000) and 91.2% for babesiosis in cattle (Bhikane et al., 2001). However similar reports for piscine hematozoans are lacking.

Isometamidium chloride (Samorin) was first described by Wag et al. (1958) and was obtained from the coupling reaction between m-amido-benzenediazonium chloride and homidium chloride. The drug is effective against African trypanosomiasis but has not been much used in the treatment of the disease in India.

Looking into the possibility that these drugs could be effective against piscine blood flagellates, the present experiment was designed to assess the efficacy of the aforementioned drugs. The parameters taken into consideration were PCV, ESR, Hb% and parasitemia.

Under parasitized conditions, the above-mentioned hematological parameters deviate from their normal values as seen in the chapter on "Hematological Invesigations." In the present experiment, the PCV values declined by 33.87% and 49.76% in ts-infected and tpl-infected fish respectively. Group I comprised of untreated fish and fish of the remaining groups (II – V) were treated with different doses of drugs as given above. Group II fish treated with 5.0 mg kg^{-1} Berenil showed an initial recovery up to 2nd WPT by 24.73% and 46.00% in Groups IIA and IIB respectively but after that, there was a relapse as the PCV values started declining again indicating that the dose concentration was less. This was further confirmed by the fact that parasitemia in Groups IIA and IIB showed a decline to 1.3×10^3 and 1.1×10^2 as compared to the untreated values in Group IB (1.5×10^5/ml) and Group IC (1.6×10^4) respectively as the parasites were not eliminated completely. Similarly, PCV in Samorin-treated fish, Group VIIA and VIIB (0.1 mg kg^{-1}) showed an initial recovery up to 2nd WPT (19.48% and 28.63% respectively), but then again, a relapse of parasitemia occurred as observed for Berenil-treated fish and the PCV started declining again and the parasites were not eliminated indicating the low dose level of the drug. However, with both the drugs, not only there was a continuous recovery in Groups IIIA (34.98%) and IIIB (55.34%) and VIIIA (41.5%) and VIIIB (54.0%), but also the parasitemia was nil at 5th WPT, thus the parasites had been completely eradicated by the drugs. In Groups IVA and IVB (Berenil-treated) although there was a recovery of 34.98% and 55.34% respectively but 3 fish of Group IVA and 2 fish of Group IVB died at 4th week WPT and 1 fish of Group IVA and 2 of IVB died on 5th WPT so that at the end of the experiment, only 2 fish of each group survived indicating that the dose level of 12.5 mg kg^{-1} was an overdose for the fish. However, in Samorin-treated fish, Groups IXA and IXB there were no casualities and all fish survived till the end of experimentation showing a recovery rate of 41.91% and 73.63% respectively. All fish of Groups VA, VB, XA and XB fish died in the first week itself suggesting that dose levels of 15.0 mg kg^{-1} Berenil and 2.0 mg kg^{-1} Samorin were lethal for the fish. Kumar et al. (2009) recorded 20% recovery rate of PCV in rats infected with *Trypanosoma evansi* and treated with 1 mg kg^{-1} body weight of isometamidium-HCl which was higher than that observed for Hb (12.7%) in the same study but this was in any case lower to the recovery rate observed in the present studies.

The ESR values showed a similar pattern of change at different dose levels of Berenil and Samorin. There was an increase in the ESR level of untreated ts- parasitized (80.89%) and tpl-infected (100.0%) fish. When these fish were treated with 5.0 mg kg^{-1} Berenil, there was an intitial recovery upto 13.04% in Group IIA and 21.91% in Group IIB upto 2nd WPT as also in Groups VIIA (13.29%) and VIIB (22.09%) but soon afterwards, a relapse occurred and the ESR values started elevating again indicating the low dose concentration of the drugs. When the concentration of the drugs was increased to 10.0 mg kg^{-1} Berenil and 0.5 mg kg^{-1} Samorin, the dose levels appeared to be optimum as not only there was a recovery of 32.90% (Group IIIA) and 32.50% (Group IIIB) but also the parasitaemia was nil showing complete control of the parasites. In Groups VIIIA and VIIIB, the results were similar, complete eradication of the parasites and recovery of 32.91% and 35.91% respectively. In Groups IVA and IVB although there was a recovery of 32.91% and 40.44% respectively up to 3rd WPT, but 4 fish of each group died during the subsequent weeks indicating that the dose level of 12.5 mg kg^{-1} Berenil was high as it could not be tolerated by all fish. Nevertheless, Group IXA and IXB fish showed a continuous recovery reaching 34.81% and 41.43% respectively up to 5th WPT. As mentioned for PCV, the dose levels of 15.0 mg kg^{-1} Berenil and 2.0 mg kg^{-1} Samorin were toxic and all the fishes of Groups V and X died during the first week of experimentation.

The Hb% values decreased in untreated ts-parasitized (29.10%) and tpl-infected (50.74%) fish. When these fish were treated with 5.0 mg kg^{-1} Berenil, there was again an initial recovery of 21.05% in Group IIA, 53.03% in Group IIB, 26.37% in Group VIIA and 57.35% in VIIB (up to 2nd WPT) at low dose levels, but then, a relapse occurred and the Hb values started declining again indicating that the dose level of the drug was low and the parasites were not completely cleared. When the concentration of the drugs was increased to 10.0 mg kg^{-1} Berenil and 0.5 mg kg^{-1} Samorin, the dose levels appeared to be optimum, there was a recovery of 29.47% (Group IIIA) and 74.24% (Group IIIB) and the parasitaemia was also nil showing complete control of the parasites. In Groups VIIIA and VIIIB, the results were similar, complete elimination of the parasites and recovery of 29.67% and 76.47% respectively. In Groups IVA and IVB although there was a recovery of 32.63% and 74.24% respectively up to 3rd WPT, but 4 fish of each group died during the subsequent weeks indicating that the dose level of 12.5 mg kg^{-1} Berenil was high as it could not be tolerated by all fish and some of them died. However, Group IXA and IXB fish showed a continuous recovery reaching 32.96% and 83.82% respectively up to 5th WPT. The dose levels of 15.0 mg kg^{-1} Berenil and 2.0 mg kg^{-1} Samorin were again found to be toxic and all the fishes of Groups V and X died during the first week of experimentation. The recovery rate during the present studies was higher to that reported by Kumar et al. (2009) (12.7% recovery rate in Hb value under treatment with isometamidium chloride which was 6.06% lower than that of the uninfected controls).

Over the years, attempts have been made to observe the therapeutic effects of certain drugs. Havelka et al. (1965) stated that *Trypanoplasma cyprini* (= *T. borreli*) is killed *in vitro* by methylene blue, gentian violet and malachite green in concentrations of 1 in 2,000 to 1 in 20,000. Exposure of infected carp fingerlings to 1 in 2000 and 1 in 5000 solutions of the above dyes resulted in the disappearance of flagellates from the peripheral blood; in field experiments, the use of these dyes increased the survival rate of heavily infected fish and reduced the parasitemia. However, using goldfish infected with *T. danilewskyi* or *T. borreli* it was found that concentrations tolerated by the goldfish did not in the least affect the flagellates.

Ardelli and Woo (1998) observed *in vitro* effects of crystal violet on *Cryptobia salmositica.* The drug did not inhibit *in vitro* multiplication of a non-pathogenic strain of *Cryptobia salmositica* at low concentrations (0.01 μM and 0.001 μM) but multiplication was inhibited at higher concentrations (> or = 0.05 μM). In contrast, the pathogenic strain of *C. salmositica* did not multiply *in vitro* when incubated with crystal violet (0.001, 0.01, and 0.05 μM). The infectivity of the pathogenic strain was significantly reduced after *in vitro* exposure to crystal violet. Crystal violet lysed *C. salmositica* (100.0 μM) and caused lesions on mitochondrial and nuclear membranes of the parasite. Pathogenic strains of *Cryptobia salmositica* and *C. bullocki* were more susceptible to lysis after *in vitro* exposure to crystal violet than the nonpathogenic strains of *C. salmositica* and *C. catostomi*.

The antiprotozoal activity of phospholipid analogues, originally developed as anti-cancer drugs, has been determined in the past decade (Croft et al., 2003). The susceptiblility of this drug was also shown for *Trypanosoma cruzi* and *T. brucei*. Miltefosine, an alkylphosphocholine, was registered for the oral treatment of visceral leishmaniasis in India in March 2002. During the present studies, Berenil and Samorin were selected as the drugs to observe their efficacy on fish flagellates. Isoun (1975) reported that Berenil, Samorin and Novidium (homidium chloride) were the most commonly used drugs against animal trypanosomiasis in Tropical Africa and a knowledge of the free drug concentration was essential to determine the therapeutic efficacy of the drugs.

In Nigeria, Berenil at 3.5–7.0 mg/kg body weight and Novidium at 1.0 mg/ kg body weight are used to treat clinical cases mainly by *T. vivax* and *T. congolense*, whereas Samorin at 0.5 mg/kg was recommended for chemoprophylaxis (Na' Isa, 1970). Macadam and Williamson (1972) investigated the trypanocidal drug action of diminazene on the fine structure of trypanosomes (*T. cruzi, T. congolense, T. brucei* and *T. vivax*). Tropical Africa, isometamidium chloride is widely used against trypanosomiasis in domestic animals (Kinabo et al., 1989) and it is also used as a prophylactic drug against bovine trypanosomiasis (Kinabo and Bogan, 1987). and Arakawa (1991) reported varied responses to Berenil and Samorin as curative drugs in trypanosome-infected mice.

In fish, the effect of Samorin (1.0 mg/kg weight) reaches peak level in the blood 2–3 weeks after intramuscular injection (Ardelli and Woo, 2000). Also, *in vitro* exposure to sublethal levels of the drug reduces infectivity of the parasite to fish and changes the surface glycoprotein antibody-receptor sites of the parasite. This alteration of surface epitopes explains the protection of some parasites from lysis by complement fixing antibodies when rainbow trout with acute infections were treated with the drug (Ardelli et al., 2000).

The drug is therapeutic against *C. salmositica* in rainbow trout during pre- and post-clinical phases of the disease. Ardelli and Woo (1999) demonstrated that 1.0 mg/kg of isometamidium chloride delivered intra-muscularly was therapeutic in experimentally infected rainbow trout (*O. mykiss*). However, it was not effective during acute disease partly as the drug "modifies" surface epitopes of the parasite so that they are not lysed by complement fixing antibodies (Ardelli and Woo, 1999). The drug was more effective in infected Atlantic salmon, and at a higher dose (2.5 mg/kg), the infection was eliminated in about 30% of adult fish and the parasitemias were significantly reduced in remaining fish. Also, all infected juvenile chinook salmon, *Oncorhynchus tshawytscha,* treated with isometamidium chloride (1.0 mg/kg) survived the disease while 100% of untreated infected fish died with massive parasitaemias. The drug has a prophylactic value, and it does not seem to affect fish growth, food consumption, blood complement levels, or hematocrit values in fish (Ardelli and Woo, 2001a).

Tettey et al. (1999) devised a procedure for the evaluation of chemical equivalence of proprietary formulations of isometamidium by combining the analysis of the principal component (isometamidium), HPLC profiling of related substances and determination of the inorganic impurity, ammonium chloride, using a modification of the Berthelot (Indophenol) reaction whereby Samorin® samples showed inter-batch consistency of composition.

There have been renewed studies on the mechanisms of action of drugs against trypanosomes *in vitro* (Williamson and Macadam, 1975; Williamson, 1976). During the last decade attention has been diverted towards the understanding of drug action and the development of resistance in relation to the mechanism by which the drug is ingested by the parasite. The primary importance of any antiparasitic drug is the efficacy by which it is taken up by the parasite. It has been proposed that this process can be due to passive diffusion, endocytosis, receptor medicated uptake or transported-mediated uptake.

The mechanisms of action of drugs against trypanosomes are generally not well understood, with the exception of eflornithine (Wang, 1995). Eflornithine is used to treat early and late stages of *Trypanosoma brucei gambiense* infections and inhibits ornithine decarboxylase (Bey et al., 1987). Isometamidium inhibits both mammalian and *Trypanosoma*l DNA polymerases and interacts with glycosomes, kinetoplast DNA, glycoprotein biosynthesis, lipid metabolism, and membrane transport systems.

Trypanocidal drugs contain strongly basic groups which bind the kinetoplast DNA by non-intercalative interaction in the parasite (Brack et al., 1972). The trypanocidal drug binds irreversibly to blood proteins in rabbit (*Alvi* et al., 1985). The latter authors incubated diminazene with blood products and found that binding to plasma and serum was 50% and 35% respectively. On examination of the red blood cells they found that 70% of diminazene was bound to purified hemoglobin and that red blood cell membranes did not show any binding. The authors concluded that diminazene binds to a number of blood proteins and can cross the red cell membrane to bind to hemoglobin.

Isometamidium chloride is an amphiphilic cationic phenanthridinium and is highly positively charged under physiological conditions (Wang, 1995). The primary actions of phenanthridiniums are to block nucleic acid synthesis by intercalation between DNA base pairs (Wagner, 1971), leading to inhibition of RNA (Richardson, 1973) and DNA polymerases (Marcus et al., 1982), and thus inhibition of incorporation of nucleic acid precursors into DNA and RNA (Lantz and Van Dyke, 1972). Isometamidium is thought to exert its cytotoxic effects on trypanosomes by modulating lipid (Dixon et al., 1971) and ATP (Frank-Henderson et al., 1977) metabolism and glycoprotein biosynthesis (Casero et al., 1982). The kinetoplast, but not nuclear DNA, was disrupted when trypanosomes were exposed to isometamidium (Newton, 1974). In addition, isometamidium can produce dyskinetoplastic trypanosomes; these are cells that retain the mitochondrial membranes but do not contain detectable kinetoplast DNA (Shapiro, 1993). Although it has been used as a trypanocidal drug over the years, the exact toxic mechanism of isometamidium has not been elucidated, although it is presumed to act in ways similar to other phenanthridiniums (Kinabo and Bogan, 1988).

The kinetoplast of *C. salmositica* is the principal target of isometamidium. Fluorescence combined with phase-contrast microscopy demonstrated the initial and rapid (within 1 min of exposure) accumulation of the drug within the kinetoplast (Ardelli and Woo, 2001a).

The authors demonstrated that isometamidium accumulated in the kinetoplast of *C. salmositica* and caused a rapid disruption of kinetoplast DNA. This is not unusual, as a number of anti-*Trypanosoma*l agents are known to bind DNA. These include ethidium bromide, a DNA intercalating agent, and the diamidines, which include Berenil and pentamidine (Shapiro, 1993). Berenil and pentamidine bind to the minor groove of A–T-rich sites with high DNA sequence specificity (Reinert, 1999). Isometamidium is produced by linking a portion of the Berenil molecule to an ethidium nucleus, and consequently it also disrupts kinetoplast DNA (Kinabo and Bogan, 1988). It was suggested by the authors that these interactions may be part of the overall mode of action of the drug.

Several anti-*Trypanosoma*l drugs are known to inhibit oxygen consumption, although this has not been demonstrated for isometamidium. For example, the two calcium channel antagonists, isradipine and lacidipine, inhibit growth and oxygen consumption of *Trypanosoma cruzi* epimastigotes at μMolar concentrations (Nunez-Vergera et al., 1998). Similarly, an antibiotic, ascofuranone, specifically inhibits respiration and *in vitro* growth of long slender bloodstream forms of *Trypansoma brucei* (Minagawa et al., 1996). Licochalcone A alters the structure of the mitochondrion of *Leishmania* promastigotes and inhibits oxygen consumption and carbon dioxide production (Zhai et al., 1995). *Cryptobia salmositica* consumes oxygen under *in vitro* conditions (Thomas et al., 1992). Ardelli and Woo (2001a) demonstrated that isometamidium inhibited oxygen consumption and carbon dioxide production by *C. salmositica* and part of this could be due to its effects on the mitochondrion.

Some of the exact actions have been elucidated by Pilch et al. (1995). Diminazene as an anti-*Trypanosoma*l agent binds to the AT-rich regions of nucleic acid duplexes. Binding occurs via complexation into the minor grooves of AT- rich domains of DNA double helices. It can bind DNA as well as RNA duplexes, while exhibiting properties characteristic of both intercalation and minor groove binding. This binding unwinds negative supercoils in plasmids and has also been found to interfere with the activities of the eukaryotic type II topo-isomerase enzymes (Portugal, 1994). The higher level of chromosomal aberrations induced by Berenil might be related to the ability of this drug to interfere with activities of eukaryotic topoisomerase (Portugal, 1994; Othman and Ahmed, 2004).

Although information available on the interaction of Berenil with proteins is scanty, speculations have been made that Berenil enhances protein synthesis in liver and inhibits enzymes involved in phosphorylation and drug detoxification (Mallick et al., 2000). The authors reported that high plasma protein–binding capacity of Berenil serves as a circulating reservoir influencing the therapeutic efficacy of the drug and provides protection for a variable period. Also, protein binding reduces the rate of loss of a drug from the body and the reversibly bound drug molecules are always in dynamic equilibrium with their free counterpart.

The infectivity of *C. salmositica* in fish blood was reduced after *in vitro* exposure to isometamidium. However, this could also be because most *Cryptobia* were killed and therefore isometamidium would seem to reduce infectivity. The number of infected fish and parasitemia was lower in fish that received the parasites exposed to isometamidium (Ardelli and Woo, 2001b); this might be because these fish received lower inocula and also some of the parasites were not infective. Isometamidium at 1 mg/ml in Iscove's medium prevented infectivity of sensitive *T. brucei* stocks, *T. brucei evansi*, and *T. vivax* (Kaminsky et al., 1990). Similarly, infectivity of *T. b. brucei* was prevented after incubation for 4 hr in plasma from cattle treated with diminazene aceturate. Isometamidium has many negatively charged groups and thus is strongly bound at physiologic pH by polyanionic molecules such as nucleic acids, mucopolysaccharides, and serum albumin. Isometamidium binds to serum proteins in rats (Philips et al., 1967), and this was confirmed by Ardelli and Woo (1999). The lytic activity of isometamidium was significantly reduced in a medium containing serum supplement and this resulted in a delayed cryptobiacidal effect. The same authors demonstrated that 1.0 mg/kg isometamidium was therapeutic in rainbow trout during preclinical and chronic disease phases. However, the drug was not effective when trout were treated at acute disease, and as a result, the infection was more severe in treated as compared to nontreated fish. Two *in vitro* experiments were performed to examine the effects of isometamidium on complement-mediated lysis. The lytic titers of immune plasma or fresh complement did not change after 3 hr incubation with isometamidium (Ardelli and Woo, 1999). Thus, isometamidium did not interfere with antibody production or function, or complement production, or ions required to initiate and regulate complement. Instead, it was suggested that isometamidium altered surface epitopes on *C. salmositica* and thus protected the parasite from complement-mediated lysis. This was confirmed by Ardelli and Woo (2001a) as *C. salmositica* maintained in drug supplemented medium were not lysed by complement-fixing antibodies nor agglutinated by immune plasma. *C. salmositica* has the first 5 enzymes of the Embden–Meyerhoff pathway, as well as the peroxisomal enzyme catalase, sequestered within glycosomes, and these enzymes were demonstrated in whole cell lysates of the parasite (Ardelli and Woo, 2000). The glycosome in *C. salmositica* may provide an advantage to the parasite by allowing the switch between glycosomal glycolysis in the absence of oxygen or inhibition of oxidative phosphorylation and mitochondrial respiration in the presence of oxygen. This may occur in the leech vector (hypoxic conditions) or after exposure to chemotherapeutic compounds, like isometamidium, that damage the mitochondrion.

A concentration-dependant inhibition of membrane Ca^{++}- ATPase activity, as well as significant secondary binding of diminazene within DNA corresponding to G+C rich sites have also been reported (Ariyibi et al., 2001). Cytogenetically, Berenil is recognized as an agent that binds to nucleic acid duplex. The generally accepted mode of Berenil binding is via complexation into the minor groove of ATP–rich domains of DNA double helices. This affinity of Berenil binding to higher ordered nucleic acid structure and complexation with a DNA and RNA triple helix was reported in many studies (Durand et al., 2002).

In rats, it has been suggested that the drug blocks neuromuscular transmission and stimulates cholinergic receptors with resulting paralysis. Apparently, these receptor sites are occupied by isometamidium when drug concentrations are relatively high in the blood. As drug level decreases, the binding of the drug to the receptors is reversed and recovery from paralysis and other acute reactions quickly ensues (Philips et al., 1967). These findings are supported by Ardelli and Woo (1999) who proposed that a similar condition may occur in the rainbow trout which survived a high dose (2.5 mg kg^{-1}) of isometamidium. In the study conducted by the latter authors, it took 1 to 3 wk after treatment before the therapeutic effects of isometamidium were observed in rainbow trout. In the present case, results could be detected at 2 WPT as noted above.

Immunity is not known to increase the prophylactic effects of isometamidium in Boran cattle (Whitelaw et al., 1986). Rainbow trout that survive a *C. salmositica* infection are protected from subsequent challenge, and this acquired immunity is nonsterile (Jones and Woo, 1987). In rainbow trout, antibodies are detectable at approximately 2 wk after infection (Woo, 1990; Sitja-Bobadilla and Woo, 1994) and this was confirmed by Ardelli and Woo (1999) who investigated the possibility that protective antibodies may increase the therapeutic effects of isometamidium during acute disease; however, parasitaemias and lytic antibody titers were higher in treated fish as compared to untreated infected controls. Generally, when the parasitaemia is high, the titre of free complement fixing antibodies are expected to be low, and it increases as the parasitaemia declines (Li and Woo, 1995). This was due to lysis of the parasite (Li and Woo, 1995) or

C. salmositica-specific antibody being bound by antigens in the blood (Ardelli and Woo, 1997).

Humoral immunity is important in protection and complement-fixing antibody is required for fish to recover from an infection (Li and Woo, 1995). In the study conducted by Ardelli and Woo (1999), antibodies were being produced in treated fish suggesting that isometamidium does not interfere with the secretion of immunoglobulins by plasma cells. It is possible that isometamidium decreased the effectiveness of antibodies by interfering with the complement cascade and as a result the parasitaemias were higher in treated fish. The drug may affect complement-mediated lysis by binding the important Mg++ and Ca++ ions required to activate the classical pathway. It is also likely that the drug alters the glycoprotein receptors (epitopes) for antibodies on the surface of the parasite and thus protects the parasite from complement fixing antibodies. Kinabo and Bogan (1988) have shown that isometamidium binds to tissue macromolecules and precipitates tissue proteins.

Ardelli and Woo (2001c) conjugated isometamidium chloride to polyclonal antibodies (from "recovered" fish) and the monoclonal antibody (mAb-001). The conjugated drug was on the entire parasite while the unconjugated drug accumulated only in the kinetoplast. Before drug conjugation, both antibodies agglutinated living parasites but they reacted differently after drug conjugation. Polyclonal antibodies-conjugated drug (PAIC) lysed most of the parasites and it no longer agglutinated the parasite. In contrast, the mAb-00-conjugated drug did not lyse *C. salmositica* but agglutinated it. After *in vitro* exposure to PAIC, the infectivity of the parasite and subsequent parasitaemias in inoculated fish were significantly lowered. Fish survival was much higher in juvenile chinook salmon infected with parasites exposed to the polyclonal antibodies-conjugated drug (PAIC) than to drug alone (Drug) or to polyclonal antibodies alone (Antibody) or to drug plus polyclonal antibody (PAI). Also, preliminary studies indicated the drug-antibody conjugate to be effective when injected into infected fish (Woo, 2010).

Kumar et al. (2009) observed the effect of Isometamidium treatment on rats infected with *T. evansi*. The infection resulted in 100% mortality by day 7 in the infected untreated rats. The drug exerted 90% trypanocidal efficacy. Hematological studies indicated that the parasites altered the values significantly however, Isometamidium treatment significantly ($P < 0.05$) improved Hb and PCV without any significant changes in TEC and TLC values.

Samorin accumulates rapidly in the kinetoplast of the parasite (Ardelli and Woo, 2001b) causes condensation of its kinetoplast DNA, forms vacuoles, and swells the mitochondrial cristae. Although the parasite normally undergoes aerobic respiration (Hontzeas et al., 2001), it also has glycolytic enzymes sequestered in microbodies called glycosomes (Ardelli et al., 2000). The *in vitro* oxygen consumption and carbon dioxide production decrease significantly after drug exposure with very significant increases in secretion of glycolytic products (lactate and pyruvate) as the parasite switches from aerobic respiration to glycolysis after its mitochondrion is damaged by the drug (Ardelli and Woo, 2001a).

Dyskinetoplasty in trypanosomes is induced by many trypanocidal cationic drugs, such as acriflavine, Surfen C (Williamson and Rollo, 1959), Prothidium (Ray and Malhotra, 1960), Berenil (Killick-Kendrick, 1964) and Ethidium (Riou, 1967), so that an attack on kinetoplast DNA is likely to be an integral part of the mode of action of cationic trypanocides. Both spontaneous and chemically induced strains of akinetoplastic *T. evansi have been reported.* Repeated administration of trypaflavin caused the complete disappearance of kinetoplast in *T. evansi* in 81 days. With other trypanosomes such as *T. equiperdum, T. brucei* and *T. gambiense* it took 54, 121 and 122 days respectively, for the complete disappearance of this organelle (Ray and Malhotra, 1960). Baker et al. (1961) demonstrated by electron microscopy the existence of a structure resembling a tube that appears to connect the kinetoplast with the exterior at the posterior end of the parasite; probably enabling drugs to penetrate rapidly into this organelle and allow basic drugs to come into contact with the band of deoxyribonucleic acid which lies therein. However, this has not been reported by other authors. Ormerod and Shaw (1963) have shown that cationic drugs enter the kinetoplast within minutes of treatment. The authors reported that the first change occurred within 1 min of intraperitoneal injection of the drug, consisted of the appearance of a slightly darker area in the region of the kinetoplast. Within 5 mins, a marked darkening of this region had occurred: this did not appear to be confined to the kinetoplast itself, but was diffuse and appeared to involve the whole of the posterior end of the trypanosome. Within 15 mins, an increase in contrast of such granules as were already present had occurred and a dark spot appeared at the center of the nucleus (apparently nucleolus); at the same time a granule began to form in the region of the kinetoplast and during the following hour, the granule extended to form an irregular scar-like mark. It was also shown that a substantial part of the trypanocidal action of the drugs maybe due to an immune response of the host to trypanosomes which have been affected (perhaps even killed) by initial activity of the drug and the new granules which appeared after administration of the drugs are related to the immune response rather than to the action of the drugs themselves (Ormerod and Shaw, 1963).

The above factors may be relevant, not only to the intra-*Trypanosoma*l loci of drug attack, but also to differential host-parasite drug sensitivity. Macadam and Williamson (1972) found the primary focus of all three diamidine trypanocides on the DNA core of the kinetoplast.

Gutteridge and Coombs (1977) proposed the following mechanisms for the selectivity of drugs:

1. Drug only permeates to or is concentrated in the pathogen.
2. Drug is metabolized to the active species only in the pathogen.
3. Drug attacks a target present only in pathogen.
4. Drug discriminates between isofunctional targets in the pathogen and the host.
5. Drug blocks a target which is of greater importance to the pathogen than to the host.

Of prime importance is the mechanism of drug resistance and cross resistance and unfortunately, little progress has been made in understanding the nature of most of the drugs. Gutteridge and Coombs (1977) also proposed the mechanism by which microorganisms can become resistant to drugs:

1. Metabolism of drug to an inactive form;
2. Alteration in permeability therefore drug no longer enters pathogen;
3. Metabolic lesion bypassed by alternative metabolic pathway;
4. Target altered so that its sensitivity to inhibition is decreased;
5. Target enzyme increased in quantity so that pathogen can survive high percentage inhibitions of activity.

During the present chemotherapeutical studies, the impact of the drugs was also observed on the parasites by examining the blood smears of treated fish. At different dose levels of the drugs, the treated parasites darkened and the parasites appeared diffused and disfigured prior to lysis of the parasite by the trypanocidal action of the drugs (Plate LXXIA-C).

The recovery of the test fish, *Clarias batrachus* was observed and compared graphically for the three parameters under study: PCV (Plate LXXIIA), ESR (Plate LXXIIB) and Hb (Plate LXXIII). On comparison of different dose levels of the drugs and different hematological parameters, it was found that Hb showed the best response to the drugs where the peak recovery was noticed at 1.0 mg kg^{-1} Samorin in tpl-infected *Clarias*. The anti-*Trypanosoma*l efficacy of Berenil for Hb was also pretty high, being 74.24% at dose levels 10.0 mg kg^{-1} and 12.5 mg kg^{-1} when the parasitemia was completely suppressed and the infection cleared. When the two parasites were compared, the trypanoplasms responded better to the drugs by showing a better recovery rate as compared to the trypanosomes (Plate LXXIV). However, with regard to achieving the normal hematological values subsequent to drug exposure, the ts-infected treated fish could reach closer to the normal values as compared to the tpl-infected, trypanocide-treated infected fish. Based on the performance of the drugs in the experiment, Berenil at dose level 10.0 mg kg^{-1} and Samorin at the dose level 1.0 mg kg^{-1} can be recommended as the dose of curative therapy for the control of piscine flagellates. Moreover, the two drugs compared, Samorin appears to be the better drug of choice. However, it may be further noted that none of the fish that had received trypanocide treatments had regained their initial PCV, ESR and Hb values up to 5th WPT. In this respect, the best response was observed for Hb where in Group IIIB and Group IVA, the Hb value was only less by 5.9% as compared to the normal Hb value but could even remain less up to a maximum of 57.30% as observed in Group IIA for ESR values (Berenil treated) (Table 6.10). It maybe recalled that in this Group, the dose level was low (5.0 mg kg^{-1}) and relapse occurred during the 3rd WPT resulting in low-level consistent parasitemia. Therefore, the graphs in Plates LXXII – LXXIV represent the recovery rate at 5 mg kg^{-1} Berenil and 0.1 mg kg^{-1} Samorin at 2 WPT whereas the recovery rate at other dose levels are represented at 5 WPT.

Drug toxicity to Berenil and Samorin appears to be unpredictable in individual animals and there are some anecdotal reports of the drug not exerting the expected anti-parasitic action. The relapses at 2–3 weeks of therapy, dictates the need for further study of the pharmacokinetics of drugs in order to enable us to better understand the cause for the variation in the clinical responses seen.

The ANOVA test for the experiments conducted above show the values of significance within the group and between groups (Tables 6.11–6.14). The results indicate that the change in hematological parameters of flagellate-infected *Clarias batrachus* are significant ($P < 0.001$). The drug-treated fish showed a recovery in their hematological parameters and the comparative recovery rates between ts- infected and tpl-infected fish were insignificant ($P > 0.01$) when compared to each other. However, the recovery rates of hematological parameters in Berenil- treated (Table 6.11) and Samorin-treated (Table 6.12) infected fish were significant ($P< 0.001$) indicating the effectivity of the tested drugs.

Table 6.10. Recovery rate (%) of Different Hematological Parameters of *Clarias batrachus* Infected with Blood Flagellates Subsequent to Treatment with Different Dose Levels of Berenil and Samorin

Haematological Parameters	**Parasites**	**Drugs**							
		Berenil				**Samorin**			
		Dose levels (mg kg^{-1})				Dose levels (mg kg^{-1})			
		5.0	10.0	12.5	15.0	0.1	0.5	1.0	2.0
PCV	Ts- infected	24.73	34.62	34.98	Lethal	19.48	41.5	41.91	Lethal
	Tpl-infected	46.00	55.34	55.34	Lethal	28.63	54.0	73.63	Lethal
ESR	Ts- infected	13.04	32.90	32.91	Lethal	13.29	32.91	34.81	Lethal
	Tpl-infected	21.91	32.50	40.44	Lethal	22.09	35.91	41.43	Lethal
Hb	Ts- infected	21.05	29.47	32.63	Lethal	26.37	29.67	32.96	Lethal
	Tpl-infected	53.03	74.24	74.24	Lethal	57.35	76.47	83.82	Lethal

Table 6.11. ANOVA Results of Hematological Parameters of Control, Infected (*Trypanosoma* and *Trypanoplasma*) and Different Treatment Regimes of *Clarias batrachus* Treated with Berenil

	Trypanosoma			***Trypanoplasma***		
	Hb%	**ESR**	**PCV**	**Hb%**	**ESR**	**PCV**
Control	13.40±0.42c	0.89±0.03a	42.80±0.23d	13.40±0.42d	0.89±0.03a	42.80±0.23d
Infected	9.50±0.22a	1.61±0.11b	28.30±0.56a	6.60±0.19a	1.78±0.11c	21.50±0.62a
Treatment 1*	9.90±0.27a	1.74±0.11b	30.30±0.63b	9.70±0.27b	1.76±0.11c	29.40±0.46b
Treatment 2**	12.30±0.16b	1.08±0.04a	38.10±0.51c	11.50±0.15c	1.21±0.09b	33.40±0.42c
Treatment 3***	12.60±0.11b	1.07±0.02a	38.20±0.49c	11.50±0.12c	1.06±0.04ab	33.40±0.06c

* 5.0 mg kg^{-1} ** 10.0 mg kg^{-1} *** 12.5 mg kg^{-1}

[a,b,c,d] Superscript showing the significance level at $P<0.001$

Table 6.12. *ANOVA results of Hematological Parameters of Control, Infected (Trypanosoma and Trypanoplasma) and Different Treatment Regimes of Clarias batrachus Treated with Samorin*

	Trypanosoma			***Trypanoplasma***		
	Hb%	**ESR**	**PCV**	**Hb%**	**ESR**	**PCV**
Control	13.60±0.20c	0.90±0.01a	44.00±0.66d	13.60±0.20d	0.90±0.01a	44.00±0.66e
Infected	9.10±0.24a	1.58±0.02d	27.20±0.12a	6.80±0.18a	1.81±0.03e	22.00±0.40a
Treatment 1*	9.22±0.12a	1.48±0.02c	31.70±0.16b	9.40±0.13b	1.59±0.03d	25.60±0.14b
Treatment 2**	11.80±0.31b	1.06±0.02b	38.50±0.17c	12.05±0.14c	1.16±0.02c	34.10±0.16c
Treatment 3***	12.13±0.17b	1.03±0.02b	38.60±0.11c	12.50±0.16c	1.06±0.01b	38.20±0.11d

** 0.1 $mg\ kg^{-1}$ ** 0.5 $mg\ kg^{-1}$ *** 1.0 $mg\ kg^{-1}$*

[a,b,c,d,e] *Superscript showing the significance level at $P<0.001$.*

Table 6.13. ANOVA Values of Different Hematological Parameters of *Clarias batrachus* Treated with Berenil

Hemoglobin (Hb)					
	Sum of Squares	**df**	**Mean Square**	**F**	**Significance**
Ts-infected Between groups	72.072	4	18.018	45.045	0.000
Within groups	10.0	25	0.4		
Total	82.072	29			
Tpl-infect Between groups	157.512	4	39.378	103.845	0.000
Within groups	9.48	25	0.379		
Total	166.992	29			
Erythrocyte Sedimentation Rate (ESR)					
Ts-infected Between groups	3.34	4	0.835	25.62	0.000
Within groups	0.815	25	3.26E-02		
Total	4.155	29			
Tpl-infected Between groups	4.007	4	1.002	25.648	0.000
Within groups	0.976	25	3.91E-02		
Total	4.983	29			
Packed Cell Volume (PCV)					
Ts-infected Between groups	877.272	4	219.318	143.834	0.000
Within groups	38.12	25	1.525		
Total	915.392	29			
Tpl-infected Between groups	1425.12	4	356.28	360.9	0.000
Within groups	24.68	25	0.987		
Total	1449.8	29			

Table 6.14. ANOVA Values of Different Hematological Parameters of *Clarias batrachus* Treated with Samorin

Hemoglobin (Hb)					
	Sum of Squares	**df**	**Mean square**	**F**	**Significance**
Ts-infected Between groups	91.981	4	22.995	80.723	0.000
Within groups	7.122	25	0.285		
Total	99.103	29			
Tpl-infected Between groups	181.368	4	45.342	275.468	0.000
Within groups	4.115	25	0.165		
Total	185.483	29			
Erythrocyte Sedimentation Rate (ESR)					
Ts-infected Between groups	2.165	4	0.541	326.812	0.000
Within groups	4.14E-02	25	1.66E-03		
Total	2.206	29			
Tpl-infected Between groups	3.494	4	0.873	339.417	0.000
Within groups	6.43E-02	25	2.57E-03		
Total	3.558	29			
Packed Cell Volume (PCV)					
Ts-infected Between groups	1037.64	4	259.41	415.721	0.000
Within groups	15.6	25	0.624		
Total	1053.24	29			
Tpl-infected Between groups	1948.608	4	487.152	619.471	0.000
Within groups	19.66	25	0.786		
Total	1968.268	29			

A Tribute

Stars in the field of Parasitology in general and Hematozoan parasites in particular have soared high in the sky, spreading their glamour to inspire other workers and carry on the relay. Their dedication, devotion and perseverance have given new dimensions to the domain of Parasitological Research.

I humbly pay tribute to the names of the researchers mentioned in the following pages whose research contributions have acted as stepping stones for me to tread on and continue the race of parasitological research. There are no words to acknowledge their valuable contributions which have perceptibly or imperceptibly initiated a flow of thought in my inquisitive mind to further their research. Each consulted research article is a pearl in itself which is capable of being strewn into a perfect necklace of knowledge.

The work presented in the thesis is a humble attempt to carry on the race started by the earlier researchers which would have been impossible without repeated consultation of their thought-provoking contributions. In the following pages, I mention the names and publications of all the worthy researchers involved in 1039 research papers who have been a constant source of inspiration and have provided the necessary guidelines for the execution of this piece of work.

I owe my deepest sense of gratitude to all the names associated with the publications.

"A salute to all those researchers Whose experiences and findings Have paved the way for fellow researchers."

PLATE I

Map showing district Bareilly and perennial rivers of Bareilly merging with river Ramganga

PLATE II

***A*. The Institute of Ichthyobiology and Aquaculture, Golysz, Poland.**

B. **Pond of the Institute of Ichthyobiology and Aquaculture, Golysz, Poland.**

C. **Carps being harvested at the Institute of Ichthyobiology and Aquaculture, Golysz, Poland.**

PLATE III

Experimental set-up of aquaria with recirculating water system for maintenance of *Cyprinus carpio* at the Institute of Ichthyobiology and Aquaculture, Chybie, Poland.

Selection of labeled *Cyprinus carpio* for blood sampling at the Institute of Ichthyobiology and Aquaculture, Chybie, Poland.

PLATE IV

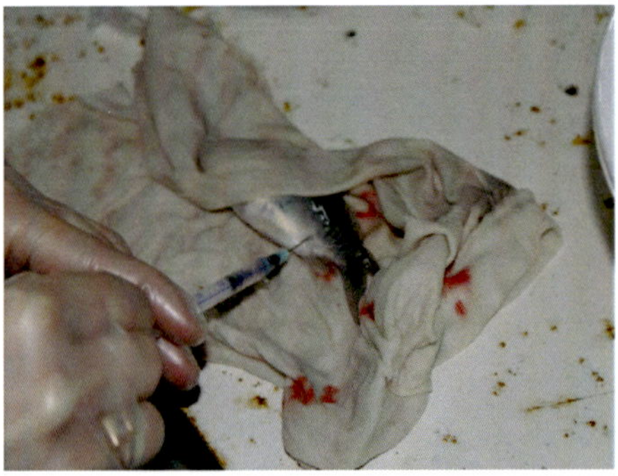

***A.* Locating the caudal vein of *Cyprinus carpio* for blood collection.**

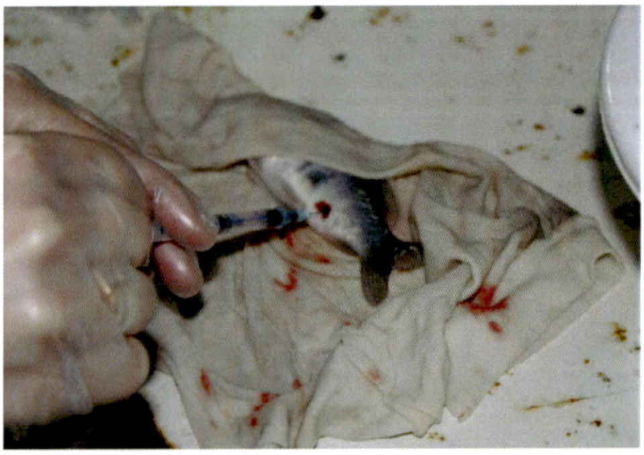

B.* Collection of blood from the caudal vein of *Cyprinus carpio.

PLATE V

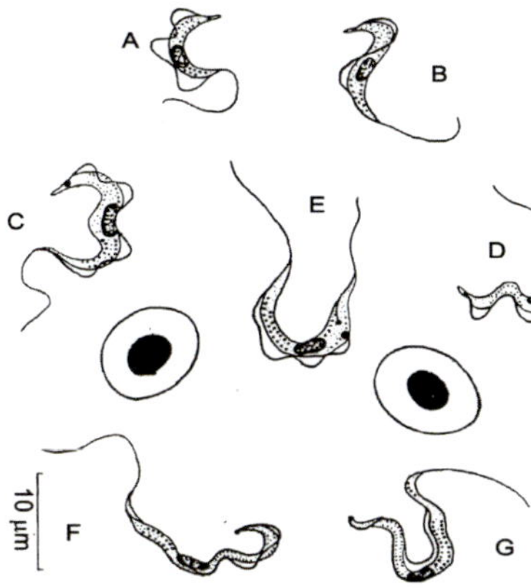

Camera lucida diagrams of *Trypanosoma colisi* (Gupta, 1986).

PLATE VI

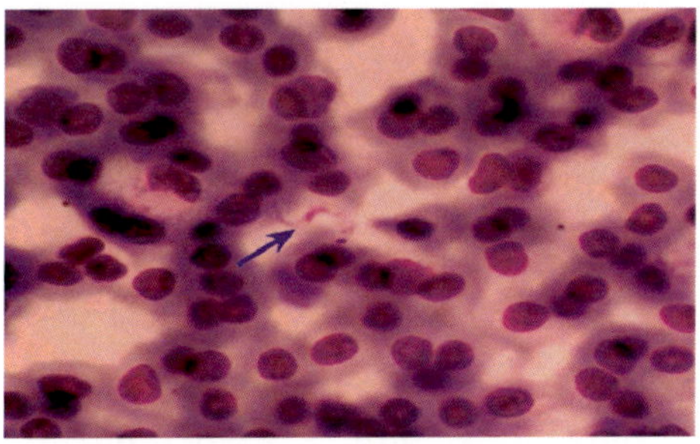

***A.* Photomicrograph of *T. colisi* (Gupta, 1986) showing stumpy form (X 1000).**

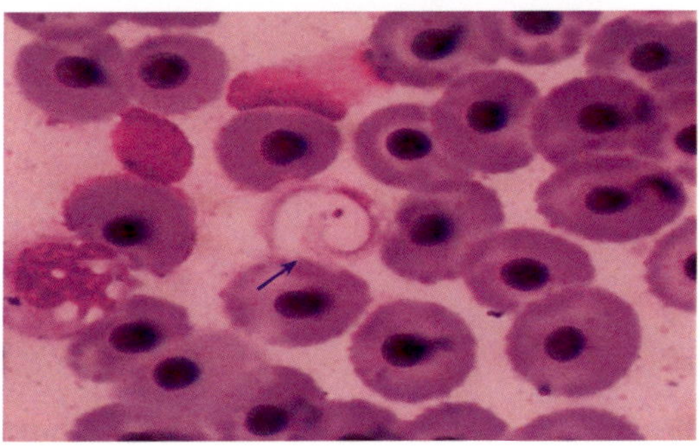

***B.* Photomicrograph of *T. colisi* (Gupta, 1986) showing intermediate form (X 1000).**

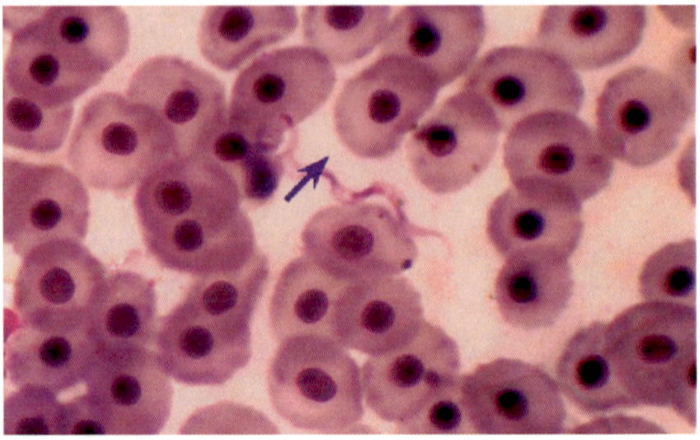

C. **Photomicrograph of *T. colisi* (Gupta, 1986) showing slender form (X 1000).**

PLATE VII

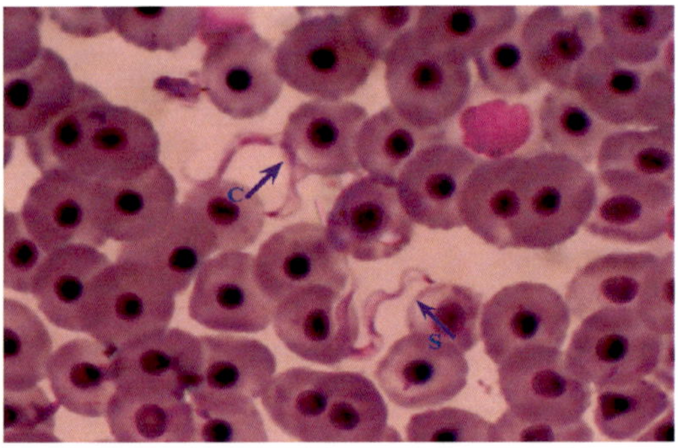

A. **Photomicrograph of *Trypanosoma kargenensis* (Gupta and Gupta, 1994) showing 'C' and 'S' shaped parasites (X1000).**

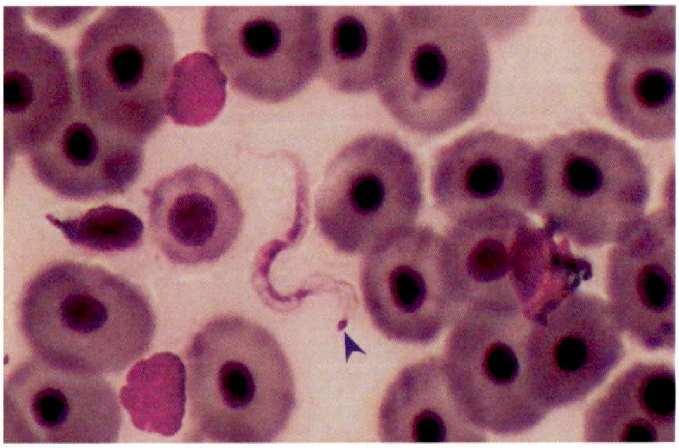

B. **Photomicrograph of *Trypanosoma kargenensis* (Gupta and Gupta, 1994) showing prominent kinetoplast (X1000).**

PLATE VIII

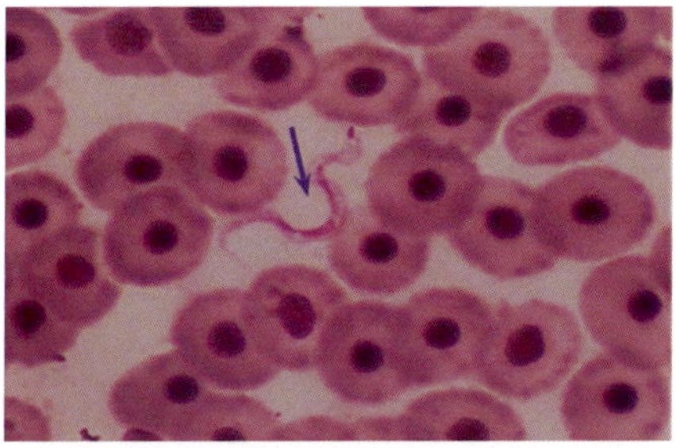

A. **Photomicrograph of *Trypanosoma bagroides* (Gupta and Gupta, 1996) showing elongated nucleus (X1000).**

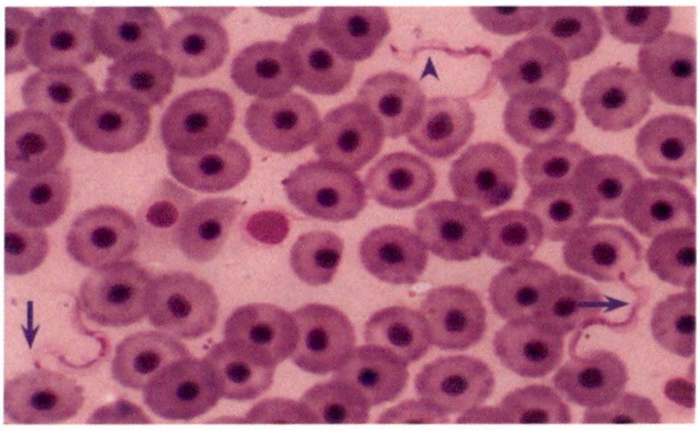

B. **Photomicrograph of *Trypanosoma bagroides* (Gupta and Gupta, 1996) showing folds in undulating membrane (X1000).**

PLATE IX

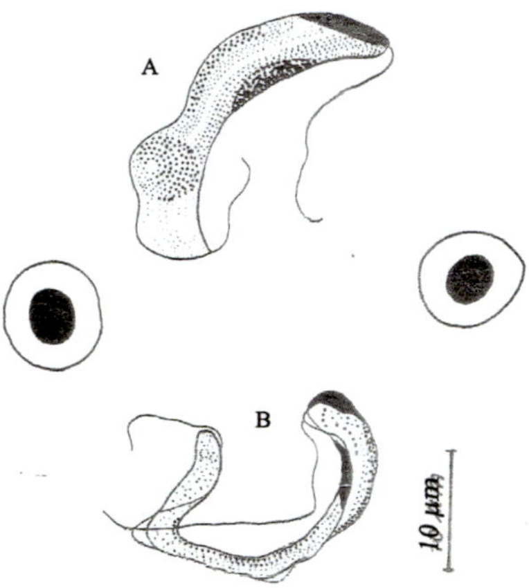

Camera lucida diagrams of *Trypanoplasma maguri* (Gupta and Gupta, 1987).

PLATE X

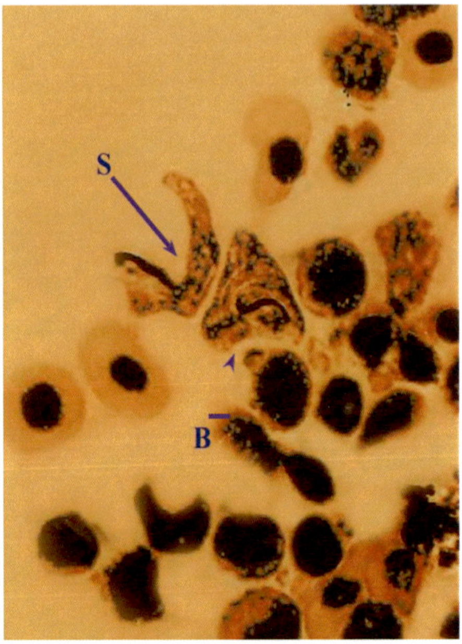

Photomicrograph of *Trypanoplasma maguri* (Gupta and Gupta, 1987). Bulky (B) and Slender (S) forms (X1000).

PLATE XI

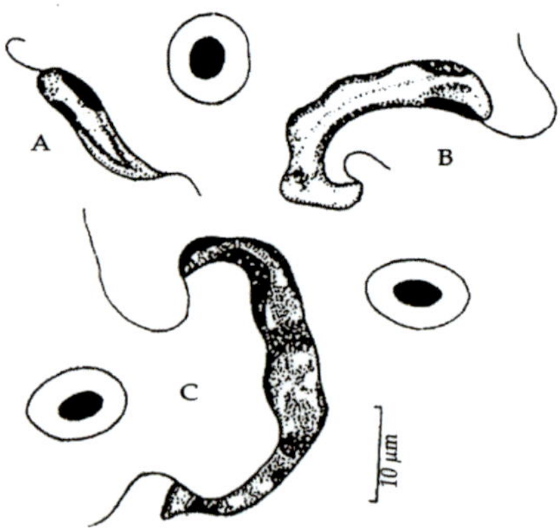

Camera lucida diagrams of *Trypanoplasma guptii* (Gupta and Gupta, 1987). (A) Minute form; (B) Intermediate form; (C) Bulky form.

PLATE XII

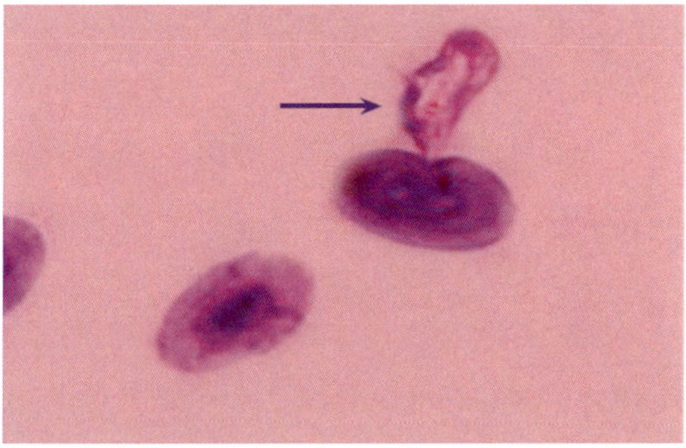

A. **Photomicrograph of minute form of *Trypanoplasma guptii* (Gupta and Gupta, 1987) having pointed end (X1000).**

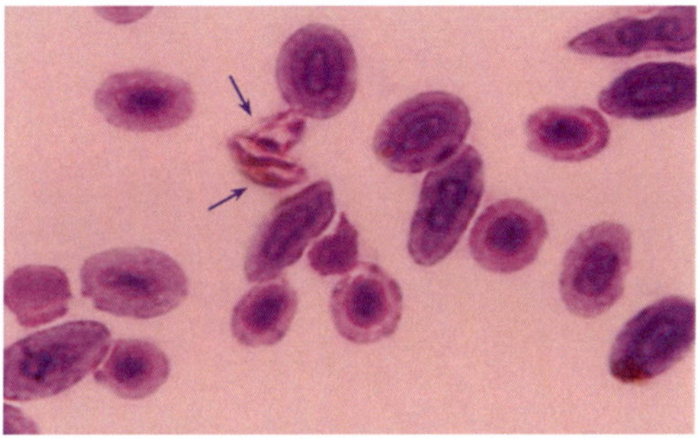

B. **Photomicrograph of minute form of *Trypanoplasma guptii* (Gupta and Gupta, 1987) having a uniformly broad body (X1000).**

PLATE XIII

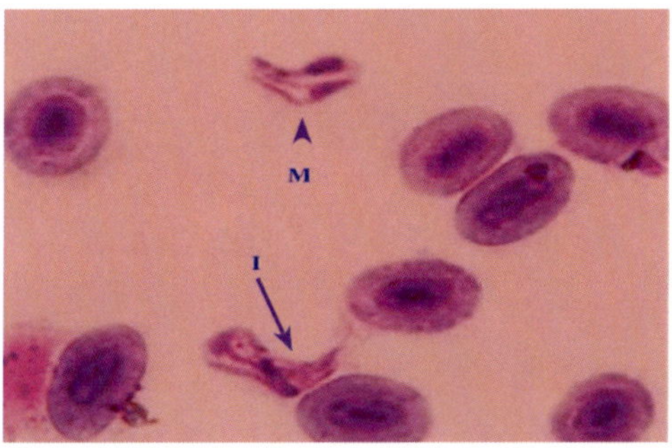

A. **Photomicrograph of minute (M) and intermediate (I) forms of *Trypanoplasma guptii* (Gupta and Gupta, 1987) in close proximity (X1000).**

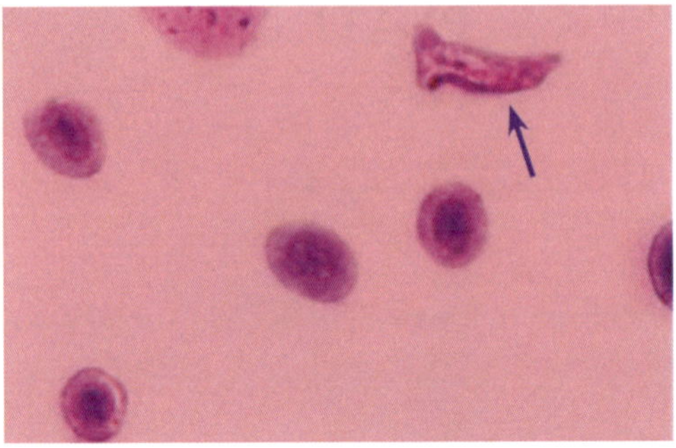

B.* Photomicrograph of intermediate form of *Trypanoplasma guptii
(Gupta and Gupta, 1987) having a straight posture (X1000).

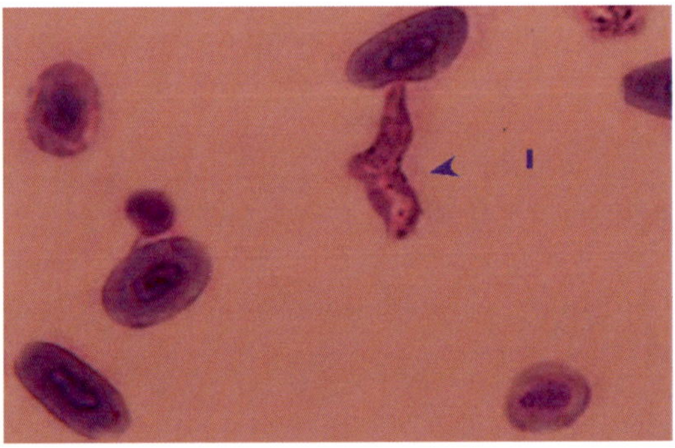

C.* Photomicrograph of intermediate form of *Trypanoplasma guptii
(Gupta and Gupta, 1987) having a notched posture (X1000).

PLATE XIV

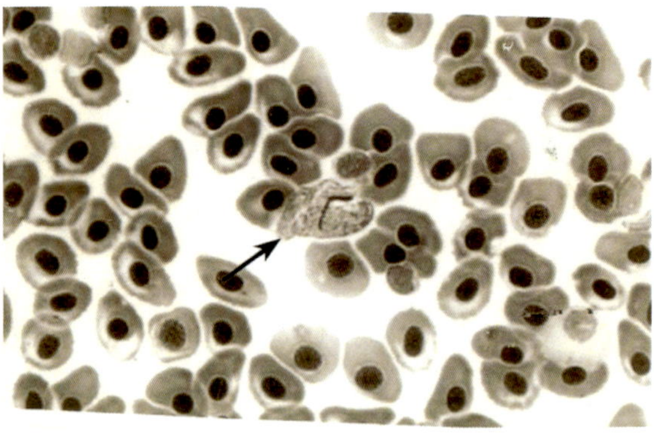

***A.* Photomicrograph of minute form of *Trypanoplasma guptii* (Gupta and Gupta, 1987) having broad anterior and narrow posterior end (X1000).**

B. Photomicrograph of intermediate form of *Trypanoplasma guptii* (Gupta and Gupta, 1987) having a curved posture (X1000).

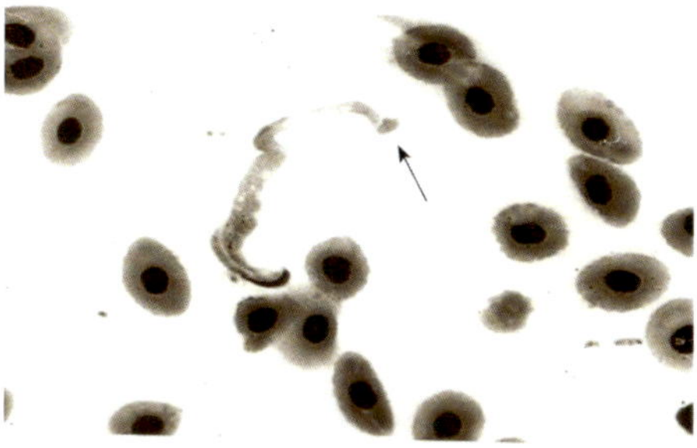

C. Photomicrograph of bulky form of *Trypanoplasma guptii* (Gupta and Gupta, 1987) (X1000).

PLATE XV

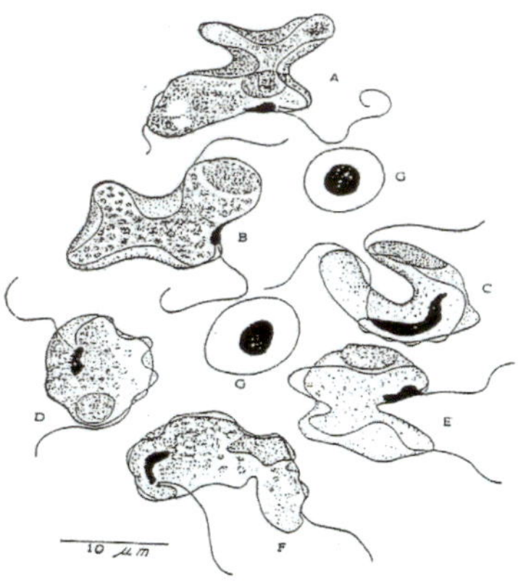

Camera lucida diagrams of *Trypanoplasma catli* (Gupta et al., 1999).

A-F – Different shapes of the parasite; G – Blood cells.

PLATE XVI

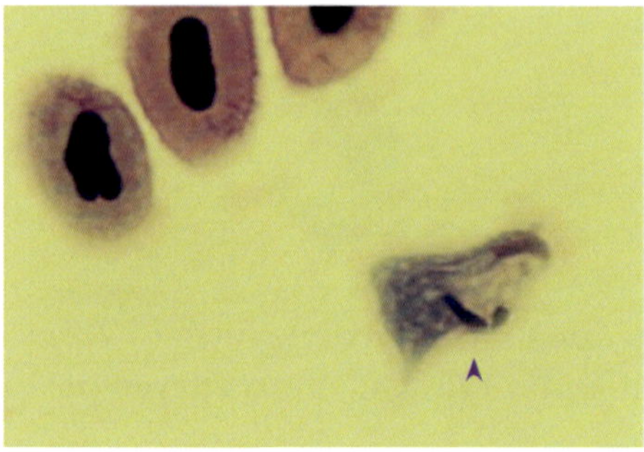

A. **Photomicrograph of *Trypanoplasma catli* (Gupta and Gupta, 1998) showing pointed end (X1000).**

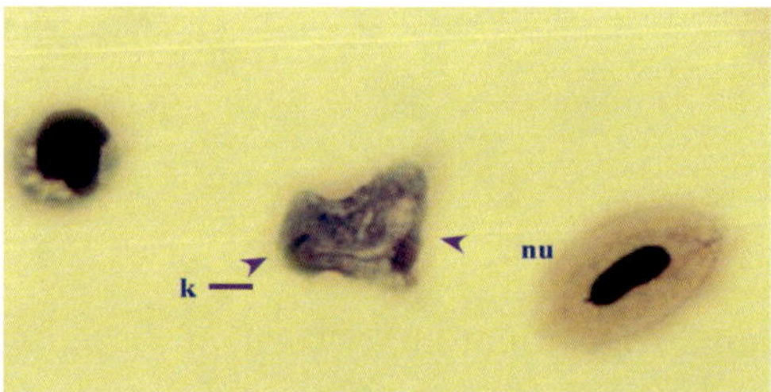

B. **Photomicrograph of *Trypanoplasma catli* (Gupta and Gupta, 1998) showing opposed nucleus and kinetoplast (X1000).**

PLATE XVII

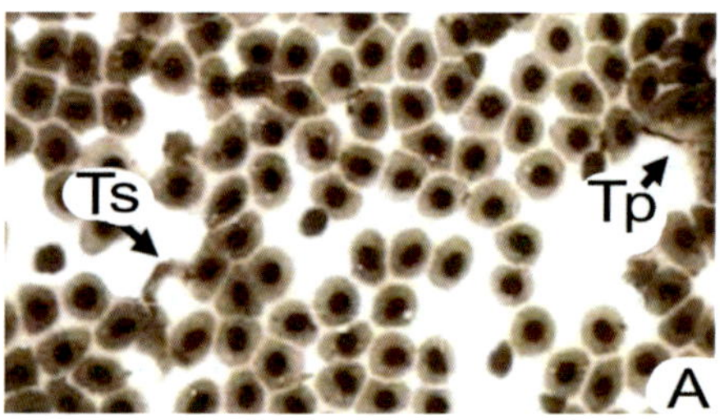

A. **Photomicrograph of the blood of *Clarias batrachus* showing concomitant infection of *Trypanosoma batrachi* (Qadri, 1962) (Ts) and *Trypanoplasma haematalis* (Gupta and Gupta, 2010). (Tp) (X 450).**

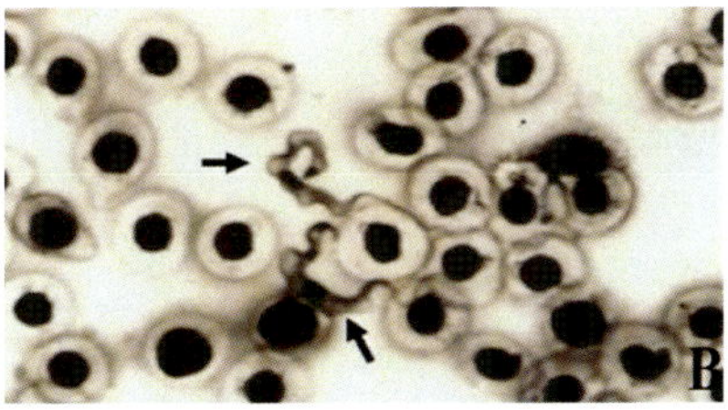

B. **Photomicrograph of the blood of *Clarias batrachus* showing *Trypanosoma batrachi* (Qadri, 1962) (X 1000).**

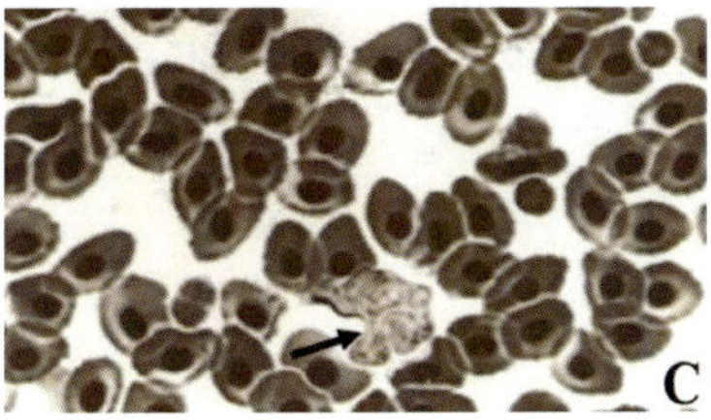

C. **Photomicrograph of the blood of *Clarias batrachus* showing *Trypanoplasma haematalis* (Gupta and Gupta, 2010) (X 1000)**

PLATE XVIII

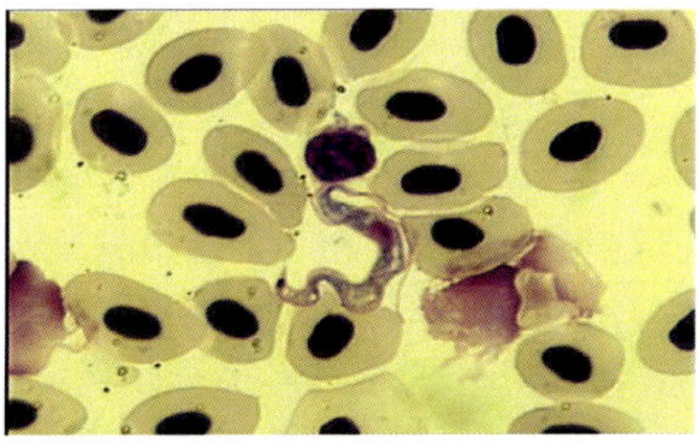

A. **Photomicrograph of *Trypanosoma danilewskyi* (Laveran and Mesnil, 1904) showing 'C' shaped body (X1000).**

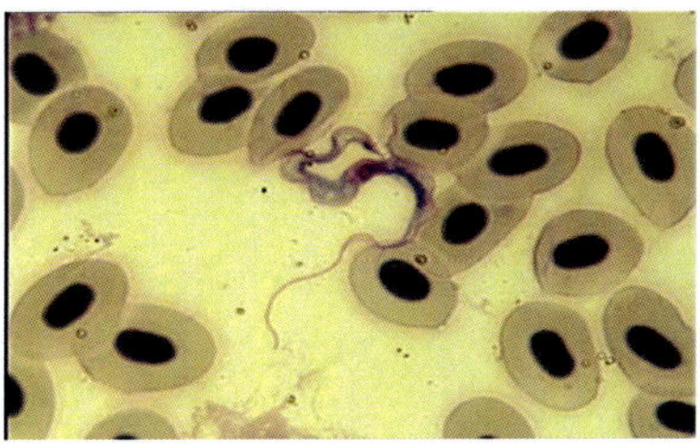

B. **Photomicrograph of *Trypanosoma danilewskyi* (Laveran and Mesnil, 1904) showing 'S' shaped body (X1000).**

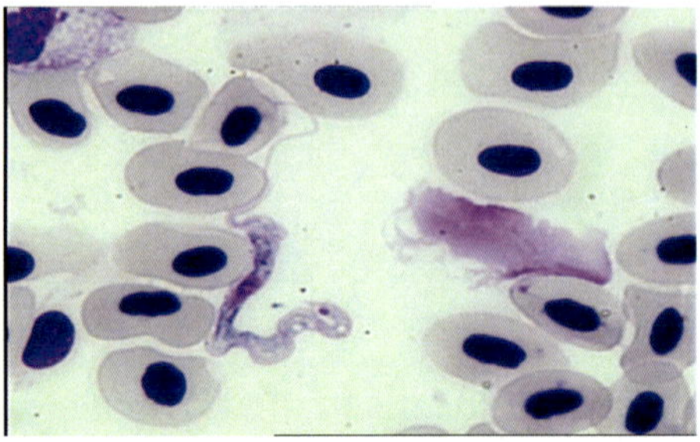

C. **Photomicrograph of *Trypanosoma danilewskyi* (Laveran and Mesnil, 1904) showing undulated body (X1000).**

PLATE XIX

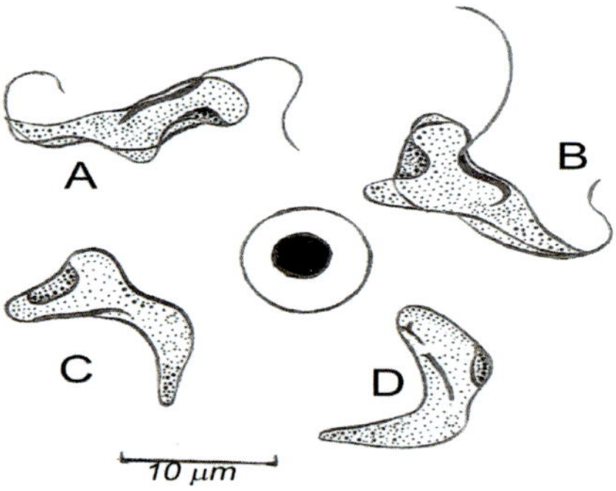

Camera lucida diagrams of *Trypanosoma golysiana* (Gupta and Pilarczyk, 1994).

PLATE XX

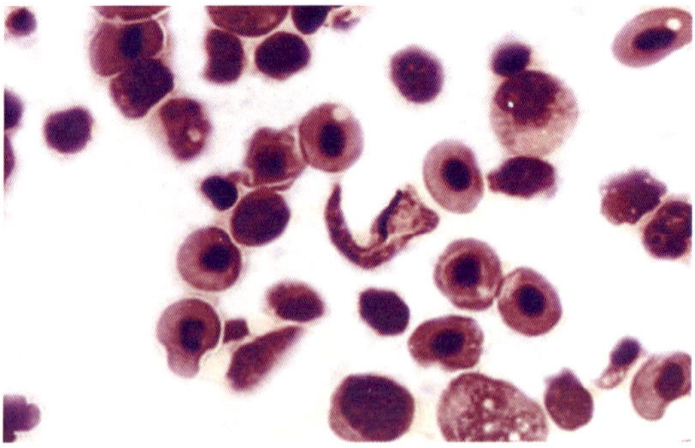

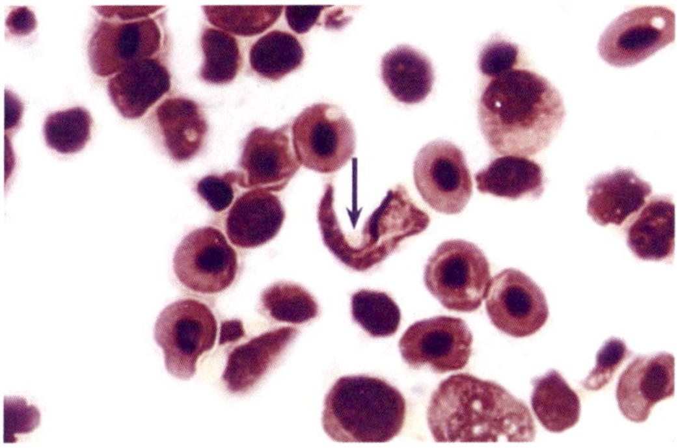

Photomicrographs of blood forms of *Trypanoplasma golysiana* (Gupta and Pilarczyk, 1994) (X1000).

PLATE XXI

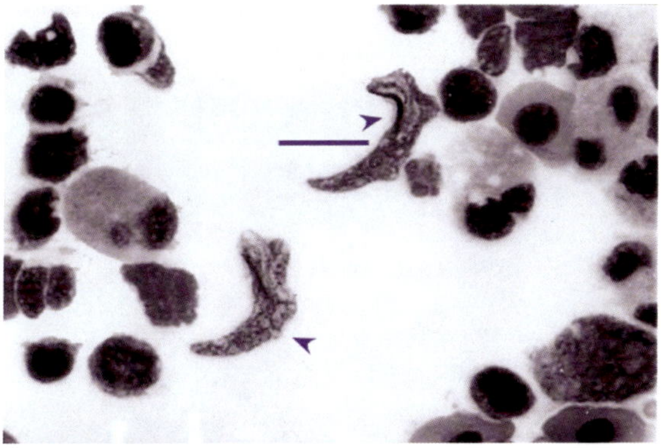

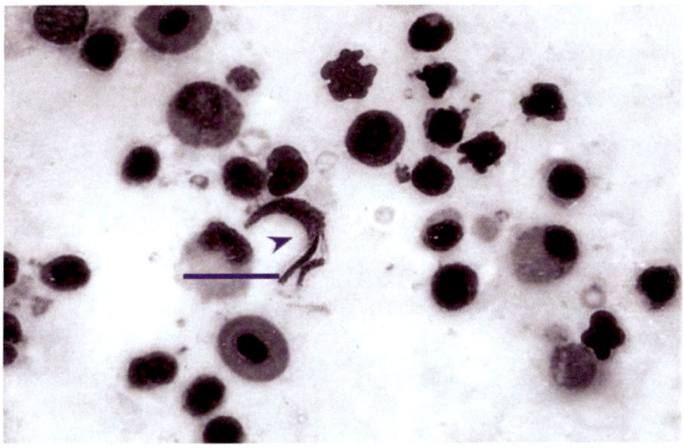

Photomicrographs of kidney forms of *Trypanoplasma golysiana* (Gupta and Pilarczyk, 1994) (X1000).

PLATE XXII

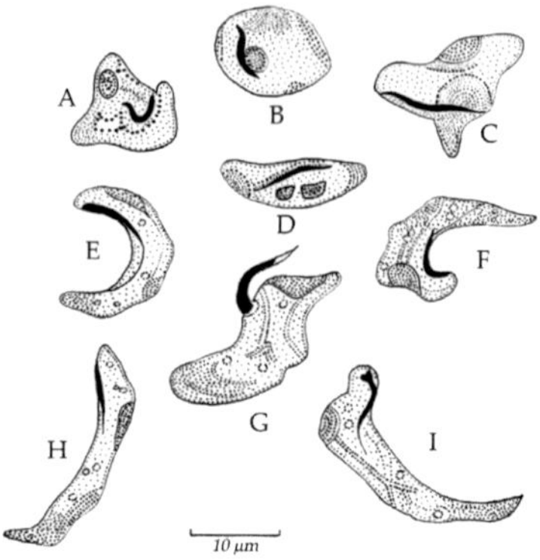

Camera lucida diagrams of *Trypanosoma polygolysiana* (Gupta and Pilarczyk, 1994) A–D: Stumpy forms; E–G: Intermediate forms; H-I: Large forms.

PLATE XXIII

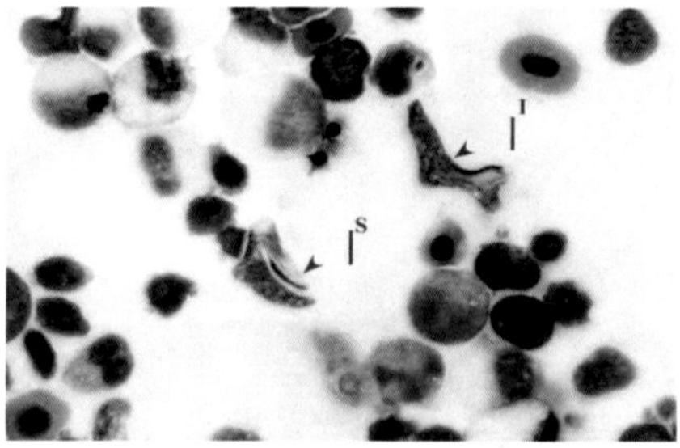

***A.* Photomicrograph of *Trypanoplasma polygolysiana* (Gupta and Pilarczyk, 1994). Stumpy (S) and Intermediate (I) forms (X1000).**

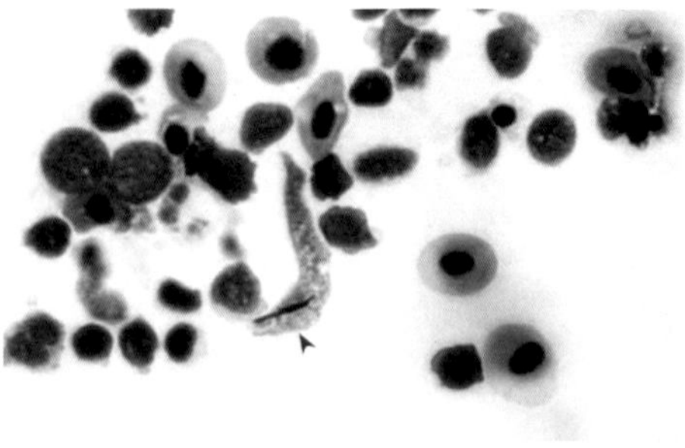

***B.* Photomicrograph of *Trypanoplasma polygolysiana* (Gupta and Pilarczyk, 1994). Large form (X1000).**

PLATE XXIV

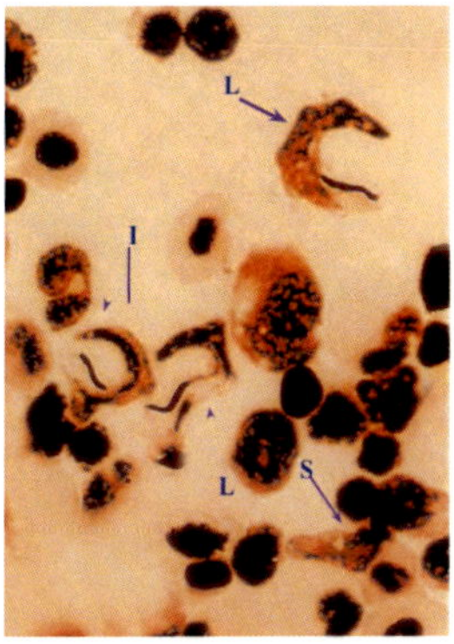

Photomicrograph of *Trypanoplasma polygolysiana* (Gupta and Pilarczyk, 1994). Stumpy (S), Intermediate (I) and Large (L) forms (X1000).

PLATE XXV

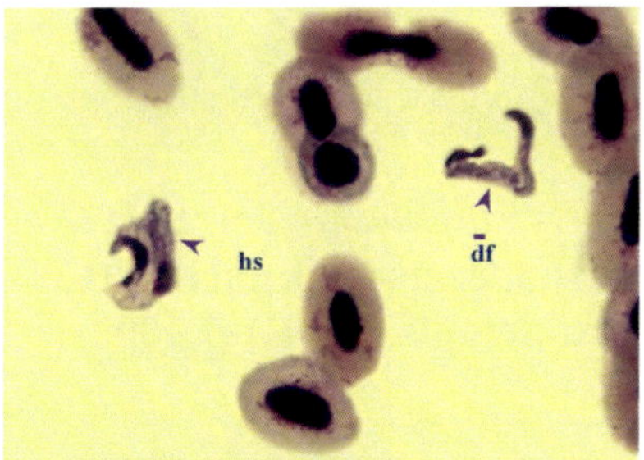

***A.* Photomicrograph of *Trypanoplasma borreli* (Laveran and Mesnil, 1901) showing horse-shoe shape (hs) and dividing form (df) (X1000).**

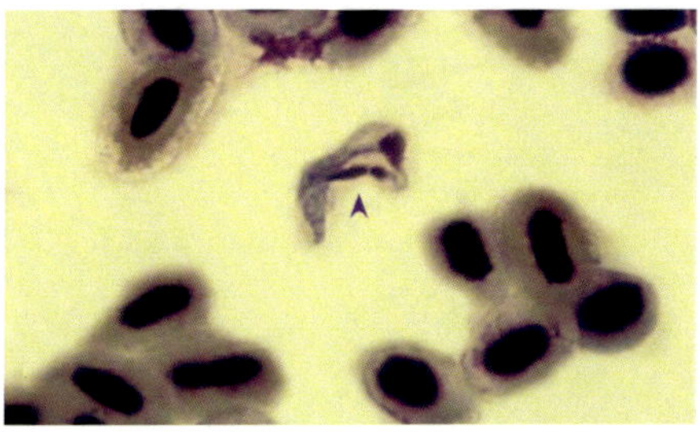

***B.* Photomicrograph of *Trypanoplasma borreli* (Laveran and Mesnil, 1901) showing C-shaped body, marginal and notched kinetoplast (X1000).**

PLATE XXVI

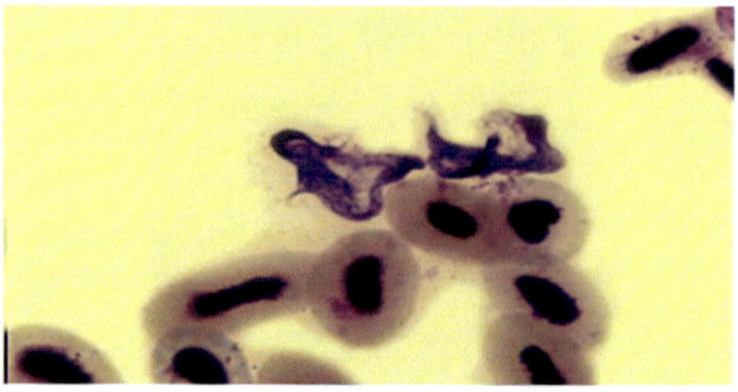

A. **Photomicrograph of *Trypanoplasma borreli* (Laveran and Mesnil, 1901) showing stumpy shape (X1000).**

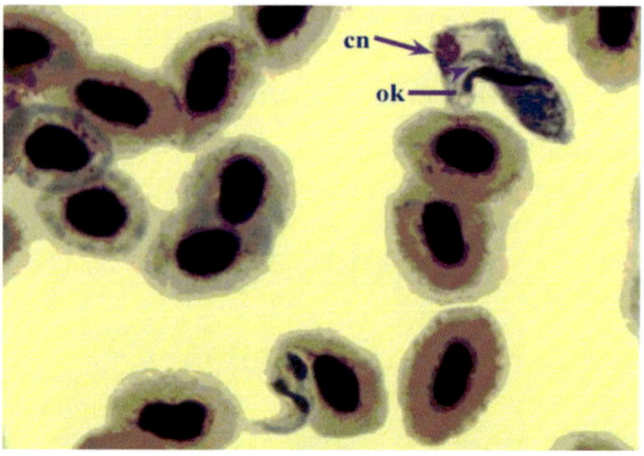

B. **Photomicrograph of *Trypanoplasma borreli* (Laveran and Mesnil, 1901) showing question mark shape, circular nucleus (CN) and obliquely placed kinetoplast (OK) (X1000).**

PLATE XXVII

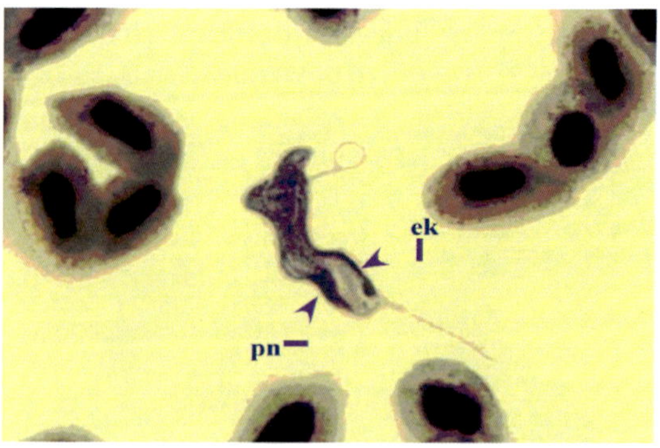

A. **Photomicrograph of *Trypanoplasma borreli* (Laveran and Mesnil, 1901) showing elongated shape, pointed nucleus at one end (PN) and elongated kinetoplast (EK) (X1000).**

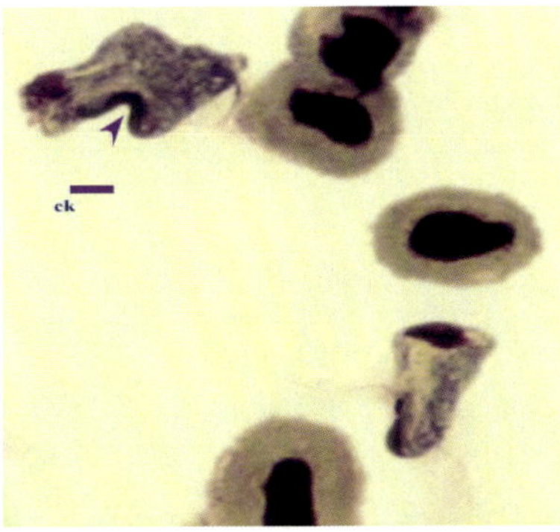

***B.* Photomicrograph of *Trypanoplasma borreli* (Laveran and Mesnil, 1901) showing curved kinetoplast (ck) and origin of flagella (X1000).**

PLATE XXVIII

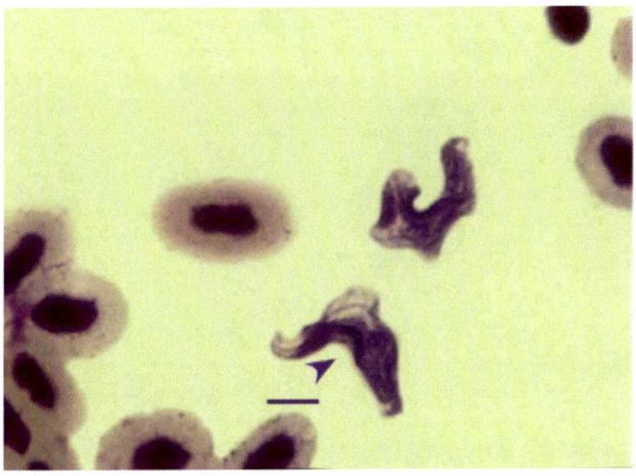

***A.* Photomicrograph of *Trypanoplasma borreli* (Laveran and Mesnil, 1901) showing partially marginal kinetoplast (X1000).**

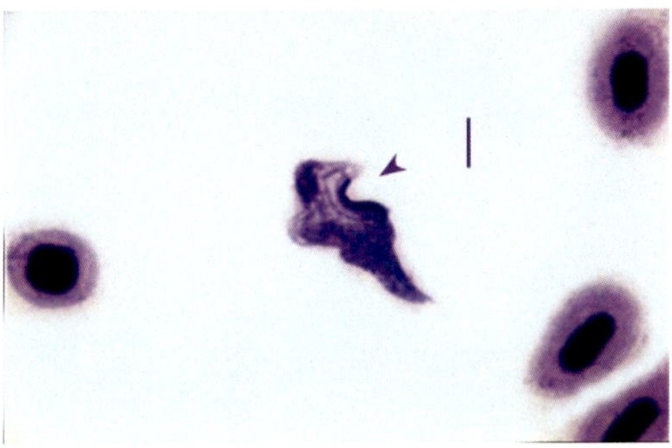

***B.* Photomicrograph of *Trypanoplasma borreli* (Laveran and Mesnil, 1901) showing posteriorly running flagellum (X1000).**

PLATE XXIX

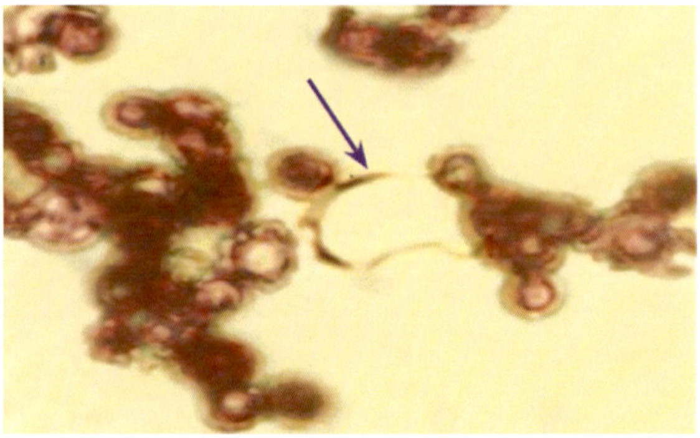

A. **Photomicrograph of *Trypanosoma lewisi* (Laveran and Mesnil, 1901) showing posterior long and thin end (X1000).**

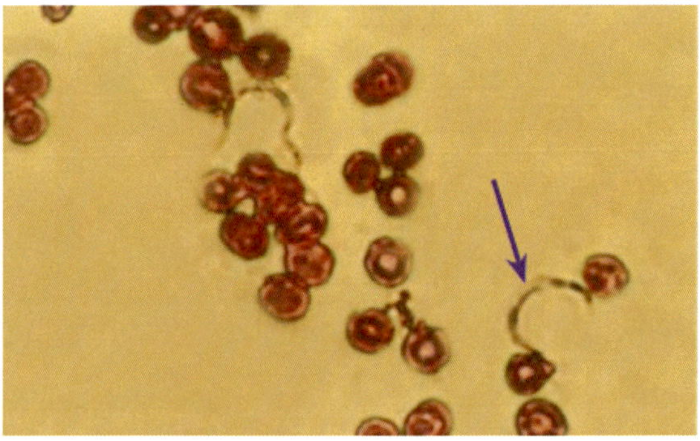

B. **Photomicrograph of *Trypanosoma lewisi* (Laveran and Mesnil, 1901) showing immature forms (X1000).**

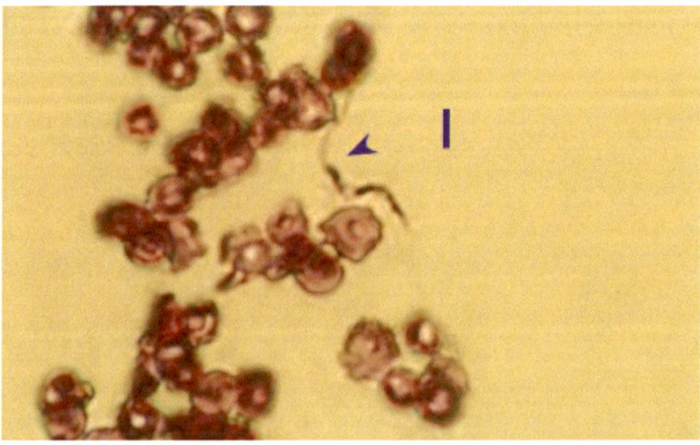

C. **Photomicrograph of *Trypanosoma lewisi* (Laveran and Mesnil, 1901) showing classical adult slender forms (X1000).**

PLATE XXX

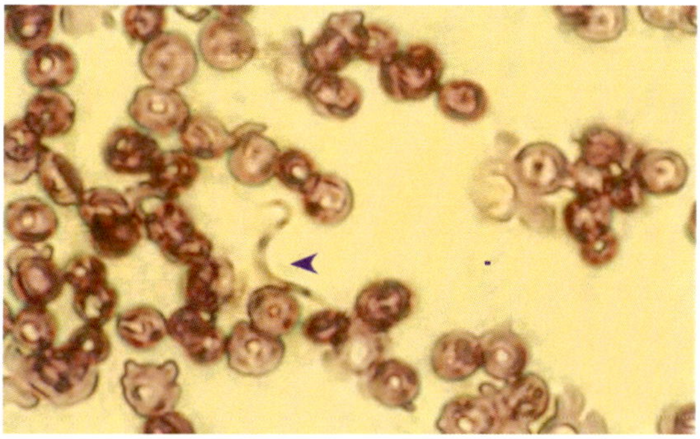

A **Photomicrograph of *Trypanosoma lewisi* (Laveran and Mesnil, 1901) showing C-shaped parasite (X1000).**

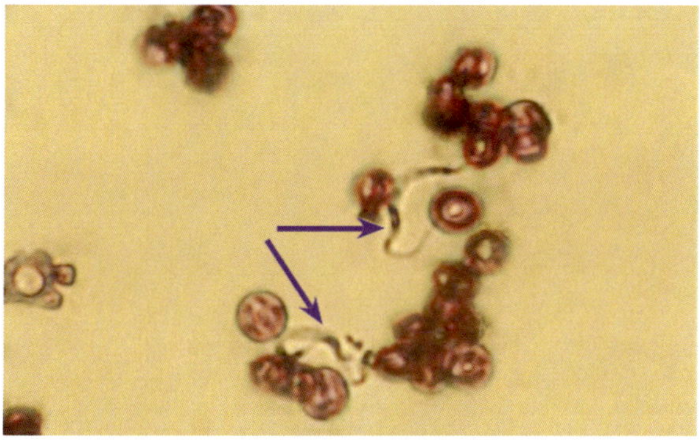

***B.* Photomicrograph of *Trypanosoma lewisi* (Laveran and Mesnil, 1901) showing S-shaped parasite (X1000).**

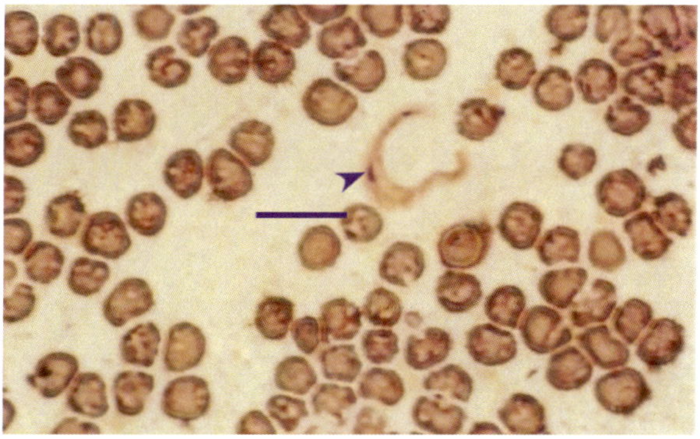

***C.* Photomicrograph of *Trypanosoma lewisi* (Laveran and Mesnil, 1901) showing oval nucleus, long axis lying parallel to the body (X1000).**

PLATE XXXI

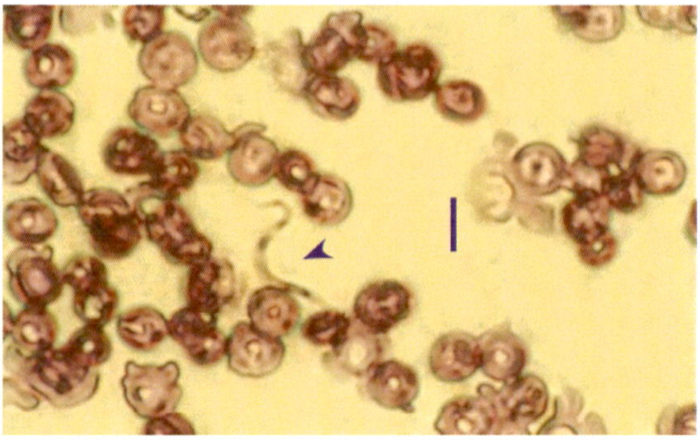

***A.* Photomicrograph of *Trypanosoma lewisi* (Laveran and Mesnil, 1901) showing oval kinetoplast situated far from the posterior end (X1000).**

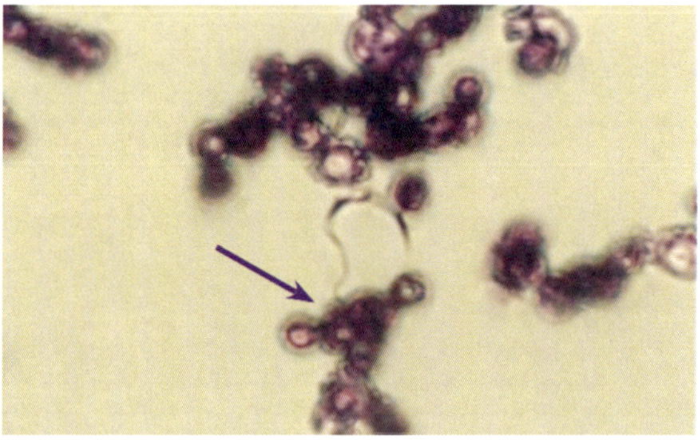

***B.* Photomicrograph of *Trypanosoma lewisi* (Laveran and Mesnil, 1901) showing well developed anteriorly directed flagellum (X1000).**

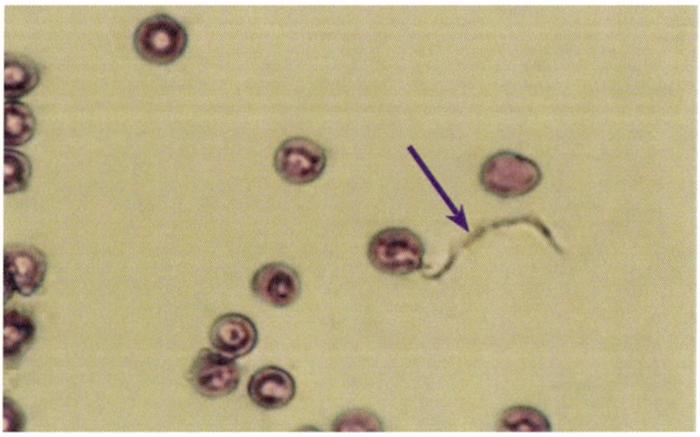

***C.* Photomicrograph of *Trypanosoma lewisi* (Laveran and Mesnil, 1901) showing prominent undulating membrane on the convex side of the parasite (X1000).**

PLATE XXXII

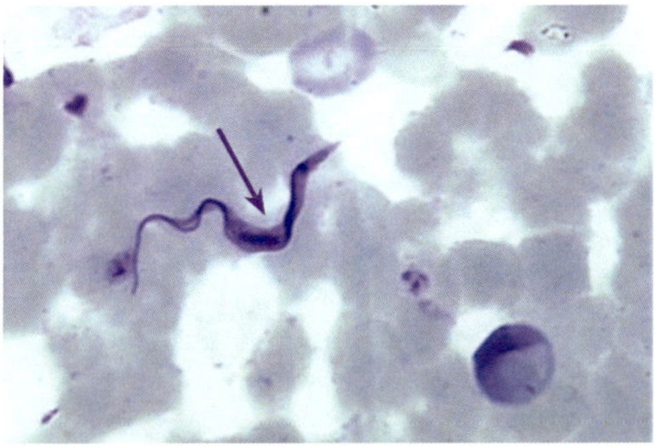

A. **Photomicrographs of curved and slender *Trypanosoma grosi kosewiense* (Karbowiak and Wita, 2004) (X1000).**

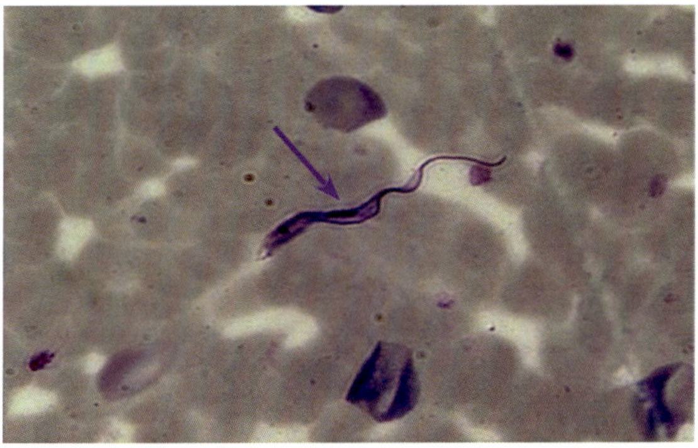

B. Photomicrographs of straight and slender Trypanosoma grosi kosewiense (Karbowiak and Wita, 2004) (X1000).

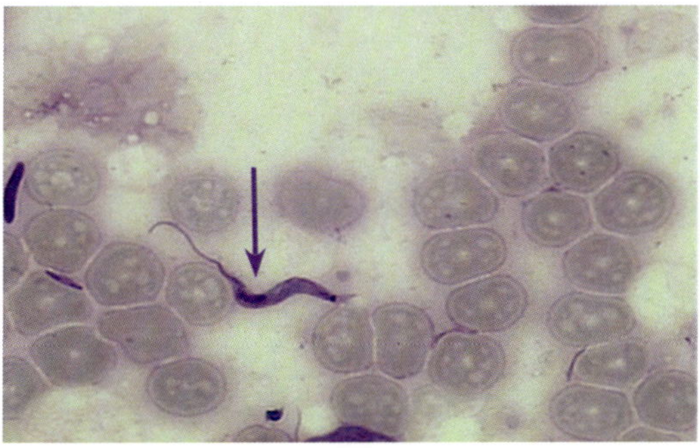

C. **Photomicrographs of slender form of *Trypanosoma grosi kosewiense* (Karbowiak and Wita, 2004) showing oval nucleus (X1000).**

PLATE XXXIII

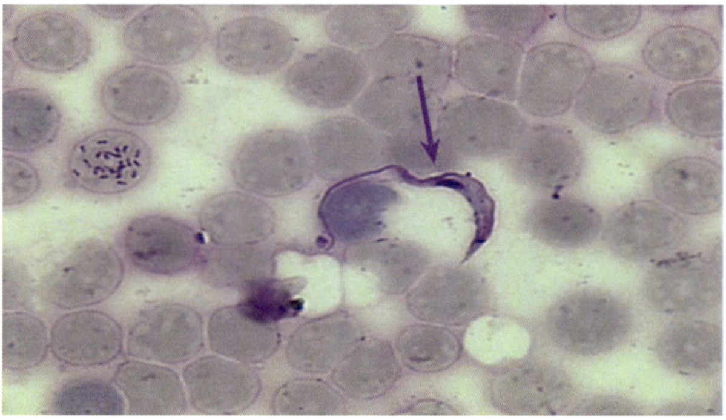

A. Photomicrographs of stout form of T*rypanosoma grosi kosewiense* (Karbowiak and Wita, 2004) showing oval nucleus (X1000).

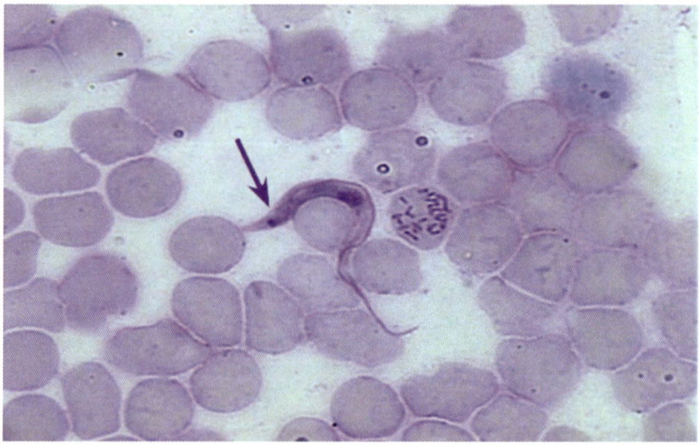

***B.* Photomicrographs of stout form of *Trypanosoma grosi kosewiense* (Karbowiak and Wita, 2004) showing oval kinetoplast (X1000).**

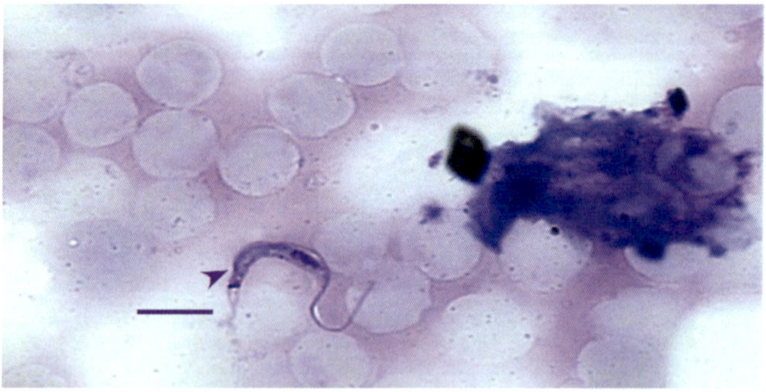

***C.* Photomicrographs of stout form of *Trypanosoma grosi kosewiense* (Karbowiak and Wita, 2004) showing perpendicular kinetoplast (X1000).**

PLATE XXXIV

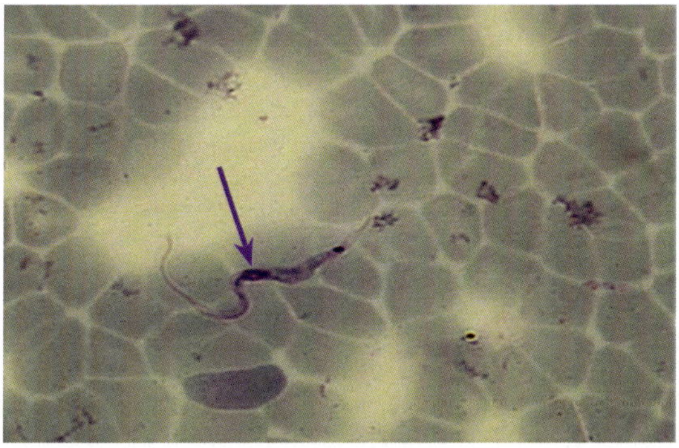

A. **Photomicrograph of *Trypanosoma evotomys* (Molyneux, 1969) showing elongated body (X1000).**

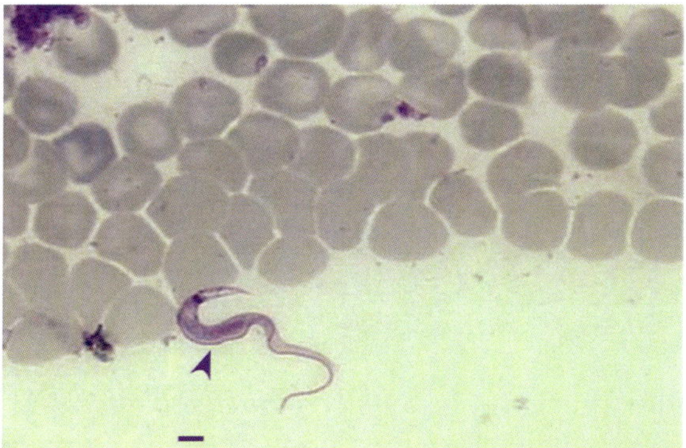

B. **Photomicrograph of *Trypanosoma evotomys* (Molyneux, 1969) showing elongated nucleus (X1000).**

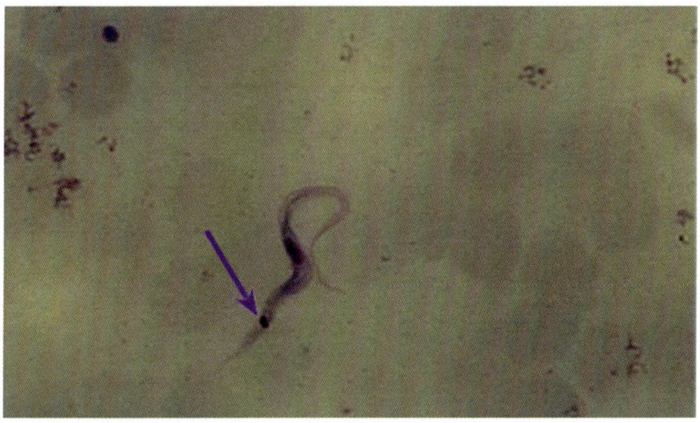

C. **Photomicrograph of *Trypanosoma evotomys* (Molyneux, 1969) showing spherical kinetoplast (X1000).**

PLATE XXXV

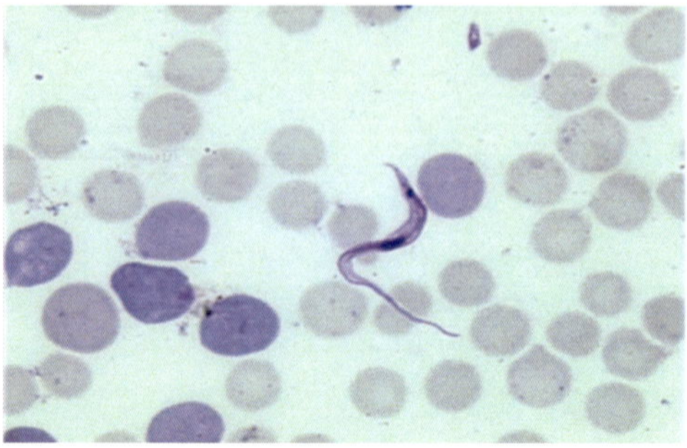

A. **Photomicrograph of *Trypanosoma microti* (Laveran and Pettitt, 1909) showing C-shaped body and extended flagellum (X1000).**

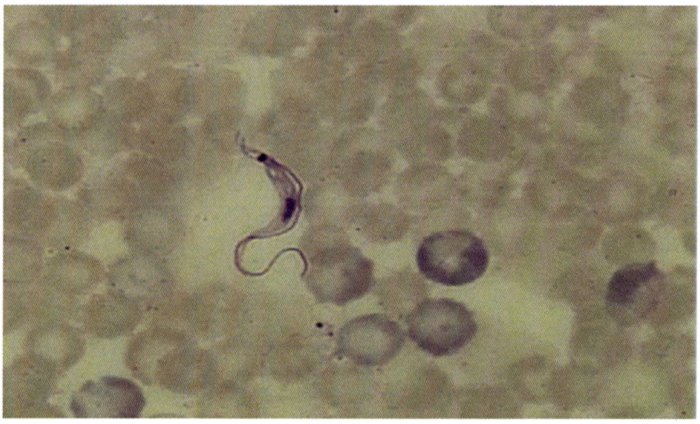

B. **Photomicrograph of small *Trypanosoma microti* (Laveran and Pettitt, 1909) showing C-shaped body and curved flagellum (X1000).**

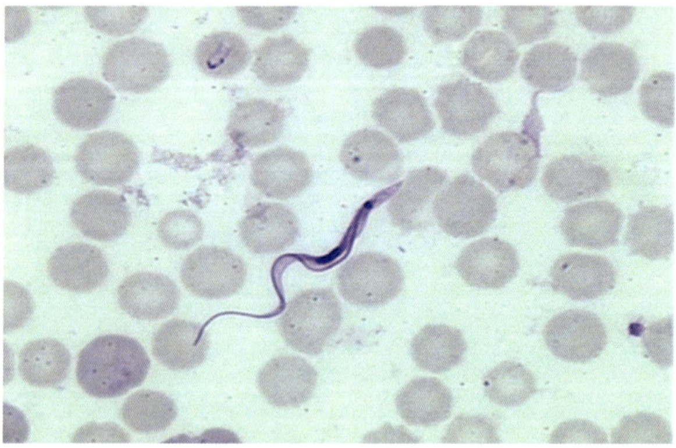

C. **Photomicrograph of *Trypanosoma microti* (Laveran and Pettitt, 1909) showing extened S-shaped body (X1000).**

PLATE XXXVI

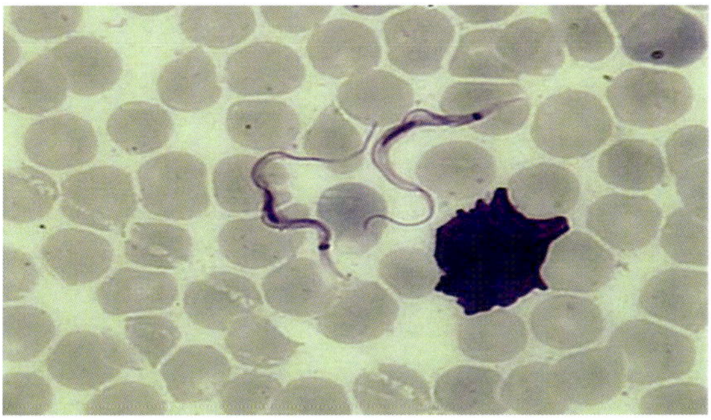

A. Photomicrograph of *Trypanosoma microti* (Laveran and Pettitt, 1909) showing sinuous body (X1000).

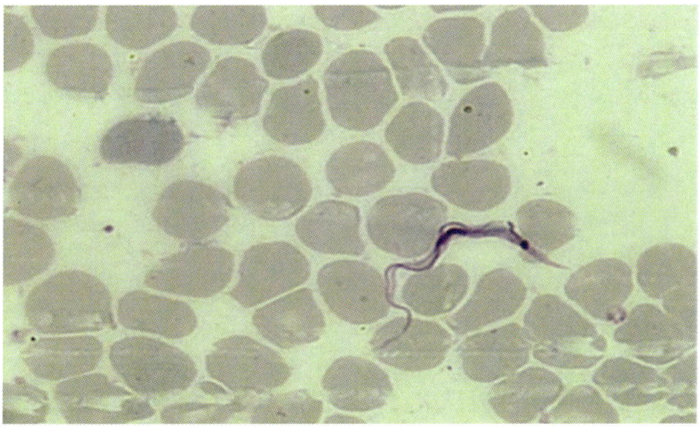

B. Photomicrograph of large *Trypanosoma microti* (Laveran and Pettitt, 1909) (X1000).

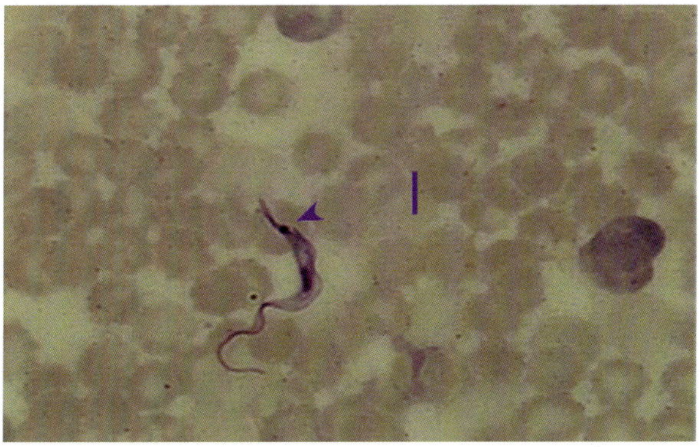

C. Photomicrograph of *Trypanosoma microti* (Laveran and Pettitt, 1909) showing round kinetoplast (X1000).

PLATE XXXVII

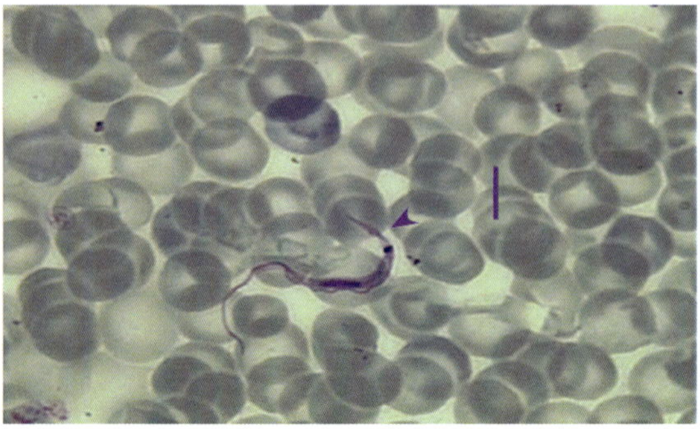

A. **Photomicrograph of *Trypanosoma microti* (Laveran and Pettitt, 1909) showing compressed kinetoplast (X1000).**

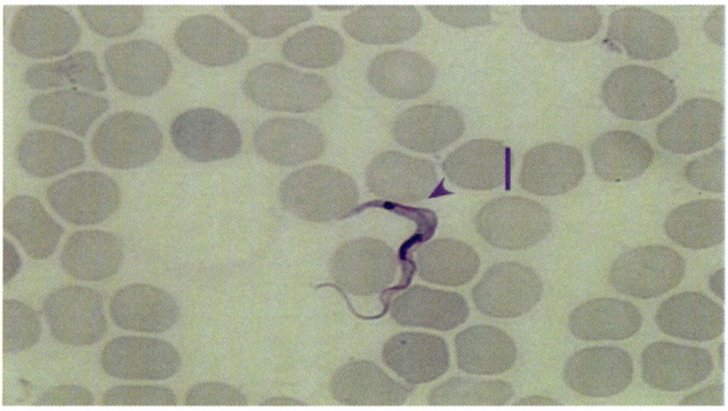

B. **Photomicrograph of *Trypanosoma microti* (Laveran and Pettitt, 1909) showing 4 folds of undulating membrane (X1000).**

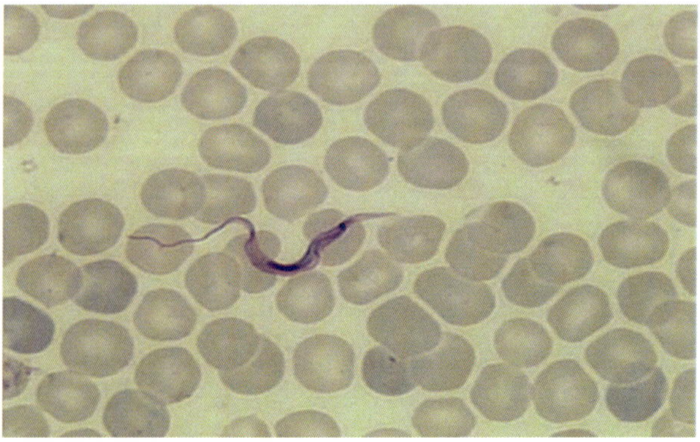

C. **Photomicrograph of *Trypanosoma microti* (Laveran and Pettitt, 1909) showing 5 folds of undulating membrane (X1000).**

PLATE XXXVIII

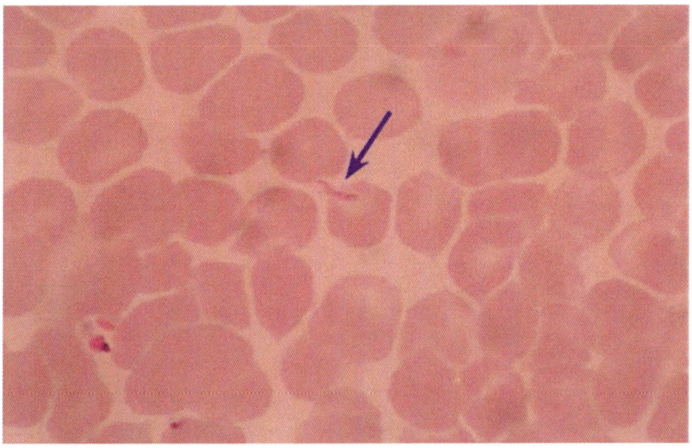

A. Photomicrograph of sporozoite of *Babesia microti* (Franca, 1912) entering the erythrocyte of *Microtus oeconomus* (X1200).

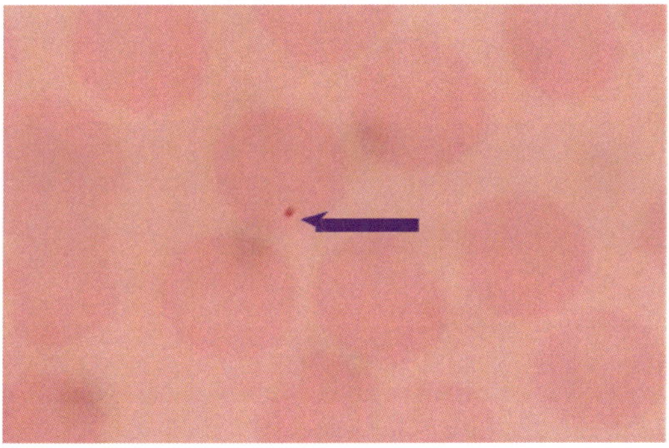

B. Photomicrograph of dot-like *Babesia microti* (Franca, 1912) with very less cytoplasm during early development (X1200).

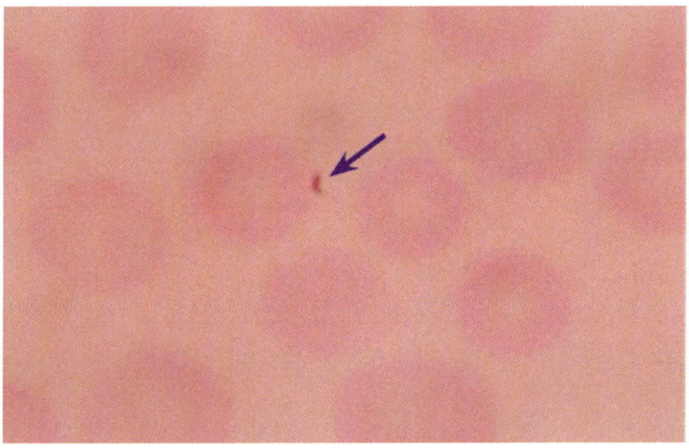

C. Photomicrograph of crescent form of *Babesia microti* (Franca, 1912) (X1200).

PLATE XXXIX

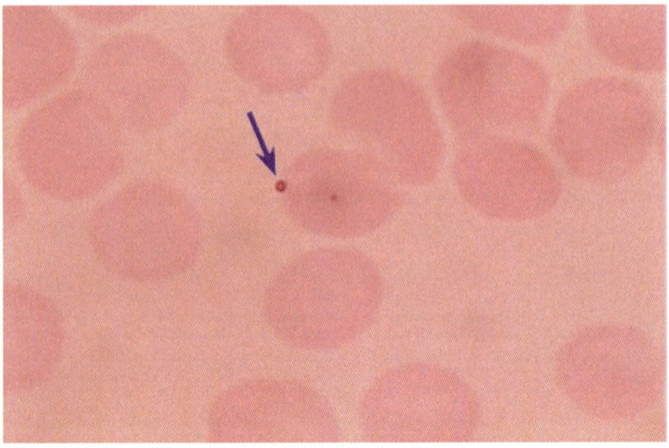

A. **Photomicrograph of micro-ring stage of *Babesia microti* (Franca, 1912) at tip of the erythrocyte (X1200).**

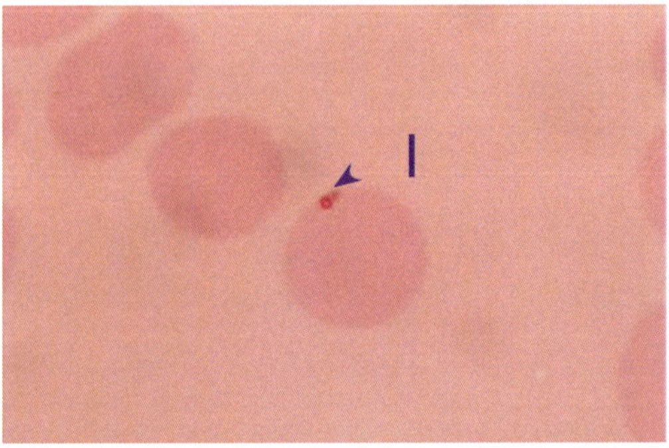

B. **Photomicrograph of micro-ring stage of *Babesia microti* (Franca, 1912) inside the erythrocyte (X1200).**

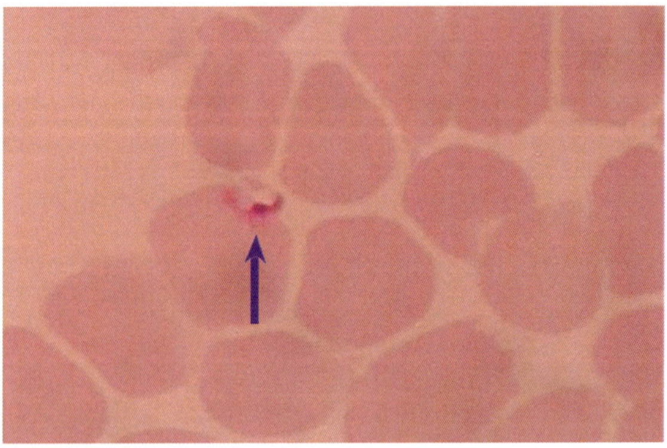

C. **Photomicrograph of ring form of *Babesia microti* (Franca, 1912) (X1200).**

PLATE XL

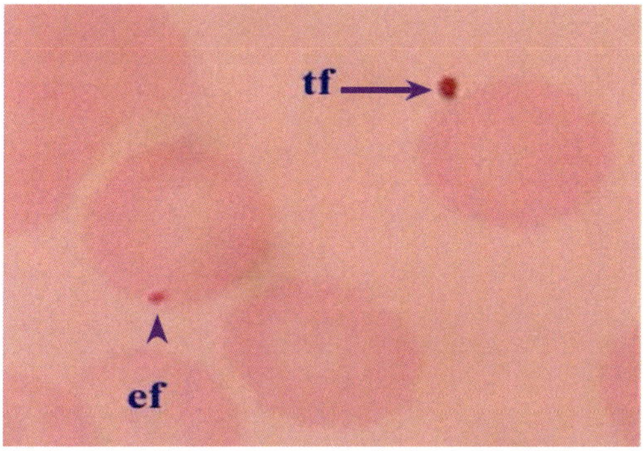

A. **Photomicrograph of tear drop form (tf) and eye form (ef) of *Babesia microti* (Franca, 1912) (X1200).**

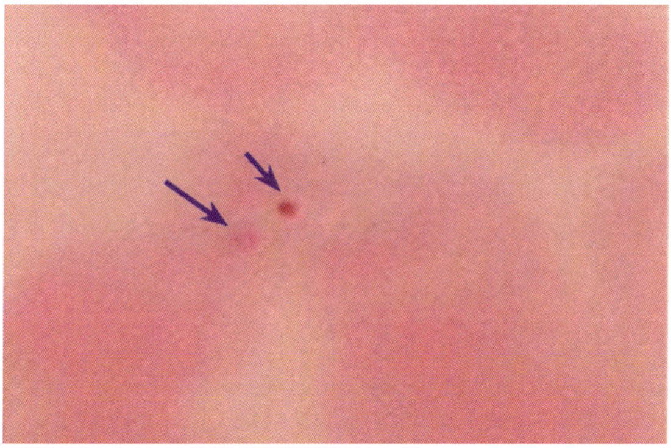

B. **Photomicrograph of double eye of *Babesia microti* (Franca, 1912) (X1200).**

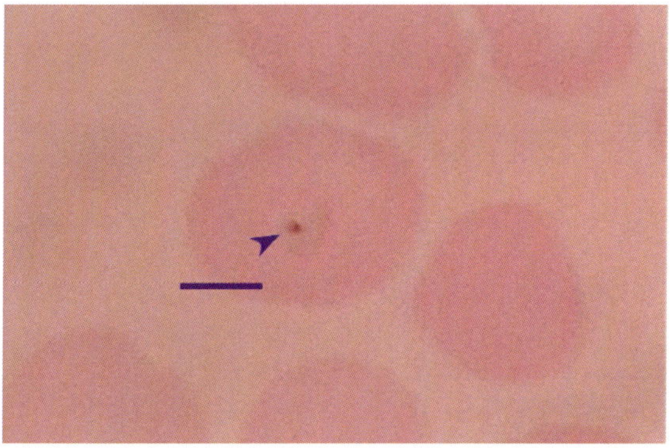

C. **Photomicrograph of trophozoite of *Babesia microti* (Franca, 1912) (X1200).**

PLATE XLI

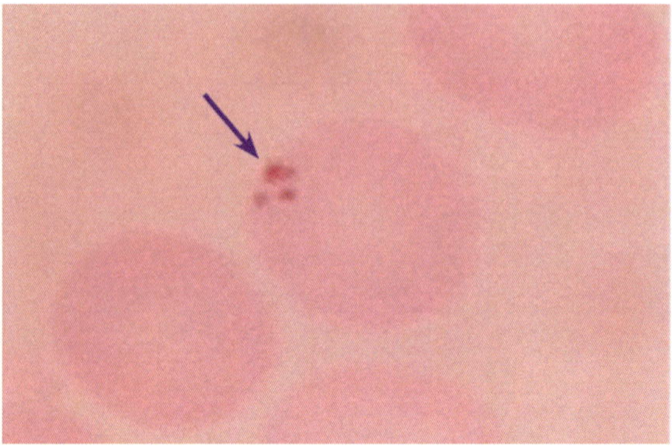

A. **Photomicrograph of tetrad dots of *Babesia microti* (Franca, 1912) (X1200).**

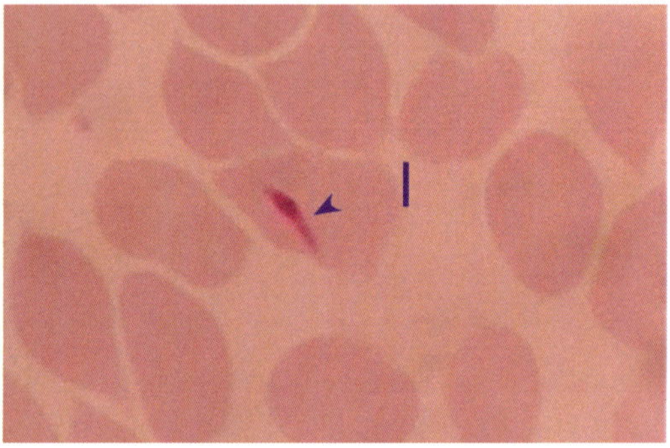

B. **Photomicrograph of thick long worm string pattern of *Babesia microti* (Franca, 1912) (X1200).**

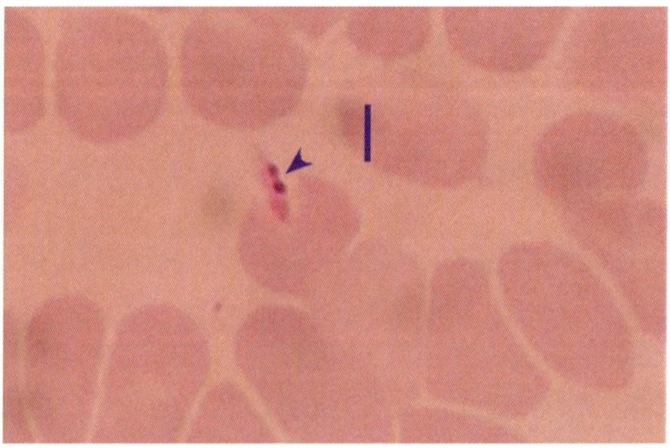

C. **Photomicrograph of merozoite of *Babesia microti* (Franca, 1912) escaping from erythrocyte to infect new cells (X1200).**

PLATE XLII

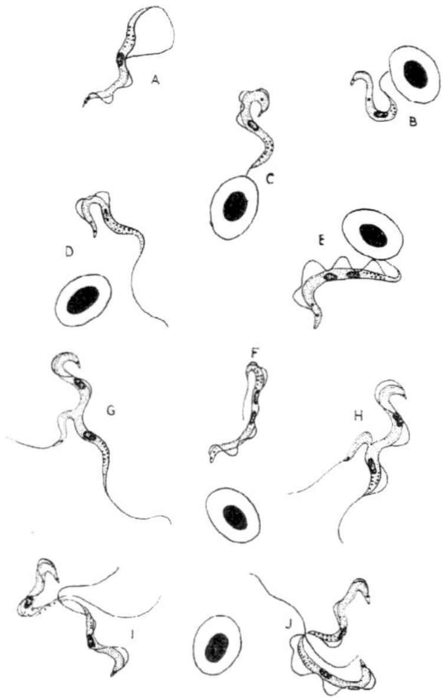

Camera lucida diagrams of divisional stages of *Trypanosoma trichogasteri* A-C – Kinetoplast division and migration, D-F – Nuclear division and separation, Cytokinesis, I – Cytokinesis completed with incomplete undulating membrane, J – Division completed prior to final separation.

PLATE XLIII

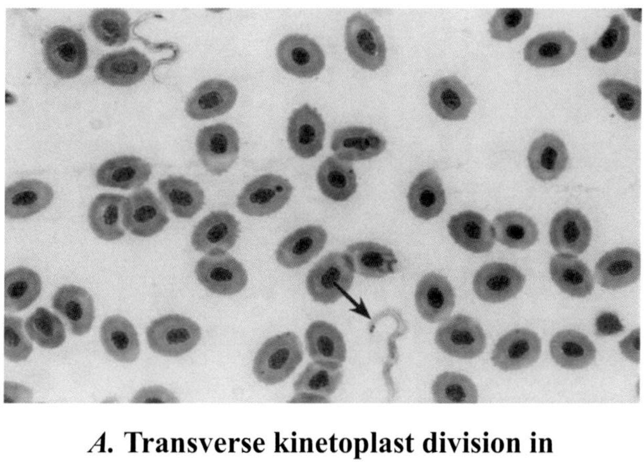

A.* Transverse kinetoplast division in *Trypanosoma trichogasteri.

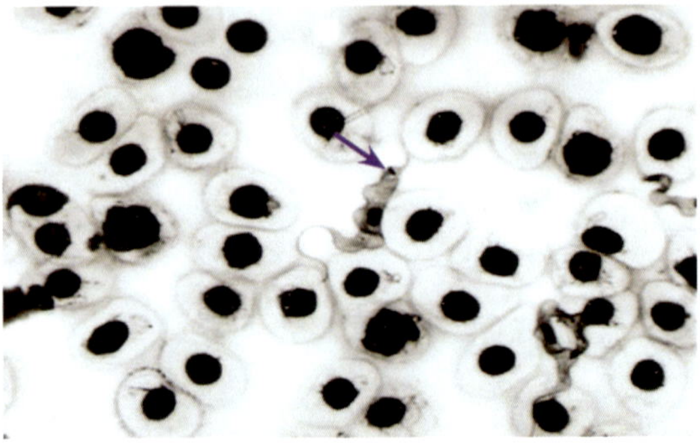

B. **Longitudinal kinetoplast division in *Trypanosoma trichogasteri.***

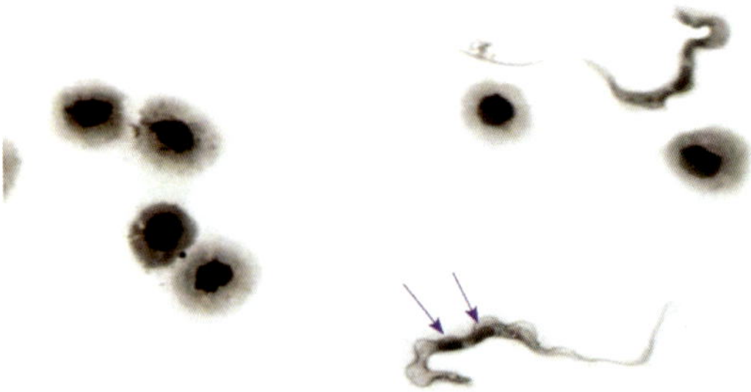

C. **Nuclear division in mature trypomastigote of *Trypanosoma trichogasteri.***

PLATE XLIV

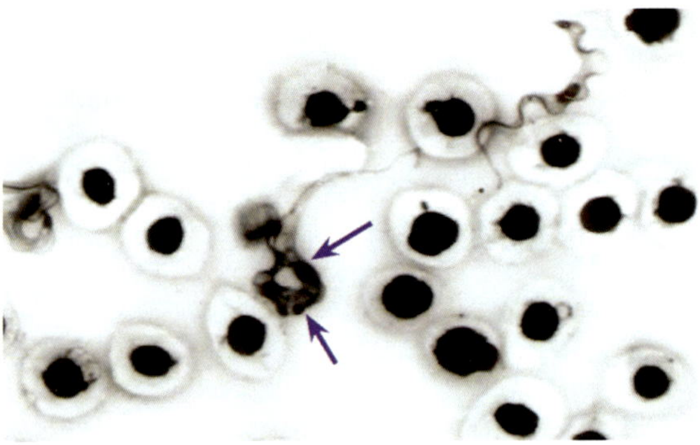

A. **Two nuclei and two kinetoplasts in *Trypanosoma trichogasteri.***

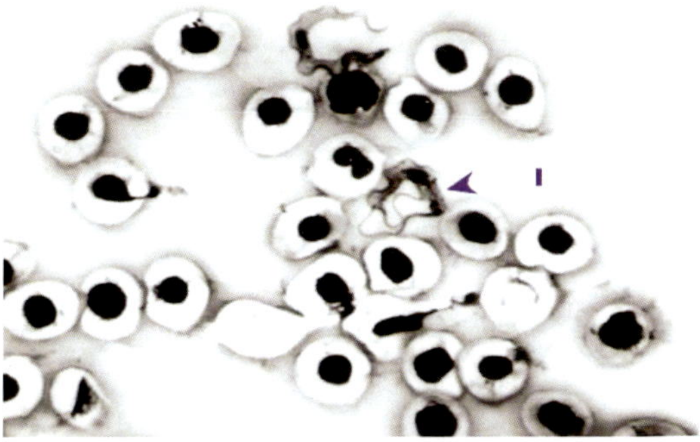

B. **One nucleus (N), two kinetoplasts (K), two flagella (F1, F2) and cytokinesis from posterior end in *Trypanosoma trichogasteri.***

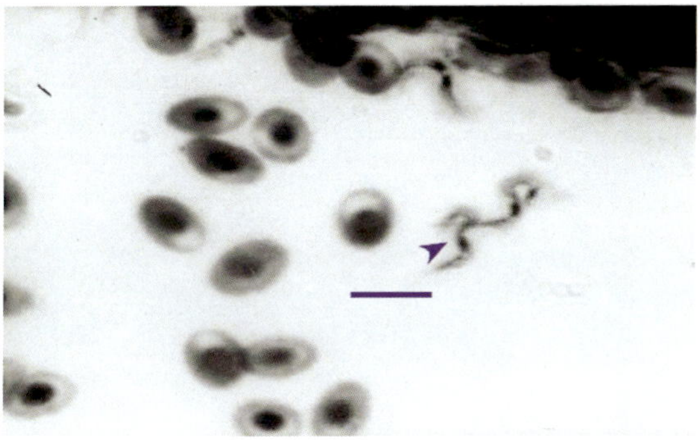

C. **Cytokinesis (from anterior end) in *Trypanosoma kargenensis.***

PLATE XLV

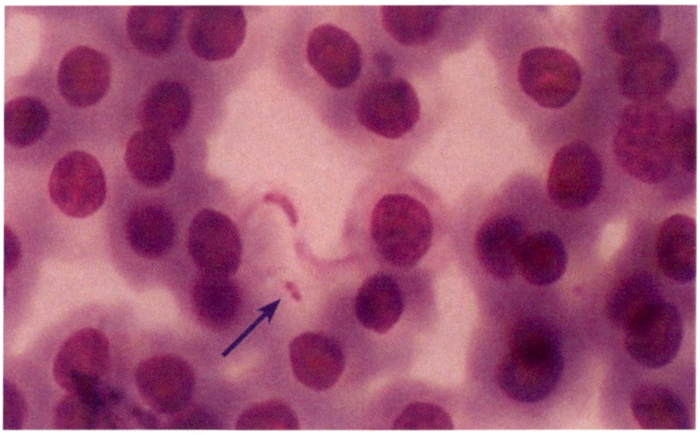

A. **Transverse kinetoplast division in in *Trypanosoma kargenensis.***

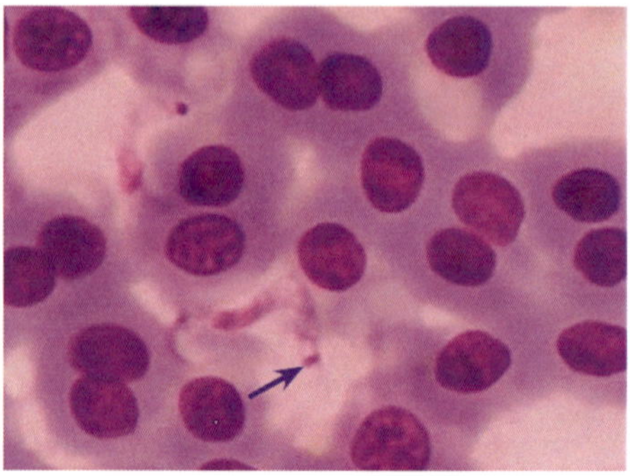

B. **Longitudinal kinetoplast division**
Trypanosoma kargenensis.

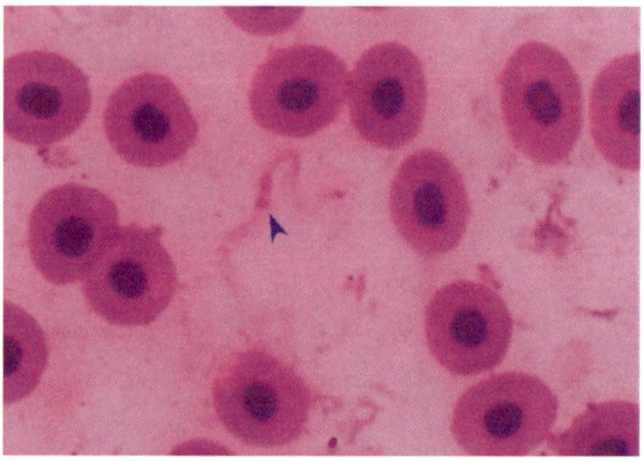

C. **Nuclear division in**
Trypanosoma kargenensis.

PLATE XLVI

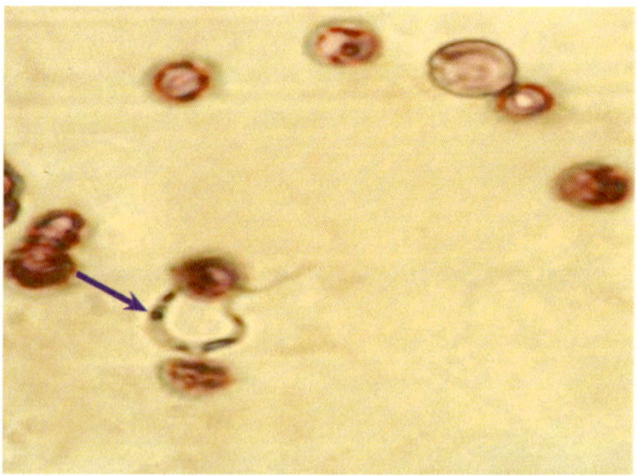

A. **Photomicrograph of *Trypanosoma lewisi* (Laveran and Mesnil, 1901) showing kinetoplast division in the immature form (X1000).**

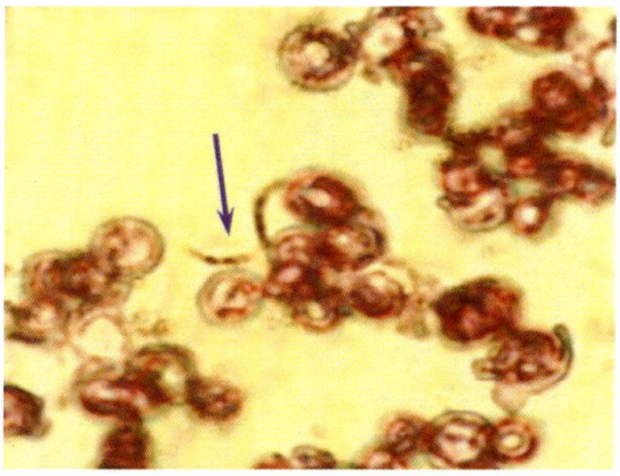

_B._ Photomicrograph of _Trypanosoma lewisi_ (Laveran and Mesnil, 1901) showing kinetoplast division in the classical adult form (X1000).

PLATE XLVII

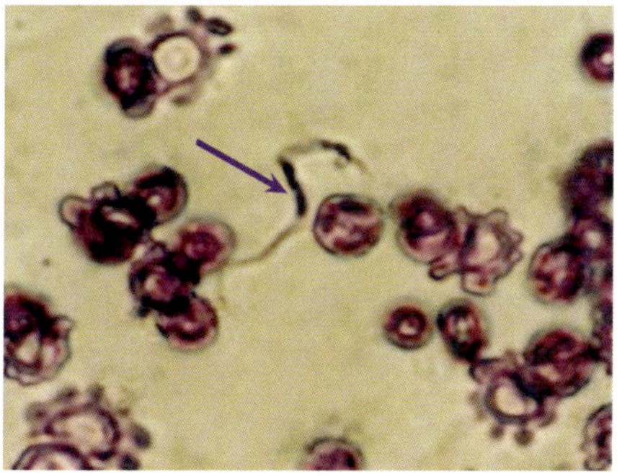

_A._ Photomicrograph of _Trypanosoma lewisi_ (Laveran and Mesnil, 1901) showing early nuclear division with two nuclei in close proximity to each other (X1000).

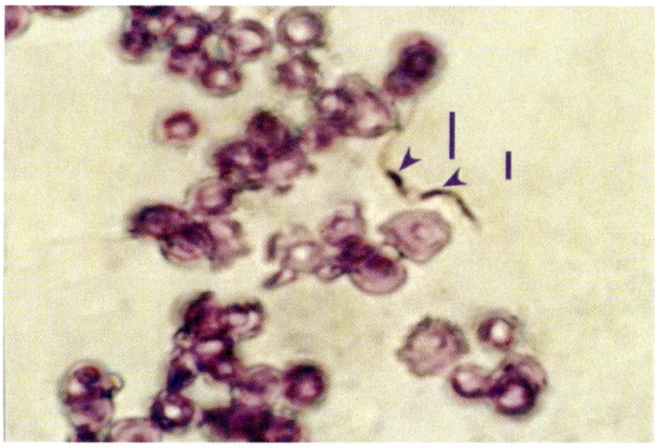

_B._ Photomicrograph of _Trypanosoma lewisi_ (Laveran and Mesnil, 1901) showing late nuclear division with two nuclei migrating (X1000).

PLATE XLVIII

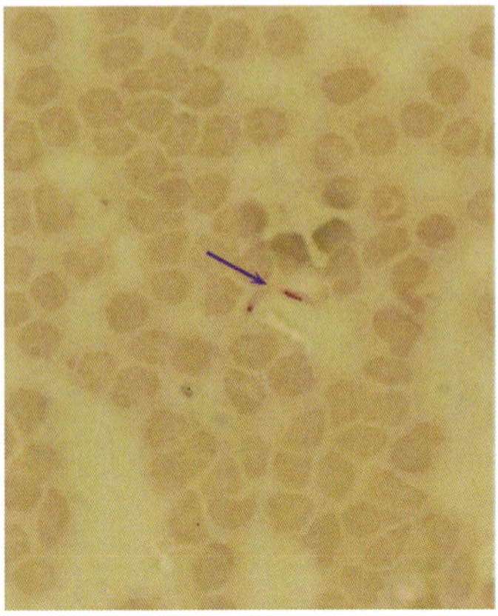

Photomicrograph of *Trypanosoma microti* (Laveran and Pettitt, 1909) showing flagellar division.

PLATE XLIX

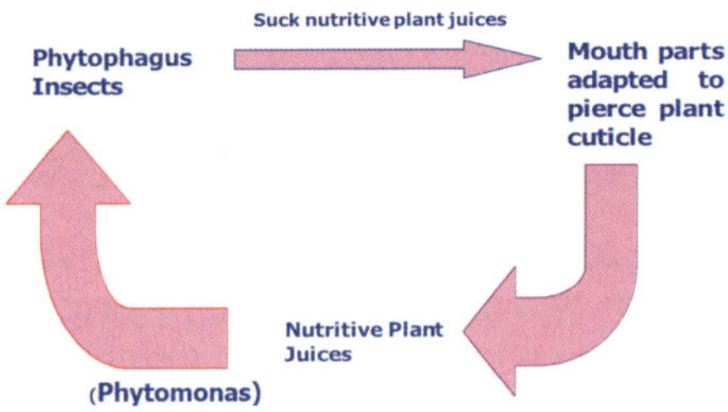

***A.* Evolution of *Trypanosoma* from a *Phytomonas*-like ancestor whose mouth parts were adapted to pierce plant cuticle.**

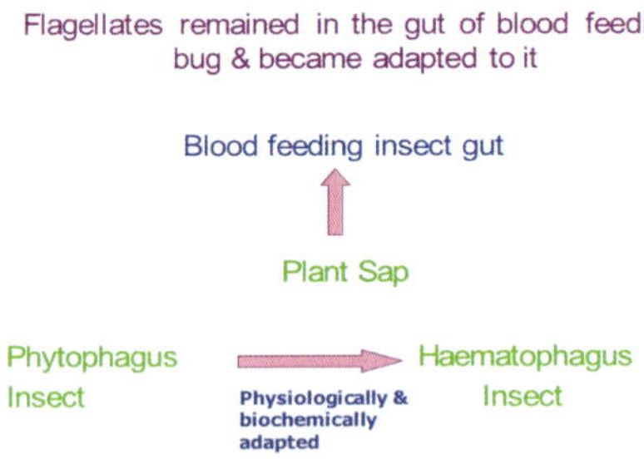

***B. Phytomonas* physiologically and biochemically adapted to hematophagous insect.**

PLATE L

A. **Evolutionary recapitulation where *Phytophagus* bug evolved to pierce skin and drink the blood of vertebrates.**

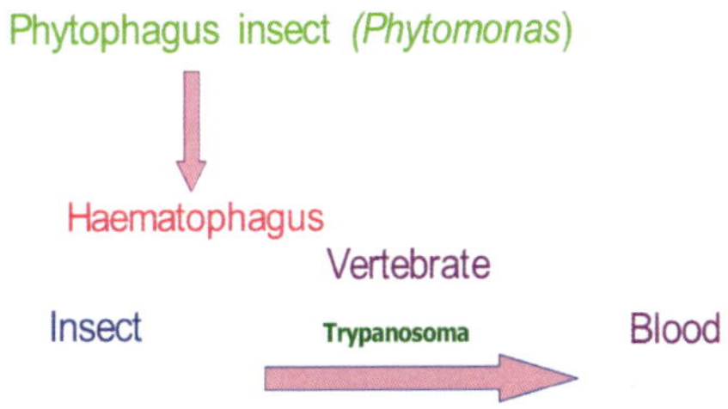

B. **From the hematophagous insect, the parasite evolved to *Trypanosoma* to feed on vertebrate blood.**

PLATE LI

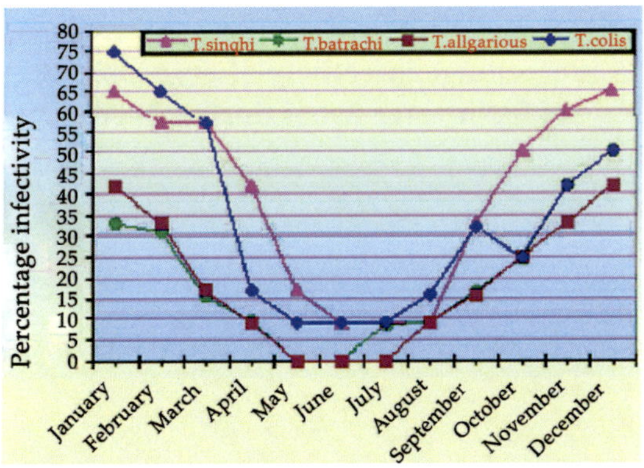

A. Percentage infectivity of Trypanosoma in freshwater fishes during 2006. Percentage infectivity of *Trypanosoma* in freshwater fishes during 2006.

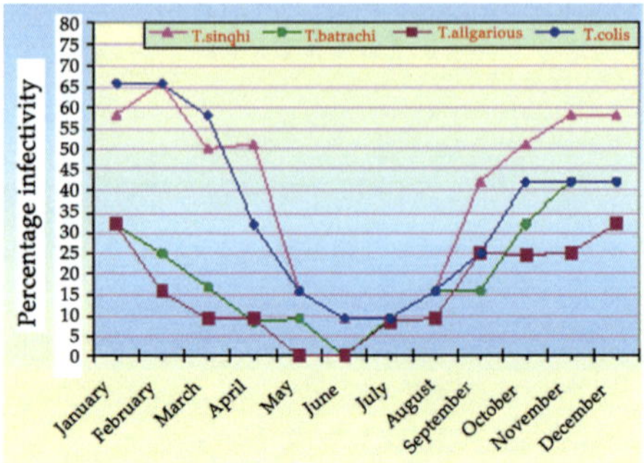

B. **Percentage infectivity of *Trypanosoma* in freshwater fishes during 2007.**

PLATE LII

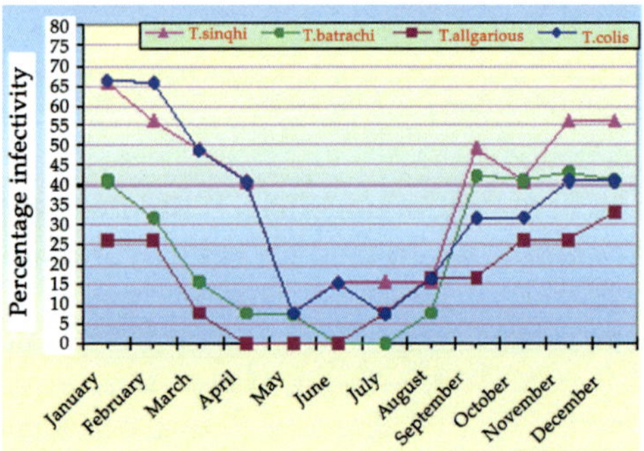

A. Percentage infectivity of *Trypanosoma* in freshwater fishes during 2008.

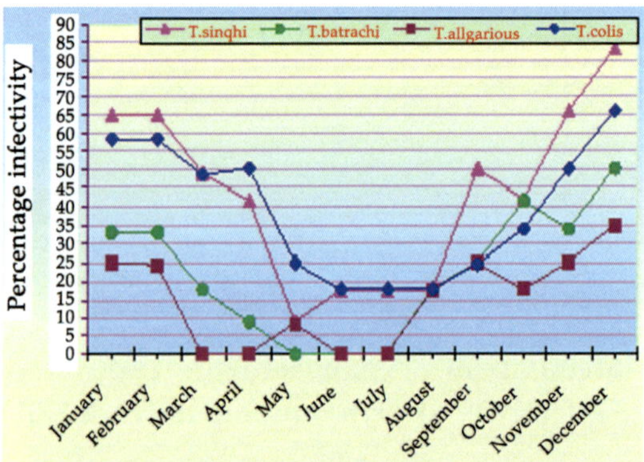

B. **Percentage infectivity of *Trypanosoma* in freshwater fishes during 2009.**

PLATE LIII

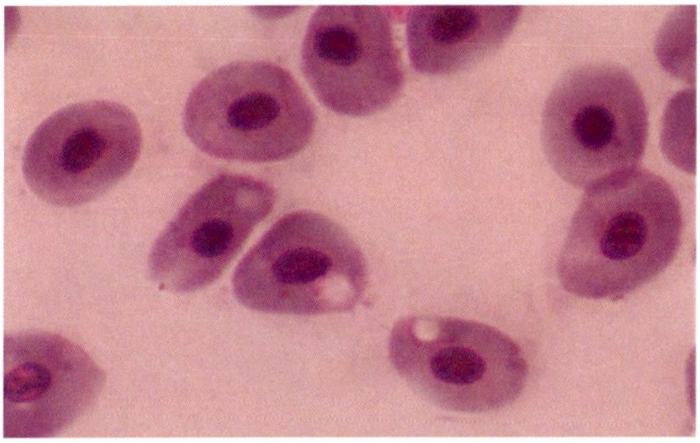

A. **Photomicrograph of blood of Ts-infected *Wallago attu* showing vacuolation of RBC's (X 1000).**

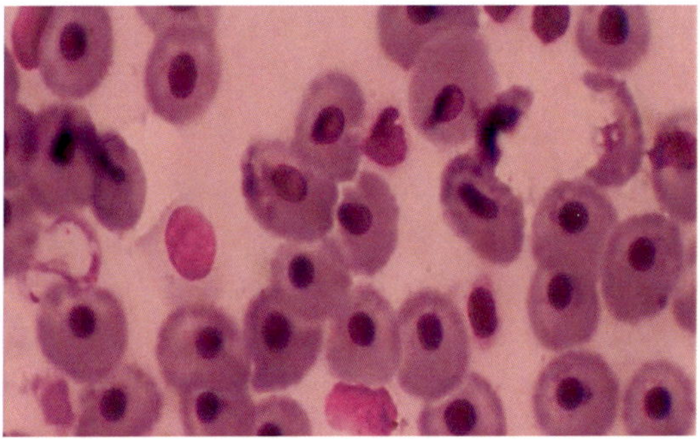

B. **Photomicrograph of blood of Ts-infected *Wallago attu* showing karyorrhexis of RBC(X 1000).**

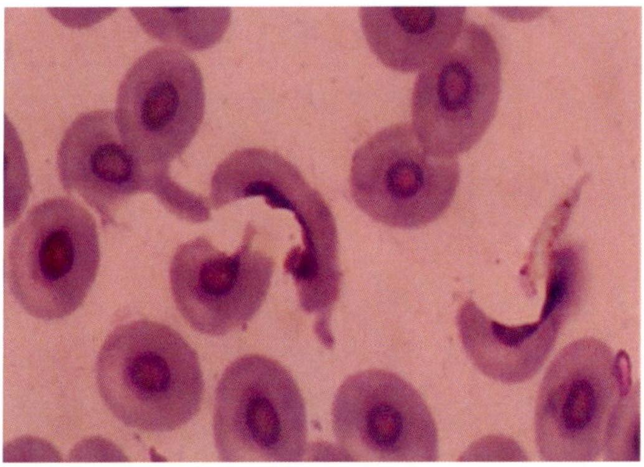

C. **Photomicrograph of blood of Ts-infected *Wallago attu* showing cytolysis and cell death.**

PLATE LIV

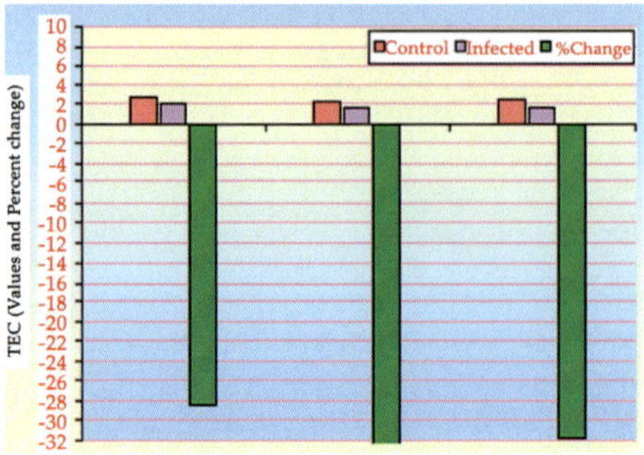

A. Total Erythrocyte Count (TEC) in freshwater fishes infected with flagellates.

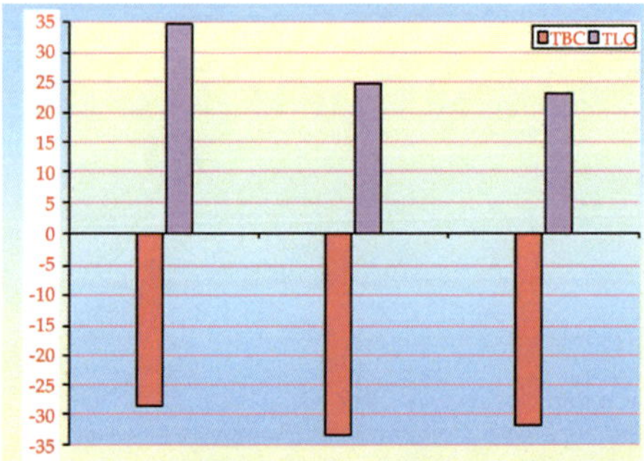

B. Percent change in TEC and TLC in freshwater fishes infected with flagellates.

PLATE LV

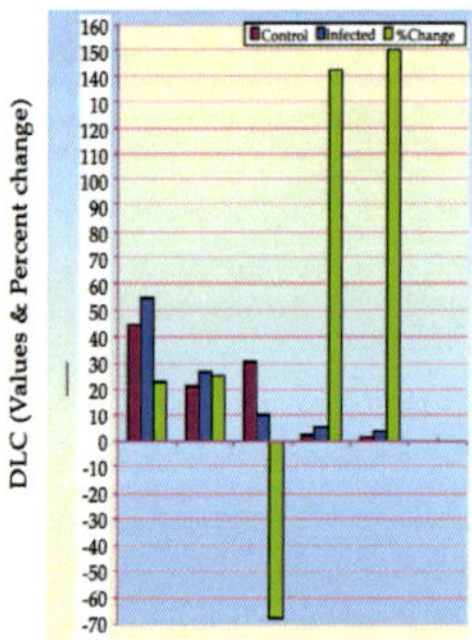

***A.* Differential Leukocyte Counts (DLC) in in *Heteropneustes fossilis* infected with trypanosomes.**

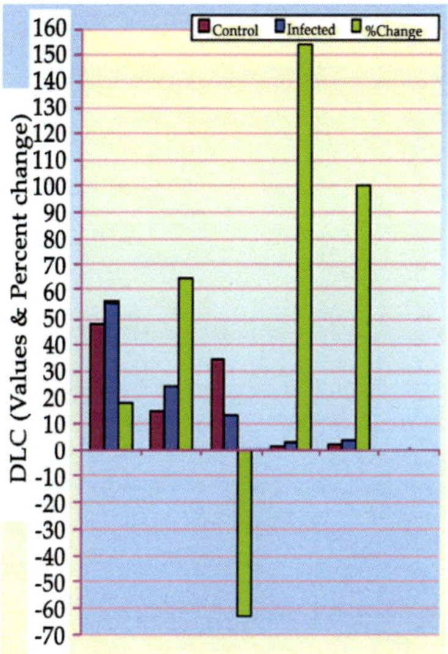

***B*. Differential Leukocyte Counts (DLC)**
***Colisa fasciata* infected with trypanosomes.**

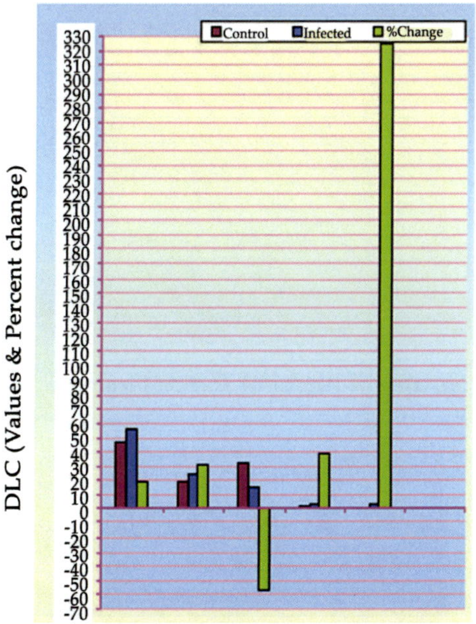

***C*. Differential Leukocyte Counts (DLC)**
***Channa punctatus* infected with Trypanosomes.**

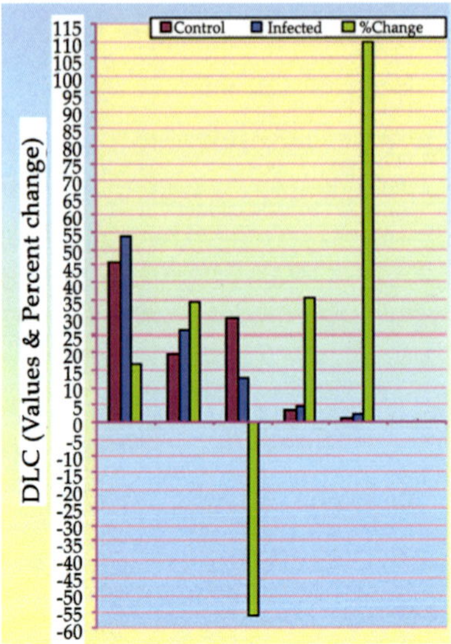

***D*. Differential Leukocyte Counts (DLC) in *Clarias batrachus* infected with trypanosomes.**

PLATE LVI

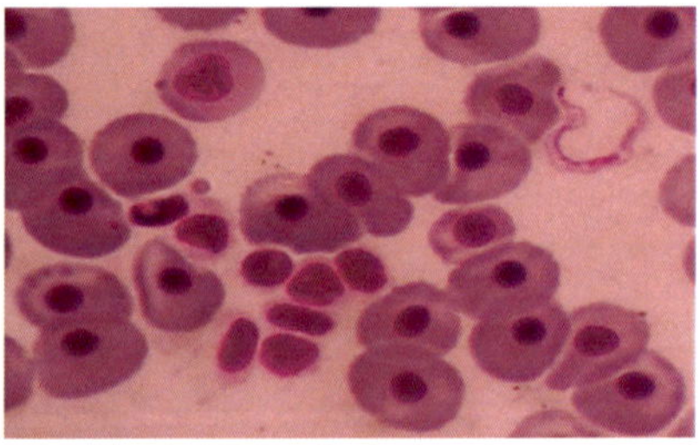

***A*. Photomicrograph of blood of *Heteropneustes fossilis* infected with *Ts kargenensis* showing lymphocytosis (X 1000).**

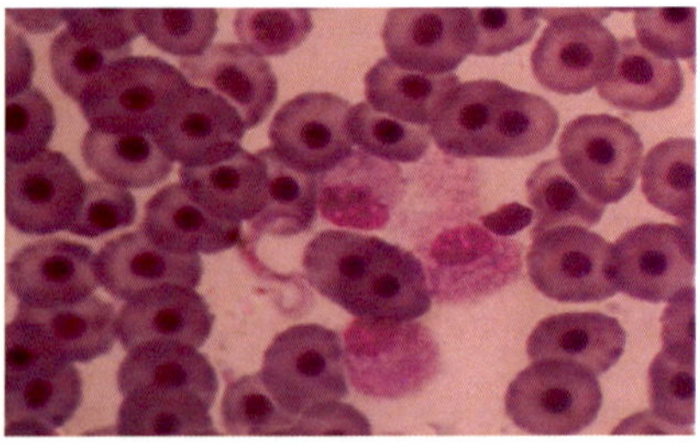

***B*. Photomicrograph of blood of *Heteropneustes* fossilis infected with *Ts kargenensis* showing myelocytes (X 1000).**

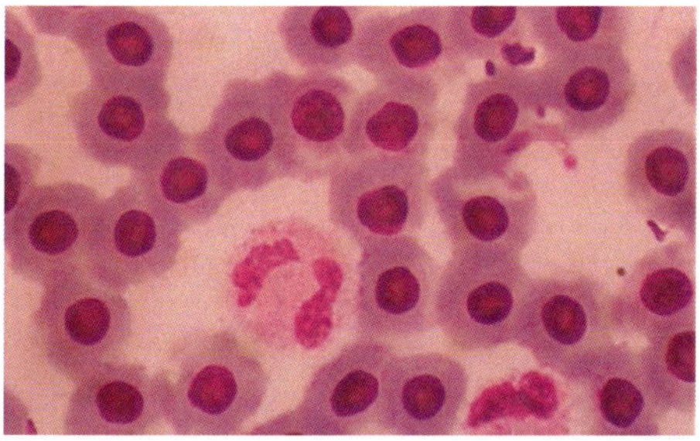

C. **Photomicrograph of blood of *Heteropneustes fossilis* infected with Ts *kargenensis* showing eosinophil (X 1000).**

PLATE LVII

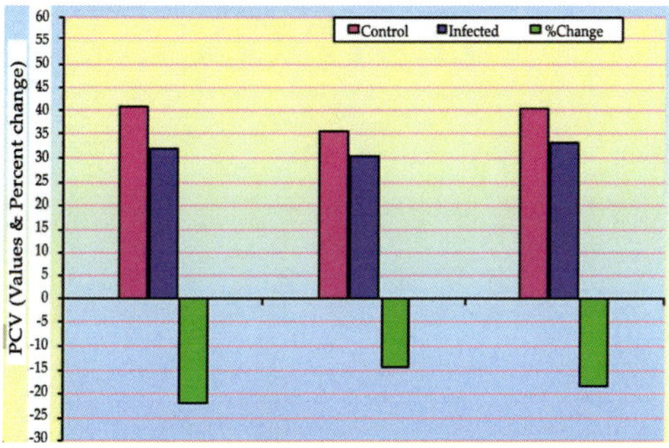

Changes in PCV in *Cyprinus carpio* infected with flagellates.

PLATE LVIII

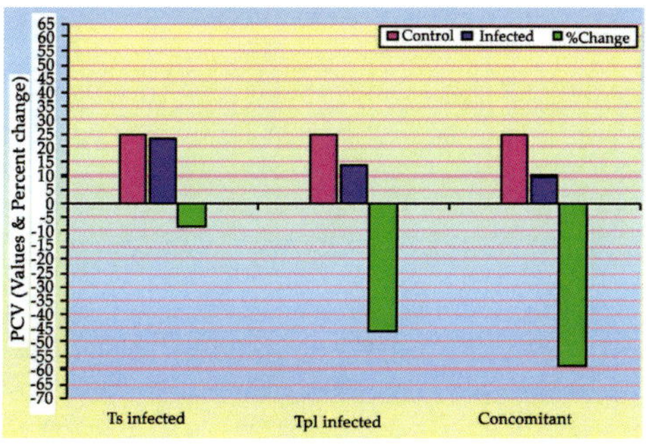

A. Changes in PCV in *Mystus vittatus* infected with flagellates.

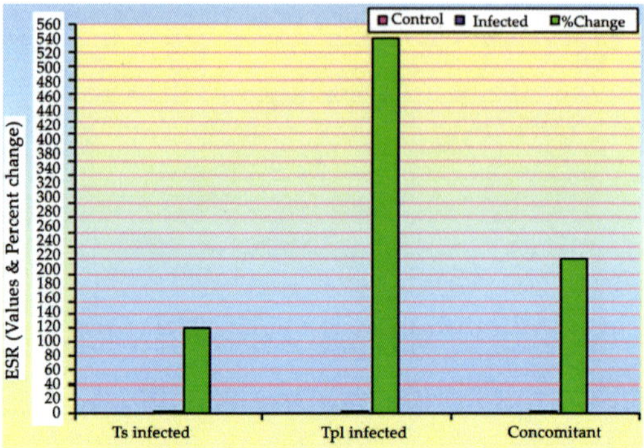

B. Changes in ESR in *Mystus vittatus* infected with flagellates.

PLATE LIX

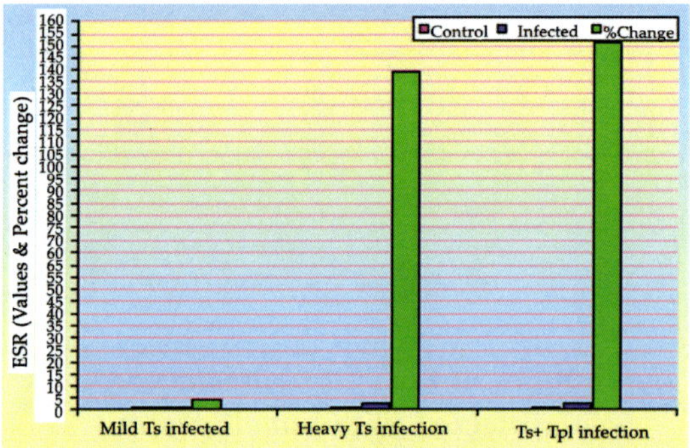

A. Changes in PCV in *Clarias batrachus* infected with flagellates (Natural infection).

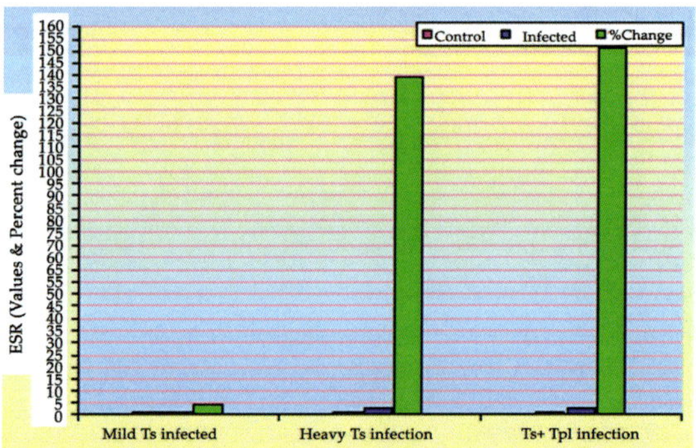

B. Changes in ESR in *Clarias batrachus* infected with flagellates (Natural infection).

PLATE LX

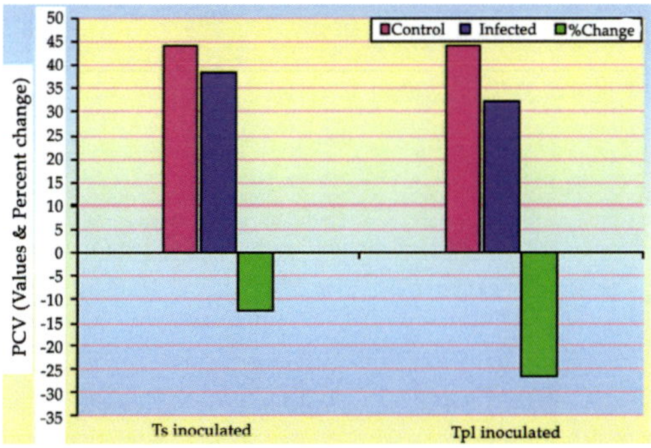

Changes in PCV in *Clarias batrachus* infected with flagellates (Experimental inoculation).

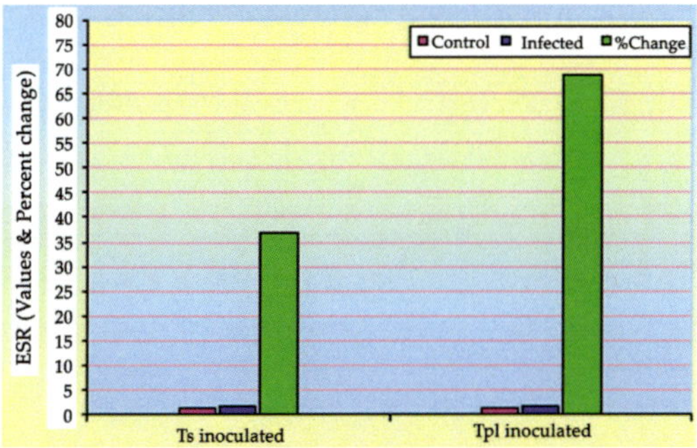

B.**Changes in ESR in *Clarias batrachus* infected with flagellates (Experimental inoculation).**

PLATE LXI

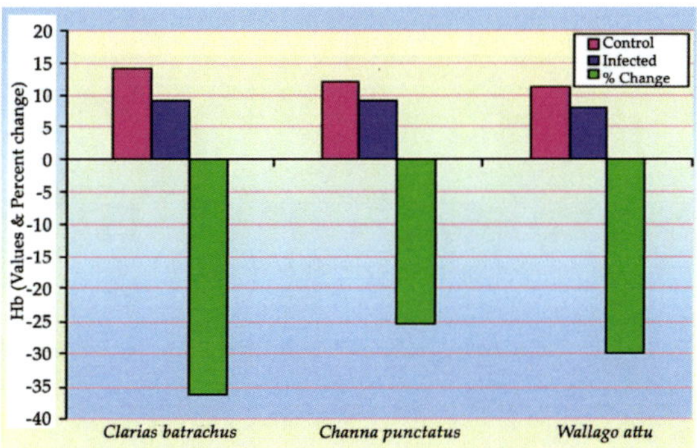

C. Hemoglobin changes in fresh water fishes infected with flagellates.

PLATE LXII

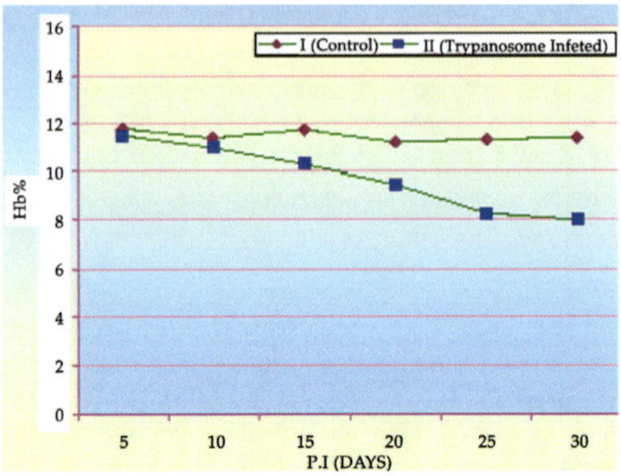

A. Changes in Hb in *Wallago attu* after *Trypanosoma atti* inoculation (at every 5 day p.i.).

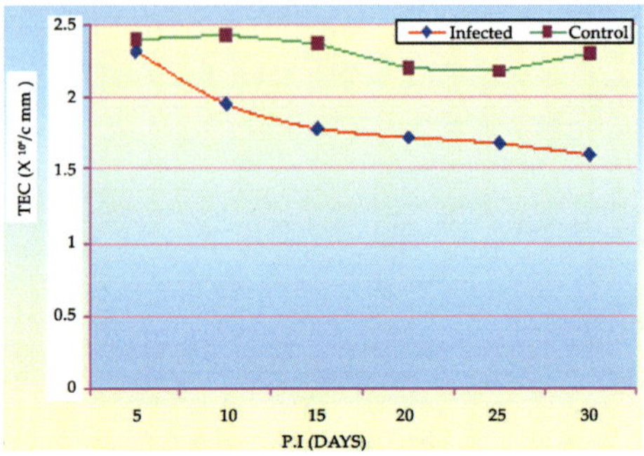

B. Changes in TEC in *Wallago attu* after *Trypanosoma atti* inoculation (at every 5 day p.i.)

PLATE LXIII

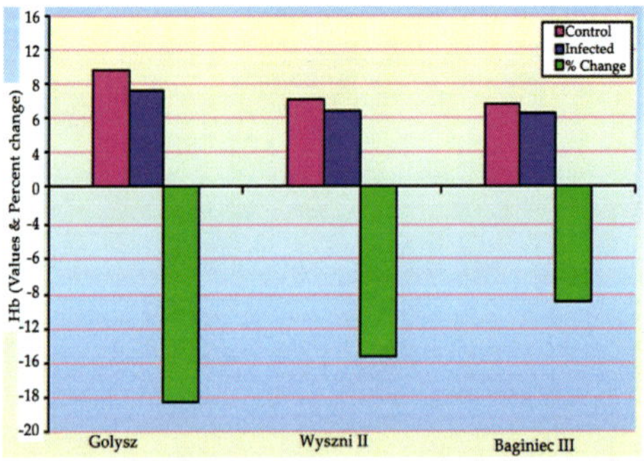

A. Hemoglobin changes in *Cyprinus carpio* infected with flagellates.

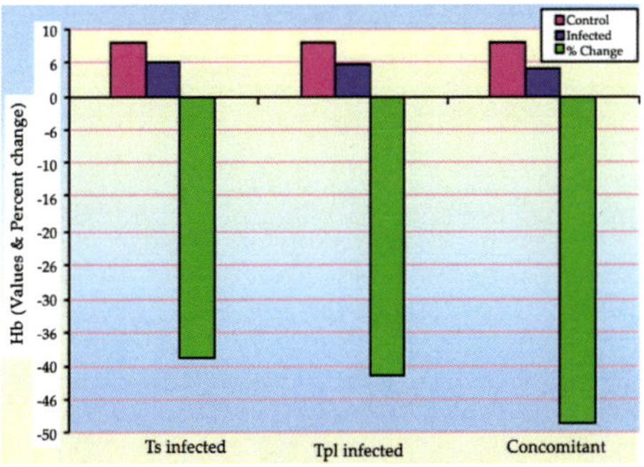

B emoglobin changes in *Mystus vittatus* nfected with flagellates.

PLATE LXIV

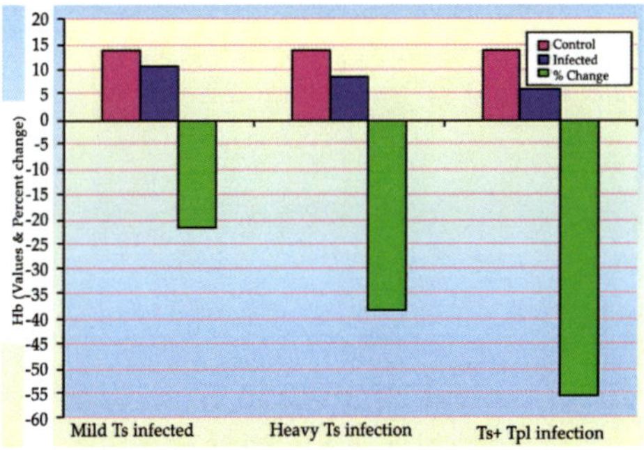

***A.* Hemoglobin changes in *Clarias batrachus* infected with flagellates (Natural infection).**

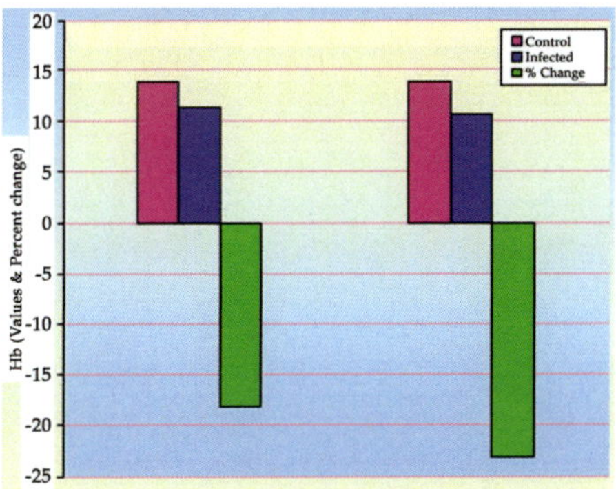

***B.* Hemoglobin changes in *Clarias batrachus* infected with flagellates (Experimental inoculation).**

PLATE LXV

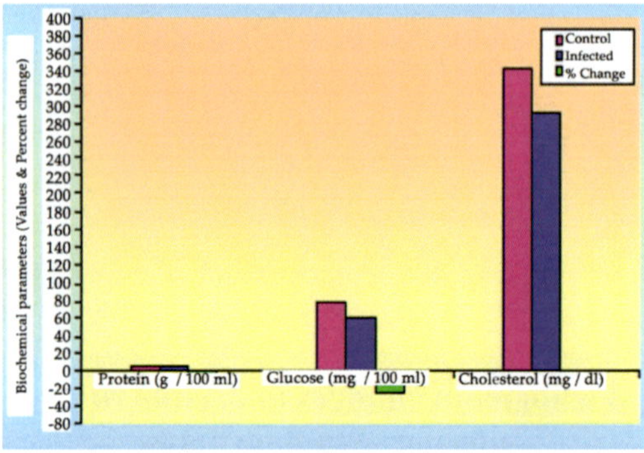

A. **Biochemical changes in *Wallago attu* infected with *T. atti.***

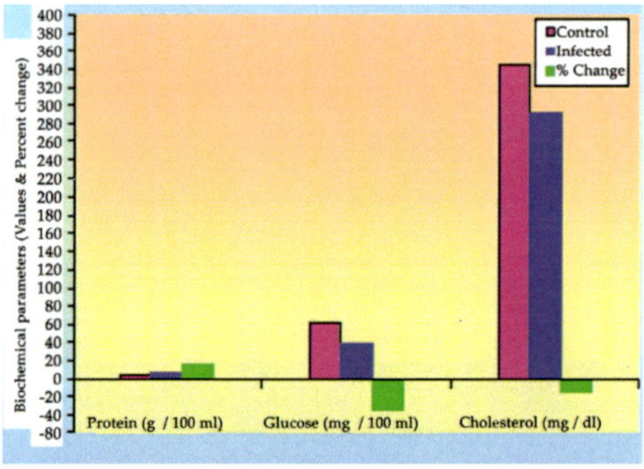

B*. Biochemical changes in *Heteropneustes fossilis* infected with *T. singhii.

PLATE LXVI

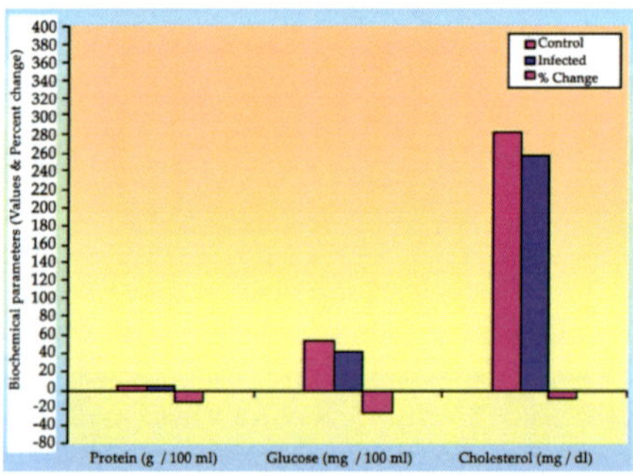

A. Biochemical changes in *Channa punctatus* infected with *Ts. Aligaricus.*

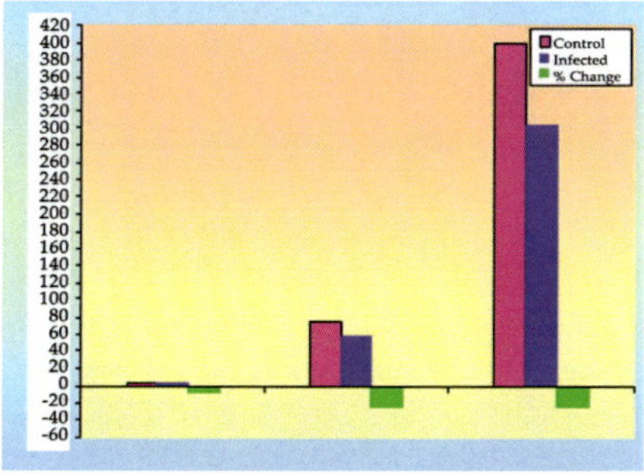

B. Biochemical changes in *Cyprinus carpio* from Golysz infected with flagellates.

PLATE LXVII

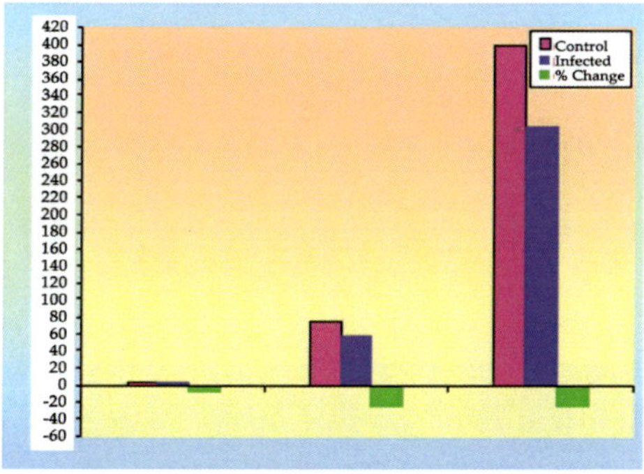

A. Biochemical changes in *Cyprinus carpio* rom Wyszni II infected with flagellates.

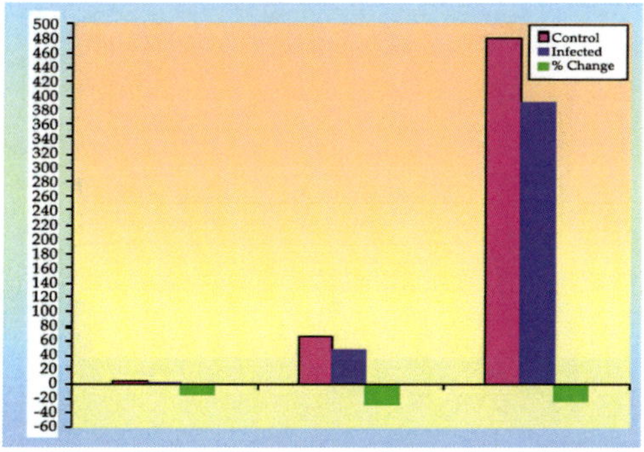

B. Biochemical changes in *Cyprinus carpio* from Baginiec III infected with flagellates.

PLATE LXVIII

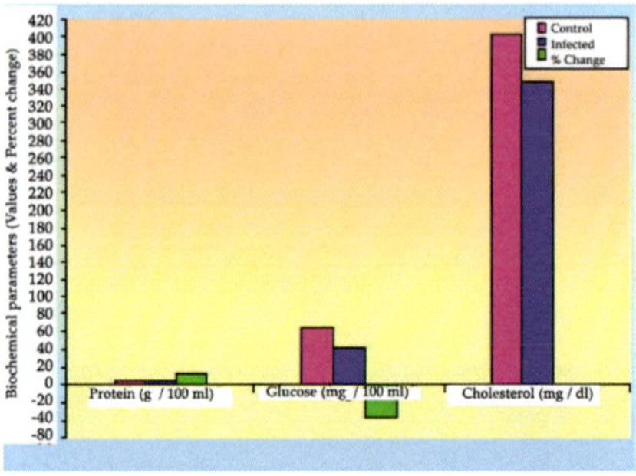

A. Biochemical changes in *Clarias batrachus* infected with *Trypanosoma.*

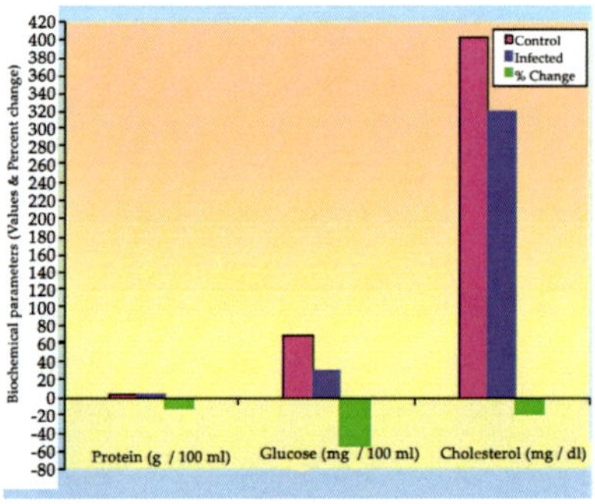

B. Biochemical changes in *Clarias batrachus* infected with *Trypanoplasma.*

PLATE LXIX

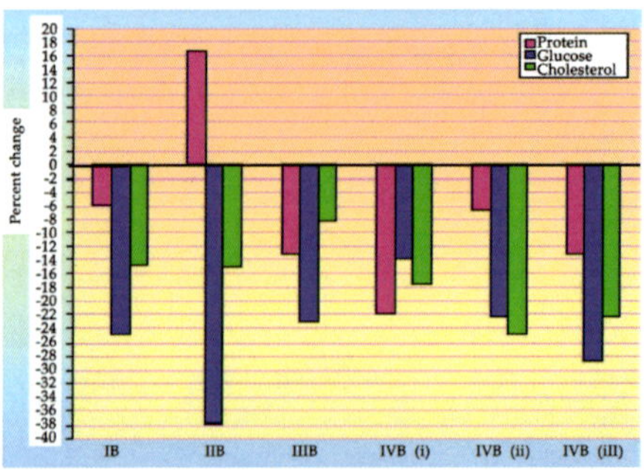

C. Percent changes in biochemical parameters in fish infected with flagellates.

PLATE LXX

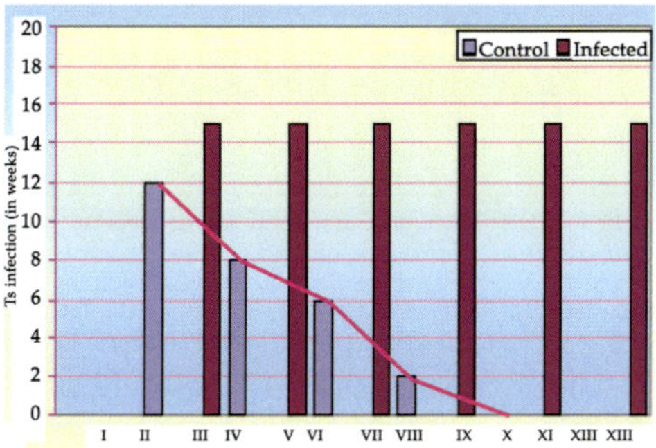

D. Immune response in *Heteropneustes fossilis* challenged with *Trypanosoma kargenesis*. Columns represent infection (weeks) (the line represents the decline in infectivity to produce immune serum).

PLATE LXXI

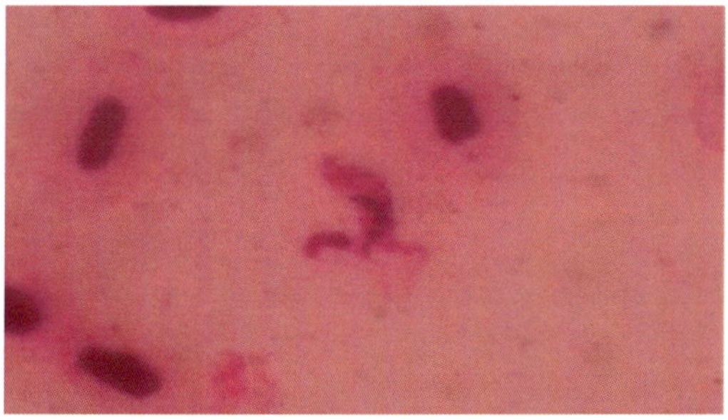

A.* Photomicrograph of blood of *Clarias batrachus* treated with 10.0 mg^{-1} Berenil showing destruction of *Trypanoplasma haematalis.

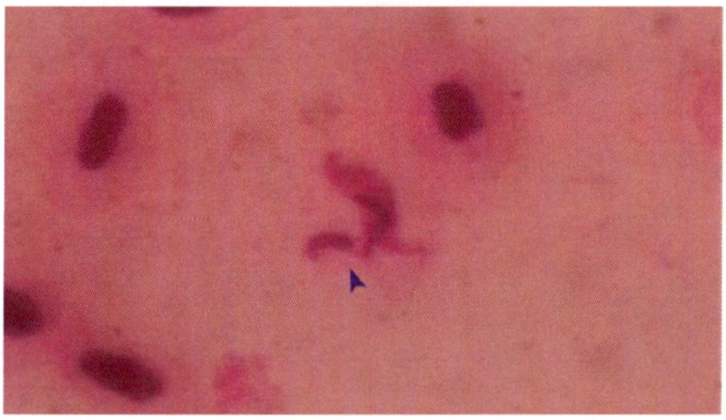

B.* Photomicrograph of blood of *Clarias batrachus* treated with 0.5 mg^{-1} Samorin showing destruction of *Trypanoplasma haematalis.

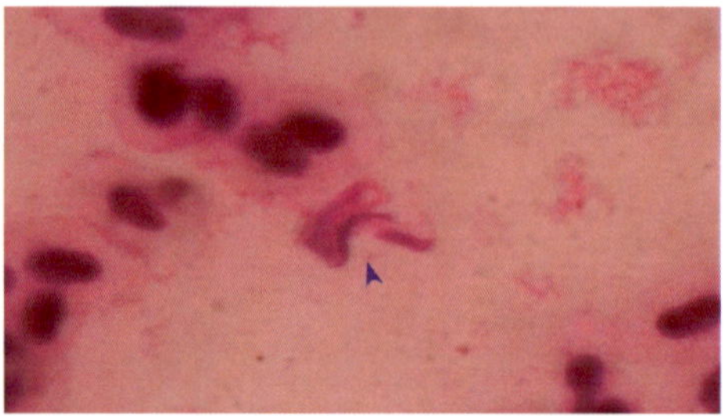

C. **Photomicrograph of blood of *Clarias batrachus* treated with 1.0 mg^{-1} Samorin showing destruction of *Trypanoplasma haematalis.***

PLATE LXXII

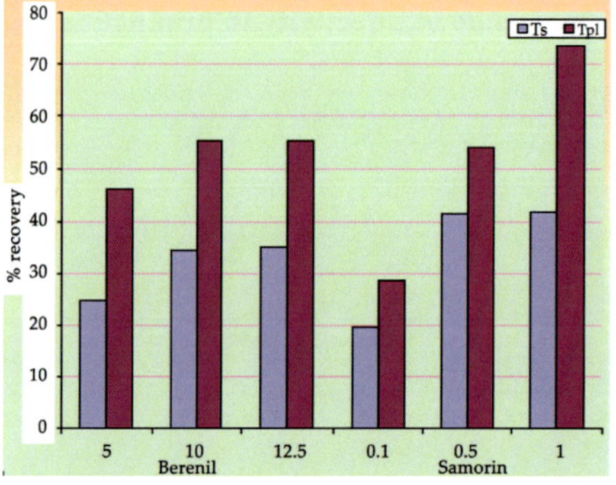

A. Recovery in PCV (%) of flagellate-infected *Clarias batrachus* treated with different doses of Berenil and Samorin.

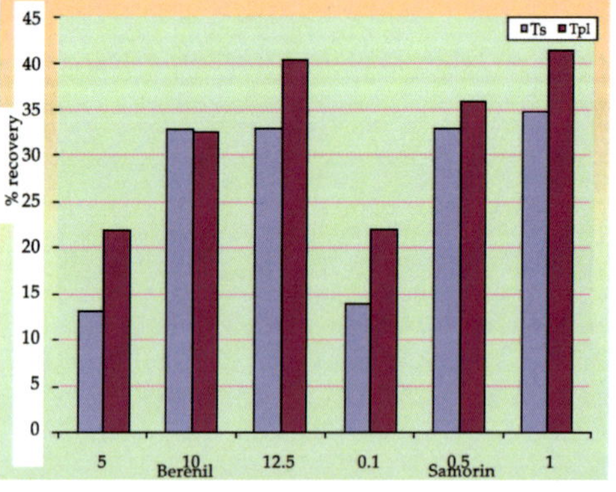

B. Recovery in ESR (%) of flagellate-infected *Clarias batrachus* treated with different doses of Berenil and Samorin.

PLATE LXXIII

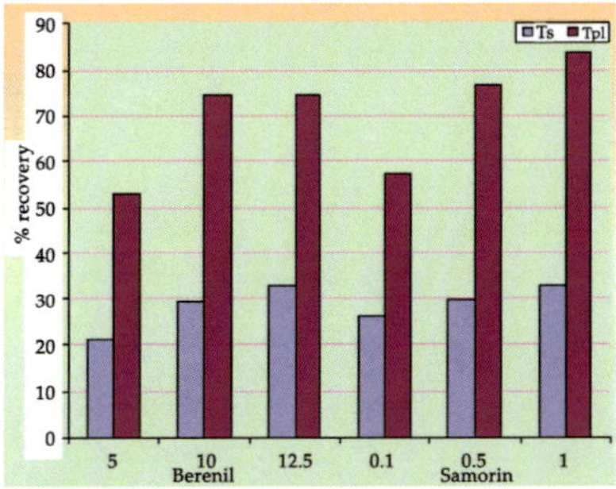

C. Recovery in Hb(%) of flagellate-infected *Clarias batrachus* treated with different doses of Berenil and Samorin.

PLATE LXXIV

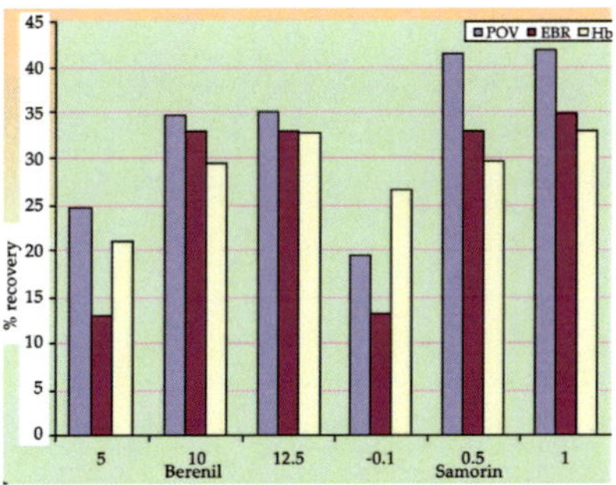

A. Recovery of hematological parameters in *Clarias batrachus* infected with *Trypanosoma batrachus* treated with Berenil and Samorin.

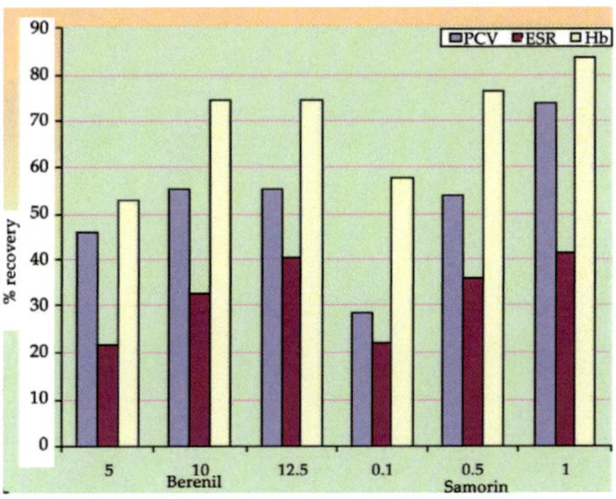

B. Recovery of hematological parameters in *Clarias batrachus* infected with *Trypanoplasma haematalis* treated with Berenil and Samorin.

Bibliography

1. Anderson, D. P., & Hennessen, W. (1981). *Fish Biologics: Serodiagnostics and Vaccines*. Developments in Biological Standards (Vol. 49). Karger.
2. Anderson, J. F., Mintz, E. D., Gadbaw, J. J., & Magnarelli, L. A. (1991). *Babesia microti*, human babesiosis, and Borrelia burgdorferi in Connecticut. *Journal of Clinical Microbiology, 29*(12), 2779–2783.
3. Anderson, K. (1972). Studies of the helminth fauna of Norway. XXI. The influence of population size (intensity of infection) on morphological characters in *Diphyllobothrium dendriticum* Nitzsch in the golden hamster (*Mesocricetus auratus* Waterhouse). *Norwegian Journal of Zoology, 20*, 1–7.
4. Anderson, T. J. C. (1990). Blood parasites of mammals from Papua New Guinea. *Journal of Wildlife Diseases, 26*(2), 291–294.
5. Appleby, B. M., Anwar, M. A., & Petty, S. J. (1999). Short-term and long-term effects of food supply on parasite burden in tawny owls, *Strix aluco*. *Functional Ecology, 13*, 315–321.
6. Archer, R. K. (1965). *Haematological Techniques for Use on Animals*. Blackwell Scientific Publications.
7. Ardelli, B. F., & Woo, P. T. K. (1997). Protective antibodies and anamnestic response in *Salvelinus fontinalis* to *Cryptobia salmositica* and innate resistance of *Salvelinus namaycush* to the haemoflagellate. *Journal of Parasitology, 83*, 943–946.
8. Ardelli, B. F., & Woo, P. T. K. (1998). The *in vitro* effects of crystal violet on the pathogenic haemoflagellate *Cryptobia salmositica* Katz, 1951. *Parasite, 5*, 27–36.
9. Ardelli, B. F., & Woo, P. T. K. (1999). The therapeutic use of isometamidium chloride against *Cryptobia salmositica* in rainbow trout (*Oncorhynchus

mykiss). *Diseases of Aquatic Organisms, 37*, 195–203.

10. Ardelli, B. F., & Woo, P. T. K. (2000). An antigen-capture enzyme-linked immunosorbent assay (ELISA) to detect isometamidium chloride in *Oncorhynchus* spp. *Diseases of Aquatic Organisms, 39*, 231–236.
11. Ardelli, B. F., & Woo, P. T. K. (2001a). The *in vitro* effects of isometamidium chloride (Samorin) on the piscine hemoflagellate *Cryptobia salmositica* (Kinetoplastida, Bodonina). *Journal of Parasitology, 87*, 194–202.
12. Ardelli, B. F., & Woo, P. T. K. (2001b). Therapeutic and prophylactic effects of isometamidium chloride (Samorin) against the hemoflagellate *Cryptobia salmositica* in chinook salmon (*Oncorhynchus tshawytscha*). *Parasitology Research, 87*, 18–26.
13. Ardelli, B. F., & Woo, P. T. K. (2001c). Conjugation of isometamidium chloride to antibodies and the use of the conjugate against the hemoflagellate, *Cryptobia salmositica*: An immuno-chemotherapeutic strategy. *Journal of Fish Diseases, 24*, 439–451.
14. Ardelli, B. F., & Woo, P. T. K. (2006). Immunocompetent cells and their mediators in fin fish. In P. T. K. Woo (Ed.), *Fish Diseases and Disorders* (Vol. 1, 2nd ed., pp. 702–724). CABI.
15. Ardelli, B. F., Witt, J. D. S., & Woo, P. T. K. (2000). The identification of glycosomes and metabolic end products in pathogenic and nonpathogenic strains of *Cryptobia salmositica* (Kinetoplastida: Bodonidae). *Diseases of Aquatic Organisms, 42*(1), 41–51.
16. Ariyibi, A., Odunuga, O., & Olorunsogo, O. (2001). Concentration-dependant inhibition of sheep erythrocyte ghost plasma membrane Ca^{2} -ATPase activity by Berenil [1,3-bis (4'-amidinophenyl) triazene] *in vitro*. *Journal of Veterinary Pharmacology and Therapeutics, 24*, 233–236.
17. Arkoosh, M. R., & Kaattari, S. L. (1991). Development of immunological memory in rainbow trout (*Oncorhynchus mykiss*). I. An immunochemical and cellular analysis of the B cell response. *Developmental and Comparative Immunology, 15*, 279–293.
18. Arneberg, P., Skorping, A., Grenfell, B. T., & Read, A. F. (1998). Host densities as determinants of abundance in parasite communities. *Proceedings of the Royal Society of London. Series B: Biological Sciences, 265*, 1283–1289.
19. Arowolo, R. O., & Ikede, B. O. (1977). Susceptibility of a rodent-adapted strain of *Trypanosoma vivax* to Berenil, Samorin and Novidium. *Acta Tropica, 34*, 61–64.
20. Arowolo, R. O., & Ikede, B. O. (1980). Susceptibility of the Y58 rodent-

adapted strain of *Trypanosoma vivax* to two trypanocides in sheep. *Veterinary Record, 106*, 59.

21. Arrea, G. C., Carmona, M. C., Bermudez, O. M. G., & Abrahams, E. (1998). Effect of *Trypanosoma lewisi* (Kinetoplastida: Trypanosomatidae) on the infection of white rats with *Toxoplasma gondii* (Eucoccidia: Sarcocystidae) oocysts. *Revista de Biología Tropical, 46*, 1121–1123.
22. Arriero, E., Moreno, J., Merino, S., & Martinz, J. (2008). Habitat effects on physiological stress response in nestling blue tits are mediated through parasitism. *Physiological and Biochemical Zoology, 81*(2), 195–203.
23. Ashman, P. U., & Seed, J. R. (1974). The effects of photoperiod and *Trypanosoma brucei gambiense* infection on the reproductive organs of male *Microtus montanus*. *Journal of Reproduction and Fertility, 40*, 51–60.
24. Asin, S., & Catala, S. (1995). Development of *Trypanosoma cruzi* in *Triatoma infestans*: Influence of temperature and blood consumption. *Journal of Parasitology, 81*, 1–7.
25. Ayala, F. J. (1972). Competition between species. *American Scientist, 60*, 348–357.
26. Babes, V. (1888). Sur l'hémoglobinurie bactérienne du boeuf. *Comptes Rendus des Séances de l'Académie des Sciences, 107*, 692–694.
27. Babigumira, B. M., Waiswa, P., Wampande, E., & Waiswa, C. (2009). Clinical manifestations and treatment of the potentially zoonotic *T. brucei* infections in domestic pigs. *African Journal of Animal and Biomedical Sciences, 5*, 1–5.
28. Bacigalupo, U., & De la Plaza, N. (1948). Presencia de tripanosomas en las rayas de Mar del Plata. *Trypanosoma morplatensis* n. sp. *Revista de la Sociedad Argentina de Biología, 24*, 269–274.
29. Bajer, A., Bednarska, M., Pawelczyk, A., Konopka, E., Karbowiak, G., & Sinski, E. (1998). Blood parasites in a wild rodent country community of Mazury Lakes District, Poland. *Wiadomości Parazytologiczne, 44*, 426.
30. Bajer, A., Pawelczyk, A., Behnke, J. M., Gilbert, F. S., & Sinski, E. (2001). Factors affecting the component community structure of haemoparasites in bank voles (*Clethrionomys glareolus*) from the Mazury Lake District region of Poland. *Parasitology, 122*, 43–54.
31. Baker, J. R. (1960). Trypanosomes and Dactylosomes from the blood of freshwater fish in East Africa. *Parasitology, 50*, 515–526.
32. Baker, J. R. (1961). Trypanosomes of African freshwater fish: An addendum. *Parasitology, 51*, 263.

33. Baker, J. R. (1966). Studies on *Trypanosoma avium* (iv). The development of infective metacyclic trypanosomes in cultures grown *in vitro*. *Parasitology, 56*, 15–16.

34. Baker, J. R. (1974). Protozoan parasites of the blood of British wild birds and mammals. *Journal of Zoology, London, 172*, 169–190.

35. Baker, J. R. (1976). Blood fluids–protozoans. In C. R. Kennedy (Ed.), *Ecological Aspects of Parasitology* (pp. 371–386). North Holland Publishing Company.

36. Baker, J. R., Chitty, D., & Phipps, E. (1963). Blood parasites of wild voles, *Microtus agrestis*, in England. *Parasitology, 53*, 297–301.

37. Bakker, B. M., Mensonides, F. I., Teusink, B., van Hoek, P., Michels, P. A., & Westerhoff, H. V. (2000). Compartmentation protects trypanosomes from the dangerous design of glycolysis. *Proceedings of the National Academy of Sciences, 97*, 2087–2092.

38. Bakker, B. M., Michels, P. A., Opperdoes, F. R., & Westerhoff, H. V. (1997). Glycolysis in bloodstream form *Trypanosoma brucei* can be understood in terms of the kinetics of the glycolytic enzymes. *Journal of Biological Chemistry, 272*, 3207–3215.

39. Bardsley, J. E., & Hamsen, R. (1973). The trypanosomes of Anura. *Advances in Parasitology, 11*, 1–73.

40. Barrow, J. H. (1954). Observation of some host specificity and immunological reactions of trypanosome infections in some freshwater fish of Europe. *Anatomical Record, 120*, 750–751.

41. Barrow, J. H. (1955). Social behavior in freshwater fish and its effect on resistance to trypanosome. *Proceedings of the National Academy of Sciences, U.S.A., 41*, 676–679.

42. Barrow, J. H. (1958). The biology of *Trypanosoma diemyctyli* Tobey III. Factors influencing the cycle of *Trypanosoma diemyctyli* in the vertebrate host *Triturus v. viridescens*. *Journal of Protozoology, 5*, 161–170.

43. Barta, D. (1983). Serum lymphocyte immunoregulatory factors (SLIF). *Veterinary Immunology and Immunopathology, 4*, 279.

44. Barton, B., & Iwama, G. K. (1991). Physiological changes in fish from stress in aquaculture with emphasis on the response and effects of corticosteroids. *Annual Review of Fish Diseases, 1*, 3–26.

45. Baska, F., & Gonda, E. (2006). Common protozoan diseases of pet fish. *Kisallat Praxis, 7*(3), 86–92.

46. Bauer, O. N. (1941). Contributions to the knowledge of fish parasites of the river Khatanga. *Transactions of the Institute of Taiga, Pol. Zeml.,*

Series Prom., Khas., 16, 84–103.

47. Bauer, O. N. (1961). Parasitic diseases of cultured fishes and methods of their prevention and treatment. In V. A. Dogiel, G. K. Petrushevki, & Yu. I. Polyanski (Eds.), *Parasitology in Fishes* (pp. 365). Oliver and Boyd Ltd.
48. Bauer, O. N. (1984). *Key to the Determination of Parasites of Freshwater Fishes of the USSR*. Nauka.
49. Beament, J. W. L. (1961). Mechanisms in biological competition. In F. L. Milthorpe (Ed.), *Biological Competition* (pp. 62–71). The University Press.
50. Beamish, F. W. H., Sitja-Bobadilla, A., Jebbink, J. A., & Woo, P. T. K. (1996). Bioenergetic cost of cryptobiosis in fish: Rainbow trout (*Oncorhynchus mykiss*) infected with *Cryptobia salmositica* and with an attenuated live vaccine. *Diseases of Aquatic Organisms, 25*(1, 2), 1–8.
51. Becker, C. D. (1967). *Trypanosoma occidentalis* n. sp. from freshwater teleosts in Washington State. *Journal of Protozoology, 14*, 153–156.
52. Becker, C. D. (1970). Haematozoa of fishes, with emphasis on North American records. In S. F. Snieszko (Ed.), *A Symposium on Diseases of Fishes and Shellfishes* (Spec. Publ. No. 5, pp. 82–100). American Fisheries Society.
53. Becker, C. D. (1977). Flagellate parasites of fish. In J. P. Krier (Ed.), *Parasitic Protozoa* (Vol. I, pp. 441). Academic Press.
54. Becker, C. D. (1980). Haematozoa from resident and anadromous fishes of the central Columbia River: A survey. *Canadian Journal of Zoology, 58*, 356–362.
55. Becker, C. D., & Katz, M. (1965a). Transmission of the hemoflagellate, *Cryptobia salmositica* (Katz, 1951), by a rhynchobdellid vector. *Journal of Parasitology, 51*, 95–99.
56. Becker, C. D., & Katz, M. (1965b). Distribution ecology and biology of the salmonid leech, *Piscicola salmositica* (Rhynchobdellae: Piscicolidae). *Journal of the Fisheries Research Board of Canada, 22*, 1175–1195.
57. Becker, C. D., & Katz, M. (1965c). Infections of the hemoflagellate, *Cryptobia salmositica* (Katz, 1951), in freshwater teleosts of the Pacific Coast. *Transactions of the American Fisheries Society, 94*, 327–333.
58. Becker, C. D., & Overstreet, R. M. (1979). Haematozoa of marine fishes from the northern Gulf of Mexico. *Journal of Fish Diseases, 2*, 469–479.
59. Bellanti, J. A. (1978). *Immunology*. W. B. Saunders Co.
60. Benigno, E., & Marges, B. (1978). *Rats and Their Control*. UP Science Education Centre.

61. Bennett, G. F. (1962). The haematocrit centrifuge for laboratory diagnosis of hematozoa. *Canadian Journal of Zoology, 40*, 124–125.

62. Berens, R. L., Brun, R., & Krassner, S. M. (1976). A simple monophasic medium for axenic culture of hemoflagellates. *Journal of Parasitology, 62*(3), 360–365.

63. Bey, P., Danzin, C., & Jung, M. (1987). Inhibition of basic amino acid decarboxylases involved in polyamine biosynthesis. In P. P. McCann & A. Sjoerdsma (Eds.), *Inhibition of Polyamine Metabolism: Biological Significance and Basis for New Therapies* (pp. 1–31). Academic Press.

64. Bhatt, S. N., & Singh, H. R. (1981). Haematological studies on a hillstream fish *Schizothorax richardsoni* (Gray). *International Journal of Academic Ichthyology, 2,* 13–15.

65. Bhikane, A. U., Narladkar, B. W., Anantwae, L. G., & Bhokre, A. P. (2001). Epidemiology, clinico-pathology, and treatment of babesiosis in cattle. *Indian Veterinary Journal, 78,* 726–729.

66. Bienek, D. R., & Belosevic, M. (1997). Comparative assessment of growth of *Trypanosoma danilewskyi* (Laveran & Mesnil) in medium containing fish or mammalian serum. *Journal of Fish Diseases, 20,* 217–221.

67. Biozzi, G., Mouton, D., Sant'Anna, O. A., Passos, H. C., Gennari, M., Reis, M. H., Ferreira, V. C. A., Heumann, A. M., Bouthillier, Y., Ibanez, O. M., Stifell, C., & Siqueira, M. (1979). Genetics of immunoresponsiveness to natural antigens in the mouse. *Current Topics in Microbiology and Immunology, 85,* 31–98.

68. Birch, L. C. (1957). The meanings of competition. *The American Naturalist, 91,* 5–18.

69. Blaxhall, P. C. (1972). The haematological assessment of the health of fresh water fish. *Journal of Fish Biology, 4,* 593–604.

70. Blaxhall, P. C. (1985). The separation and cultivation of fish lymphocytes. In M. J. Manning & M. F. Tatner (Eds.), *Fish Immunology* (pp. 1–374). Academic Press.

71. Bloom, B. R. (1971). *In vitro* approaches to the mechanisms of cell-mediated immune reactions. *Advances in Immunology, 13,* 101–208.

72. Bone, G. J., & Parent, G. (1963). Stearic acid, an essential growth factor for *Trypanosoma cruzi*. *Journal of General Microbiology, 31,* 261–266.

73. Borah, S., & Yadav, R. N. S. (1996). Biochemical and haematological changes in a freshwater fish *Heteropneustes fossilis* (Bloch): Impact of crude oil and some pesticides commonly used in tea gardens (Ph.D. thesis). Dibrugarh University, Assam, India.

74. Botham, J. W., Grace, M. F., & Manning, M. J. (1980). Ontogeny of first and second set alloimmune reactivity in fish. In M. M. Manning (Ed.), *Phylogeny of Immunological Memory* (p. 83). Elsevier/North-Holland Biomedical Press.

75. Bower, S. M., & Evelyn, T. P. T. (1988). Acquired and innate resistance to the haemoflagellate *Cryptobia salmositica* in sockeye salmon (*Oncorhynchus nerka*). *Developmental and Comparative Immunology, 12,* 749–760.

76. Bower, S. M., & Margolis, L. (1983). Direct transmission of the haemoflagellate *Cryptobia salmositica* among Pacific salmon (*Oncorhynchus spp.*). *Canadian Journal of Zoology, 61,* 1242–1250.

77. Bower, S. M., & Margolis, L. (1984). Detection of infection and susceptibility of different Pacific salmon stocks (*Oncorhynchus spp.*) to the haemoflagellate *Cryptobia salmositica*. *Journal of Parasitology, 70,* 273–278.

78. Bower, S. M., & Margolis, L. (1985). Effects of temperature and salinity on the course of infection with the haemoflagellate *Cryptobia salmositica* in juvenile Pacific salmon (*Oncorhynchus spp.*). *Journal of Fish Diseases, 8,* 25–33.

79. Bower, S. M., & Woo, P. T. K. (1977a). *Cryptobia catastomi*: Incubation in plasma of susceptible and refractory fishes. *Experimental Parasitology, 43,* 63–68.

80. Bower, S. M., & Woo, P. T. K. (1977b). Morphology and host specificity of *Cryptobia catostomi* n. sp. (Protozoa: Kinetoplastida) from white sucker (*Catostomus commersoni*) in southern Ontario. *Canadian Journal of Zoology, 55,* 1082–1092.

81. Bower, S. M., & Woo, P. T. K. (1979). The prevalence of *Trypanosoma catostomi* in white sucker (*Catostomus commersoni*) from southern Ontario. *Journal of Wildlife Diseases, 15,* 429–431.

82. Brack, C., Delain, E., & Riou, G. (1972). Replicating, covalently closed, circular DNA from kinetoplasts of *Trypanosoma cruzi*. *Proceedings of the National Academy of Sciences of the United States of America, 69,* 1642–1646.

83. Brain, M. V. (1956). Segregation of species of the ant genus *Myrmica*. *Journal of Animal Ecology, 25,* 319–337.

84. Brands, S. J. (1989–2005). *Systema Naturae 2000*. Amsterdam, The Netherlands. http://sn2000.taxonomy.nl/ (accessed on 30 October 2024).

85. Brasseur, P., & Gorenflot, A. (1992). Human babesiosis in Europe. *Memorias do Instituto Oswaldo Cruz, 87,* Suppl. 3, 131–132.

86. Brasseur, P., & Gorenflot, A. (1996). Human babesial infections in Europe. *Roczniki Akademii Medycznej im. Juliana Marchlewskiego w Białymstoku, 41,* 117–122.
87. Briendl, V. (1911). Trypanosomes and *Trypanoplasma* of some Czechoslovakian fishes. *Sitzungsberichte der Böhmischen Gesellschaft der Wissenschaften, 33,* 1–34.
88. Briendl, V. (1915). Study on the blood parasites of freshwater fishes. *Rozpravy České Akademie císaře Františka Josefa pro vědy, slovesnost a umění, 18,* 1–29.
89. Brooke, H. E. (1964). Blood serum protein and calcium levels in yearlings brook trout. *Progressive Fish-Culturist, 26,* 107–110.
90. Brumpt, E. (1906). Sur quelques espèces nouvelles de trypanosomes parasites des poisons d'eau douce: leur mode d'évolution. Mode de transmission et évolution des trypanosomes des poissons etc. *Comptes Rendus des Séances de la Société de Biologie, 60,* 160–164.
91. Brun, R. (1982). Cultivation of procyclic trypomastigates of *Trypanosoma congolense* in a semi-defined medium with direct adaptation from blood stream forms. *Zeitschrift für Parasitenkunde, 67,* 129–135.
92. Brun, R., Jenni, L., Schonenberger, M., & Schell, K. F. (1981). *In vitro* cultivation of blood stream forms of *Trypanosoma brucei, T. rhodesiense,* and *T. gambiense*. *Journal of Protozoology, 28*(4), 470–479.
93. Bunnajirakul, S. (1998). Histopathological observation on the common carp (*Cyprinus carpio*) infected with the blood flagellate *Trypanoplasma borreli* (Dissertation). Tierärztliche Hochschule, Hannover.
94. Bunnajirakul, S., Steinhagen, D., Hetzel, U., Korting, W., & Drommer, W. (2000). A study of sequential histopathology of *Trypanoplasma borreli* (Protozoa: Kinetoplastida) in susceptible common carp, *Cyprinus carpio*. *Diseases of Aquatic Organisms*, *39*, 221–229.
95. Burreson, E. M. (1982). The life cycle of *Trypanoplasma bullocki* (Zoomastigophorea: Kinetoplastida). *Journal of Parasitology*, *29*(1), 72–77.
96. Burreson, E. M. (1983). Trypanoplasmiasis in flounder along the Atlantic coast of the United States. In D. P. Anderson, M. Dorson, & P. H. Dubourget (Eds.), *Antigens of Fish Pathogens: Development and Production for Vaccines and Serodiagnostics* (pp. 273). Collection Foundation Marcel Merieux.
97. Burreson, E. M. (2007). Haemoflagellates of Oregon marine fishes with the description of new species of *Trypanosoma* and *Trypanoplasma*. *Journal of Parasitology*, *93*(6), 1442–1451.

98. Burreson, E. M., & Frizzell, L. J. (1986). The seasonal antibody response in juvenile summer flounder (*Paralichthys dentatus*) to the haemoflagellate *Trypanoplasma bullocki*. *Veterinary Immunology and Immunopathology*, *12*, 395–402.

99. Burreson, E. M., & Karlsbakk, E. (2007). Multiplication of *Trypanosoma pacifica* (Euglenozoa: Kinetoplastea) in English sole (*Parophrys vetulus*) from Oregon coastal waters. *Journal of Parasitology*, *93*, 932–933.

100. Burreson, E. M., & Pratt, I. (1972). *Trypanosoma pacifica* sp. n. from the English sole *Parophrys vetulus* Girard from Oregon. *Journal of Protozoology*, *19*, 555–556.

101. Burreson, E. M., & Sypek, J. P. (1981). *Cryptobia* sp. (Mastigophora: Kinetoplastida) from the gills of marine fishes in the Chesapeake Bay. *Journal of Fish Diseases*, *4*, 519–522.

102. Burreson, E. M., & Zwerner, D. E. (1982). The role of host biology, vector biology and temperature in the distribution of *Trypanoplasma bullocki* infections in the lower Chesapeake Bay. *Journal of Parasitology*, *68*, 306–313.

103. Bush, A. O., Lafferty, K. D., Lotz, J. M., & Shostak, A. W. (1997). Parasitology meets ecology on its own terms: Margolis et al. revisited. *Journal of Parasitology*, *83*, 575–583.

104. Bykhovski, B. E. (1957). *Monogenoidea, Their Systematics and Phylogeny*. Izd. Akad. Nauk SSSR. Moscow and Leningrad.

105. Caira, J. N., Jensen, K., & Holsinger, K. E. (2003). On a new index of host specificity. In C. Combes & J. Jourdane (Eds.), *Taxonomy, Ecology and Evolution of Metazoan parasites* (pp. 111–120). Presses Universitaires de Perpignan.

106. Callero, C. (1952). Incidence of *Trypanosoma lewisi*, *Sarcocystis muris*, species of *Spirochaeta* and microfilarial larvae in rats in Panama city and suburbs. *Journal of Parasitology*, *38*, 369.

107. Calvo, M. M., de L., Nogueda, B., & Alejandre, R. (1992). La vía oral: Una puerta de acceso para *Trypanosoma cruzi*. *Revista Latinoamericana de Microbiología*, *39–42*.

108. Camardella, L., Caruso, C., D'Avino, R., di Prisco, G., Rutigliano, B., Tamburrini, M., Fermi, G., & Perutz, M. F. (1992). Haemoglobin of the Antarctic fish *Pagothenia bernacchii*: Amino acid sequence, oxygen equilibria and crystal structure of its carbonmonoxy derivative. *Journal of Molecular Biology*, *20*, 449–460.

109. Cannon, L. R. G. (1973). Diet and intestinal helminths in a population of perch, *Perca flavescens*. *Journal of Fish Biology*, *5*, 447–457.

110. Carton, Y. (1973). La réponse immunitaire chez les agnathes et les poissons: Structure des immunoglobulines. *Annales de Biologie*, *12*, 139.

111. Casero, R. A. J., Porter, C. W., & Bernacki, R. J. (1982). Activity of tunicamycin against *Trypanosoma brucei in vitro* and *in vivo*. *Antimicrobial Agents and Chemotherapy*, *22*, 1008–1011.

112. Castellani, A., & Willey, A. (1905). Observations on the haematozoa of vertebrates in Ceylon. *Quarterly Journal of Microscopical Science*, *49*, 39.

113. Cavalier-Smith, T. (1993). Kingdom Protozoa and its 18 phyla. *FEMS Microbiology Reviews (Amsterdam)*, *57*, 953–994.

114. Cazzulo, J. J. (1994). Intermediate metabolism in *Trypanosoma cruzi*. *Journal of Bioenergetics and Biomembranes*, *26*(2), 157–165.

115. Chagas, C. (1909). Ueber eine neue Trypanosomiasis des Menschen. Studien über Morphologie und Entwicklungszyklus des *Schizotrypanum cruzi* n. gen., n. sp., Erreger einer neuen Krankheit des Menschen. *Memórias do Instituto Oswaldo Cruz*, *1*, 159–218.

116. Chaka, H., & Abebe, G. (2003). Drug-resistant trypanosomes: A threat to cattle production in the Southwest of Ethiopia. *Revue Élevage Médecine Vétérinaire des Pays Tropicaux*, *56*, 33–36.

117. Chakravarty, A. K. (1996). *Immunology*. Tata McGraw-Hill Publishing Co. Ltd.

118. Chalachnikov, A. P. (1888). Recherches sur le sang des animaux à sang chaud et à sang froid. *Archives de Naturopathie*, *1*, 65.

119. Chanchal, A. K., Pandey, B. N., Singh, S. B., & Prashad, S. (1979). Cyclic changes in haematological values of *Anabas testudineus* (Bloch). *Annales Zoologicae*, 111–124.

120. Chandler, A. C., & Read, C. P. (1970). *Introduction to Parasitology*. Wiley Eastern Private Ltd.

121. Channon, J. Y., & Kasper, L. H. (1996). *Toxoplasma gondii*: Induced immunosuppression by human peripheral blood monocytes: Role of gamma interferon. *Infection and Immunity*, *64*, 1181–1189.

122. Chappell, L. H. (1969). Competitive exclusion between two intestinal parasites of the three-spined stickleback, *Gasterosteus aculeatus* L. *Journal of Parasitology*, *55*, 775–778.

123. Chappell, L. H. (1980). *Physiology of Parasites*. John Wiley and Sons.

124. Chatterji, A. (1960). The liver in *Trypanosoma evansi* infection of white mice. *Bulletin of the Calcutta School of Tropical Medicine*, *8*, 119–121.

125. Chen, C. L., & Hsieh, Sh. R. (1964). Parasitic flagellates of fishes from

Hwama lake. *Acta Hydrobiologica*, *51*, 37–49.

126. Cheng, T. C. (1986). Physiology and biochemistry of parasitic flagellates. In *General Parasitology* (pp. 139–151). Academic Press.
127. Chitambo, H., & Arakawa, A. (1991). Therapeutic effect of Berenil and Samorin in mice infected with four species of trypanosome population isolated from Zambian cattle. *Veterinary Parasitology*, *39*, 43–52.
128. Christensen, G. M., Kim, M. C., Brungs, W. A., & Hund, R. (1972). Changes in the blood of the brown bull head, *Ameiurus nebulosus* (Le Suer), following short and long-term exposure to copper (II). *Toxicology and Applied Pharmacology*, *23*, 417–427.
129. Clark, T. B., & Wallace, F. G. (1960). A comparative study of kinetoplast ultrastructure in the *Trypanosomatidae*. *Journal of Protozoology*, *7*, 115–124.
130. Clarkson, M. J. (1968). Blood and plasma volumes in sheep infected with *Trypanosoma vivax*. *Journal of Comparative Pathology*, *78*, 189–193.
131. Clayton, C. E., & Michels, P. (1996). Metabolic compartmentation in African trypanosomes. *Parasitology Today*, *12*, 465–471.
132. Cleaveland, S., Hess, G. R., Dobson, A. P., Laurenson, K. M., McCallum, H. I., Roberts, M. G., & Woodroffe, R. (2002). The role of pathogens in biological conservation. In P. J. Hudson, A. Rizzoli, B. T. Grenfell, H. Heesterbeek, & A. P. Dobson (Eds.), *The Ecology of Wildlife Diseases* (pp. 139–150). Oxford University Press.
133. Cochrane, J. C., Sagorin, L., & Wilcock, J. C. (1975). *Capillaria hepatica* infection in man. *South African Medical Journal*, *31*, 751–755.
134. Cohen, S. (1982). Survival of parasites in the immunocompetent host. In S. Cohen & K. S. Warren (Eds.), *Immunology of Parasitic Infections* (pp. 138–161). Blackwell Scientific Publications.
135. Cooper, G. R., & McDaniel, N. (1970). The determination of glucose by the ortho-toluidine method. In R. MacDonald (Ed.), *Standard Methods of Clinical Chemistry* (Vol. 6, p. 159). Academic Press.
136. Cooper, H. L. (1971). Biochemical alterations accompanying initiation of growth in resting cells. In R. Baserga (Ed.), *Cell Cycle and Cancer* (pp. 1–197). Dekker.
137. Corliss, J. O. (1994). An interim utilization ('user-friendly') hierarchical classification and characterization of the protists. *Acta Protozoologica*, *33*, 1–51.
138. Cornell, H. V., & Lawton, J. H. (1992). Species interaction, local and regional processes, and limits to the richness of ecological communities:

A theoretical perspective. *Journal of Animal Ecology*, *61*, 1–12.

139. Cornille, R., & Hornung, M. (1968). Determination of serum IgM levels for the diagnosis of *Trypanosoma rhodesiense* infection. *American Journal of Tropical Medicine and Hygiene*, *17*, 522–527.
140. Cossarini-Dunier, M. (1986). Protection against enteric redmouth disease in rainbow trout, *Salmo gairdneri* Richardson, after vaccination with *Yersinia ruckeri* bacterin. *Journal of Fish Diseases*, *9*, 27–33.
141. Cottrell, B. J. (1976). The immune response of plaice, *Pleuronectes platessa*, to tissue parasites. *Parasitology*, *24*, 74.
142. Cottrell, B. J. (1977). A trypanosome from plaice, *Pleuronectes platessa* L. *Journal of Fish Biology*, *11*, 35–47.
143. Cox, F. E. (1970). Parasitic protozoa of British wild mammals. *Mammal Review*, *1*, 1–28.
144. Cranney, J. S., & Heckmann, R. A. (1996). *Trypanosoma ataria* sp. n. (Kinetoplastida: Bodonidae) in fishes from the Sevier River drainage, Utah. *Great Basin Naturalist*, *56*(2), 142–149.
145. Crawley, H. (1909). Studies on blood parasites, II. The priority of *Cryptobia* Leidy, 1846 over *Trypanoplasma* (Laveran and Mesnil, 1901). *Bulletin of the US Bureau of Animal Industry*, *119*, 16–20.
146. Crawshaw, G. J. (1983). Amphibian medicine. In M. Fowler (Ed.), *Zoo and Wild Animal Medicine* (pp. 131–139). Saunders.
147. Croft, S. L., Seifert, K., & Duchêne, M. (2003). Antiprotozoal activities of phospholipid analogues. *Molecular and Biochemical Parasitology*, *126*(2), 165–172.
148. Crofton, H. D. (1971a). A quantitative approach to parasitism. *Parasitology*, *62*, 179–193.
149. Crofton, H. D. (1971b). A model of host-parasite relationship. *Parasitology*, *63*, 343–364.
150. Crombie, A. C. (1946). Further experiments on insect competition. *Proceedings of the Royal Society of London Biological Sciences*, *133*, 76–109.
151. Cross, S. X. (1934). A probable case of non-specific immunity between two parasites of ciscoes of the Trout Lake region of northern Wisconsin. *Journal of Parasitology*, *20*, 244–245.
152. Currie, J. L. M., & Woo, P. T. K. (2007). Susceptibility of sexually mature rainbow trout, *Oncorhynchus mykiss*, to experimental cryptobiosis caused by *Cryptobia salmositica*. *Parasitology Research*, *101*(4), 1057–1067.
153. Czichos, J., Nonnengaesser, C., & Overpath, P. (1986). *Trypanosoma*

brucei: Cis-aconitate and temperature reduction as triggers of synchronous transformation of bloodstream to procyclic trypomastigotes *in vitro*. *Experimental Parasitology*, *62*, 283–291.

154. D'Alesandro, P. A. (1975). Ablastin: The phenomenon. *Experimental Parasitology*, *38*, 303–308.
155. Dabral, R. (1980). Some aspects of haematological studies in a new species of freshwater teleosts (Doctoral dissertation, Rohilkhand University, Bareilly).
156. Daly, J. J., & Degiusti, D. L. (1971). *Trypanosoma catostomi* (n.sp.) from the white sucker *Catostomus c. commersoni* (Lacepède). *Journal of Protozoology*, *18*, 414–417.
157. Damian, R. T. (1964). Molecular mimicry: Antigen sharing by parasite and host and its consequences. *American Naturalist*, *98*, 129–149.
158. Daniela, G., Alessandro, V. H., Caroline, L., Jack, P., Marco, B., Lelio, M., Elio, P., Guido di Prisco, & Cinzia, V. (2007). Hemoglobin structure/function and globin-gene evolution in the Arctic fish *Liparis tunicatus*. *Gene*, *30*, 58–68.
159. Danilewskyi, B. (1885). Zur Parasitologie des Blutes. *Biologisches Zentralblatt*, *5*, 529.
160. Danilova, N., Bussmann, J., Jekosch, K., & Steiner, L. A. (2005). The immunoglobulin heavy chain locus in zebra fish: Identification and expression of a previously unknown isotype, immunoglobulin. *Z. Nat. Immunol., 6,* 295–302.
161. Dantrakool, A., Sombon, P., Hashimoto, T., & Saito-Ito, A. (2004). Identification of a new type of *Babesia* species in wild rats (*Bandicoota indica*) in Chiang Mai Province, Thailand. *Journal of Clinical Microbiology, 42,* 850–854.
162. Darji, A. A., Lucas, S., Magez, E., Torreele, J., Palacios, M., Sileghem, E., Bajyana Songa, R., Hamers, & De Baetselier, P. (1992). Mechanisms underlying trypanosome-elicited immunosuppression. *Annales de la Société Belge de Médecine Tropicale, 72*(Suppl.), 27–38.
163. Darwin, C. (1859). *The Origin of Species by Means of Natural Selection.* Harvard University Press. (Original work published 1859)
164. Davies, A. J., Mastri, C., Ariyanayagam, R., Sappal, G. K., & Wilkins, A. (1992). Simple methods for culturing and preserving *Trypanosoma granulosum* Laveran and Mesnil from European eels (*Anguilla anguilla* L.). *Bulletin of the European Association of Fish Pathologists, 12,* 8–10.
165. Davies, A. J., Mastri, C., Thorborn, D. E., & Mackintosh, D. (1995).

Experiments with growth of the eel trypanosome, *Trypanosoma granulosum* (Laveran and Mesnil, 1902), in semi-defined and defined media. *Journal of Fish Diseases, 18,* 599–608.

166. Davila, A. M. R., Ramirez, L., & Silva, R. A. M. S. (1998). Biometrical alteration of *Trypanosoma evansi* isolate in laboratory rodents. *Veterinary Parasitology, 76,* 149–152.
167. Davis, A. K., Maney, D. L., & Maerz, J. C. (2008). The use of leukocyte profiles to measure stress in vertebrates: A review for ecologists. *Functional Ecology, 22*(5), 760–772.
168. Davis, B. S. (1952). Studies on the trypanosomes of some Californian mammals. *University of California Publications in Zoology, 57,* 145–250.
169. Davis, J. R., & Huffman, D. G. (1978). Some factors associated with the distribution of helminths among individual mosquito fish, *Gambusia affinis*. *Texas Journal of Science, 30,* 43–53.
170. Davoust, B., Boni, M., Branquet, D., Ducos de Lahitte, J., & Martet, G. (1997). Research on three parasitic infestations in rats captured in Marseille: Evaluation of the zoonotic risk. *Bulletin de l'Académie Nationale de Médecine, 181,* 887–895.
171. De Souza, W., & Meyer, H. (1974). On the fine structure of the nucleus in *T. cruzi* in tissue culture forms spindle fibres in the dividing nucleus. *Journal of Protozoology, 21,* 48–52.
172. DeBach, P. (1966). The competitive displacement and coexistence principle. *Annual Review of Entomology, 2,* 183–212.
173. Delain, E., Brack, C. H., Riou, G., & Festy, B. (1971). Ultrastructural alterations of *Trypanosoma cruzi* kinetoplast induced by the interaction of trypanocidal drug (Hydroxy stilbamidine) with the kinetoplast DNA. *Journal of Ultrastructural Research, 37,* 200–218.
174. De-Mello, I. F., & Valles, C. (1936). On the *Trypanosoma* found in the blood of Indian freshwater fish *C. batrachus* (Linn). *Proceedings of the Indian Academy of Sciences, 3,* 12–124.
175. Denman, A. M. (1973). Methods of separation of human blood lymphoid cell population. *Journal of Immunological Methods, 2,* 331–351.
176. Desquesnes, M., Ravel, S., & Cuny, G. (2002). PCR identification of *Trypanosoma lewisi*, a common parasite of laboratory rats. *Kinetoplastid Biology and Disease, 1,* 2–7.
177. Desvaux, F. X., & Charlemagne, J. (1981). Characterization of the humoral response to particulate antigens. *Immunology, 43,* 755–762.
178. Dhandt, A. A. (1977). Interspecific competition between great and blue tit.

Nature, 268, 521–523.

179. Diamant, A. (1990). Morphology and ultrastructure of *Cryptobia eilatica* n. sp. (Bodonidae: Kinetoplastida), an ectoparasite from the gills of marine fish. *Journal of Protozoology, 37,* 482–489.
180. Dias, J. A. T. S. (1955). Estudos sobre os hematozóarios dos peixes de águas doces de Moçambique. I. Descrição de algumas novas espécies de tripanossomas, parasites de *Tilapia mossambica* (Peters, 1952). *Moçambique: Doc Trimestral, 82,* 47.
181. Dixon, H., & Williamson, J. (1970). The lipid composition of blood and culture forms of *Trypanosoma lewisi* and *Trypanosoma rhodesiense* compared with that of their environment. *Comparative Biochemistry and Physiology, 33,* 111–128.
182. Dixon, H., Ginger, C. D., & Williamson, J. (1971). The lipid metabolism of blood and culture forms of *Trypanosoma lewisi* and *Trypanosoma rhodesiense*. *Comparative Biochemistry and Physiology, 39B,* 247–266.
183. Dobson, A. P. (1998). The population biology of parasite-induced changes in host behaviour. *Quarterly Review of Biology, 63,* 139–165.
184. Dobson, A. P., & May, R. M. (1987). Effects of parasites on fish populations: Theoretical aspects. *International Journal for Parasitology, 17,* 363–370.
185. Dobson, A. P., & Pacala, S. W. (1992). The parasites of *Anolis* lizards in northern Antilles. II. The structure of the parasite community. *Oecologia, 91,* 118–125.
186. Dobson, A. P., & Roberts, M. (1995). The population dynamics of parasitic helminth communities. *Parasitology, 109,* S97–S108.
187. Docampo, R., & Moreno, S. N. (1990). The metabolism and mode of action of gentian violet. *Drug Metabolism Reviews, 22,* 161–178.
188. Doflein, F. (1901). *Die Protozoen als Parasiten und Krankheitserreger.* 1st ed. Fischer.
189. Doflein, F. (1916). *Lehrbuch der Protozenkunde* (4th ed.). Fischer, Jena.
190. Dogiel, V. A. (1958). Ecology of the parasites of freshwater fishes. In V. A. Dogiel, G. K. Petrushevski, & Y. I. Polyanski (Eds.), *Parasitology of Fishes* (Translated by Z. Kabata, 1961, pp. 1–47). Leningrad University Press.
191. Dogiel, V. A. (1964). *General Parasitology*. Oliver and Boyd.
192. Dogiel, V. A. (1970). Ecology of the parasites of freshwater fishes. In V. A. Dogiel, G. K. Petrushevski, & Yu I. Polyanski (Eds.), *Parasitology of Fishes* (Translated by Z. Kabata, pp. 1–47). T.F.H. Publications, Inc.
193. Dore, A. B. (1918). Rat trypanosomes in New Zealand. *New Zealand*

Journal of Science and Technology, 1(4), 200.

194. Douglas, S. D. (1971). Human lymphocyte growth *in vitro*: Morphological, biochemical and immunological significance. *International Review of Experimental Pathology, 10,* 41.

195. Dowell, S. F. (2001). Seasonal variation in host susceptibility and cycles of certain infectious diseases. *Emerging Infectious Diseases, 7.* Retrieved from: http://www.cdc.gov/ncidod/eid/vol7no3/dowell.htm (accessed on 30 October 2024).

196. Dube, S. C., & Munshi, J. S. D. (1973). A quantitative study of the erythrocytes and haemoglobin in the blood of an air-breathing fish, *Anabas testudineus* (Bloch), in relation to its body size. *Folia Haematologica, 100,* 436–446.

197. Dubey, N. K., & Pandey, P. K. (1985). Trypanosomiasis in *Heteropneustes fossilis* (Bloch) and its effect on the plasma electrolytes. *Uttar Pradesh Journal of Zoology, 5*(1), 69–71.

198. Duh, D., Petrovec, M., & Avsic-Zupanc, T. (2001). Diversity of *Babesia* infecting European sheep ticks (*Ixodes ricinus*). *Journal of Clinical Microbiology, 39,* 3395–3397.

199. Durand, M., Seche, E., & Maurizot, J. C. (2002). Effect of strand orientation on the interaction of Berenil with DNA triple helices. *European Biophysics Journal, 30,* 625–630.

200. Dutton, J. E., Todd, J. L., & Tobey, E. N. (1907). Concerning certain parasitic protozoa observed in Africa. Part II. *Annals of Tropical Medicine and Parasitology, 1,* 287.

201. Dvorak, A. M. (1982). Anaphylactic degranulation of guinea pig basophilic leukocytes II. Evidence for regranulation of mature basophils during recovery from degranulation *in vitro. Laboratory Investigation, 56,* 461.

202. Dykova, I., & Lom, J. (1979). Histopathological changes in *Trypanosoma danilewskyi* (Laveran and Mesnil, 1904), and *Trypanoplasma borreli* (Laveran and Mesnil, 1902), infections of goldfish, *Carassius auratus* (L.). *Journal of Fish Diseases, 2,* 381–390.

203. Eiras, J. C. (1988). *Trypanosoma granulosum* Laveran and Mesnil (1902) (Protozoa: Kinetoplastida) infecting *Anguilla anguilla* L. in Portugal. *Ciencia Biologica Ecol. Sistemas (Portugal), 8*(1/2), 8–10.

204. Elliot, J. W., Fowler, L. G., & Burrows, R. B. (1966). Effects of age, growth, and diet on characteristics of salmon fingerlings. *Technical Paper, 8,* Bureau of Sport Fish and Wildlife, 3–11.

205. Ellis, A. E. (1982). Differences between the immune mechanisms of fish

and higher vertebrates. In R. J. Roberts (Ed.), *Microbial Diseases of Fish* (pp. 1–29). Academic Press.

206. Ellis, A. E. (1985). Development of fish vaccines: Strategies and future considerations. In A. E. Ellis (Ed.), *Fish and Shellfish Pathology* (pp. 412). Academic Press.

207. Ellsaesser, C. F., & Clem, L. W. (1986). Haematological and immunological changes in channel catfish stressed by handling and transport. *Journal of Fish Biology, 28,* 511–521.

208. Ellsaesser, C. F., & Clem, L. W. (1987). Cortisol-induced haematological and immunological changes in channel catfish (*Ictalurus punctatus*). *Comparative Biochemistry and Physiology, 87A*(2), 405–408.

209. El-Sayed, N. M., Myler, P. J., Bartholomeu, D. C., et al. (2005). The genome sequence of *Trypanosoma cruzi*, etiologic agent of Chagas disease. *Science, 309,* 409–415.

210. Elton, C., Ford, E. B., Baker, J. R., & Gardner, A. D. (1931). The health and parasites of a wild mouse population with a bacteriological note by A. D. Gardner. *Proceedings of the Zoological Society of London, 3,* 657–721.

211. Engel, D. W., & Davis, E. M. (1964). Relationship between activity and blood composition in certain teleosts. *Copeia, 31,* 586–587.

212. Erondu, E. S., Nubia, C., & Nwadukwe, F. O. (1993). Haematological studies on four catfish species raised in freshwater ponds in Nigeria. *Journal of Applied Ichthyology, 9,* 250–256.

213. Esch, G. W., Hazen, T. C., & Aho, J. M. (1977). Parasitism and r- and K-selection. In G. W. Esch (Ed.), *Regulation of Parasitic Populations* (pp. 9–62). Academic Press.

214. Evans, D. A., & Brown, R. C. (1972). The utilization of glucose and proline by culture forms of *Trypanosoma brucei*. *Journal of Protozoology, 19,* 686–690.

215. Ewers, W. H. (1973). A host parasite list of the protozoan and helminth parasites of New Guinea animals. *International Journal for Parasitology, 3,* 89–110.

216. Eyles, D. E. (1952). Incidence of *Trypanosoma lewisi* and *Hepatozoon muris* in the Norway rat. *Journal of Parasitology, 38,* 222–225.

217. Fabrikant, J. L. (1972). Cell population kinetics in the seminiferous epithelium under low dose irradiation. *American Journal of Roentgenology, 114,* 792.

218. Facer, A. C. (1976). Blood hyperviscosity during *Trypanosoma (Trypanozoan) brucei* infection of rabbits. *Journal of Comparative*

Pathology, 86, 393–407.

219. Fairlamb, A. H., & Cerami, A. (1985). Identification of a novel, thiol-containing co-factor essential for glutathione reductase enzyme activity in Trypanosomatids. *Molecular and Biochemical Parasitology, 14,* 187–198.
220. Fairlamb, A. H., & Opperdoes, F. R. (1986). Carbohydrate metabolism in African trypanosomes, with special reference to the glycosome. In M. J. Morgan (Ed.), *Carbohydrate Metabolism in Cultured Cells* (pp. 183–224). Plenum Publishing.
221. Faisal, M., Perkins, P. S., & Cooper, E. L. (1990). Infestation by the pennellid copepod *Phrixocephalus cincinnatus* modulates cell-mediated immune responses in the Pacific arrowtooth flounder, *Atheresthes stomias*. In F. O. Perkins & T. C. Cheng (Eds.), *Pathology in Marine Science* (pp. 471–478). Academic Press.
222. Fantham, H. B. (1919). Some parasitic protozoa found in South Africa. II. *South African Journal of Science, 16,* 185–191.
223. Fantham, H. B., Porter, A., & Richardson, L. R. (1942). Some haematozoa observed in vertebrates in eastern Canada. *Parasitology, 34,* 194–205.
224. Fattohy, Z. I. (1978). *Trypanosoma neinevana* n. sp. from the fish *Barbus grypus* Heckel in Iraq. *Current Science, 47,* 33–35.
225. Fay, F. H., & Rausch, R. L. (1969). Parasitic organisms in the blood of arvicoline rodents in Alaska. *Journal of Parasitology, 55*, 1258–1265.
226. Feng, S., & Woo, P. T. K. (1997). Complement fixing antibody production in thymectomized *Oncorhynchus mykiss* vaccinated against or infected with the pathogenic haemoflagellate *Cryptobia salmositica*. *Folia Parasitologica, 44*(3), 188–194.
227. Feng, S., & Woo, P. T. K. (1998). *In vitro* and *in vivo* effects of rabbit anti-thymocyte serum on circulating leukocytes and production of complement fixing antibodies in thymectomized *Oncorhynchus mykiss* (Walbaum) infected with *Cryptobia salmositica* Katz, 1951. *Journal of Fish Diseases, 21*(4), 241–248.
228. Feng, S., & Woo, P. T. K. (2001). Cell membrane glycoconjugates on virulent and avirulent strains of the haemoflagellate *Cryptobia salmositica* (Kinetoplastida). *Journal of Fish Diseases, 24*, 23–32.
229. Fernando, F. S., Benigno, E. A., Hoque, M. M., et al. (1985). *Rodent Biology and Control*. College of Agriculture, University of the Philippines.
230. Ferrante, A., Allison, A. C., & Hirumi, H. (1982). Polyamine oxidase-mediated killing of African trypanosomes. *Parasitology and Immunology, 4*, 349–354.

231. Fiennes, R. N. T. W. (1954). Haematological studies in trypanosomiasis of cattle. *Veterinary Research, 66*, 423–434.
232. Figueroa, F., Mayer, W. E., Lom, J., Dykova, I., Weller, M., Peckova, H., & Klein, J. (1999). Fish trypanosomes: Their position in kinetoplastid phylogeny and variability as determined from 12S rRNA kinetoplast sequences. *Journal of Eukaryotic Microbiology, 46*(5), 473–481.
233. Fijan, N. (1984). Vaccination of fish in European pond culture: Prospects and constraints. In J. Olah (Ed.), *Fish, Pathogens and Environment in European Polyculture* (pp. 1–265). Akademiai Kiado Budapest.
234. Findlay, G. M. (1950). *Recent Advances in Chemotherapy* (3rd ed.). Churchill Livingstone.
235. Firdaus, S., Jafri, A. K., & Rahman, N. (1994). Effects of iron-deficient diet on the growth and haematological characteristics of the catfish, *Heteropneustes fossilis* Bloch. *Journal of Aquatic Tropical, 9*, 179.
236. Fischer, D. B., & Mueller, G. C. (1968). An early alteration in the phospholipid metabolism of lymphocytes by PHA. *Proceedings of the National Academy of Sciences, 60*, 1396.
237. Fish, W. R., Nelson, R. T., & Hirumi, H. (1987). Cell adhesion in *Trypanosoma*: *In vitro* studies of the interaction of *Trypanosoma vivax* with immobilized organic dyes. *Journal of Protozoology, 34*, 457–464.
238. Ford, B. J. (1990). *Trypanosoma* spp. pathological to freshwater fish. *Microscope, 38*, 15–22.
239. Forward, G. M., Ferguson, M. M., & Woo, P. T. K. (1995). Susceptibility of brook charr, *Salvelinus fontinalis*, to the pathogenic haemoflagellate, *Cryptobia salmositica*, and the inheritance of innate resistance by the progenies of resistant fish. *Parasitology, 111*, 337–345.
240. Franjola, R., Sato, G., & Montefusco, A. (1995). Prevalencia de infeccion por protozoos en roe dores sinantropicos de la ciudad de Valdivia, Chile. *Boletín Chileno de Parasitología, 50*, 66–72.
241. Frank, C. (1978). Kleinsaugerprotozoen im Neusiedlerseegebiet. *Angewandte Parasitologie, 19*, 137–154.
242. Frank, S. A. (1996). Models of parasite virulence. *Quarterly Review of Biology, 71*, 37–78.
243. Frank-Henderson, J., Battell, M. L., Zombor, G. I., & Khoo, M. K. Y. (1977). Effects of ethidium catabolism and purine nucleotide synthesis in Ehrlich ascites tumor cells *in vitro*. *Cancer Research, 37*, 3434–3441.
244. Freeman, B. J., Payne, K. R., Cornford, E. M., & MacInnis, A. J. (1973). Effects of *Trypanosoma lewisi* on concurrent infections with *Hymenolepis*

diminuta. *Journal of Protozoology, 20*, 512.

245. Freeman, T., Smithers, S. R., Targett, G. A. T., & Walker, P. J. (1970). Specificity of immunoglobulin G in Rhesus monkeys infected with *Schistosoma mansoni*, *Plasmodium knowlesi*, and *Trypanosoma brucei*. *Journal of Infectious Diseases, 121*, 401–406.
246. Fuller, C. A. (1996). Population dynamics of two species of *Eimeria* (Apicomplexa: Eimeriidae) in deer mice (*Peromyscus maniculatus*): Biotic and abiotic factors. *Journal of Parasitology, 82*, 220–225.
247. Fussganger, R. (1995). Berenil in veterinary medicine: Report from the Chemotherapeutic Institute of Faberwerke Hoechst, AG.
248. Fussganger, R., & Bauer, F. (1962). Berenil, a new chemotherapeutic agent in veterinary medicine. The chemotherapeutic and the parasitological laboratory of Farbwerke Hoechst (pp. 504–531).
249. Galuzo, I. G., & Novinskaya, W. F. (1961). Animal trypanosomes from Kazakhstan. II. Trypanosomes of rodents. *Prirodnaya Ochagovost' Boleznei i Voprosy Parazitologii, 3*, 151–172.
250. Galvani, A. P. (2003). Epidemiology meets evolutionary ecology. *Trends in Ecology & Evolution, 18*, 132–139.
251. Gangwar, R. (2001). Haemoprotozoan infectivity in fishes and anurans and their impact on biochemical indices (Ph.D. thesis). M. J. P. Rohilkhand University, Bareilly.
252. George, J. N., Stokes, E. F., Wicker, D. J., & Conrad, M. E. (1966). c.f. von Brand, T. (1973). Biochemistry of parasites. Academic Press.
253. Gibson, W. C., Lom, J., Peckova, H., Ferris, V. R., & Hamilton, P. B. (2005). Phylogenetic analysis of freshwater fish trypanosomes from Europe using ssu rRNA sequences and random amplification of polymorphic DNA. *Parasitology, 130*, 405–412.
254. Goethert, H. K., & Telford, S. R. (2003). What is *Babesia microti*? *Parasitology, 127*, 301–309.
255. Gordon, M. S. (1972). *Animal Physiology: Principles and Adaptation* (2nd ed.). Macmillan.
256. Gorenflot, A., Moubri, K., Precigout, E., Carcy, B., & Schetters, T. P. (1998). Human babesiosis. *Annals of Tropical Medicine and Parasitology, 92*, 489–501.
257. Grace, F., & Manning, J. (1980). Histogenesis of the lymphoid organs in rainbow trout, *Salmo gairdneri* Rich, 1836. *Developmental and Comparative Immunology*, *4*, 255–264.
258. Grant, P. T., & Fulton, J. D. (1957). The catabolism of glucose by strain of

Trypanosoma rhodesiense. *Biochemical Journal*, *66*, 242–250.

259. Grant, P. T., & Sargent, J. R. (1960). Properties of 1-a glycerophosphate oxidase and its role in the respiration of *Trypanosoma rhodesiense*. *Biochemical Journal*, *76*, 229–237.
260. Greaney, G. S., & Powers, D. A. (1977). Cellular regulation of an allosteric modifier of fish haemoglobin. *Nature*, *270*, 73–74.
261. Griffin, B. R. (1983). Opsonic effect of rainbow trout (*Salmo gairdneri*) antibody on phagocytosis of *Yersinia ruckeri* by trout leukocytes. *Developmental and Comparative Immunology*, *7*, 253–259.
262. Grogl, M., Marinkelle, J. C., Suarez, F. M., Sanchez, N., & Guhl, F. (1980). *Trypanosoma magdulenae* n. sp. (Protozoa: Kinetoplastida) from fresh water teleost, *Petenia karussii* in Colombia. *Journal of Parasitology*, *66*, 1022–1026.
263. Gruby, D. (1843). Recherches et observations sur une nouvelle espèce d'hématozoaire, *Trypanosoma sanguinis*. *Comptes Rendus de l'Académie des Sciences*, *17*, 1134–1136.
264. Grunwaldt, E. (1977). Babesiosis on Shelter Island. *New York State Journal of Medicine*, *8*, 1320–1321.
265. Grutter, A. S., & Poulin, R. (1998). Intraspecific and interspecific relationships between host size and the abundance of parasitic larval gnathiid isopods on coral reef fishes. *Marine Ecology Progress Series*, *164*, 263–271.
266. Gu, Z., Wang, J., Ke, X., Liu, Y., Liu, X., Gong, X., & Li, A. (2010). Phylogenetic position of the freshwater fish trypanosome, *Trypanosoma ophiocephali* (Kinetoplastida) inferred from the complete small subunit ribosomal RNA gene sequence. *Parasitology Research*, *106*(5), 1039–1042.
267. Gu, Z., Wang, J., Li, M., Zhang, J., & Gong, X. (2007). Redescription of *Trypanosoma siniperca* Chang, 1964 from fresh water fish of China based on morphological and molecular data. *Parasitology Research*, *100*, 395–400.
268. Gu, Z., Wang, J., Zhang, J., & Gong, X. (2006). Redescription of *Trypanosoma ophiocephali* Chen, 1964 (Kinetoplastida: Trypanosomatina: Trypanosomatidae) and first record from the blood of dark sleeper (*Odontobutis obscura* Temminck et Schlegel) in China. *Parasitology Research*, *100*, 149–154.
269. Guerrero, O. M., Chinchilla, M., & Abrahams, E. (1997). Increasing of *Toxoplasma gondii* (Coccidia, Sarcocystidae) infections by *Trypanosoma*

lewisi (Kinetoplastida, *Trypanosomae*) in white rats. *Revista de Biología Tropical*, *45*(2), 877–882.

270. Guimares, A. C. R., Otto, T. D., Alves-Ferreira, M., Miranda, A. B., & Degrave, W. M. (2008). *In silico* reconstruction of the amino acid metabolic pathways of *Trypanosoma cruzi*. *Genetics and Molecular Research*, *7*, 872–882.

271. Gupta, D. K., & Gupta, Neelima. (1994). A comparative study of culture media for *Trypanosoma kargenensis* n. sp. from *Heteropneustes fossilis* and its effect on the biochemical composition of fish. In A. F. S. Philippines (Eds.), *Proceedings of the Third Asian Fish Forum* (pp. 320–323).

272. Gupta, D. K., & Gupta, Neelima. (1998). A new record of *Trypanoplasma* (Kinetoplastida: Cryptobiidae) inhabiting the blood of carp, *Catla catla* in India. *Bulletin of the NRC Egypt*, *24*, 75–83.

273. Gupta, D. K., Gupta, Neelima, & Gangwar, R. (2001). Dynamics of infection of *Trypanosoma* in *Clarias batrachus* and their role in inducing protein and haemoglobin alterations. In S. K. Malhotra & V. S. Chauhan (Eds.), *V. N. Capoor Commemoration Volume* (pp. 35–38).

274. Gupta, D. K., Gupta, Neelima, & Gangwar, R. (2006). Two new species of *Trypanosoma* from fresh water fishes (*Heteropneustes fossilis* and *Channa punctatus*) from Bareilly, India. *Journal of Parasitic Diseases*, *30*(1), 58–63.

275. Gupta, D. K., Gupta, Neelima, & Gupta, A. (2001). Infectivity and pathogenicity in *Clarias batrachus* parasitized by haemoflagellates. *Applied Fisheries and Aquaculture*, *1*(1), 75–77.

276. Gupta, D. K., Gupta, Neelima, & Gupta, A. (2002a). Haemoglobin values in Bagrid and Saccobranchid fishes due to *Trypanosoma* infection. *Himalayan Journal of Environmental Zoology*, *16*, 23–26.

277. Gupta, D. K., Gupta, Neelima, & Gupta, A. (2002b). Dimorphism of *Trypanosoma* in *Heteropneustes fossilis* and its role in inducing haematological aberrations. *Flora and Fauna*, *8*, 3–6.

278. Gupta, D. K., Gupta, Neelima, & Yadav, P. (1987). New polymorphic trypanosomes from two freshwater fishes. *Himalayan Journal of Environmental Zoology*, *1*, 93–97.

279. Gupta, D. K., Gupta, Neelima, & Yadav, P. (1988). On a new trypanoplasm from an Indian freshwater fish. *Revista Iberica de Parasitología*, *48*(3), 227–232.

280. Gupta, D. K., Gupta, Neelima, & Yadav, P. (2000). Dynamics of infection of *Trypanosoma monomporpha* var. nov. catlae from *Catla catla*. *Advances*

in Biosciences, *19*, 67–76.

281. Gupta, D. K., Gupta, Neelima, & Yadav, P. (2003). *Trypanosoma piscidium* n. sp. and its role in inducing anemia in *Colsia fasciatus*. In B. N. Pandey, R. K. Choudhary, & B. K. Singh (Eds.), *Biodiversity Conservation, Environmental Pollution and Ecology* (Vol. II, pp. 127–133). A.P.H. Publishing Corporation, New Delhi.
282. Gupta, D. K., Yadav, P., & Gupta, Neelima. (1990). Pathogenicity in catfishes infected with *Trypanoplasma* (Cryptobia) (Mastigophora: Kinetoplastida). In R. Hirano & I. Hanyu (Eds.), *Proceedings of the Second Asian Fish Forum* (pp. 713–715).
283. Gupta, N., & Jairajpuri, D. S. (1984). Effect of locality of the host and seasonal variations on the incidence of infection of trypanosomes. *Comparative Physiology and Ecology*, *9*, 373–375.
284. Gupta, Neelima, & Gupta, D. K. (1985). Haematological changes due to *Trypanosoma batrachi* and *T. aligaricus* infection in two fresh water teleosts. *Angewandte Parasitologie*, *26*, 193–196.
285. Gupta, Neelima, & Gupta, D. K. (1986). Trypanosome infectivity and changes in the glucose level of two fresh water fishes. *Indian Journal of Parasitology*, *10*, 213–215.
286. Gupta, Neelima, & Gupta, D. K. (1987a). Polymorphism in *Trypanoplasma guptii* n. sp. (Kinetoplastida: Cryptobiidae). *Rivista di Parasitologia*, *48*, 319–325.
287. Gupta, Neelima, & Gupta, D. K. (1987b). Dimorphism in *Trypanoplasma (Cryptobia) maguri* n. sp.: Effect on blood glucose level of host. *Rivista Iberica de Parasitología*, *47*, 317–324.
288. Gupta, Neelima, & Gupta, D. K. (1988). Prevalence of trypanosome infections in freshwater fishes of northern India: Cross transmission experiments on host specificity. *Archiv für Protistenkunde*, *136*, 203–209.
289. Gupta, Neelima, & Gupta, D. K. (1989). Sterol variations in two teleosts due to trypanosome infectivity. *Indian Journal of Parasitology*, *13*, 229–230.
290. Gupta, Neelima, & Gupta, D. K. (1990a). Trypanosomes as etiological agents of anemia in fresh water siluroid *Wallago attu*. In R. Hirano & I. Hanyu (Eds.), *Proceedings of the Second Asian Fisheries Forum* (pp. 709–711). A.F.S.
291. Gupta, Neelima, & Gupta, D. K. (1990b). Macro-environmental influence on trypanosome infectivity of some fresh water fishes of Northern India. *Advances in Biosciences*, *9*, 29–36.

292. Gupta, Neelima, & Gupta, D. K. (1990c). Concomitant haemoflagellates in *Mystus vittatus*: Degree of infectivity, inhibition and host specificity. *Rivista Iberica de Parasitología*, *50*, 9–13.

293. Gupta, Neelima, & Gupta, D. K. (1998). Hypocholesterolemia as a consequential event of haemoprotozoan infection in a catfish. *Flora and Fauna*, *3*, 59–60.

294. Gupta, Neelima, & Gupta, D. K. (1999). PHA and Con A as mitogenic factors for the cultivation of lymphocytes of catfishes. *Asian Fisheries Science*, *12*, 9–14.

295. Gupta, Neelima, & Jairajpuri, D. S. (1981a). *Trypanosoma trichogasteri* n. sp. from a freshwater telecost, *Trichogaster fasciata*. *Indian Journal of Parasitology*, *5*, 35–36.

296. Gupta, Neelima, & Jairajpuri, D. S. (1981b). *Trypanosoma singhii* n. sp. from the freshwater teleost, *Heteropneustes fossilis*. *Indian Journal of Parasitology*, *5*, 251–253.

297. Gupta, Neelima, & Jairajpuri, D. S. (1982a). *Trypanosoma aligaricus* n. sp. from the freshwater murrel, *Ophiocephalus punctatus* Bloch. *Archiv für Protistenkunde*, *125*, 109–114.

298. Gupta, Neelima, & Jairajpuri, D. S. (1982b). A new polymorphic trypanosome from an Indian freshwater fish, *Wallago attu* (Bl. & Schn.). *Rivista di Parasitologia*, *43*, 287–293.

299. Gupta, Neelima, & Jairajpuri, D. S. (1983). *Trypanosoma batrachi* Qadri, 1962 and its effect on the biochemical composition of the blood of *Clarias batrachus*. *Acta Protozoologica*, *22*, 79–85.

300. Gupta, Neelima, & Jairajpuri, D. S. (1985). Two new trypanosomes from Indian freshwater fishes. *Indian Journal of Parasitology*, *9*, 207–211.

301. Gupta, Neelima, & Pilarczyk, A. (1994). Effect of flagellates on the growth of table carp, *Cyprinus carpio*, in ponds from Poland. In *The Third Asian Fisheries Forum* (pp. 365–368). Asian Fisheries Society.

302. Gupta, Neelima, & Saraswat, H. (1991). *Trypanosoma rohilkhandae* n. sp. from freshwater teleost fish *Channa (Ophiocephalus) punctatus*. *Himalayan Journal of Environmental Zoology*, *5*, 49–53.

303. Gupta, Neelima, & Saraswat, H. (1993). Haemoprotozoosis in *Clarias batrachus* (Linn.) from freshwater ponds of Rohilkhand region, India. *Himalayan Journal of Environmental Zoology*, *7*, 47–52.

304. Gupta, Neelima, Gupta, D. K., & Jairajpuri, D. S. (1983). Effect of age of freshwater fishes on the incidence of trypanosomes. *Indian Journal of Parasitology*, *7*, 101–102.

305. Gupta, Neelima, Gupta, D. K., & Saraswat, H. (2000). Hypoglycemia in *Heteropneustes fossilis* parasitized by two new species of parasites (*Trypanosoma karelensis* n. sp. and *Myxosoma fossili* n. sp.). In U. P. Singh, R. S. Chauhan, & A. P. Sharma (Eds.), *Proceedings of the National Symposium on Fish Health Management & Sustainable Aquaculture* (pp. 179–186). Pantnagar.

306. Gupta, Neelima, Gupta, D. K., & Sharma, R. K. (1987). On the division of trypanosomes in *Colisa fasciata in vivo*. *Angewandte Parasitologie*, *28*, 15–19.

307. Gupta, Neelima, Gupta, D. K., & Yadav, P. (1998). New records of some haemoparasites affecting *Puntius ticto* (Pisces: Cyprinidae) in India: Observations on interaction and displacement of species. *Research and Reviews in Parasitology*, *58*, 103–108.

308. Gupta, Neelima, Saraswat, H., & Gupta, D. K. (1998). A new variety of *Trypanosoma* (Kinetoplastida: Trypanosomatidae) from the blood of *Colisa fasciatus* and its role in inducing haematological changes. *Journal of Parasitology and Applied Animal Biology*, *7*, 33–40.

309. Gupta, Neelima. (1981). *Trypanosome Parasites of Some Fishes of Aligarh* (Doctoral dissertation, Aligarh Muslim University, Aligarh). 137 pp.

310. Gupta, Neelima. (1986a). *Trypanosoma colisi* sp. n. from a freshwater fish *Colisa fasciata* Bloch. *Rivista Iberica de Parasitología*, *46*, 1–6.

311. Gupta, Neelima. (1986b). Trypanosome infectivity in relation to environmental factors. In *Proceedings of the National Symposium on Fish* (pp. 38–41).

312. Gupta, Neelima. (2003). Impact of concurrent infectivity on haemato-indices of cold blooded vertebrates. In B. N. Pandey & P. S. Murthy (Eds.), *Contributions to Biosciences* (pp. 1–14). ZSI Pub.

313. Gupta, Neelima. (2006). Historical review of piscine trypanosomiasis and survey of Indian *Trypanosoma*. *Journal of Parasitic Diseases*, *30*(2), 101–115.

314. Gutteridge, W. E., & Coombs, G. H. (1977). *Biochemistry of Parasitic Protozoa*. University Park Press.

315. Guttman, H. N. (1963). Experimental glimpses at the lower Trypanosomatidae. *Experimental Parasitology*, *14*, 129–142.

316. Guttowa, A. (1967). Experimental co-infection of Copepoda naturally infected with *Proteocephalus* sp. with the larvae of *Diphyllobothrium latum* (L). *Acta Parasitologica*, *14*, 399–404.

317. Haas, W. (1994). Physiological analysis of host-finding behaviour in

trematode cercariae: Adaptations for transmission success. *Parasitology, 109*, S15–S29.

318. Hadden, J. W. (1977). Cyclic nucleotides in lymphocyte proliferation and differentiation. In J. Hadden, R. Coffrey, & F. Spreafico (Eds.), *Immunopharmacology* (pp. 1–28). Plenum Press.
319. Hadden, J. W. (1978). The action of immunopotentiators *in vitro* on lymphocyte and macrophage activation. In G. H. Werner & F. Flocks (Eds.), *The Pharmacology of Immunoregulation* (pp. 321–350). Academic Press.
320. Hafner, M. S., Sudman, P. D., Villablance, F. X., Spradling, T. A., Demastes, J. W., & Nadler, S. A. (1994). Disparate rates of molecular evolution in co-speciating hosts and parasites. *Science, 265*(5179), 1087–1090.
321. Haldar, D. P., & Mukherjee, M. (1979). An annotated list of protozoan parasites from *Ophicephalus punctatus* Bloch, a common food-fish. *Archiv für Protistenkunde, 121*, 392–400.
322. Halevy, S. (1962). Comparative studies of lipid metabolism of Trypanosomatidae. *Bulletin of the Research Council of Israel, 10E2*, 65–68.
323. Hall, R. P. (1953). *Protozoology*. Prentice Hall, Inc.
324. Halvorsen, O. (1976). Negative interaction amongst parasites. In C. R. Kennedy (Ed.), *Ecological Aspects of Parasitology* (pp. 99–114). North-Holland Publishing Company.
325. Halvorsen, O., & Anderson, K. (1974). Some effects of population density in infections of *Diphyllobothrium dendriticum* (Nitzsch) in golden hamster (*Mesocricetus auratus* Waterhouse) and common gull (*Larus canus* L.). *Parasitology, 69*, 149–160.
326. Halvorsen, O., & Macdonald, S. (1972). Studies of the helminth fauna of Norway XXVI: The distribution of *Cyathocephalus truncatus* (Pallas) in the intestine of brown trout (*Salmo trutta* L.) Norway. *Journal of Zoology, 20*, 265–272.
327. Hamers, R. (1995). Studies on carp lymphoid cells from kidney, peripheral blood and spleen stimulated *in vitro* with blood parasites and super-antigen. *Bulletin of the European Association of Fish Pathologists, 14*, 191–194.
328. Hamilton, P. B., Stevens, J. R., Gaunt, M. W., Gidley, J., & Gibson, W. C. (2004). Trypanosomes are monophyletic: Evidence from genes for glyceraldehyde phosphate dehydrogenase and small subunit rRNA. *International Journal for Parasitology, 34*, 1393–1404.
329. Hamilton, W. D., & Zuk, M. (1982). Heritable true fitness and bright birds:

A role for parasites. *Science, 218*, 384–387.

330. Hamilton, W. D., Axelrod, R., & Tanese, R. (1990). Sexual reproduction as an adaptation to resist parasites (a review). *Proceedings of the National Academy of Sciences, U.S.A., 87*, 3566–3573.
331. Hansen, J. D., Landis, E. D., & Philips, R. B. (2005). Discovery of a unique Ig heavy-chain isotype (IgT) in rainbow trout: Implications for a distinctive B cell developmental pathway in teleost fish. *Proceedings of the National Academy of Sciences, U.S.A., 102*, 6919–6924.
332. Harris, J. W., & Kellermeyer, R. W. (1970). In *The Red Cell: Production, Metabolism, Destruction: Normal and Abnormal* (p. 98). Harvard University Press.
333. Harvey, S. C. (1949). The carbohydrate metabolism of *Trypanosoma lippicum*. *Journal of Biological Chemistry, 179*, 435–453.
334. Hasan, R., & Qasim, S. Z. (1962). *Trypanosoma punctati* n. sp. from the fish *Ophicephalus punctatus* Bloch, a fresh water murrel of India. *Z. Parasitenkd, 22*, 118–122.
335. Hassam, A. A., Bamidele, A., & Adegbaju, W. A. (2007). Haemoparasites of *Clarias gariepinus* and *Syndontis clarias* from Lekki lagoon, Lagos, Nigeria. *Journal of American Science, 3*(3), 61–67.
336. Havelka, J., Tesarcik, J., & Volf, F. (1965). Research on antiparasitic and antifungal control measurements. I. Investigation of the action of new antiparasitic medicaments on *Cryptobia cyprini* Plehn, 1903 (syn.: *Trypanoplasma cyprini* Plehn, 1903). *Papers of the Vyzkumny ustav rybarsky a hydrobiologicky, Vodnany, 5*, 67–87.
337. Hawking, F. (1963). Drug resistance of *Trypanosoma congolense* and other trypanosomes to quinapyramine, phenanthridines, berenil and other compounds in mice. *Annals of Tropical Medicine and Parasitology, 57*, 262–267.
338. Healing, T. D. (1981). Infections with blood parasites in the small British rodents *Apodemus sylvaticus*, *Clethrionomys glareolus* and *Microtus agrestis*. *Parasitology, 83*, 179–180.
339. Hedin, S. G. (1891). The Haematokrit: A new apparatus for the investigation of blood. *Skandinavisches Archiv für Physiologie, 2*, 134–140.
340. Helfert, S., Estévez, A. M., Bakker, B., Michels, P., & Clayton, C. (2001). Roles of triosephosphate isomerase and aerobic metabolism in *Trypanosoma brucei*. *Biochemical Journal, 357*, 117–125.
341. Henry, H. (1913a). A new haemosporidian parasite from *Scomber scomber*, the common mackerel. *Journal of Pathology and Bacteriology,*

18, 228–231.

342. Henry, H. (1913b). A summary of the blood parasites of British sea-fish. *Journal of Pathology and Bacteriology, 18*, 228–231.

343. Herbert, W. J., & Inglis, M. D. (1973). Immunization of mice against *T. brucei* infection, by the administration of released antigen adsorbed to erythrocytes. *Transactions of the Royal Society of Tropical Medicine and Hygiene, 67*, 268.

344. Herbert, W. J., & Lumsden, W. H. R. (1976). *Trypanosoma brucei*: A rapid matching method for estimating the host's parasitaemia. *Experimental Parasitology, 40*, 427–431.

345. Herrera, L., & Urdaneta-Morales, S. (1997). Synantropic rodent reservoirs of *Trypanosoma (Schizotrypanum) cruzi* in the valley of Caracas, Venezuela. *Revista do Instituto de Medicina Tropical de São Paulo, 39*, 279–282.

346. Hewitt, R. I., Gumble, A. R., Taylor, L. H., & Wallace, W. S. (1956). In *Antibiotics, Annual* (p. 722).

347. Hickey, C. R. (1976). Fish haematology, its uses and significance. *New York Fish and Game Journal, 23*(2), 170–175.

348. Hickman, C. P. Sr., Hickman, C. P. Jr., Hickman, F. M., & Roberts, L. S. (1979). *Integrated Principles of Zoology*. C. V. Mosby Company.

349. Hirono, I., Nam, B. H., Enomoto, J., Uchino, K., & Aoki, T. (2003). Cloning and characterization of a cDNA encoding Japanese flounder *Paralichthys olivaceus* IgD. *Fish & Shellfish Immunology, 15*, 63–70.

350. Hirumi, H., Doyle, J. J., & Hirumi, K. (1977). African trypanosomes: Cultivation of animal-infective *Trypanosoma brucei in vitro*. *Science, 196*, 992–994.

351. Hoare, C. A. (1955). Intraspecific biological groups in pathogenic Protozoa. *Refuah Veterinar, 12*, 263–270.

352. Hoare, C. A. (1964). Morphological taxonomic studies on mammalian trypanosomes. X. Revision of the systematics. *Journal of Protozoology, 11*, 200–207.

353. Hoare, C. A. (1972). *The Trypanosomes of Mammals: A Zoological Monograph*. Blackwell.

354. Hoare, C. A., & Wallace, F. G. (1966). Developmental stages of Trypanosomatid flagellates: A new terminology. *Nature, 212*, 1385–1386.

355. Hochberg, M. E., Michalakis, Y., & De Meeus, T. (1992). Parasitism as a constraint on the rate of life history evolution. *Journal of Evolutionary Biology, 5*, 491–504.

356. Hoffman, G. L., & Meyer, F. P. (1974). *Parasites of Freshwater Fishes: A Review of Their Control and Treatment*. T.F.H. Publications, Inc.
357. Hoffmeyer, I. (1973). Interaction and habitat selection in the mice *Apodemus flavicollis* and *A. sylvaticus*. *Oikos, 24*, 108–116.
358. Hogstedt, G. (1980). Prediction and test of the effects of interspecific competition. *Nature, 283*, 64–66.
359. Hollande, A. (1952). Ordre des Bodonides (Bodonidea ord. nov.). In P. P. Grasse (Ed.), *Traite de Zoologie* (Vol. I, Fasc. I, pp. 1–1071). Masson.
360. Holmes, J. C. (1961). Effects of concurrent infections on *Hymenolepis diminuta* (Cestoda) and *Moniliformis dubius* (Acanthocephala). I. General effects and comparison with crowding. *Journal of Parasitology, 47*, 209–216.
361. Holmes, J. C. (1962). Effects of concurrent infections on *Hymenolepis diminuta* (Cestoda) and *Moniliformis dubius* (Acanthocephala). II. Growth. *Journal of Parasitology, 48*, 87–96.
362. Holmes, J. C. (1973). Site selection by parasitic helminths: Interspecific interactions, site segregation, and their importance to the development of helminth communities. *Canadian Journal of Zoology, 51*, 333–347.
363. Holmes, J. C. (1986). The structure of helminth communities. In M. J. Howell (Ed.), *Parasitology-Quo vadit*: Proceedings of the 6th International Congress of Parasitology (pp. 203–208). Australian Academy of Sciences.
364. Holmes, J. C., & Zohar, S. (1990). Pathology and host behaviour. In C. J. Barnard & J. M. Behnke (Eds.), *Parasitism and Host Behavior* (pp. 34–64). Taylor and Francis.
365. Homer, M. J., Aguilar, D. I., Telford, S. R. I., Krause, P. J., & Persing, D. H. (2000). Babesiosis. *Clinical Microbiology Reviews, 13*, 451–469.
366. Hontzeas, N., Feng, S., & Woo, P. T. K. (2001). Inhibitory effects of a monoclonal antibody (MAb-001) on *in vitro* oxygen consumption and multiplication of the pathogenic haemoflagellate, *Cryptobia salmositica* Katz. *Journal of Fish Diseases, 24*(7), 391–398.
367. Hoppe, J. O., & Chapman, C. W. (1947). Role of glucose in acute parasitaemic death of the rat infected with *Trypanosoma equiperdum*. *Journal of Parasitology, 33*, 509–516.
368. Houston, A. H. (1997). Review: Are the classical haematological variables acceptable indicators of fish health? *Transactions of the American Fisheries Society, 126*, 879–894.
369. Hughes, A. L., & Piontkivska, H. (2003). Phylogeny of *Trypanosomatidae* and *Bodonidae* (Kinetoplastida) based on 18S rRNA: Evidence for

paraphyly of *Trypanosoma* and six other genera. *Molecular Biology and Evolution, 20*, 644–652.

370. Hughes, K. P., & Smith, S. A. (2003). Common and emerging diseases in commercially-cultured summer flounder, *Paralichthys dentatus*. *Journal of Applied Aquaculture, 14*(2/3), 163–178.

371. Humiczewska, M., & Kuzna-Grygiel, W. (1995). Przypadki babesiosis u czzelowieka. Zarazenie Mieszkanca Szczecina przez *Babesia microti*. In *Borrelioza z Lyme i Inne Choroby Przenoszone Przez Kleszcze*: Międzynarodowe Sympozium (Streszczenia) (pp. 132–133). Białowieża, Poland.

372. Humiczewska, M., & Kuzna-Grygiel, W. (1997). Przypadek importowanej ludzkiej babesiozy w Polsce. *Wiadomości Parazytologiczne, 43*, 227–229.

373. Hurd, H., & Fogo, S. (1991). Changes induced by *Hymenolepis diminuta* (Cestoda) in the behaviour of intermediate host *Tenebrio molitor* (Coleoptera). *Canadian Journal of Zoology, 69*, 2291–2294.

374. Hutchinson, G. E. (1957). Concluding remarks. *Cold Spring Harbor Symposia on Quantitative Biology, 22*, 415–427.

375. Ijagbone, I. F., & Agbede, S. A. (2000). A case of congenital transmission of *Trypanosoma brucei* in mice. *Tropical Veterinarian, 18*, 37–38.

376. Ikede, B. O., & Losos, G. J. (1972). Spontaneous canine trypanosomiasis caused by *T. brucei*: Meningo-encephalomyelitis with extra-vascular localization of trypanosomes in the brain. *Bulletin of Epizootic Diseases of Africa, 20*, 221–228.

377. Ikede, B. O., Lule, M., & Jerry, R. J. (1977). Anemia in trypanosomiasis: Mechanism of erythrocyte destruction in mice infected with *Trypanosoma congolense* or *T. brucei*. *Acta Tropica, 34*, 53–60.

378. Ikeziani, O. (1946). Studies in trypanosomiasis III: The plasma whole blood and erythrocyte potassium of rats during the course of infection with *T. brucei* and *T. equiperdum*. *Journal of Parasitology, 32*, 379–382.

379. Ingles, L. G. (1936). Worm parasite of California amphibia. *Transactions of the American Microscopical Society, 55*, 73–92.

380. Ioff, I. G., Bozhenko, V. B., & Levashev, M. M. (1926). *Trypanoplasma* acipenseris sp. n.–a new parasite of sterlet blood. *Russkhyi Gidrobiologhicheskyi Zhurnal, 5*, 225–233.

381. Islam, A. K. M. N., & Woo, P. T. K. (1991a). Anemia and its mechanism in goldfish *Carassius auratus* infected with *Trypanosoma danilewskyi*. *Dis. Aquat. Org., 11*, 37–44.

382. Islam, A. K. M. N., & Woo, P. T. K. (1991b). Anorexia in goldfish

Carassius auratus infected with *Trypanosoma danilewskyi*. *Dis. Aquat. Org.*, *11*, 45–48.

383. Islam, A. K. M. N., & Woo, P. T. K. (1991c). *Trypanosoma danilewskyi* in *Carassius auratus*: The nature of protective immunity in recovered goldfish. *J. Parasitol.*, *77*, 258–262.

384. Islam, A. K. M. N., & Woo, P. T. K. (1992). Effects of temperature on the *in vitro* and *in vivo* multiplication of *Trypanosoma danilewskyi*, Laveran and Mesnil. *Folia Parasitol.*, *39*, 1–12.

385. Isoun, T. T. (1975). The histopathology of experimental disease produced in mice infected with *Trypanosome vivax*. *Acta Trop. (Basel)*, *32*, 267–272.

386. Jackson, J. A. (1999). Analysis of parasite host-switching: Limitations on the use of phylogenies. *Parasitology*, *119*, S111–S123.

387. Jaffe, J. J. (1967). Sensitivity of *Trypanosoma equiperdum* to the action of tumor-inhibitory antibiotics *in vivo*. *Nature*, *London*, *213*, 704–705.

388. Jain, N. C. (1986). *Schalm's Veterinary Haematology* (4th ed.). K. M. Verghese Comp.

389. Jalali, B., & Barzegar, M. (2006). Fish parasites of Zarivar lake. *J. Agr. Sci. Tech.*, *8*(1), 47–59.

390. Janovy, J. J., Clopton, R. E., & Percival, T. J. (1992). The roles of ecological and evolutionary influences in providing structure to parasite species assemblages. *J. Parasitol.*, *78*, 630–640.

391. Jansen, M. E., & Burreson, E. M. (1990). Parasites of summer flounder, *Paralichthys dentatus*, in the Chesapeake Bay. *J. Helminthol. Soc. Wash.*, *57*(1), 31–39.

392. Jastrzebski, M. (1984). Polymorphism of flagellates *Trypanosoma* occurring in the blood of crucian carp (*C. auratus*), perch (*P. fluviatilis*) and stone perch (*Acerina cereena*). *Wiad. Parazytol.*, *30*(2), 172–182.

393. Jayasri, M. Y., & Mathur, B. N. (1993). *Myxosoma aravalea*, a new myxosporean parasite in kidney of fresh water fish, *Puntius sophore* (Ham) from Rajasthan (India). *Indian J. Parasitol.*, *17*, 121–124.

394. Jesudhasan, P. R. R., Tan, C. W., & Woo, P. T. K. (2007). A metalloproteinase gene from the pathogenic piscine haemoflagellate, *Cryptobia salmositica*. *Parasitol. Res.*, *100*(4), 899–904.

395. Johnson, T. O., & Ekanem, J. T. (2003). Effect of Tetracycline on late stages African trypanosomiasis in rats. *Biokemistry*, *14*(1), 51–56.

396. Johnston, T. H. (1909). On some haemogregarines from Australian reptiles. *Proc. Linn. Soc. N.S.W.*, *34*(3), 400–410.

397. Jones, S. R. M., & Woo, P. T. K. (1987). The immune response of rainbow

trout, *Salmo gairdneri* Richardson, to the haemoflagellate, *Cryptobia salmositica* Katz, 1951. *J. Fish Dis.*, *10*, 395–402.

398. Jones, S. R. M., & Woo, P. T. K. (1990). *Cryptobia cataractae* from the blood of *Semotilus atromaculatus*: Structure and division in the fish. *J. Protozool.*, *76*, 645–649.

399. Jones, S. R. M., & Woo, P. T. K. (1991). Culture characteristics of *Trypanosoma catostomi* and *T. phaleri* from North American freshwater fishes. *Parasitology*, *103*, 237–243.

400. Jones, S. R. M., & Woo, P. T. K. (1992). Vector specificity of *Trypanosoma catostomi* and its infectivity to fresh water fishes. *J. Parasitol.*, *78*, 87–92.

401. Jones, S. R. M., & Woo, P. T. K. (1993). Biology of *Trypanosoma burresoni* n.sp. from the American eel, *Anguilla rostrata*. *J. Parasitol.*, *79*, 542–547.

402. Jones, S. R. M., Palmen, M., & van Muiswinkel, W. B. (1993). The effects of inoculum route and dose on the immune response of common carp, *Cyprinus carpio* to the blood parasite, *Trypanoplasma borreli*. *Vet. Immunol. Immunopathol.*, *36*, 369–378.

403. Jones, S. R. M., Wiegertjes, G. F., & Van-Muiswinkel, W. B. (1995). Modulation of carp (*Cyprinus carpio* L.) cellular immune function *in vitro* caused by the blood parasite, *Trypanoplasma borreli* Laveran and Mesnil. *Fish Shellfish Immunol.*, *5*(5), 381–383.

404. Jones, S. R. M., Woo, P. T. K., & Stevenson, R. M. (1986). Immunosuppression in *Salmo gairdneri* Richardson, caused by the haemoflagellate *Cryptobia salmositica* Katz, 1951. *J. Fish Dis.*, *9*, 431–438.

405. Joshi, B. D. (1973). Studies on blood of freshwater fishes of India (Ph.D. thesis, University of Lucknow, Lucknow, India).

406. Joshi, B. D. (1976a). Two new species of trypanosomes from fresh water teleosts. *Indian J. Zoot.*, *17*, 5–10.

407. Joshi, B. D. (1976b). On two new species of trypanosomes from two freshwater teleosts. *J. Anim. Morphol. Physiol.*, *25*, 1–7.

408. Joshi, B. D. (1978). Two new species of trypanosomes from two fresh water teleosts. *Indian J. Zoot.*, *27*, 5–10.

409. Joshi, B. D. (1979a). On the occurrence of trypanosomes in the blood of some freshwater teleosts of Lucknow (UP, India). *Proc. Indian Acad. Sci.*, *88B*, 59–63.

410. Joshi, B. D. (1979b). Effects of starvation on some haematological values in some freshwater fishes. *Indian J. Exp. Biol.*, *17*, 968–969.

411. Joshi, B. D. (1979c). On the histomorphological variation of the circulating

blood cells of *Channa punctatus* infected with trypanosomes. *Indian J. Zoot.*, *20*, 133–136.

412. Joshi, B. D. (1980). Arneth counts of the granulocytes, leukocytes of 33 species of freshwater teleosts. *Indian J. Anim. Res.*, *14*, 43–47.
413. Joshi, B. D. (1981). On some biochemical changes in the serum and liver of a fresh-water fish infected with trypanosomes. *Acta Parasitol.*, *28*, 197–201.
414. Joshi, B. D. (1982a). Changes in the blood glucose and liver glycogen contents of healthy and trypanosome-infected fish, *Clarias batrachus*, following intramuscular injection of glucose solution. *Angewandte Parasitologie, 23*, 121–124.
415. Joshi, B. D. (1982b). Three new species of hematozoans from freshwater teleosts. *Proceedings of the Indian Academy of Sciences, 91B*, 397–406.
416. Joshi, B. D. (1983). *Trypanosoma rupicola* (sp. nov.) from a hill stream fish *Nemacheilus rupicola* (Hora). *Current Science, 52*, 141–142.
417. Joshi, B. D. (1985). On the occurrence of trypanosomes from certain hill-stream fishes of Almora and related alterations in some blood values of a fish *Nemacheilus rupicola* (Hora). *Uttar Pradesh Journal of Zoology, 5*(2), 199–203.
418. Joshi, B. D. (1986). Effect of desiccation on some haematological values of freshwater catfish *Heteropneustes fossilis*. *Comparative Physiology and Ecology, 11*, 65–67.
419. Joshi, B. D. (1989a). Physio-pathological studies on the blood of a few hill-stream teleosts. *Proceedings of the National Symposium, 127–137.*
420. Joshi, B. D. (1989b). Haematological studies on *Clarias batrachus* infected with trypanosomes. *Proceedings of the National Symposium on Animal Protection and Changing Environment, 203–210.*
421. Joshi, B. D., & Dabral, R. (1981). Some haematological changes in a freshwater catfish, *Heteropneustes fossilis*, infected with the trypanosome *Trypanosoma maguri*. *Proceedings of the Indian Academy of Sciences, 90*, 295–301.
422. Joshi, B. D., & Sharma, T. (1982). Sex-related variations in some haematological values of the hillstream fish, *Tor putitora*, during the pre-spawning period. *Proceedings of Life Science and Human Well-Being, 16, 17.*
423. Joshi, B. D., & Tandon, R. S. (1977). Seasonal variations in the haematological values of freshwater fishes, *Heteropneustes fossilis* and *Mystus vittatus*. *Comparative Physiology and Ecology, 2*, 22–26.

424. Joshi, B. D., & Tandon, R. S. (1980). Effects of trypanosome infection on some haematological values in ten species of freshwater teleosts. *Journal of Animal Morphology and Physiology, 27*, 39–44.

425. Joshi, B. D., Chaturvedi, L. D., & Dabral, R. (1980). Some haematological values of *Clarias batrachus* following its sudden transfer to varying temperatures. *Indian Journal of Experimental Biology, 18*(1), 76–77.

426. Joshi, B. D., Dabral, R., & Sharma, T. (1982). Haematological response of a freshwater teleost, *Clarias batrachus*, following partial hepatectomy. *Journal of Animal Morphology and Physiology, 29*, 149–156.

427. Joshi, B. D., Sharma, T., & Giri, G. S. (1983). On some haematological values of *Nemacheilus rupicola* heavily infested with helminthic cysts. *Indian Journal of Physiology and Natural Sciences, 3*, 38–39.

428. Joshi, S. S., & Singh, B. (2000). Evaluation of therapeutic and chemoprophylactic efficacy of certain drugs against clinical surra in buffaloes. *Indian Veterinary Journal, 77*, 895–897.

429. Kabata, Z. (1985). *Parasites and Diseases of Fish Cultured in the Tropics*. Taylor and Francis Ltd.

430. Kalman, D., Sreter, T., Szell, Z., & Egyed, L. (2003). *Babesia microti* infection of anthropophilic ticks (*Ixodes ricinus*) in Hungary. *Annals of Tropical Medicine and Parasitology, 97*, 317–319.

431. Kaminsky, R., Gumm, I. D., Zweygarth, E., & Chuma, F. (1990). A drug incubation infectivity test (DIIT) for assessing resistance in trypanosomes. *Veterinary Parasitology, 34*, 335–343.

432. Kang, Y. S., & Park, E. H. (1975). Leukocyte culture of the eel without autologous serum. *Japanese Journal of Genetics, 50*, 159–161.

433. Kapila, R., Kapila, S., & Basade, Y. (2000). Sex-related haematological variations in Himalayan golden mahseer, *Tor putitora* (Ham.). *Indian Journal of Fisheries, 47*(10), 81–85.

434. Kapoor, V. C. (1998). *Theory and Practice of Animal Taxonomy*. Oxford and IBH Publishing Co. Pvt. Ltd.

435. Karbowiak, G. (2004). Zoonotic reservoir of *Babesia microti* in Poland. *Polish Journal of Microbiology, 53*(Suppl), 61–65.

436. Karbowiak, G., & Sinski, E. (1994). Rola kleszczsy i drobnych ssakow w szerzeniu sie zakazen *Borrelia burgdorferi* i *Babesia microti*. *Przeglad Epidemiologiczny, 48*, 219–224.

437. Karbowiak, G., & Sinski, E. (1995). Ticks (*Ixodidae*) in zoonotic propagation of piroplasms *Babesia microti* and *Babesia divergens*. *Wiadomosci Parazytologiczne, 41*(3), 321–327.

438. Karbowiak, G., & Sinski, E. (1996a). Occurrence and morphological characteristics of a *Trypanosoma evotomys* strain from North Poland. *Acta Parasitologica, 41*(3), 105–107.

439. Karbowiak, G., & Sinski, E. (1996b). The finding of *Babesia microti* in bank vole *Clethrionomys glareolus* in the district of Mazury lakes (Poland). *Acta Parasitologica, 41*, 50, 51.

440. Karbowiak, G., & Wita, I. (1998). The occurrence of *Herpetosoma* trypanosomes in some parts of Poland. *18 Parasitologische Tagung, Deutsche Gesellschaft für Parasitologie e.V.*, Drezno, 24–28.

441. Karbowiak, G., & Wita, I. (2001a). Przypadki zarazenia szczurow wedrownych *Rattus norvegicus* (Berkenhout, 1769) swidrowcem *Trypanosoma lewisi* (Kent, 1880) Laveran i Mesnil, 1901. Na terenie aglomeracji warszawskiej. *Wiadomości Parazytologiczne, 47*(3), 377–382.

442. Karbowiak, G., & Wita, I. (2001b). Ekologiczne aspekty zarazenia nornicy rudej *Clethrionomys glareolus* (Schreber, 1780) swidrowcem *Trypanosoma (Herpetosoma) evotomys* Haswen, 1912. *Wiadomości Parazytologiczne, 47*(4), 789–795.

443. Karbowiak, G., & Wita, I. (2001c). Parazytofauna krwi drobnych gryzoni lesnych w Polsce. *Polska Akademia Nauk, Dzialalnosc Naukowa. Wybrane Zagadnienia, 11*, 55–57.

444. Karbowiak, G., & Wita, I. (2004). *Trypanosoma (Herpetosoma) grosi* kosewiense sub sp.n., the parasite of the yellow necked mouse *Apodemus flavicollis* (Melchior, 1834). *Acta Protozoologica, 43*, 173–178.

445. Karbowiak, G., Rychlik, L., Nowakowski, W., & Wita, I. (2005). Natural infections of small mammals with blood parasites on the borderland of boreal and temperate forest zones. *Acta Theriologica, 50*, 31–42.

446. Karbowiak, G., Slemenda, S., Pieniazek, N. J., Rychlik, L., Stanko, M., Sinski, E., Wita, I., & Czaplinska, U. (2004). Babesiosis of wild animals in Poland. *IX European Multicolloquium of Parasitology*, Valencia, Spain.

447. Karbowiak, G., Stanko, M., Fricova, J., Wita, I., & Czaplinska, U. (2003). The communities of blood parasites in field mouse *Apodemus agrarius*. *Wiadomosci Parazytologiczne, 49*, 407.

448. Karbowiak, G., Stanko, M., Rychlik, L., Nowakowski, W., & Siuda, K. (1999). The new data about the zoonotic reservoir of *Babesia microti* in small mammals in Poland. *Acta Parasitologica, 44*, 142–144.

449. Karbowiak, G., Wita, I., & Czaplinska, U. (2002). Protozoan parasites in the blood of the social vole (*Microtus socialis*) in Askania Nova

Reserve, Ukraine. In *Abstracts of the XII Conference of the Ukrainian Parasitological Society* (p. 131).

450. Karbowiak, G., Wita, I., & Czaplinska, U. (2004). Protozoan parasites in blood of the social vole (*Microtus socialis*) in Askania-Nova Reserve Ukraine. *Vestnik Zoologii, Suppl.*, 18, 186–187.
451. Karbowiak, G., Wita, I., Czaplinska, U., & Polishchuk, I. K. (1998). *Trypanosoma (Herpetosoma) sp.* in the social vole (*Microtus socialis*) for first time in Ukraine. *Wiadomości Parazytologiczne, 44*, 466.
452. Karbowiak, G., Wita, I., Czaplinska, U., Slemenda, S., Norman, I., & Pieniazek, J. (2004). Canine babesiosis in Poland. *Wiadomości Parazytologiczne, 50*, 52–54.
453. Karbowiak, G., Wita, I., Rychlik, L., & Gadomska, K. (1998). The morphology of *Herpetosoma* trypanosomes in small rodents in Poland. *Wiadomości Parazytologiczne, 44*, 467.
454. Karlsbakk, E., & Nylund, A. (2006). Trypanosomes infecting cod *Gadus morhua* L. in the North Atlantic: A resurrection of *Trypanosoma pleuronectidium* Robertson, 1906 and delimitation of *T. murmanese* (Nikitin, 1927), with a review of other trypanosomes from North Atlantic and Mediterranean teleosts. *Systematic Parasitology, 65*, 175–203.
455. Kartman, L. (1954). Observations on *Trypanosoma lewisi* and *Grahamella* sp. in the blood of rats from the Hamakua District, Island of Hawaii. *Journal of Parasitology, 40*, 571–579.
456. Katz, M. (1951). Two new haemoflagellates (Genus *Cryptobia*) from some western Washington teleosts. *Journal of Parasitology, 37*, 245–250.
457. Kaur, M. (1996). Haematological studies in a freshwater fish *Rita rita* (Ham.). (Ph.D. thesis). Magadh University, Bodh Gaya, Bihar.
458. Kellner, H. M., Eckert, H. G., & Volz, H. M. (1985). Studies in cattle on the disposition of the anti-*Trypanosoma*l drug diminazene diaceturate (Berenil). *Tropical Medicine and Parasitology, 36*, 199–204.
459. Kennedy, C. R. (1968). Population biology of the cestode *Caryophyllaeus laticeps* (Pallas, 1781) in dace, *Leuciscus leuciscus* L. of the river Avon. *Journal of Parasitology, 54*, 538–543.
460. Kennedy, C. R. (1970). The population biology of helminths of British freshwater fish. *Symposia of the British Society for Parasitology, 9*, 145–159.
461. Kennedy, C. R. (1975). *Ecological Animal Parasitology*. Blackwell Scientific Publications.
462. Kent, W. S. (1880). *A Manual of the Infusoria*. Bogue.

463. Kesseylitz, G. (1906). Generations und Wirtwechsel von *Trypanoplasma borreli*, Laveran und Mesnil. *Archiv für Protistenkunde, 7*, 1–74.

464. Khan, A. R., Chishti, M. Z., Ahmad, F., Rashid, M., & Bakshi, S. (2004). Seasonal occurrence of helminth parasites in *Schizothorax* in Dal Lake in Kashmir. *Journal of Parasitic Diseases, 28*, 23–28.

465. Khan, R. A. (1972). On a *Trypanosoma* from the Atlantic cod, *Gadus morhua*. *Canadian Journal of Zoology, 50*, 1051–1054.

466. Khan, R. A. (1976). The life cycle of *Trypanosoma murmanensis* Nikitin. *Canadian Journal of Zoology, 54*, 1840–1849.

467. Khan, R. A. (1977a). Susceptibility of marine fish to trypanosomes. *Canadian Journal of Zoology, 55*, 1235–1241.

468. Khan, R. A. (1977b). Blood changes in Atlantic cod (*Gadus morhua*) infected with *Trypanosoma murmanensis*. *Journal of the Fisheries Research Board of Canada, 34*, 2193–2196.

469. Khan, R. A. (1978). Longevity of *Trypanosoma murmanensis* in the marine leech, *Johanssonia* sp. *Canadian Journal of Zoology, 56*, 2061–2063.

470. Khan, R. A. (1985). Pathogenesis of *Trypanosoma murmanensis* in marine fish of the northwestern Atlantic following experimental transmission. *Canadian Journal of Zoology, 63*, 2141–2144.

471. Khan, R. A. (1987). Effects of chronic exposure to petroleum hydrocarbons on two species of marine fish infected with a haemoprotozoan, *Trypanosoma murmanensis*. *Canadian Journal of Zoology, 65*, 2703–2709.

472. Khan, R. A. (1991). Prevalence of trypanosome infection in the marine fish leech *Johanssonia arctica* off eastern Newfoundland (Canada). *Canadian Journal of Zoology, 69*, 1194–1198.

473. Khan, R. A., & Lacey, D. (1986). Effect of concurrent infections of *Lernaeocera branchialis* (Copepoda) and *T. murmanensis* (Protozoa) on Atlantic cod, *Gadus morhua*. *Journal of Wildlife Diseases, 22*, 201–208.

474. Khan, R. A., & Newman, M. W. (1982). Blood parasites from fish of the Gulf of Maine to Cape Hatteras, northwest Atlantic Ocean with notes on the distribution of fish haematozoa. *Canadian Journal of Zoology, 60*, 396–402.

475. Khan, R. A., Barrett, M., & Murphy, J. (1980). Blood parasites of fish from the Northwestern Atlantic Ocean. *Canadian Journal of Zoology, 58*(5), 770–781.

476. Khan, R. A., Murphy, J., & Taylor, D. (1980). Prevalence of a trypanosome in Atlantic cod (*Gadus morhua*) especially in relation to stocks in the Newfoundland area. *Canadian Journal of Fisheries and Aquatic Sciences,*

37, 1467–1475.

477. Khan, S. H. (1974). Variations in the blood cell counts of the catfish, *Clarias batrachus* during starvation. *Indian Journal of Zoology, 2*, 59–64.
478. Khan, S. H., Siddique, A. Q., & Hasnain, A. (1972). The hypotonic fragility of some fish erythrocytes. *Proceedings of the Indian Academy of Sciences, 75B*, 266–270.
479. Khaybulyayev, K. Kh. (1970). On the role of leeches in the developmental cycle of blood parasites of fishes. *Parazitologiya, 4*, 13–17.
480. Khaybulyayev, K. Kh. (1976). On the specificity of trypanosomes and cryptobiids from the blood of fishes. *Proceedings of the II All-Union Congress of Soviet Protozoologists, Naukova Dumka, Kiev, 1*, 143–144.
481. Kierszenbaum, F., & Howard, J. G. (1976). Mechanisms of resistance against experimental *Trypanosoma cruzi* infection: The importance of antibodies and antibody-forming capacity in the Biozzi high and low responder mice. *Journal of Immunology, 116*, 1208–1211.
482. Killick-Kendrick, R. (1964). The apparent loss of the kinetoplast of *Trypanosoma evansi*. *Annals of Tropical Medicine and Parasitology, 58*, 481–490.
483. Kinabo, L. D. B., & Bogan, J. A. (1987). Binding of isometamidium to calf thymus DNA and lipids: Pharmacological implications. *Journal of Veterinary Pharmacology and Therapeutics, 10*, 357–362.
484. Kinabo, L. D. B., & Bogan, J. A. (1988). The pharmacology of isometamidium. *Journal of Veterinary Pharmacology and Therapeutics, 11*, 233–245.
485. Kinabo, L. D. B., Bogan, J. A., McKellar, Q. A., & Murray, M. (1989). Relay bioavailability and toxicity of isometamidium residues: A model for human risk assessment. *Veterinary Human Toxicology, 31*, 417–421.
486. Kingston, N. (1991). A brief review of *Trypanosoma* (Megatrypanum) infections in ruminants in North America and Europe. *Wiadomości Parazytologiczne, 37*(2), 211–218.
487. Kingston, N., & Morton, J. K. (1973). Trypanosomes from elk (*Cervus canadensis*) in Wyoming. *Journal of Parasitology, 59*, 1132–1133.
488. Kingston, N., & Morton, J. K. (1975). *Trypanosoma cervi* sp. n. from elk (*Cervus canadensis*) in Wyoming. *Journal of Parasitology, 61*, 17–23.
489. Kingston, N., Bobek, B., Perzanowski, K., Wita, I., & Maki, L. (1992). Description of *Trypanosoma* (Megatrypanum) *stefanskii* sp. n. from roe deer (*Capreolus capreolus*) in Poland. *Journal of the Helminthological Society of Washington, 59*, 89–95.

490. Kingston, N., Drozdz, J., & Rutkowska, M. (1985). *Trypanosoma* sp. in Red Deer (*Cervus elaphus*) and Elk (*Alces alces*) in Poland. *Proceedings of the Helminthological Society of Washington, 51*(1), 144–145.

491. Kingston, N., Drozdz, J., Rutkowska, M., Wita, I., & Maki, L. (1992). Redescription of *Trypanosoma* (Megatrypanum) *wrublewskii* Wladimiroff et Yakimoff, 1909 from the European bison (*Bison bonasus* L.) from Puszcza Bialowieska, Poland. *Acta Parasitologica, 37*(4), 163–168.

492. Kipp, H. (1968). Untersuchungen zum Vorkommen, zur Artsspezifitat und Pathogenitat der Cryptbein und Trypanosomen bei der Scheie (*Tinca tinca* L.). Inaugural-dissertation, Tierarztliche Fakultat der Ludwig-Maxmilians-Universitat, München. 115 pp.

493. Kirshenblat, Ya. D. (1941). Specificity of parasites towards their hosts. *Usp. Sovrem. Biol., 14*, 271–294.

494. Kisielewska, K. (1970). On the theoretical foundations of Parasito-synecology. *Bulletin of the Polish Academy of Sciences. Series of Biological Sciences, 18*, 103–106.

495. Klatt, P., & Hadju, P. (1976). Pharmacokinetic investigation on diminazene and rolitetracycline in comparison to a combination of both. *Veterinary Record, 99*, 372–374.

496. Klontz, G. W. (1972). Haematological techniques and the immune response in rainbow trout. In *Diseases of Fish Symposium Zoological Society London, No. 30* (L. E. Mawdesley-Thomas, Ed.), Academic Press, London. pp. 89–99.

497. Koella, J. C., Sørensen, F. L., & Anderson, R. A. (1998). The malaria parasite, *Plasmodium falciparum* increases the frequency of multiple feeding of its mosquito vector, *Anopheles gambiae*. *Proceedings of the Royal Society of London. Biological Sciences, 265*, 763–768.

498. Korn, E. D., Greenblatt, C. L., & Lees, A. M. (1965). Synthesis of unsaturated fatty acids in the slime mould *Physarum polycephalum* and the zooflagellates *Leishmania tarentolae*, *Trypanosoma lewisi*, and *Crithidia* sp.: A comparative study. *Journal of Lipid Research, 6*, 43–50.

499. Koumansvandiepen, J. C. E., Egberts, E., Peixoto, B. R., Taverne, N., & Rombout, J. H. W. M. (1995). B-cell and immunoglobulin heterogeneity in carp (*Cyprinus carpio* L.)-an immuno(cyto)chemical study. *Developmental and Comparative Immunology, 19*, 97–108.

500. Kozloff, E. N. (1948). The morphology of *Cryptobia helices* Leidy, with an account of the fate of the extranuclear organelles in division. *Journal of Morphology, 8*, 253–279.

501. Krampitz, H. E. (1959). Über das europäische Waldmaustrypanosom, *Trypanosoma grosi* Laveran et Pettit, 1909 (Protomonadina, Trypanosomatidae). *Zeitschrift für Parasitenkunde, 19*, 232–258.

502. Krampitz, H. E. (1961). Kritisches zur taxonomischen und systematischen parasitischen Säugetiertrypanosomen mit besonderer Beachtung einiger der in Wühlmausen verbreiteten spezifischen Formen. *Zeitschrift für Tropenparasitologie, 12*, 117–138.

503. Krampitz, H. E., & Baumler, W. (1978). Occurrence, host range and seasonal prevalence of *Babesia microti* (Franca, 1912) in rodents of southern Germany. *Zeitschrift für Parasitenkunde, 58*, 15–33.

504. Krampitz, H. E., Buschmann, H., & Munchhoff, P. (1986). Gibt es latente Babesien-Infektionen beim Menschen in Süddeutschland? *Mitteilungen der Österreichischen Gesellschaft für Tropenmedizin und Parasitologie, 8*, 233–243.

505. Krebs, C. J. (1972). *Ecology: The Experimental Analysis of Distribution and Abundance*. Harper and Row, New York.

506. Krettli, A. U., & Pereira, F. E. L. (1981). Immunosuppression in protozoal infections. In M. Levandowsky & S. H. Hutner (Eds.), *Biochemistry and Physiology of Protozoa* (2nd ed., pp. 431). Academic Press.

507. Krier, J. P. (1977). *Parasitic protozoa* Vol. I. Taxonomy, kinetoplastids, and flagellates of fish. Academic Press.

508. Krier, J. P., & Baker, J. R. (1987). *Parasitic Protozoa*. Allen and Unwin, Inc.

509. Krishnamurthy, R., & Wahul, M. A. (1986). Studies on the morphology of some trypanoplasms from the blood of freshwater fishes in Maharashtra II. From fishes of the genus *Labeo*. *Archiv für Protistenkunde, 131*, 319–325.

510. Kruse, P., Steinhagen, D., & Korting, W. (1989). Development of *Trypanoplasma borreli* (Mastigophora: Kinetoplastida) in the leech vector, *Piscicola geometra*, and its infectivity for the common carp, *Cyprinus carpio*. *Journal of Parasitology, 75*(4), 527–530.

511. Kruse, P., Steinhagen, D., & Korting, W. (1991). The division process of *Trypanoplasma borreli (Laveran and Mesnil, 1901)* (Mastigophorea, Kinetoplastida) in experimentally infected common carp (*Cyprinus carpio* L.). *Archiv für Protistenkunde, 140*, 349–352.

512. Kruse, P., Steinhagen, D., Korting, W., & Friedhoff, K. T. (1989). Morphometrics and redescription of *Trypanoplasma borreli (Laveran and Mesnil, 1901)* (Mastigophora, Kinetoplastida) from experimentally infected common carp (*Cyprinus carpio* L.). *Journal of Protozoology, 36*,

408–411.

513. Kruszewicz, A. G. (1994). Związek parazytologii z ornitologią Część II. Przeglad systematyczny pasożytów i możliwości diagnozowania inwazji. *Notatki Ornithologiczne, 35*, 421–430.

514. Krylov, M. V. (1981). Piroplasmids. In *Protozoa* (Vol. 2). Izd Nauka.

515. Kumar, A., Singh, H. R., & Chopra, A. K. (1984). Haematological changes in coldwater fish *Schizothorax agiostomus* (Heckel) infected with black spot disease. *Indian Journal of Parasitology, 8*, 129–131.

516. Kumar, U., Jas, R., & Ghosh, J. D. (2009). Effect of Isometamidium hydrochloride on *Trypanosoma evansi* infections in rat. *Journal of Parasitology and Diseases, 33*, 36–41.

517. Kundu, T. K. (1985). *Myxosoma indirae* n. sp. (Myxozoa: Myxosomatidae) from the head cartilage, scale, and tail fin of *Cirrhina mrigala* (Ham.). *Acta Protozoologica, 24*, 333–337.

518. Kundu, T. K., & Chatterjee, M. (2005). Observations on a trypanosome, *Trypanosoma notopterae* n. sp. from a freshwater teleost fish, *Notopterus notopterus* (Pallas). *Journal of Current Science, 7*, 489–494.

519. Labbe, A. (1896). Researches zoologiques, cytologiques, et biologiques sur les coceidies. *Archives de Zoologie Expérimentale et Générale, 4*, 517–654.

520. LaCount, D. J., Gruszynski, A. E., Grandgenett, P. M., Bangs, J. D., & Donelson, J. E. (2003). Expression and function of the *Trypanosoma brucei* major surface protease (GP63) genes. *Journal of Biological Chemistry, 278*(27), 24658–24664.

521. Ladle, R. J. (1992). Parasites and sex: Catching the Red Queen. *Trends in Ecology & Evolution, 7*, 405–408.

522. Laha, R., Hemaprasanth, H., & Bhattacharya, D. (1997). Observations on the prevalence of *Trypanosoma lewisi* infection in wild rats and a trial on its adaptation in unnatural host. *Journal of Parasitology and Applied Animal Biology, 6*, 5–8.

523. Laidley, C. W., Woo, P. T. K., & Leatherland, J. F. (1988). The stress response of rainbow trout to experimental infection with the blood parasite *Cryptobia salmositica* Katz, 1951. *Journal of Fish Biology, 32*, 253–261.

524. Laird, M. (1948). *Trypanosoma hepatretae* n.sp. a blood parasite of hagfish. *Nature, 161*, 440–441.

525. Laird, M. (1951). Blood parasites of mammals in New Zealand. *Zoological Publications, Victoria University College, 9*, 1–14.

526. Lamers, C. H. J., & Pilarczyk, A. (1982). Immune response and antigen

localization in carp (*Cyprinus carpio*) after administration of *Yersinia ruckeri* O-antigen. *Developmental and Comparative Immunology, Supplement, 2*, 107–114.

527. Lamers, C. H. J., De Haas, M. J. H., & Van Muiswinkel, W. B. (1985). Humoral response and memory formation in carp after injection of *Aeromonas hydrophila* bacterin. *Developmental and Comparative Immunology, 9*, 65–75.
528. Lanham, S. R., & Godfrey, D. C. (1970). Isolation of salivarian trypanosomes from man and other mammals using DEAE cellulose. *Experimental Parasitology, 28*, 521–534.
529. Lantz, C., & Van Dyke, K. (1972). *Plasmodium berghei*: Inhibited incorporation of AMP-8-3H into nucleic acids of erythrocyte-free malarial parasites by acridines, phenanthridiniums, and 8-amino-quinolines. *Experimental Parasitology, 31*, 255–261.
530. Larsh, J. E. Jr., & Donaldson, A. W. (1944). The effect of concurrent infection with *Nippostrongylus* on the development of *Hymenolepis* in mice. *Journal of Parasitology, 30*, 18–20.
531. Laveran, A., & Mesnil, F. (1901). Sur les flagelles à membrane ondulante des poissons (genres *Trypanosoma* Gruby et *Trypanoplasma* n.gen.). *Comptes Rendus Hebdomadaires des Séances de l'Académie des Sciences, Paris, 133*, 670–675.
532. Laveran, A., & Mesnil, F. (1902). Sur le mode de multiplication des trypanosomes des poissons. *Comptes Rendus Hebdomadaires des Séances de l'Académie des Sciences, Paris, 134*, 1405–1409.
533. Laveran, A., & Mesnil, F. (1904). *Trypanosomes and Trypanosomiases* (Translated by D. Nabarro). W. T. Keener and Co.
534. Laveran, A., & Mesnil, F. (1907). *Trypanosomes and Trypanosomiasis*. Bailliere.
535. Laveran, A., & Mesnil, F. (1912). *Trypanosomes et Trypanosomiasis* (2nd ed.). Masson.
536. Laveran, A., & Pettit, A. (1909). Sur un trypanosome d'un campagnol Microtus arvalis Pallas. *Comptes Rendus des Séances de la Société de Biologie*, *67*, 798–800.
537. Lavier, G. (1921). Haemogregarines, Grahamella, spirochete et trypanosome du campagnol indigène Microtus arvalis Pallas. *Bulletin de la Société de Pathologie Exotique*, *14*, 569.
538. Lavier, G. (1936). Sur quelques flagelles intestinaux de poissons marins. *Annales de Parasitologie Humaine et Comparée, 14*, 67–68.

539. Lebailly, C. (1905). Recherches sur les hématozoaires parasites des téléostéens marins. *Archives de Parasitologie*, *10*, 348–404.

540. Lebailly, C. (1906). Les hématozoaires des téléostéens marines. *Archives de Parasitologie*, *10*, 392.

541. Leidy, J. (1846). Description of a new genus and species of Entozoa. *Proceedings of the Academy of Natural Sciences of Philadelphia*, *3*, 100, 101.

542. Leonard, G., Soanes, D. M., & Stevens, J. R. (2011). Resolving the question of trypanosome monophyly: A comparative genomics approach using whole genome data sets with low taxon sampling. *Infection, Genetics and Evolution*, *11*, 955–959.

543. Letch, C. A. (1977). *Studies on Trypanosomes of Small Fishes from the River Lee* (Ph.D. thesis). Council for National Academic Awards.

544. Letch, C. A. (1979). Host restriction, morphology and isoenzymes among trypanosomes of some British freshwater fishes. *Parasitology*, *79*, 107–117.

545. Letch, C. A., & Ball, S. J. (1979). Prevalence of *Trypanosoma cobitis* Mitrophanow, 1883 in fishes from the river Lee. *Parasitology*, *79*, 119–124.

546. Levin, S. A. (1970). Community equilibria and stability, and an extension of the Competitive Exclusion Principle. *American Naturalist*, *104*, 413–423.

547. Levine, N. D., Corliss, J. O., Cox, F. E. G., Deroux, G., Grain, J., Honigberg, B. M., Leedale, G. F., Loeblich III, A. R., Lom, J., Lynn, D. H., Meninfield, E. G., Page, F. C., Poljansky, G., Sprague, V., Vavra, J., & Wallace, F. G. (1980). A newly revised classification of the Protozoa (The Committee on Systematics and Evolution of the Society of Protozoologists). *Journal of Protozoology*, *27*, 37–58.

548. Lewington, S., Whitlock, G., Clarke, R., Sherliker, P., Emberson, J., Halsey, J., Qizilbash, N., Peto, R., & Collins, R. (2007). Blood cholesterol and vascular mortality by age, sex, and blood pressure: A meta-analysis of individual data from 61 prospective studies with 55,000 vascular deaths. *Lancet*, *370*(9602), 1829–1839.

549. Lewis, J. W., & Ball, S. J. (1981). Observations on the division of the epimastigotes of *Trypanosoma cobitis*. *Archiv für Protistenkunde*, *124*, 337–344.

550. Li, S., & Woo, P. T. K. (1995). Efficacy of a live *Cryptobia salmositica* vaccine, and the mechanism of protection in vaccinated *Oncorhynchus*

mykiss (Walbaum) against cryptobiosis. *Veterinary Immunology and Immunopathology*, *48*, 343–353.

551. Li, S., & Woo, P. T. K. (1997). Vaccination of rainbow trout, *Oncorhynchus mykiss* (Walbaum), against cryptobiosis: Efficacy of the vaccine in fresh and sea water. *Journal of Fish Diseases*, *20*(5), 369–374.
552. Li, S., Cowey, C. B., & Woo, P. T. K. (1996). The effects of dietary ascorbic acid on *Cryptobia salmositica* infection and on vaccination against cryptobiosis in *Oncorhynchus mykiss*. *Diseases of Aquatic Organisms*, *24*, 11–16.
553. Linardi, P. M., & Botelho, J. R. (2002). Prevalence of *Trypanosoma lewisi* in *Rattus norvegicus* from Belo Horizonte, State of Minas Gerais, Brazil. *Memórias do Instituto Oswaldo Cruz*, *97*(3), 411–414.
554. Lingard, A. (1904). A short account of the various *Trypanosoma* found to date in the blood of some of the lower animals and fish. *Indian Medical Gazette*, *39*, 445.
555. Linton, R. W. (1930). The blood chemistry of acute trypanosome infection. *Journal of Experimental Medicine*, *52*, 103–111.
556. Linxian, H., Wenchuan, Y., Wenfeng, P., & Yuguang, L. (1994). A survey on the parasitic protozoa of rodents from Xiamen, Fujian province and Tacheng of Xinjiang Autonomous Region, China. *Wuyi Science Journal*, *12c*, 144–148.
557. Lippi, M., & Sebastiani, A. (1958). *Archives Italiennes de Sciences Médicales Tropiques de Parasitologie*, *5*, 327–335. c.f. von Brand, T. (1973). *Biochemistry of Parasites*. Academic Press.
558. Lo, C. M., Morand, S., & Galzin, R. (1998). Parasite diversity/host age and size relationship in three coral-reef fishes from French Polynesia. *International Journal for Parasitology*, *28*, 1695–1708.
559. Lom, J. (1969). Cold blooded vertebrate immunity to protozoa. In G. J. Jackson, R. Herman, & I. Singer (Eds.), *Immunity to Parasitic Animals* (Vol. 1, pp. 249–265). Appleton-Century-Crofts.
560. Lom, J. (1973). Experimental infection in goldfish with blood flagellates. In *Progress in Protozoology*. 4th International Congress in Protozoology, Clermont-Ferrand, France, p. 225 (Abstract).
561. Lom, J. (1979). Biology of trypanosomes and trypanoplasms of fish. In W. H. R. Lumsden & D. E. Evans (Eds.), *Biology of the Kinetoplastida* (Vol. 2, pp. 269–337). Academic Press.
562. Lom, J. (1980). *Cryptobia branchialis* Nie from fish gills: Ultra structural evidence of ectocommensal function. *Journal of Fish Diseases*, *3*, 427–436.

563. Lom, J., & Dykova, I. (1984). Pathogenicity of some protozoan parasites of cyprinid fishes. In *Fish, Pathogens and Environment in European Polyculture*. Proceedings of the International Seminar (pp. 23–27).

564. Lom, J., & Dykova, I. (1992). Protozoan parasites of fishes. In *Developments in Aquaculture and Fisheries Science* (Vol. 26, pp. 1–315). Elsevier Science Publishing.

565. Lom, J., & Laird, M. (1976). Parasitic protozoa from marine and euryhaline fish of Newfoundland and New Brunswick. II. Microsporidia. *Transactions of the American Microscopical Society*, *95*, 569–580.

566. Lom, J., & Machackova, B. (1986). Experimental evidence of pathogenicity of *Trypanoplasma borreli* and *Trypanosoma danilewskyi* for carp fingerlings. *Bulletin of the European Association of Fish Pathologists*, *6*, 87–88.

567. Lom, J., Dykova, I., & Machakova, B. (1986). Experimental evidence of pathogenicity of *Trypanosoma borelli* and *Trypanosoma danilewskyi* for carp fingerlings. *Bulletin of the European Association of Fish Pathologists*, *6*, 87–88.

568. Lopes, R. A., Paula, Lopes, O. V., Satake, T., & Garavelo, J. C. (1991). Trypanosomes of Brazilian fishes I. *T. vialeri* sp. n. from *Pterodoras granulosus* Valenciennes, 1833. *Naturalia* (Brazil), *16*, 19–29.

569. Love, R. M. (1980). *The Chemical Biology of Fishes*. Academic Press.

570. Lowry, O. H., Rosenbrough, N. J., Farr, A. L., & Randall, R. J. (1951). Protein measurement with Folin-Phenol reagent. *Journal of Biological Chemistry*, *193*, 265–275.

571. Lubega, G. W., Ochola, D. O. K., & Prichard, R. K. (2002). *Trypanosoma brucei*: Anti-tubulin antibodies specifically inhibit trypanosome growth in culture. *Experimental Parasitology*, *102*, 134–142.

572. Lucky, Z. (1980). Methods for the diagnosis of fish diseases. Amerind Publishing Co. Pvt. Ltd.

573. Lumsden, W. H. R., & Evans, D. A. (1976). *Biology of Kinetoplastida* (Vol. 1). Academic Press.

574. Lymbery, J. A. (1989). Host specificity, host range and host preference. *Parasitol. Today*, *5*, 298.

575. Lynch, J. E., Grodone, P. N., & Bomforth, B. J. (1956). *Antibiotics Annual* (1955–1956).

576. Macadam, R. F., & Williamson, J. (1972). Drug effects on the fine structure of *Trypanosoma rhodesiense*: Diamidines. *Transactions of the Royal Society of Tropical Medicine and Hygiene*, *66*, 897–904.

577. Macdonald, D. W., Anwar, M., Newman, C., Woodroffe, R., & Johnson, P. J. (1999). Inter-annual differences in the age-related prevalences of *Babesia* and *Trypanosoma* parasites of European badgers (*Meles meles*). *Journal of Zoology*, *247*, 65–70.
578. Macdonald, G. (1957). *The Epidemiology and Control of Malaria*. Oxford University Press.
579. MacKenzie, K., & Gibson, D. I. (1970). Aspects of fish parasitology. In A. E. R. Taylor & R. Muller (Eds.), *Fish Parasitology* (pp. 1–42). Blackwell Scientific Publications.
580. Mackenzie, R. B. (1972). Public health importance of rodents in South America. *Bulletin of the World Health Organization*, *47*, 161–169.
581. Mackerras, J. E. (1959). The haematozoa of Australian mammals. *Australian Journal of Zoology*, *7*, 105–135.
582. Mackintosh, D., Coleman, K., & Davies, A. J. (2003). Mitogenic activity of insulin in the culture of a trypanosome: Duration and dose response. *Cell Biology International*, *27*(1), 75–78.
583. Maegraith, B. G. (1969). *Military Medicine*, *134*, 1129–1131. c.f. von Brand, T. (Ed.). (1973). *Biochemistry of Parasites*. Academic Press.
584. Mafiana, C. F., Osho, M. B., & Sam-Wobo, S. (1997). Gastrointestinal helminth parasites of the black rat (*Rattus rattus*) in Abeokuta, southwest Nigeria. *Journal of Helminthology*, *71*, 217–220.
585. Mahnert, V. (1972). Grahamella and Sporozoa as blood parasites of Alpine small mammals. *Acta Tropica*, *29*, 88–100.
586. Malhotra, S. K. (1981). Systems models for parasite pathways in ichthyoparasitology of the Garhwal Himalayan ecosystem. *Current Science*, *50*, 874–875.
587. Malhotra, S. K. (1985). Systems models in ichthyoparasitology of the Garhwal Himalayan riverine ecosystems. *Bulletin of the Botanical Society of Saugar*, *32*, 159–163.
588. Mallick, T. K., Sardar, K. K., Parija, S. C., Patel, M. B., & Misra, S. N. (2000). Plasma-protein binding of Berenil in black Bengal goats. *Indian Veterinary Journal*, *77*, 492–494.
589. Mamman, M., Mckeever, D. J., Aliu, Y. O., & Peregrine, A. S. (1996). Pharmacokinetics of diminazene in plasma and lymph of goats. *American Journal of Veterinary Research*, *57*, 710–714.
590. Mamo, E., & Holmes, P. H. (1975). The erythrokinetics in zebu cattle chronically infected with *Trypanosoma congolense*. *Research in Veterinary Science*, *18*, 105.

591. Mandal, A. K. (1975). Two new trypanosomes from Indian fresh water fishes. *Angewandte Parasitologie*, *16*, 87–93.

592. Mandal, A. K. (1977). *Trypanosoma choudhuryi* sp. nov. from *Tilapia mossaambica* (Peters). *Acta Protozoologica*, *16*, 1–4.

593. Mandal, A. K. (1978a). Two new species of trypanosomes from Indian fresh water fishes. *Angewandte Parasitologie*, *19*, 158–162.

594. Mandal, A. K. (1978b). Studies on the haematozoa of some fishes of commercial importance from India. *Asian Congress of Parasitology*, 185–186.

595. Mandal, A. K. (1979). Studies on the haematozoa of catfishes belonging to the genus *Mystus* Scopoli from India. *Bulletin of the Zoological Survey of India*, *2*, 12–23.

596. Mandal, A. K. (1980). *Trypanosoma tandoni* sp. nov. from fresh water shark, *Wallago attu* (Schn). *Bulletin of the Zoological Survey of India*, *2*, 183–186.

597. Mandal, A. K. (1984). A study on the haematozoa of fishes of commercial importance from India. (Tech. monograph). *Zoological Survey of India*, *9*, 51 pp.

598. Maraghi, S., & Molyneux, D. H. (1989). Studies on cross immunity in *Herpetosoma* trypanosomes of *Microtus*, *Clethrionomys* and *Apodemus*. *Parasitology Research*, *75*, 175–177.

599. Maraghi, S., Molyneux, D. H., & Wallbanks, K. R. (1989). Differentiation of rodent of the subgenus *Herpetosoma* by lectins. *Tropical Medicine and Parasitology*, *40*, 273–278.

600. Maraghi, S., Wallbanks, K. R., & Molyneux, D. H. (1995). Oral transmission of trypanosomes of subgenus *Herpetostoma* from small mammals. *Parasitology Research*, *81*, 693–695.

601. Marciacq, Y., & Seed, J. R. (1970). Reduced levels of glycogen and glucose-6-phosphate in liver of guinea pigs infected with *Trypanosoma gambiense*. *Journal of Infectious Diseases*, *121*, 653–655.

602. Marcus, S. L., Kopelman, R., Koll, B., & Bacchi, C. J. (1982). Effects of exogenous polyamine and trypanocides on the DNA polymerase activities from *Trypanosoma brucei*, mouse thymus and murine leukemia virus. *Parasitology*, *5*, 231–243.

603. Martin, C. (1910). Observations on *Trypanoplasma congeri*. Part I. The division of the active form. *Quarterly Journal of Microscopic Science*, *55*, 485–496.

604. Martin, C. (1913). Further observations on the intestinal trypanoplasm of

fishes, with a note on the division of *Trypanoplasma cyprini* in the crop of a leech. *Quarterly Journal of Microscopic Science*, *59*, 175–195.

605. Maruta, H., & Mizuno, D. (1971). Selective recognition of various erythrocytes in endocytosis by mouse peritoneal macrophages. *Nature*, *234*, 246.

606. Maslov, D. A., Podlipaev, S. A., & Lukes, J. (2001). Phylogeny of the Kinetoplastida: Taxonomic problems and insights into the evolution of parasitism. *Memórias do Instituto Oswaldo Cruz*, *96*, 397–402.

607. Maslov, D. A., Yasuhira, S., & Simpson, L. (1999). Phylogenetic affinities of *Diplonema* within the Euglenozoa as inferred from the SSU rRNA gene and partial COI protein sequences. *Protist*, *150*, 33–42.

608. Masoumian, M., & Pazooki, J. (1999). Myxosporean parasites from Mesopotamian part of Iran. *Iranian Journal of Fisheries Sciences*, *1*(1), 35–46.

609. Mathews, M. J., Kingston, N., & Morton, J. K. (1977). *Trypanosoma cervi* Kingston and Morton, 1975 from the mule deer, *Odocoileus hemionus* in Wyoming. *Journal of Wildlife Diseases*, *13*, 33–39.

610. Mathis, C., & Leger, M. (1911). Trypanosomes des batrachiens du Tonkin. *Annales de l'Institut Pasteur*, *25*, 671–681.

611. Mavor, J. W. (1915). On the occurrence of the trypanoplasm probably *T. borreli* of the common sucker, *Catostomus commersonii*. *Journal of Parasitology*, *2*, 1–6.

612. McDonald, L. R., & Roberson, E. L. (1988). Antiprotozoan drugs. In N. H. Booth & L. E. McDonald (Eds.), *Veterinary Pharmacology and Therapeutics* (pp. 963). Iowa State University Press.

613. McDougald, L. R., & Hansen, M. F. (1970). *Histomonas meleagridis*: Effect on plasma enzymes in chickens and turkeys. *Experimental Parasitology*, *27*, 229–235.

614. McGhee, R. B. (1964). *American Journal of Tropical Medicine and Hygiene*, *13*, 219–224. c.f. von Brand, T. (Ed.). (1973). *Biochemistry of Parasites*. Academic Press.

615. Mdachi, R. E., Murilla, G. A., Omukuba, J. N., & Cagnolati, V. (1995). Disposition of diminazene aceturate (Berenil) in trypanosome-infected pregnant and lactating cows. *Veterinary Parasitology*, *58*, 215–225.

616. Mehlhorn, H. (1988). *In vitro* cultivation of certain parasitic protozoa. In *Parasitology in Focus* (pp. 702–718). Springer-Verlag.

617. Mehlhorn, H., & Schein, E. (1984). The piroplasms: Life cycle and sexual stages. *Advances in Parasitology*, *23*, 37–103.

618. Mehta, M., & Woo, P. T. K. (2002). Acquired cell-mediated protection in rainbow trout, *Oncorhynchus mykiss*, against the haemoflagellate, *Cryptobia salmositica*. *Parasitology Research*, *88*(11), 956–962.

619. Mercado, T. I., Butany, J. W., & Ferrans, V. J. (1986). *Trypanosoma cruzi*: Ultrastructural changes produced by an anti-*Trypanosomal* factor from *Pseudomonas fluorescens*. *Experimental Parasitology*, *61*, 65–75.

620. Michels, P. A., Bringaud, F., Herman, M., & Hannaert, V. (2006). Metabolic functions of glycosomes in Trypanosomatids. *Biochimica et Biophysica Acta*, *1763*, 1463–1477.

621. Miletti, L. C., Koerich, L. B., Pacheco, L. K., Steindel, M., & Stambuk, B. U. (2006). Characterization of D-glucose transport in *Trypanosoma rangeli*. *Parasitology*, *133*, 721–727.

622. Miller, R. S. (1967). Advances in ecological research. In J. B. Cragg (Ed.), *Advances in Ecological Research* (pp. 1–74). Academic Press.

623. Milne, A. (1961). Mechanisms in biological competition. In F. L. Milthorpe (Ed.), *Biological Competition* (pp. 40–61). The University Press.

624. Minchella, D. J. (1985). Host life-history variation in response to parasitism. *Parasitology*, *90*, 205–216.

625. Minchin, E. A. (1909). Observations on the flagellates parasitic in the blood of freshwater fishes. *Proceedings of the Zoological Society of London. Biological Sciences*, *2*, 30.

626. Mingawa, N., Yabu, Y., Kita, K., Nagai, K., Ohta, N., Meguro, K., Sakajo, S., & Yoshimoto, A. (1996). An antibiotic, ascofuranone, specifically inhibits respiration and *in vitro* growth of long slender bloodstream forms of *Trypanosoma brucei*. *Molecular and Biochemical Parasitology*, *81*, 127–136.

627. Misra, K. K., Chandra, A. K., & Choudhary, A. (1973). *Trypanosoma gachuii* n.sp. from a fresh water teleost fish *Ophicephalus gachua* Ham. *Archiv für Protistenkunde*, *115*, 18–21.

628. Miyata, A. (1978). Anuran trypanosomes in Kyushu and Ryukyu islands with descriptions of six new species. *Tropical Medicine*, *20*, 51–80.

629. Mohamed, H. A., & Molyneux, D. H. (1987). *In vitro* cultivation of *Herpetosoma* trypanosomes in insect cell tissue culture media. *Parasitology Research*, *73*, 9–14.

630. Mohamed, H. A., Molyneux, D. H., & Scott, C. M. (1987). Isoenzyme characterization of trypanosomes of the subgenus *Herpetosoma*. *Parasitology*, *94*, 39–48.

631. Mohan, H. (2000). *Pathology Practical Book*. Jaypee Brothers Medical

Publishers Pvt. Ltd.

632. Mokhayer, B., Konneshari, M., & Malaki, M. (1976). Occurrence of *Trypanosoma percae* in perches of Southern Caspian sea. In L. A. Page (Ed.), *Wildlife Disease* (pp. 555–559). Plenum Press.
633. Molnar, K., Hanek, G., & Fernand, C. H. (1974). Parasites of fish from Laurel Creek, Ontario. *Journal of Fish Biology*, *6*, 717–728.
634. Molyneux, D. H. (1969). The morphology and life history of *Trypanosoma (Herpetosoma) microti* of the field vole, *Microtus agrestis*. *Annals of Tropical Medicine and Parasitology*, *63*, 229–244.
635. Molyneux, D. H. (1970). Development patterns in trypanosomes of the subgenus *Herpetostoma*. *Annales de la Société Belge de Médecine Tropicale*, *50*, 229–238.
636. Molyneux, D. H. (1976). Biology of *Trypanosomes* of the subgenus *Herpetosoma*. In W. H. R. Lumsden & D. A. Evans (Eds.), *Biology of the Kinetoplastida* (pp. 285–325). Academic Press.
637. Molyneux, D., & Ashford, R. (1983). *The Biology of Trypanosoma and Leishmania, Parasites of Man and Domestic Animals*. Taylor and Francis.
638. Monroy, F. P., Dusanic, D. G., & Minning, T. A. (1998). *Trypanosoma musculi* survival in the kidneys of chronically infected mice: Kidney from ultra structure, surface characteristics, and serological interactions. *Journal of Parasitology*, *84*, 914–919.
639. Montel, R. (1905). Trypanosome d'un poisson de Cochinchine. *Comptes Rendus des Séances de la Société de Biologie de Paris*, *58*, 1016.
640. Moon, A. P., Williams, J. S., & Witherspoon, C. (1968). Serum biochemical changes in mice infected with *Trypanosoma rhodesiense* and *Trypanosoma duttoni*. *Experimental Parasitology*, *22*, 112–121.
641. Moore, J. (1984). Altered behavioural responses in intermediate hosts: An acanthocephalan parasite strategy. *American Naturalist*, *123*, 572–577.
642. Moore, J., & Gotelli, N. J. (1990). A phylogenetic perspective on the evolution of altered host behaviours: A critical look at the manipulation hypothesis. In C. J. Barnard & J. M. Behnke (Eds.), *Parasitism and Host Behaviour* (pp. 193–233). Taylor and Francis.
643. Morand, S., Simkova, A., Matejusova, I., Plaisance, L., Verneau, O., & Desdevises, Y. (2002). Investigating patterns may reveal processes: Evolutionary ecology of ectoparasitic monogeneans. *International Journal for Parasitology*, *32*, 111–119.
644. Morillas, J., George-Nescimen, T. M., Valeria, H., & Khan, R. A. (1987). *Trypanosoma humboldti* n. sp. from the Chilean catshark. *Journal of*

Protozoology, *34*(3), 342–344.

645. Morsy, T. A., Babrawy, A. A., Dakhil, M. M., & Abdel, Malwa M. M. (1994). *Babesia* infections in rodents trapped in Riyadh Region, Saudi Arabia, with general discussion. *Journal of the Egyptian Society of Parasitology*, *24*(1), 177–185.
646. Mukherjee, M., & Haldar, D. P. (1982). Host susceptibility experiments on a new fish trypanosome in India. *Angewandte Parasitologie*, *23*, 69–73.
647. Mulligan, H. W. (1970). *The African Trypanosomiases*. George Allen and Unwin/Ministry of Overseas Development.
648. Murad, A., Houston, A. H., & Samson, L. (1990). Haematological response to reduced oxygen carrying capacity, increased temperature and hypoxia in goldfish *Carassius auratus* L. *Journal of Fish Biology*, *36*, 289–305.
649. Murilla, G. A., Mdachi, R. E., & Karanja, W. M. (1996). Pharmacokinetics, bioavailability and tissue residues of (14C) isometamidium in non-infected and *Trypanosoma congolense*-infected Boran cattle. *Acta Tropica*, *61*, 277–293.
650. Murray, M., Huan, C. N., Lambert, P. H., & Gerber, H. (1975). The anaemia of African trypanosomiasis: Demonstration of a hemolytic factor. Proceedings of the International Scientific Council for Trypanosomiasis Research and Control. The Gambia, 460–469.
651. Na'Isa, B. (1970). Observations on the use of chemotherapeutic agents against trypanosomiasis. *Report of the Federal Ministry of Agriculture and Natural Resources, Tsetse and Trypanosomiasis Division*, Kaduna, Nigeria.
652. Nalinikumari, K., Radhika, M., Prameela, E., & Choudhuri, P. C. (2000). Mixed infection of Theileriosis and Trypanosomiasis in a cross bred cow: A case report. *Indian Veterinary Journal*, *77*, 615–616.
653. Nandi, N. C., Nandi, R., & Mandal, A. K. (1983). In: Index catalogue and bibliography of protozoan parasites from Indian fishes. *Miscellaneous Publication of the Zoological Survey of India*, *41*, 46 pp.
654. Narasimhamurti, C. C., & Saratchandra, B. (1980). Two new species of trypanosomes, *T. channai* n. sp. from *Channa punctatus* and *T. qadrii* n. sp. from *Clarias batrachus*. *Proceedings of the Indian Academy of Sciences*, *89B*, 371–376.
655. Nayler, D. C. (1971). The haematology and histopathology of *Trypanosoma congolense* infection in cattle. VI: Haematology (including symptoms). *Tropical Animal Health and Production*, *3*, 159–168.
656. Neave, S. (1906). Report of traveling pathologist and naturalist. *Second*

Report of the Welcome Tropical Laboratory, Khartoum, 183–204.

657. Nebe, O. J., Adeoye, G. O., & Agomo, P. U. (2002). Prevalence and clinical profile of malaria among the coastal dwellers of Lagos state, Nigeria. *Nigerian Journal of Parasitology*, *23*, 61–68.
658. Needham, E. A. (1969). Protozoa Parasite in fish. Ph.D. Dissertation, University of London.
659. Negm-Eldin, M. M. (1997). *Trypanosoma mukasai* (Hoare, 1932) in its biological vector *Batracobdelloides tricarinata* (Blanchard, 1897) and their life cycles. *Deutsche Tierärztliche Wochenschrift*, *104*, 215–219.
660. Negm-Eldin, M. M. (1998). Further studies on *Trypanosoma mukasai* Hoare, 1932 and its biological vector *Batracobdelloides tricarinata* (Blanchard, 1897). *Deutsche Tierärztliche Wochenschrift*, *105*, 175–181.
661. Neumann, R. O. (1909). Studien über protozoische Parasiten im Blut von Meeresfischen. *Zeitschrift für Hygiene und Infektionskrankheiten: Medizinische Mikrobiologie, Immunologie und Virologie*, *64*, 1–112.
662. Newton, B. A. (1974). The chemotherapy of trypanosomiasis and leishmaniasis: Towards a more rational approach. In *Trypanosomiasis and Leishmaniasis* (pp. 285–307). Elsevier.
663. Nico, S. J., Van As, J. G., & Davies, A. J. (2004). Fish trypanosome from the Okavango Delta, Botswana. *Folia Parasitologica*, *51*, 299–303.
664. Nie, N. H., Bent, D. H., & Hull, C. H. (1970). *SPSS: Statistical Package for the Social Sciences*. McGraw-Hill Pub. Co.
665. Nielsen, K., Sheppard, J., Holmes, W., & Tizard, I. (1978). Increased susceptibility of *Trypanosoma lewisi*-infected or decomplemented rats to *Salmonella typhimurium*. *Experientia, 34*, 118, 119.
666. Nilsson, N. A. (1967). The biological basis of freshwater fish production. In S. D. Gerkin (Ed.), *Blackwell Scientific Publications* (pp. 295–313).
667. Nmorsi, O. P. G., & Egwunyenga, A. O. (1999). Gastrointestinal helminthes of three small wild mammals in a rain forest zone of southern Nigeria. *Journal of Parasitology, 23*, 59–60.
668. Noble, E. R. (1955). Morphology and life cycles of trypanosomes. *Rev. Biol., 30*, 1–28.
669. Noble, E. R. (1968). The flagellate *Cryptobia* in two species of deep-sea fishes from the Eastern Pacific. *Journal of Parasitology, 54*, 720–724.
670. Noble, E. R., & Noble, G. A. (1982). *Parasitology* (5th ed.). Lea and Febiger.
671. Nonaka, M., & Smith, S. L. (2000). Complement system of bony and cartilaginous fish. *Fish & Shellfish Immunology, 10*, 215–228.

672. Nordling, D., Anderson, M., Zohari, S., & Gustafsson, L. (1998). Reproductive effort reduces specific immune response and parasite resistance. *Proceedings of the Royal Society of London. Biological Sciences, 265*, 1291–1298.

673. Norris, K., Anwar, M., & Read, A. F. (1994). Reproductive effort influences the prevalence of haematozoan parasitized by blood parasites. *Functional Ecology, 11*, 652–655.

674. Nowell, F. (1969). The blood picture resulting from *Nuttalia* (= *Babesia*) rodhaini and *Nuttalia* (= *Babesia*) *microti* infections in rats and mice. *Parasitology, 59*, 991–1004.

675. Noyes, H. A., Ambrose, P., Barker, P., Begon, M., Bennett, M., Bown, K. J., & Kemp, S. J. (2002). Host specificity of *Trypanosoma (Herpetosoma)* species: Evidence that bank voles (*Clethrionomys glareolus*) carry only one *T. (H.) evotomys* 18S rRNA genotype but wood mice (*Apodemus sylvaticus*) carry at least two polyphyletic parasites. *Parasitology, 124*, 185–190.

676. Nunez-Vergera, L. J., Squella, J. A., Bollo-Dragnic, S., Marin-Catalan, R., Pino, L., Diaz-Araya, & Letelier, M. E. (1998). Isradipine and lacidipine: Effects *in vivo* and *in vitro* on *Trypanosoma cruzi* epimastigotes. *General Pharmacology, 30*, 85–87.

677. Ochiogu, I. S., Uchendu, C. N., Ihedioha, J. I., & Shoyinka, S. V. O. (2008). Experimental *Trypanosoma brucei* infection in rats (*Rattus norvegicus*): Effects on different stages of gestation and the neonatal period. *Animal Reproduction, 5*(1/2), 39–49.

678. Odum, E. P. (1971). *Fundamentals of Ecology*. W. B. Saunders Company.

679. Ogawa, M., & Uegaki, J. (1927). Beobachtungen über die Blutprotozoen bei Tieren Formosais. *Archiv für Protistenkunde, 57*, 14–30.

680. Oliver, M., & Viens, P. (1985). *Trypanosoma musculi* persisting in the kidneys of mice are responsible for maintaining the host immunity. *Transactions of the Royal Society of Tropical Medicine and Hygiene, 79*, 516.

681. Olson, R. E. (1998). Discovery of the lipoproteins, their role in fat transport and their significance as risk factors. *Journal of Nutrition, 128*(2 Suppl), 439S–443S.

682. Oppenhein, J. J., & Rosenstreich, D. L. (1976). *Mitogens in Immunology*. Academic Press.

683. Opperdoes, F. R. (1987). Compartmentation of carbohydrate metabolism in trypanosomes. *Annual Review of Microbiology, 41*, 127–151.

684. Opperdoes, F. R. (1991). Glycosomes. In G. Coombs & M. North (Eds.), *Biochemical Protozoology* (pp. 134–144).

685. Opperdoes, F. R., Nohynkova, E., Van Schaftingen, E., Lambeir, A. M., Veenhuis, M., & Roy, J. V. (1988). Demonstration of glycosomes (microbodies) in the bodonid flagellate *Trypanoplasma borelli* (Protozoa, Kinetoplastida). *Molecular and Biochemical Parasitology, 30*, 155–163.

686. Opperdoes, F. R., Wierenga, R. K., Noble, M. E. M., Hol, W. G. J., Wilson, M., Kuntz, D. A., Callens, M., & Perie, J. (1990). Unique properties of glycosomal enzymes. In N. Agabian & A. Cerami (Eds.), *Parasites: Molecular Biology, Drug and Vaccine Design* (pp. 233–246). Wiley-Liss.

687. Orecka-Grabda, T., & Wierzbicka, J. (1996). Observations on *Trypanosoma granulosum* Laveral et Mesnil, 1902 (Protozoa, Kinetoplastida). A blood parasite of eel. *Acta Ichthyologica et Piscatoria, 26*, 39–47.

688. Ormerod, W. E., & Shaw, J. J. (1963). A study of granules and other changes in phase-contrast appearance produced by chemotherapeutic agents in trypanosomes. *British Journal of Pharmacology, 21*, 259–272.

689. Oser, B. L. (1965). *Hawk's Physiological Chemistry*. Tata McGraw Hill Publishing Company.

690. Othman, O. E., & Ahmed, S. (2004). Cytogenetic effect of the trypanocidal drug Berenil in blood cultures of river buffalo. *Journal of Biological Sciences, 4*, 180–184.

691. Overath, P., Ruoff, J., Stierhof, Y. D., Haag, J., Tichy, H., Dykova, I., & Lom, J. (1998). Cultivation of bloodstream forms of *Trypanosoma carassii*, a common parasite of freshwater fish. *Parasitology Research, 84*, 343–347.

692. Pandey, K. C., & Pandey, A. K. (1976). New species of *Trypanosoma* Gruby from fishes and its effect on piscine host. *Indian Journal of Zoology, 15*, 15–18.

693. Pandey, N. N., & Mishra, S. S. (1978). Comparative efficacy of berenil, acriflavin and sulfadimethoxine in *Babesia bigemina* infection in indigenous cow calves. *Indian Veterinary Journal, 55*, 144.

694. Paperna, I. (1964). Competitive exclusion of *Dactylogyrus extensus* by *Dactylogyrus vastator* (Trematoda, Monogenea) on the gills of reared carp. *Journal of Parasitology, 50*, 94–98.

695. Papovsky, M. A., Linberg, L. E., Syrek, L. A., & Page, P. L. (1988). Prevalence of *Babesia* antibody in a selected blood donor population. *Transfusion, 28*, 59–61.

696. Parker, J. W., & Lukes, R. J. (1971). A microculture method for lymphocyte

transformation studies in the clinical laboratory. *American Journal of Clinical Pathology, 56*, 174–180.

697. Parry, M. F., Fox, M., Burka, S. A., & Richar, W. J. (1977). *Babesia microti* infection in man. *Journal of the American Medical Association, 238*, 1282–1283.

698. Pasquier, L. D. (1989). Evolution of the immune system. In W. E. Paul (Ed.), *Fundamental Immunology* (pp. 139–142). Raven Press.

699. Paterson, W. B., & Woo, P. T. K. (1984). Ultrastructural studies on mitosis in *Trypanosoma danilewskyi* (Mastigophora: Zoomastigophorea). *Canadian Journal of Zoology, 62*, 1167–1171.

700. Paulin, J. J. (1975). Intranuclear microtubules in dividing *T. equiperdum*. *Biosystems, 7*, 308.

701. Pavlovskii, E. N. (1964). *Key to the Parasites of Freshwater Fish of the USSR*. Academy of Sciences of the USSR, Zoological Institute. Translated from Russian by Israel Program for Scientific Translations.

702. Pawelczyk, A., Bajer, A., Behnke, J. M., Gilbert, F. S., & Sinski, E. (2004). Factors affecting the component community structure of haemoparasites in common voles (*Microtus arvalis*) from the Mazury Lake District region of Poland. *Parasitology Research, 92*, 270–284.

703. Pazooki, J., & Masoumian, M. (2004). *Cryptobia acipenseris* and *Haemogregarina acipenseris* infections in *Acipenser gueldenstaedti* and *A. persicus* in the southern part of the Caspian Sea. *Journal of Agricultural Science and Technology, 6*(3/4), 95–101.

704. Pearse, A. S. (1933). Parasites of Siamese fishes and crustaceans. *Journal of the Siam Society Natural History Supplement, 9*, 179–191.

705. Peckova, H., & Lom, J. (1990). Growth, morphology and division of flagellates of the genus *Trypanoplasma* (Protozoa, Kinetoplastida) *in vitro*. *Parasitology Research, 76*, 553–558.

706. Pennycuick, L. (1971a). Quantitative effects of three species of parasites on population of three-spined sticklebacks *Gasterosteus aculeatus* L. *Journal of Zoology, 165*, 143–162.

707. Pennycuick, L. (1971b). Seasonal variation in the parasite infections in a population of three-spined sticklebacks *Gasterosteus aculeatus* L. *Parasitology, 63*, 373–388.

708. Pennycuick, L. (1971c). Frequency distribution of parasites in a population of three-spined sticklebacks *Gasterosteus aculeatus* L., with particular reference to the negative binomial distribution. *Parasitology, 63*, 389–406.

709. Pennycuick, L. (1971d). Differences in the parasite infections in three-

spined sticklebacks (*Gasterosteus aculeatus* L.) of different sex, age and size. *Parasitology, 63*, 407–418.

710. Peregrine, A. S. (1990). *Chemotherapy for Trypanosomiasis: Proceedings of a Workshop*. The International Laboratory for Research on Animal Diseases.

711. Perez-Reyes, R. (1967). Estudio taxonómico, biológico y fisiológico de algunos tripanosomas parásitos de ranas. Ph.D. thesis, Instituto Politécnico Nacional, Mexico.

712. Persing, D. H., Herwaldt, B. L., Glaser, C., Lane, R. S., Thomford, J. W., Mathiesen, D., Krause, P. J., Phillip, D. F., & Conrad, P. A. (1995). Infection with a babesia-like organism in northern California. *New England Journal of Medicine, 332*, 298–303.

713. Petrie, G. F. (1905). Observations relating to the structure and geographical distribution of certain trypanosomes. *Journal of Hygiene, Cambridge, 5*, 191.

714. Philips, F. S., Sternberg, S. S., Cronin, A. P., Sodergren, J. E., & Vidal, P. M. (1967). Physiologic disposition and intracellular localization of isometamidium. *Cancer Research, 27*, 333–349.

715. Pilch, D. S., Kirolos, M. A., Liu, X., Plum, G. E., & Breslaver, K. J. (1995). Berenil [1,3-bis(4-amidinophenyl)triazene] binding to DNA duplexes and to a RNA duplex: Evidence for both intercalative and minor groove binding properties. *Biochemistry, 34*, 9962–9976.

716. Poche, F. (1913). Das system der Protozoa. *Archiv für Protistenkunde, 30*, 125–321.

717. Polyansky, Y. I., & Bychowsky, B. E. (1963). Parasite fauna of sea fish. Results and perspectives of investigations by Soviet parasitologists on fish parasites in seas of the USSR. *Jerusalem: Israel Program for Scientific Translations*, 187–193.

718. Ponselle, A. (1913). Culture *in vitro* du *Trypanoplasma varium* Léger. *C. R. Séances Soc. Biol., 74*, 685–688.

719. Pontin, A. J. (1961). Population stabilization and competition between the ants *Lasius flavus* (F.) and *L. niger* (L.). *Journal of Animal Ecology, 30*, 47–54.

720. Pontin, A. J. (1982). *Competition and Coexistence of Species*. Pitman Advanced Publications Programme.

721. Portugal, J. (1994). Berenil acts as a poison of eukaryotic topoisomerase II. *FEBS Letters, 344*, 136–138.

722. Poulin, R. (1995). Phylogeny, ecology and richness of parasite communities

in vertebrates. *Ecological Monographs, 65*, 283–302.

723. Poulin, R., & Mouillot, D. (2003). Parasite specialization from a phylogenetic perspective: A new index of host specificity. *Parasitology, 126*, 473–480.

724. Poulin, R., & Rohde, K. (1997). Comparing the richness of metazoan ectoparasite communities of marine fishes: Controlling for host phylogeny. *Oecologia, 110*, 278–283.

725. Pound, C. J. (1905). On *Trypanosoma* and their presence in the blood of Brisbane rats. *Proceedings of the Royal Society of Queensland, 19*(1), 33–38.

726. Pradhan, S. V. (1961). A study on blood of few Indian fishes. *Proceedings of the Indian Academy of Sciences, 54B*, 251–256.

727. Preston, A. (1960). Red blood values in the plaice. *Journal of Marine Biological Association of the United Kingdom, 39*, 681–687.

728. Priotto, S., Sartori, M. J., Repossi, G., & Valentich, M. A. (2009). *Trypanosoma cruzi*: Participation of cholesterol and placental alkaline phosphatase in the host cell invasion. *Experimental Parasitology, 122*(1), 70–73.

729. Pucek, Z. (1981). *Keys to Vertebrates of Poland. Mammals.* PWN-Polish Scientific Publishers.

730. Puentes-Zambrano, J. L., Silva-Rojas, C., & Reyes-Leon, Y. (2003). Parasites in juveniles of *Lutjanus griseus* (Pisces: Lutjanidae) at La Restinga Lagoon, Margarita Island, Venezuela. *Interciencia, 28*(8), 463–468.

731. Putz, E. R. (1972a). *Cryptobia cataractae* sp.n. (Kinetoplastida, Cryptobiidae), a haemoflagellate of some cyprinid fishes of West Virginia. *Proceedings of the Helminthological Society, 39*, 18–22.

732. Putz, E. R. (1972b). Biological studies on the haemoflagellate *Cryptobia cataractae* and *Cryptobia salmositica*. *Bureau of Sport Fish and Wildlife (U. S.) Technical Paper, 630*, 3–25.

733. Qadri, S. S. (1955). The morphology of *Trypanosoma striati* n.sp. from an Indian freshwater fish. *Parasitology, 45*, 79–85.

734. Qadri, S. S. (1962a). On three trypanosomes from freshwater fishes. *Parasitology, 52*, 221–228.

735. Qadri, S. S. (1962b). An experimental study of the life cycle of *Trypanosoma danilewskyi* in the leech, *Hemiclepsis marginata*. *Journal of Protozoology, 9*, 254–258.

736. Qadri, S. S. (1962c). The development in culture of *Trypanosoma striati*

from an Indian fish. *Parasitology, 52*, 229–235.

737. Qadri, S. S. (1962d). *Trypanoplasma willoughbii* n. sp. from British freshwater fish, *Salvelinus willoughbii*. *Rivista di Parassitologia, 23*, 1–9.

738. Qayyum, A., & Qasim, S. Z. (1964). Studies on the biology of some freshwater fishes. *Journal of Bombay History Society, 61*, 74–98.

739. Rai, V., Gupta, A. K., Niyogi, A., & Agarwal, S. M. (1984). Parasitic effect on the levels of alkaline and acid phosphatase in *Channa punctatus* (Bloch) and *Clarias batrachus* (Linn.). *Geobios, 11*, 34–37.

740. Raina, A. K., Singh, R. P., Rajora, V. S., Sharma, R. D., & Banerjee, D. P. (1986). Haemato-biochemical changes in induced trypanosomiasis in buffalo calves. *Rivista di Parasitologia, 67*, 263–270.

741. Raizada, M. N., & Singh, C. P. (1982). Observations on haematological values of freshwater fish, *Cirrhinus mrigala* (Ham.). *Comparative Physiology and Ecology, 7*, 34–36.

742. Ramaswamy, M., & Reddy, T. G. (1978). A comparative study of haematology of three air-breathing fishes. *Proceedings of the Indian Academy of Sciences, 87B*(12), 381–385.

743. Randall, R. (1934). *Philippine Journal of Science, 53*, 97–105. c.f. von Brand, T. (ed.). (1973). *Biochemistry of Parasites*. Academic Press.

744. Rankin, J. S. (1945). An ecological study of the helminth parasites of amphibians and reptiles of Western Massachusetts and its vicinity. *Journal of Parasitology, 31*, 142–150.

745. Rao, K. B. V. (1975). Preliminary experiment on the infection of *Trypanosoma saccobranchi* in freshwater fish, *Saccobranchus fossilis*. *Indian Journal of Zoology, 3*, 51–53.

746. Rao, T. B., Rao, S. B., & Israel, K. (1984). Haemoglobin level of fish *Heteropneustes fossilis* due to *Trypanosoma* infection. *Current Science, 53*, 340.

747. Rauthan, J. V. S., Grover, S. P., & Jaiwal, P. (1995). Studies on some haematological changes in a hill stream fish *Barilius bendelisis* (Hamilton) infected with trypanosomes. *Flora and Fauna, 1*(2), 165, 166.

748. Ray, H. N., & Malhotra, M. N. (1960). Akinetoplastic strain of *Trypanosoma evansi* produced by 'Prothidium'. *Nature, 188*, 870.

749. Ray, R. (1992). Studies on the anuran blood parasites of the sub-Himalayan West Bengal, India. *Journal of Bengal Natural History Society, 11*, 4–8.

750. Raychaudhuri, S., & Misra, K. K. (1973). Two new fish trypanosomes from India. *Archiv für Protistenkunde, 115*, 10–17.

751. Read, C. P. (1959). The role of carbohydrates in the biology of cestodes.

VIII. Some conclusions and hypotheses. *Experimental Parasitology, 8*, 365–382.

752. Read, C. P., & Phifer, K. (1959). The role of carbohydrates in the biology of cestodes. VII. Interaction between individual tapeworms of the same and different species. *Experimental Parasitology, 8*, 46–50.

753. Real, L. A. (1996). Sustainability and the ecology of infectious diseases. *Bioscience, 46*, 88–97.

754. Reardam, J. T., & Norbury, G. (2004). Ectoparasites and haemoparasite infection in a diverse temperate lizard assemblage at Macraes flat, South Island, New Zealand. *Journal of Parasitology, 90*, 1274–1278.

755. Rehulka, J. (1985). Notes on the occurrence patterns and pathogenicity of undetermined blood protozoans (UBP) and *Trypanoplasma borreli* in carp in the second growing season. *Prace Vurh Vodnony Pap. Rih. Vothany, 14*, 43–52.

756. Reilly, B. O., & Woo, P. T. K. (1982). The *in vivo* and *in vitro* development of *Trypanosoma andersoni* Reilly and Woo, 1982 and *Trypanosoma grylli* Nigrelli, 1944. *Canadian Journal of Zoology, 60*, 124–133.

757. Reinert, K. E. (1999). DNA multimode interaction with berenil and pentamidine; double helix stiffening, unbending, and bending. *Journal of Biomolecular Structure and Dynamics, 17*, 311–331.

758. Ribeiro, R. D., Lopes, R. A., Satake, T., Nuti, S. A., & Britski, H. A. (1990). Trypanosomes of Brazilian fishes: IX *Trypanosoma pintoi*, new species from armoured catfish *Hypostomus* spp. (Pisces: Loricariidae), from Mogi-Guacy river, municipality of Bitangueiras, São Paulo, Brazil. *Rivista Brasileira de Biologia, 50*, 559–562.

759. Richardson, J. P. (1973). Mechanism of ethidium bromide inhibition of RNA polymerase. *Journal of Molecular Biology, 78*, 703–714.

760. Richardson, U. F., & Kendall, S. B. (1957). Trypanosomiasis. In *Veterinary Protozoology* (pp. 36–38). Oliver and Boyd.

761. Rijkers, G. T., Frederix-Wolters, E. M. H., & Van Muiswinkel, W. B. (1980). The immune system of cyprinid fish: The effect of antigen dose and route of administration on the development of immunological memory in carp (*Cyprinus carpio*). In M. J. Manning (Ed.), *Phylogeny of Immunological Memory* (pp. 93–102). Elsevier/North-Holland.

762. Riou, G. (1967). Induction d'une "dysyskinetoplastie" par le bromure d'ethidium clez *Trypanosoma* cruzi cultiv'e *in vitro. C. R. Acad. Sci., Hebd. Seances Acad. Sci. D., 265*, 2004–2007.

763. Roberts, L. S., & Mong, F. N. (1968). Developmental physiology of

cestodes. III. Development of *Hymenolepis diminuta* in super-infections. *J. Parasitol.*, *54*, 55–62.

764. Roberts, S. C., Swihart, K. G., Agey, M. W., Ramamoorthy, R., Wilson, M. E., & Donelson, J. E. (1993). Sequence diversity and organization of the *msp* gene family encoding *gp63* of *Leishmania chagasi*. *Mol. Biochem. Parasitol.*, *62*(2), 157–172.
765. Robertson, M. (1906). Notes on certain blood inhabiting protozoa. *Proc. R. Phy. Soc. Edinburgh*, *16*, 232.
766. Robertson, M. (1909). Further notes on a trypanosome found in the alimentary tract of *Pontobdella muricata*. *J. Microsc. Sci.*, *54*, 119.
767. Robertson, M. (1911). Transmission of flagellates living in the blood of fresh water fishes. *Trans. R. Soc. Lond.*, *202*, 29–50.
768. Rodhain, J. (1942). A props de trypanosomes de poisons du bașin du fleuve Congo. *Rev. Zool. Bot. Afr.*, *36*, 411.
769. Roitt, I. M. (1994). *Essential Immunology* (8th ed.). Blackwell Science Ltd.
770. Rose, S. M., & Rose, F. C. (1961). Mechanisms in biological competition. In F. L. Milthrope (Ed.), *The University Press* (pp. 207–218).
771. Rosenthal, H. L., Pfluke, M. L., & Buscaghia, S. (1960). A stable ion reagent for determination of cholesterol. *J. Lab. Clin Med.*, *50*, 318–322.
772. Rosenthal, P. J. (1999). Proteases of protozoan parasites. *Adv. Parasitol.*, *43*, 105–159.
773. Ross, M. H., & Reith, E. J. (1985). Connective Tissue Cells. In M. H. Ross & E. J. Reith (Eds.), *Histology, a Text and Atlas* (pp. 92–100). Harper and Row Publishers.
774. Rouskova, H., & Kulda, J. (1983). Growth and transformation of *Trypanosoma danilewskyi* in a liquid medium. *Folia Parasitol.*, *30*, 295–301.
775. Russell, D. W. (2000). Oxysterol biosynthetic enzymes. *Biochim. Biophys. Acta*, *1529*, 126–135.
776. Ruszczyk, A., Forlenza, M., Joerink, M., Ribeiro, C. M. S., Jurecka, P., & Wiegertjes, G. F. (2008). *Trypanoplasma borreli* cysteine proteinase activities support a conservation of function with respect to digestion of host proteins in common carp. *Develop. Comp. Immunol.*, *32*(11), 1348–1361.
777. Ryce, E. K. N., Zale, A. V., & MacConnell, E. (2004). Effects of fish age and parasite dose on the development of whirling disease in rainbow trout. *Dis. Aquat. Org.*, *59*(3), 225–233.

778. Ryce, E. K. N., Zale, A. V., MacConnell, E., & Nelson, M. (2005). Effects of fish age versus size on the development of whirling disease in rainbow trout. *Dis. Aquat. Org.*, *63*(1), 69–76.

779. Saeij, J. P. J., Groeneveld, A., Rooijen, N. van, Haenen, O. L. M., & Wiegertjes, G. F. (2003). Minor effect of depletion of resident macrophages from peritoneal cavity on resistance of common carp *Cyprinus carpio* to blood flagellates. *Dis. Aquat. Org.*, *57*(1/2), 67–75.

780. Saeij, J. P. J., Verburg-van-Kemenade, L. B. M., Muiswinkel, W. B. van, & Wiegertjes, G. F. (2003). Daily handling stress reduces resistance of carp to *Trypanoplasma borreli*: *In vitro* modulatory effects of cortisol on leukocyte function and apoptosis. *Dev. Comp. Immunol.*, *27*(3), 233–245.

781. Saha, S. K., Homechaudhury, S., & Banerjee, S. (1997). Natural prevalence of *Trypanosoma mukundi* in *Heteropneustes fossilis* and related haematological analysis. *J. Aqua. Trop.*, *12*, 79–87.

782. Sahlab, A. A. N., Mahmood, N. A. M., & Ibrahim, M. M. (1990). Haematological changes due to *Trypanosoma mukassai* infection in Nile cat fish, *Clarias lazera*. *Egypt. J. Comp. Pathol. Clin. Pathol.*, *3*, 451–453.

783. Sakai, D. K. (1984). Opsonization by fish antibody and complement in the immune phagocytosis by peritoneal exudate cells isolated from salmonid fishes. *J. Fish Dis.*, *7*, 29–38.

784. Salvador, A., Veiga, J. P., Martin, J., & Lopez, P. (1997). Testosterone supplementation in subordinate, small male lizards: Consequences for aggressiveness, color development and parasite load. *Behav. Ecol.*, *8*, 135–139.

785. Sanchez, S., Cuervo, J. J., & Moreno, E. (2007). Does habitat structure affect body condition of nestling? A case study with woodland Great Tits *Parus major*. *Acta Ornithol.*, *42*(2), 200–204.

786. Sandness, K., Lie, O., & Waagbo, R. (1988). Normal ranges of some blood chemistry parameters in adult formed Atlantic salmon, *Salmo salar*. *J. Fish Biol.*, *32*, 129–136.

787. Santos-Gomes, G. M., Abranches, P., Maraghi, S., Dirie, M. F., Silva Pereira, M. C., Valverde, D., & Molyneux, D. H. (1993). Laboratory and field studies on *Herpetosoma* trypanosomes from Portugal. *Ann. De Parasitol. Humaine et Comp.*, *68*, 163–168.

788. Saratchandra, B., & Jayaramaraju, P. (1981). *Trypanosoma godavariensis* n. sp. from the fresh water fish, *Channa punctatus* (Bloch). *Proc. Indian Acad. Parasitol.*, *2*(2), 53–56.

789. Sarkar, A. (2010). *Trypanosoma*. In *Understanding Parasitology* (pp. 48–

61). Discovery Publishing House Pvt. Ltd.

790. Sarkar, N. K., Mazumdar, S. K., & Pramanik, A. (1985). Observation on 4 new species of myxosporidia (*Myxozoa*) from channid (*Ophicephalid*) fishes of West Bengal, India. *Arch. Protistenkd., 130*, 289–296.
791. Saunders, D. C. (1959). *Trypanosoma balistes* n.sp. from *Balistes capriscus* Gouelin, the common trigger fish from the Florida Keys. *J. Parasitol., 45*, 623–626.
792. Savan, R., Aman, A., Sato, K., Yamaguchi, R., & Sakai, M. (2005). Discovery of a new class of immunoglobulin heavy chain from Fugu. *European Journal of Immunology, 35*, 3320–3331.
793. Sawhney, A. K., & Johal, M. S. (2000). Erythrocyte alterations induced by malathion in *Channa punctatus* (Bloch). *Bulletin of Environmental Contamination and Toxicology, 64*, 398–405.
794. Schad, G. A. (1966). Immunity, competition and natural regulation of helminth populations. *The American Naturalist, 100*, 359–364.
795. Schall, J. J., & Pearson, A. R. (2000). Body condition of a Puerto Rican anole, *Anolie gundlachi*: Effect of a malaria parasite and weather variation. *Journal of Herpetology, 34*, 489–491.
796. Schall, J. J., Pearson, A. R., & Perkins, S. L. (2000). Prevalence of malaria parasites (*Plasmodium floridense* and *Plasmodium azurophilum*) infecting a Puerto Rican lizard (*Anolie gundlachi*): A nine-year study. *Journal of Parasitology, 86*, 511–515.
797. Schaller, J. (2008). *A Laboratory Guide to Human Babesia Haematology Forms*. Hope and Academic Press.
798. Schalm, O. W. (1962). Leukocyte responses to disease in various domestic animals. *Journal of the Veterinary Medical Association, 140*, 557.
799. Schalm, O. W. (1967). *Veterinary Haematology* (2nd ed.). Lee and Febiger.
800. Scharfman, W. B., & Taft, E. G. (1977). Nantucket fever: An additional case of babesiosis. *Journal of the American Medical Association, 238*, 1281–1282.
801. Scharsack, J. P., Steinhagen, D., Kleczka, C., Schmidt, J. O., Korting, W., Michael, R. D., Leibold, W., & Schuberth, H. J. (2003a). Head kidney neutrophils of carp (*Cyprinus carpio* L.) are functionally modulated by the haemoflagellate *Trypanoplasma borreli*. *Fish and Shellfish Immunology, 14*(5), 389–403.
802. Scharsack, J. P., Steinhagen, D., Kleczka, C., Schmidt, J. O., Korting, W., Michael, R. D., Leibold, W., & Schuberth, H. J. (2003b). The haemoflagellate *Trypanoplasma borreli* induces the production of nitric

oxide, which is associated with modulation of carp (*Cyprinus carpio* L.) leukocyte functions. *Fish and Shellfish Immunology*, *14*(3), 207–222.

803. Scharsack, J. P., Steinhagen, D., Korting, W., Wanger, B., Leibold, W., & Schuberth, H. J. (2004). Some immune parameters in carp *Cyprinus carpio* susceptible and resistant to the haemoflagellate *Trypanoplasma borreli*. *Diseases of Aquatic Organisms*, *60*(1), 41–48.

804. Schiller, E. L. (1959). Experimental studies on morphological variations in the cestode genus *Hymenolepis*. IV. Influence of the host on variation in *H. nana*. *Experimental Parasitology*, *8*, 581–590.

805. Schilling, C. (1936). Versuche zur Schutzimpfung gegen Tsetsekrankheit. *Zeitschrift für Immunitätsforschung und Experimentelle Therapie*, *87*, 482–518.

806. Schonhofer, P. S., & Peters, H. D. (1977). Role of cyclic nucleotides in cultured cells. In Cramer, H., & Schultz, J. (Eds.), *Cyclic 3, 5- Nucleotides: Mechanisms of Action* (pp. 107–131). J. Wiley and Sons.

807. Schumacher, R. E., Hamilton, C. H., & Longtin, E. J. (1956). Blood sedimentation rates of brook trout as affected by furunculosis. *Progressive Fish-Culturist*, *18*, 147, 148.

808. Schuster, F. L. (2002). Cultivation of *Babesia* and *Babesia*-like blood parasites: Agents of emerging zoonotic disease. *Clinical Microbiology*, *15*, 365–373.

809. Sebek, Z. (1960). Nektere nove poznatky o krevnich parasitech nasich hmyzozravcu a mysovitych. *Vlastivedny Sbornik Vysočiny. Oddíl Věd Přírodních*, 75–80.

810. Sebek, Z. (1975). Blood parasites of small wild animals in Czechoslovakia. *Folia Parasitologica*, *22*, 11–20.

811. Sebek, Z., Rosicky, B., & Sixl, W. (1977). The occurrence of babesiosis affecting small terrestrial mammals and the importance of this zoonosis in Europe. *Folia Parasitologica*, *24*, 211–228.

812. Sebek, Z., Sixl, W., Stunzner, D., Valova, M., Hubalek, Z., & Troger, H. (1980). Blood parasites of small wild mammals in Steiermark and Burgenland. *Folia Parasitologica*, *27*, 295–301.

813. Seed, J. R., Bryam, J., & Gam, A. A. (1967). Characterization and localization of acid phosphatase activity of *Trypanosoma gambiense*. *Journal of Protozoology*, *14*, 117–125.

814. Segel, G. B. (1981). Membrane alterations in lymphocyte proliferation. *American Journal of Pediatric Hematology/Oncology*, *3*, 433.

815. Seshadri, C., Shetty, R., Gauri, N., Venkataraghavan, S., & Chari, M. V.

(1981). Serum cholesterol and total lipids in *Plasmodium vivax* malaria–A preliminary study. *Indian Journal of Medical Research*, *74*, 513–517.

816. Seyfang, A., & Duszenko, M. (1991). Specificity of glucose transport in *Trypanosoma brucei*. *European Journal of Biochemistry*, *202*, 191–196.
817. Shanavas, K. R., Ramachandran, P., & Janardanan, K. P. (1989). *Trypanosoma ompoki* n. sp. from freshwater fishes in Kerala, India, with observations on its vector-phase development and transmission. *Acta Protozoologica*, *28*(3, 4), 293–302.
818. Shapiro, T. A. (1993). Inhibition of topoisomerases in African trypanosomes. *Acta Tropica*, *54*, 251–260.
819. Sharma, P., & Ratnu, L. S. (1984). *An Introduction to Parasitology*. S. Chand and Co.
820. Sharma, T., & Joshi, B. D. (1984). Hematological studies on a hillstream fish *Nemacheilus rupicola*. *Comparative Physiology and Ecology*, *9*, 67–79.
821. Sharma, T., & Joshi, B. D. (1985). Effect of asphyxiation on some hematological values of *Nemacheilus rupicola*. *International Journal of Academic Ichthyology*, *6*, 17–22.
822. Sharma, T., & Joshi, B. D. (1986a). On some hematological values of the fish *Nemacheilus rupicola* as affected by a sudden change in its ambient water salinity. *Proceedings of the National Symposium on Fish and Environment*, 11–15.
823. Sharma, T., & Joshi, B. D. (1986b). Hematological alteration in hillstream fish *Tor putitora* in relation to body size. *Journal of Advanced Zoology*, *7*, 93–96.
824. Sharma, T., & Joshi, B. D. (1991). Haematological alterations in a hillstream fish, *Garra gotyla*, due to trypanosome infection. *Himalayan Journal of Environment and Zoology, 5*, 57–60.
825. Shaw, D. J., & Dobson, A. P. (1995). Patterns of macroparasite abundance and aggregation in wildlife populations: A quantitative review. *Parasitology, 111*(Suppl.), S111–S133.
826. Shell, E. W. (1961). Chemical composition of blood of small mouth bass. *Research Report U. S. Fish and Wildlife Service, 57*, 1–36.
827. Sherburne, S. W. (1973). Cell types, differential cell counts, and blood cell measurements of a Portuguese shark, *Centroscymnus coelolepis*, captured at 700 fathoms. *Fishery Bulletin, 71*, 435.
828. Shih, C. M., Liu, L. P., Chung, W. C., Ong, S. J., & Wang, C. C. (1997). Human babesiosis in Taiwan: Asymptomatic infection with a *Babesia*

microti-like organism in a Taiwanese woman. *Journal of Clinical Microbiology, 35*(2), 450–454.

829. Shiota, T., Kurimoto, H., Haguma, N., & Yoshida, Y. (1984). Studies on *Babesia* first found in murine in Japan: Epidemiology, morphology, and experimental infection. *Zentralblatt für Bakteriologie, Mikrobiologie und Hygiene, 256*, 347–355.
830. Shulman, S. S. (1958). Specificity of fish parasites. *Parasitology of Fishes*. University of Leningrad.
831. Siddique, S. F., & Naz, S. S. (2009). Studies on some haematological changes in fishes infected with helminth parasites of rivers Betwa, Pahunch, and other reservoirs of Jhansi. In B. N. Pandey, S. P. Trivedi, K. Jaiswal, & N. Gupta (Eds.), *World of Pests and Parasites* (pp. 91–95). Sarup Book Publishers.
832. Siddiqui, A. O., & Naseem, S. M. (1971). Seasonal variations in the biochemical composition of blood serum of *Heteropneustes fossilis* (Bloch) (Teleostomi, Heteropneustidae). *Kashmir Science, 8*, 41–50.
833. Simmons, V., Knight, R. H., & Humphyres, K. C. (1964). Separation of trypanosomes from rat blood components. In *10th Meeting of the International Scientific Committee for Trypanosomiasis Research*, Kampala (pp. 81–82). CCTA.
834. Simpson, A. G. B., Stevens, J. R., & Lukes, J. (2006). The evolution and diversity of kinetoplastid flagellates. *Trends in Parasitology, 22*, 168–174.
835. Simpson, L. (1972). The kinetoplast of the hemoflagellates. *International Review of Cytology, 32*, 139–207.
836. Simpson, M. A., Hughes, D., & Simpson, L. (1985). *Trypanosoma brucei*: Differentiation of *in vitro* grown blood stream *Trypanosoma*stigotes into procyclic forms. *Journal of Protozoology, 32*, 672–677.
837. Singh, R., & Singh, R. K. (1997). Toxic effects of fertilizer di-ammonium phosphate on serum cholesterol levels of fish *Channa punctatus*. *Advances in Biosciences, 16*, 87–91.
838. Singh, S., Singh, C., Kumar, A., Sinha, K. K., & Mishra, P. C. (2000). Certain haematological and biochemical profiles of a white tigress (*Panthera tigris Linnaeus*) suffering from trypanosomiasis. *Zoos' Print Journal, 15*, 207–208.
839. Sinha, C. K. (1986). Occurrence of *Trypanosoma mukasai* Hoare, 1932 in *Tilapia mossambica* (Peters) from India. *Acta Protozoologica, 25*, 449–452.
840. Sinski, E. (1999). Enzootic reservoir for new *Ixodes ricinus*-transmitted

infections. *Wiadomości Parazytologiczne, 45*, 135–142.

841. Sinski, E., & Karbowiak, G. (1995). Kleszcze (*Ixodidae*) w Zoonotycznym rozprzestrzenianiu piroplazm. *Wiadomości Parazytologiczne, 41*, 321–327.

842. Sitja-Bobadilla, A., & Woo, P. T. K. (1994). An enzyme-linked immunosorbent assay (ELISA) for detecting antibodies against the pathogenic haemoflagellate, *Cryptobia salmositica* Katz, and protection against cryptobiosis in juvenile rainbow trout, *Oncorhynchus mykiss* (Walbaum) inoculated with a live vaccine. *Journal of Fish Diseases, 17*, 399–408.

843. Skotarczak, B., & Cichocka, A. (2001a). Isolation and amplification by polymerase chain reaction (PCR) of *Babesia microti* and *Babesia divergens* DNA in ticks in Poland. *Annals of Agricultural and Environmental Medicine, 8*, 187–189.

844. Skotarczak, B., & Cichocka, A. (2001b). Occurrence of *Babesia microti* DNA in *Ixodes ricinus* ticks in the forest areas of Szczecin. *Folia Biologica, 49*, 247–250.

845. Skotarczak, B., Rymaszewska, A., Wodecka, B., & Sawczuk, M. (2003). Molecular evidence of co-infection with *Borrelia burgdorferi* sensu lato, human granulocytic ehrlichiosis agent, and *Babesia microti* in ticks from northwestern Poland. *Journal of Parasitology, 89*, 194–196.

846. Skotarczak, B., Wodecka, B., & Cichocka, A. (2002). Coexistence of *Borrelia burgdorferi* sensu lato and *Babesia microti* DNA in *Ixodes ricinus* ticks from northwestern Poland. *Annals of Agricultural and Environmental Medicine, 9*, 25–28.

847. Sleigh, M. A. (1991). The nature of protozoa. In J. P. Krier & J. O. Baker (Eds.), *Parasitic Protozoa* (Vol. 1, pp. 1–53). Academic Press.

848. Smallridge, C. J., & Bull, C. M. (2000). Transmission of the blood parasite *Haemolivia mariae* in a field population of the skink, *Tiliqua rugosa*. *Parasitology Research, 86*, 563–569.

849. Smirnova, L. I. (1970). *Trypanosoma* in the blood of *Lota lota* (*T. lotai* sp. nov.). *Parasitologie, 4*, 296–297.

850. Smit, N. J., Van As, J. G., & Davies, A. J. (2004). Fish trypanosomes from the Okavango Delta, Botswana. *Folia Parasitologica, 51*, 293–303.

851. Smith Trail, D. R. (1980). Behavioural interactions between parasites and hosts: Host suicide and the evolution of the complete life cycle. *The American Naturalist, 116*, 77–91.

852. Smolikova, V. (1977). Observations on the biology of the blood flagellates

of fishes (Unpublished master's thesis). Department of Parasitology, Charles University, Prague.

853. Smyth, J. D. (1996). *Animal Parasites*. Cambridge University Press.
854. Smyth, J. D., & Haslewood, G. A. D. (1963). The biochemistry of bile as a factor in determining host specificity in intestinal parasites, with particular reference to *Echinococcus granulosus*. *Annals of the New York Academy of Sciences, 113*, 234–260.
855. Snedecor, G. W., & Cochran, W. G. (1989). *Statistical Methods*. Oxford and IBH Publishing Co.
856. Sobecka, E., & Sominsha, M. (2007). Species richness, diversity, and specificity of the parasites of bream *Abramis brama* (L.) and perch *Perca fluviatilis* (L.) in the estuary of the Odra River, Poland. *Helminthologia, 44*(4), 188–192.
857. Sogin, M. L., & Silberman, J. D. (1998). Evolution of the protists and protistan parasites from the perspective of molecular systematics. *International Journal of Parasitology, 28*, 11–20.
858. Sol, D., Jovani, R., & Torrese, J. (2000). Geographical variation in blood parasites in feral pigeons: The role of vectors. *Ecography, 23*, 307–314.
859. Soulsby, E. J. L. (1976). Pathophysiology of parasitic infection. In *Pathophysiology of Parasitic Infection* (pp. 55–61). Academic Press.
860. Sousa, W. P. (1994). Patterns and processes in communities of helminth parasites. *Trends in Ecology & Evolution, 9*, 52–57.
861. Spielman, A. (1988). Lyme disease and human babesiosis: Evidence incriminating vector and reservoir hosts. In P. T. Englund & A. Sher (Eds.), *The Biology of Parasitism* (Vol. 9, pp. 147–165). Alan R. Liss Inc.
862. Srivastava, C. B. (1975). Fish pathological studies in India: A brief review. *Dr. B. S. Chauhan Commemoration Volume*, 349–358.
863. Srivastava, K., & Bhattacharya, S. M. (2003). Tetracyclines as a tool for transmission blocking of *Brugia malayi* in *Mastomys coucha*. *Current Science, 85*(5), 588–589.
864. Srivastava, S., & Choudhary, S. K. (2010). Effect of artificial photoperiod on the blood cell indices of the catfish, *Clarias batrachus*. *Journal of Stress Physiology & Biochemistry, 6*(1), 22–32.
865. Stanczak, J., Gabre, R. M., Kruminis-Lozowska, W., Racewicz, M., & Kubica-Biernat, B. (2004). *Ixodes ricinus* as a vector of *Borrelia burgdorferi* sensu lato, *Anaplasma phagocytophilum* and *Babesia microti* in urban and suburban forests. *Annals of Agricultural and Environmental Medicine, 11*, 109–114.

866. Steinhagen, D., Kruse, P., & Korting, W. (1989). Effects of immunosuppressive agents on common carp infected with the haemoflagellate *Trypanoplasma borreli*. *Diseases of Aquatic Organisms, 7*(1), 67–69.

867. Steinhagen, D., Kruse, P., & Korting, W. (1990). Some haematological observations on common carp, *Cyprinus carpio* L., experimentally infected with *Trypanoplasma borreli* Laveran and Mesnil 1901. *Journal of Fish Diseases, 13*, 157–162.

868. Steketee, R. W., Eckman, M. R., Burgess, E. C., Kuritsky, J. N., Dickerson, J., Schell, W. D., Godsey, M. S., & Davis, J. D. (1985). Babesiosis in Wisconsin: A new focus of disease transmission. *Journal of the American Medical Association, 253*, 2675–2678.

869. Stevens, D. R., & Moulton, E. (1978). Ultrastructural and immunological aspects of the phagocytosis of *Trypanosoma brucei* by mouse peritoneal macrophages. *Infection and Immunity, 19*, 972–982.

870. Stevens, J. R., Hamilton, P. B., & Brisse, S. (2011). The molecular evolution of trypanosomes. In G. Hide (Ed.), *The Molecular Epidemiology of Trypanosomes and Leishmania*. Springer.

871. Stevens, J. R., Teixeira, M. M. J., Bingle, L. E. H., & Gibson, W. C. (1999). The taxonomic position and evolutionary relationships of *Trypanosoma rangeli*. *International Journal of Parasitology, 29*, 749–757.

872. Stoskopf, M. K. (1993). Clinical pathology. In M. K. Stoskopf (Ed.), *Fish Medicine* (pp. 113–131). Saunders.

873. Strout, R. G. (1962). A method for concentrating haemoflagellates. *Journal of Parasitology, 48*, 100.

874. Strout, R. G. (1965). A new haemoflagellate (Genus *Cryptobia*) from marine fishes of Northern New England. *Journal of Parasitology, 51*, 654–659.

875. Suchankova, E., & Lom, J. (1977). Growth of the carp trypanosome, *T. danilewskyi*, in fish tissue culture. *Journal of Protozoology, 24*, 54A.

876. Sudarto, M. W., Tabel, M., & Haines, D. M. (1990). Immunohistochemical demonstration of *Trypanosoma evansi* in tissue of experimentally infected rats and naturally infected water buffalo (*Bubalus bubalis*). *Journal of Parasitology, 76*(2), 162–167.

877. Sullivan, J. J. (1979). The role of eosinophils in inflammatory reactions. *Progress in Hematology, 11*, 65.

878. Svobodova, Z., & Kolarova, J. (2004). A review of the diseases and contaminant-released mortalities of tench (*Tinca tinca* L.). *Veterinarni*

Medicina, 49(1), 19–34.

879. Swezy, O. (1919). The occurrence of *Trypanoplasma* as an ectoparasite. *Transactions of the American Microscopical Society, 38*, 20–24.

880. Sypek, J. P., & Burreson, E. M. (1983). Influence of temperature on the immune response of juvenile summer flounder, *Paralichthys dentatus*, and its role in the elimination of *Trypanoplasma bullocki* infections. *Developmental & Comparative Immunology, 7*, 277–286.

881. Szulman, S. S. (1984). Protozoa. In *Opredelitel Parazitov Presnovodnych ryb SSSR* (pp. 252–280). Moskwa-Leningrad.

882. Tait, A. (1980). Evidence for diploidy and mating in trypanosomes. *Nature, 283*, 536–538.

883. Takayanagi, T., & Nakataki, Y. (1977). *Trypanosoma gambiense*: The binding activity of antiserum to macrophages. *Experimental Parasitology, 42*, 21–26.

884. Tanabe, M. (1925). Studies on the haemoflagellate of the loach, *Misgurnus anguillicaudatus*. *Kitasato Archives of Experimental Medicine, 6*, 121–138.

885. Tandon, R. S. (1973). Studies on "Crowding Effect" on *Gastrothylax crumenifer* and *Fischoederius elongatus*, the common amphistome parasites of ruminants, observed under natural conditions. *Research Bulletin of the Meguro Parasitological Museum, 7*, 12–14.

886. Tandon, R. S. (1986). Biochemical changes in fish (*Wallago attu*) infected with *Trypanosoma*. *Indian Journal of Parasitology, 10*, 87–91.

887. Tandon, R. S., & Chandra, S. (1977a). Physiology of host-parasite relationships: Effect on serum alkaline phosphatase levels of fish hosts parasitized by trypanosomes. *Z. Parasitenkd., 52*, 195–198.

888. Tandon, R. S., & Chandra, S. (1977b). Studies on the ecophysiology of fish parasites: Effects of trypanosome infection on serum cholesterol levels of fish. *Z. Parasitenkd., 52*, 199–202.

889. Tandon, R. S., & Chandra, S. (1978). Physiology of host-parasite relationships: Effect on serum acid phosphate levels of fish hosts parasitized by trypanosomes. *J. Inland Fish. Soc. India, 10*, 159–169.

890. Tandon, R. S., & Joshi, B. D. (1973). Studies on the physiopathology of blood of freshwater fishes infected with two new forms of trypanosomes. *Z. wiss. Zool., 185*, 207–221.

891. Tandon, R. S., & Joshi, B. D. (1974). Effect of trypanosome infection on blood glucose levels of some freshwater teleosts. *J. Inl. Fish. Soc. India, 6*, 81–82.

892. Tandon, R. S., & Masih, B. P. (1981). Effect of trypanosome infection on blood ascorbic acid levels of freshwater fishes. *Indian J. Parasitol.*, *5*, 189–190.

893. Tandon, R. S., & Masih, B. P. (1983). Levels of serum cholinesterase in freshwater fishes infected with trypanosomes. *Indian J. Parasitol.*, *7*, 195–197.

894. Tavares-Dias, M., & Moraes, F. R. (2007). Leukocyte and thrombocyte reference values for channel catfish (*Ictalurus punctatus* Raf.), with an assessment of morphological, cytochemical and ultrastructural features. *Vet. Clin. Pathol.*, *36*, 49–54.

895. Taylor, A. E. R., & Baker, J. R. (1978). *Methods of Cultivating Parasites in vitro*. Academic Press.

896. Tella, J. L., Blanco, G., Forero, M. G., Gajon, A., Donazar, J., & Hiraldo, F. (1999). Habitat world geographic range and embryonic development of hosts explain the prevalence of avian hematozoa at small spatial and phylogenetic scales. *Proc. Natl. Acad. Sci.*, *96*, 1785–1789.

897. Tettey, J. N. A., Skellern, G. G., Grant, M. H., & Midgley, J. M. (1999). Investigation of the chemical equivalence of the trypanocidal products, Samorin® and Veridium®. *J. Pharm. Biomed. Anal.*, *21*, 1–7.

898. Thelohan, P. (1892). Observation sur les myxosporidies et essai de classification de ces organisms. *Bull. Soc. Phil.*, *4*, 165–178.

899. Thiroux, A. (1905). Recherches morphologiques et expérimentales sur *Trypanosoma duttoni* (Thiroux). *Ann. Inst. Pasteur*, *19*, 565–572.

900. Thomas, J. D. (1958). Studies on *Crepidostomum metoecus* (Braun) and *C. farionis* (Muller), parasitic in *Salmo trutta* L. and *S. salar* L. in Britain. *Parasitology*, *48*, 336–352.

901. Thomas, P. T., & Woo, P. T. K. (1990). Dietary modulation of humoral immune response and anemia in *Oncorhynchus mykiss* (Walbaum) infected with *Cryptobia salmositica* Katz, 1951. *J. Fish Dis.*, *13*, 435–446.

902. Thomas, P. T., & Woo, P. T. K. (1991). *In vitro* and *in vivo* effects of antimicrobial agents on viability of *Cryptobia salmositica* (Sarcomastigophora: Kinetoplastida). *Dis. Aquat. Org.*, *10*, 7–11.

903. Thomas, P. T., & Woo, P. T. K. (1995). Immunological approaches and techniques. In P. T. K. Woo (Ed.), *Fish Diseases and Disorders* (pp. 751–771). CAB International.

904. Thomas, P. T., Ballantyne, J. S., & Woo, P. T. K. (1992). *In vitro* oxygen consumption and motility of *Cryptobia salmositica*, *Cryptobia bullocki*, and *Cryptobia catostomid* (Sarcomastigophora: Kinetoplastida). *J.*

Parasitol., *78*, 747–749.

905. Thomaz, N., & Deane, M. P. (1990). Continuous growth and differentiation of *Trypanosoma (Megatrypanum) freitasi* (Rego, Magalhaes, and Siqueira, 1957), *in vitro*. *J. Protozool.*, *37*, 40–43.

906. Thompson, G. E. (1977). Physiological effect of cold. In D. Robertshaw (Ed.), *Environmental Physiology* (pp. 29–69). Baltimore, M.D.: University Park.

907. Thompson, J. D. (1908). Cultivation of the trypanosome found in the blood of the goldfish. *J. Hyg.*, *8*, 75–82.

908. Tierney, J. F., Huntingford, F. A., & Crompton, D. W. T. (1993). The relationship between infectivity of *Schistocephalus solidus* (Cestoda) and anti-predator behaviour of its intermediate host, the three-spined stickleback, *Gasterosteus aculeatus*. *Animal Behaviour*, *46*, 603–605.

909. Tizard, I. R. (2009). *Immunology: An Introduction* (3rd Indian reprint). Cengage Learning India Pvt. Ltd.

910. Tobie, E. J. (1954). Present status of chemotherapy and chemoprophylaxis of animal trypanosomiasis in the eastern hemisphere. *J. Parasitol.*, *43*, 291.

911. Tobie, E. J., von Brand, T., & Mehlman, B. (1950). Cultural and physiological observations on *Trypanosoma rhodesiense* and *Trypanosoma gambiense*. *J. Parasitol.*, *36*, 48–54.

912. Tort, L., Balasch, J. C., & Mackenzie, S. (2003). Fish immune system: A crossroads between innate and adaptive responses. *Immunologia*, *22*, 277–286.

913. Tuntasuvan, D., Jarabrum, W., Viseshakul, N., Mohvaew, K., Borisutsuwan, S., Theeraphan, A., & Kongkanjana, N. (2003). Chemotherapy of surra in horses and mules with dimenazene aceturate. *Vet. Parasitol.*, *110*, 227–233.

914. Turner, C. M. R. (1986). Seasonal and age distribution of *Babesia*, *Hepatozoon*, *Trypanosoma*, and *Grahamella* species in *Clethrionomys glareolus* and *Apodemus sylvaticus* populations. *Parasitology*, *93*, 279–289.

915. Turner, C. M. R., & Cox, F. E. G. (1985). Interspecific interactions between blood parasites in wild rodent communities. *Ann. Trop. Med. Parasitol.*, *79*, 463–465.

916. Ugbomoiko, U. S. (1997). Factors affecting the prevalence of protozoan parasites of small mammals in southern Nigeria. *Parasitica*, *53*, 5–13.

917. Ugbomoiko, U. S., & Obimiwe, B. A. (1991). Helminths of small mammals

in a forest zone of southern Nigeria. *Helminthologica*, *28*, 19–22.

918. Urdaneta-Morales, S. (1983). Pleomorphism in trypomastigotes of *Trypanosoma cruzi* from blood and cell culture. *Tropical Medicine & Parasitology*, *34*, 225–228.
919. Valentin, G. G. (1841). Ube rein Entozoon in Blute Von, *Salmo fario*. *Archives of Anatomy and Physiology*, 435.
920. Valenzuela, A. E., Silva, V. M., & Klempau, A. E. (2008). Effects of different artificial photoperiods and temperatures on haematological parameters of rainbow trout (*Oncorhynchus mykiss*). *Fish Physiology and Biochemistry*, *34*(2), 159–167.
921. Valli, V. E. O., & Mills, J. N. (1980). The quantitation of *T. congolense* in calves I. Haematological changes. *Parasitology*, *31*, 215–231.
922. Van Dallen, R. G. (1981). The eosinophil: Friend or foe? Editorial. *Mayo Clinic Proceedings*, *56*, 395.
923. Van den Bossche, P., & Doran, M. (2004). *Trypanocidal Drugs: Use and Misuse?* Harare, Zimbabwe: WHO/Regional Tsetse and Trypanosomiasis Control Programme.
924. Van Muiswinkel, W. B., & Vervoorn-Van Der Wal, B. (2006). The immune system of fish. In P. T. K. Woo (Ed.), *Fish Diseases and Disorders* (Vol. 1, pp. 678–701). CABI.
925. Van Muiswinkel, W. B., Enderson, D. P., Lamers, C. H. J., Egberts, E., Van Loon, J. J. A., & Ijssel, J. P. (1985). Fish immunology and fish health. In M. J. Manning & M. F. Tatner (Eds.), *Fish Immunology* (pp. 1–8). Academic Press.
926. van Peenen, P. F., Chang, S. J., Banknieder, A. R., & Santana, F. J. (1977). Piroplasms from Taiwanese rodents. *Journal of Protozoology*, *24*, 310–312.
927. Vassella, E., Oberle, M., Urwyler, S., Renggli, C. K., Studer, E., et al. (2009). Major surface glycoproteins of insect forms of *Trypanosoma brucei* are not essential for cyclical transmission by tsetse. *PLoS ONE*, *4*(2), e4493. https://doi.org/10.1371/journal.pone.0004493.
928. Vaughn, J. (1953). The function of the eosinophil leukocyte. *Blood*, *8*, 1.
929. Veiga, J. P., Salvador, A., Merino, S., & Puerta, M. (1998). Reproductive efforts affect immune response and parasite infection in a lizard: A phenotypic manipulation using testosterone. *Oikos*, *82*, 313–318.
930. Verlinde, C. L., Hannaert, V., Blonski, C., Willson, M., Périé, J. J., Fothergill-Gilmore, L. A., Opperdoes, F. R., Gelb, M. H., Hol, W. G., & Michels, P. A. (2001). Glycolysis as a target for the design of new anti-

trypanosome drugs. *Drug Resistance Updates*, *4*, 50–65.

931. Verma, B. B., Gautam, O. P., & Malik, P. D. (1973). The influence of *Trypanosoma congolense* infection on the disposition kinetics of diminazene aceturate in dogs. *Veterinary Record*, *93*, 465.
932. Vickerman, K. (1973). The mode of attachment of *T. vivax* in the proboscis of the tsetse fly *Glossina fuscipes*: An ultrastructural study of the epimastigote stage of the trypanosome. *Journal of Protozoology*, *20*, 394–404.
933. Vickerman, K. (1976). The diversity of the kinetoplastid flagellates. In W. H. R. Lumsden & D. A. Evans (Eds.), *Biology of the Kinetoplastida* (Vol. 1, pp. 1–34). Academic Press.
934. Vickerman, K., & Preston, T. M. (1970). Spindle microtubules in the dividing nuclei of trypanosomes. *Journal of Cell Science*, *6*, 365–383.
935. Vickerman, K., & Preston, T. M. (1976). Comparative cell biology of the kinetoplastid flagellates. In W. H. R. Lumsden & D. A. Evans (Eds.), *Biology of the Kinetoplastida* (Vol. 1, pp. 35–130). Academic Press.
936. Vickery, W. L., & Poulin, R. (1998). Parasite extinction and colonization and the evolution of parasite communities: A simulation study. *International Journal for Parasitology*, *28*, 727–737.
937. Viens, P., Targett, G. A. T., Leuchars, E., & Davies, A. J. D. (1974). The immunological response of CBA mice to *Trypanosoma musculi* I. Initial control of infections and the effect of T cell deprivation. *Clinical and Experimental Immunology*, *16*, 279–294.
938. von Brand, T. (1973). *Biochemistry of Parasites*. Academic Press.
939. Wag, W. R., Washbourn, K., Brown, K. N., & Hill, J. (1958). Structure of Isometamidium. *Nature*, *182*, 10.
940. Wagner, T. E. (1971). Physical studies on the interaction of lysergic acid diethylamide and trypanocidal dyes with DNA and DNA-containing genetic material. In F. E. Hahn (Ed.), *Progress in Molecular and Subcellular Biology* (Vol. 2, pp. 400). Springer-Verlag.
941. Wahul, M. A. (1985). Four new species of trypanoplasms from the fresh water fishes of the genus *Mystus* in Maharashtra. *Proceedings of the Indian Academy of Sciences*, *94*, 25–35.
942. Wahul, M. A. (1986a). Morphology of three new trypanosomes from fresh water fishes of the genus *Puntius*. *Archives of Protistenkunde*, *132*, 101–111.
943. Wahul, M. A. (1986b). *T. wallagoi* sp. nov. from the fresh water shark *Wallago attu* of Maharashtra (India). *Indian Journal of Parasitology*, *10*,

229–233.

944. Wahul, M. A. (1986c). Studies on the morphology of some *Trypanoplasma* from the blood of freshwater fishes in Maharashtra. I. From fishes of the genus *Puntius*. *Archives of Protistenkunde*, *131*, 309–318.
945. Wahul, M. A. (1987). Studies on the morphology of three new species of trypanosomes from the fresh water fish *Mastaembetus armatus* in Maharashtra, India. *Archives of Protistenkunde*, *133*, 111–124.
946. Waksman, S. A., Robinson, H., Metzger, H. J., & Woodruff, H. E. (1941). Toxicity of actinomycin. *Proceedings of the Society for Experimental Biology and Medicine*, *47*, 261.
947. Walker, R. I., & Willemze, R. (1980). Neutrophil kinetics and the regulation of granulopoiesis. *Reviews of Infectious Diseases*, *2*, 282.
948. Walter, G., & Liebisch, A. (1980). Studies of the ecology of some blood protozoa of wild mammals in North Germany. *Acta Tropica*, *37*, 31–40.
949. Wang, C. C. (1995). Molecular mechanisms and therapeutic approaches to the treatment of African trypanosomiasis. *Annual Review of Pharmacology and Toxicology*, *35*, 93–127.
950. Wang, R., & Belosevic, M. (1994). Cultivation of *Trypanosoma danilewskyi* (Laveran and Mesnil, 1904) in serum-free medium and assessment of the course of infection in goldfish, *Carassius auratus*. *Journal of Fish Diseases*, *17*, 47–56.
951. Warr, G. W. (1995). The immunoglobulin genes of fish. *Developmental and Comparative Immunology*, *19*, 1–12.
952. Warsi, A. A., & Fattohy, Z. I. (1976). *Trypanosoma acanthobramae* n.sp. from a fresh water fish *Acanthobrama marmid* (Heckel) (Family: Cyprinidae) from the river Tigris, Iraq. *Current Science*, *215*, 838–839.
953. Watkins, R. A., Moshier, S. E., O'Dell, W. D., & Pinter, A. J. (1991). Splenomegaly and reticulocytosis caused by *Babesia microti* infections in natural populations of the montane vole, *Microtus montanus*. *Journal of Protozoology*, *38*, 573–576.
954. Weatherhead, P. J., & Bennett, G. F. (1991). Ecology of red-winged blackbird parasitism by hematozoa. *Canadian Journal of Zoology*, *69*, 2352–2359.
955. Wedemeyer, G. A., & McLeay, D. J. (1973). c.f. von Brand, T. (ed.), (1973). *Biochemistry of Parasites* (499 pp.). Academic Press.
956. Wehnert, S. D., & Woo, P. T. K. (1980). *In vivo* and *in vitro* studies on the host specificity of *Trypanoplasma salmositica*. *Journal of Wildlife Diseases*, *16*, 183–187.

957. Wehnert, S. D., & Woo, P. T. K. (1981). The immune response of *Salmo gairdneri* during *Trypanoplasma salmositica* infection. Abstract, Canadian Society of Zoologists Annual Meeting, Waterloo, Ontario.

958. Weir, D. M. (1983). *Immunology* (255 pp.). ELBS, Edinburgh.

959. Wendelaar Bonga, S. E. (1997). The stress response in fish. *Physiological Reviews*, *77*, 591–625.

960. Wenyon, C. M. (1926). *Protozoology* (Vol. 1). Wm. Wood, New York (reprinted by Hafner).

961. Whitelaw, D. D., Bell, I. R., Holmes, P. H., Moloo, S. K., Hirumi, H., Urquhart, G. M., & Murray, M. (1986). Isometamidium chloride prophylaxis against *Trypanosoma congolense* challenge and the development of immune responses in Boran cattle. *Veterinary Record*, *118*, 722–726.

962. Wiegertjes, G. F., & Joerink, M. (2004). Macrophage polarization in the immune response to parasites. *Bulletin of the European Association of Fish Pathologists*, *24*(1), 5–10.

963. Wiegertjes, G. F., Groeneveld, A., & van Muiswinkel. (1995). Genetic variation in susceptibility of *Trypanoplasma borreli* infection in common carp (*Cyprinus carpio* L.). *Veterinary Immunology and Immunopathology*, *47*, 153–161.

964. Williams, C. B. (1951). Intrageneric competition as illustrated by Moreau's records of East African bird communities. *Journal of Animal Ecology*, *20*, 246–253.

965. Williams, H. H. (1968). *Phyllobothrium piriei* sp. nov. (Cestoda: Tetraphyllidea) from *Raja naevus* with a comment on its habitat and mode of attachment. *Parasitology*, *58*, 929–937.

966. Williamson, J. (1970). Review of chemotherapeutic and chemoprophylactic agents. In H. W. Mulligan (Ed.), *The African Trypanosomiasis* (Ch. 7). G. Allen & Unwin.

967. Williamson, J. (1976). Chemotherapy of African trypanosomiasis. *Transactions of the Royal Society of Tropical Medicine and Hygiene*, *70*, 117–119.

968. Williamson, J., & Macadam, R. F. (1975). Drug-induced lesions in trypanosome fine structure: A guide to modes of trypanocidal action. *Biochemical Pharmacology*, *24*, 147–151.

969. Williamson, J., & Rollo, I. M. (1959). Drug resistance in trypanosomes: Cross-resistance analyses. *British Journal of Pharmacology and Chemotherapy*, *14*, 423–430.

970. Wilson, C. B. (1916). Copepod parasites of fresh-water fishes and their economic relation to mussel glochidia. *Bulletin of the U.S. Bureau of Fisheries*, *34*, 131–374.

971. Wilson, E. O., & Bossert, W. H. (1964). Chemical communication among animals. *Recent Progress in Hormone Research*, *19*, 673–716.

972. Wilson, V. C., Ellis, D., & Upton, C. P. (1971). Preliminary observations on the fine structure of the trypomastigote form of *Trypanosoma (Herpetosoma) musculi* in the bloodstream phase. *Transactions of the Royal Society of Tropical Medicine and Hygiene*, *65*, 17.

973. Wintrobe, M. M. (1981). *Clinical Haematology* (8th ed., 2021 pp.). Lea and Febiger.

974. Wita, I., & Kingston, N. (1993). *Trypanosoma (Megatrypanum)* infections in wild ruminants in Poland. Abstracts IX International Congress of Parasitology, July 25–31. Berlin, Germany, 271.

975. Wita, I., & Kingston, N. (1999). *Trypanosoma cervi* in red deer. *Cervus elaphus* in Poland. *Acta Parasitologica*, *44*, 93–98.

976. Wita, I., & Ovchrenko, I. M. (1997). Occurrence of trypanosomes, *Trypanosoma carassi* in blood of roach, *Rutilus rutilus* and tench, *Tinca tinca*. *Wiadomości Parazytologiczne*, *43*, 385–392.

977. Wita, I., Karbowiak, G., & Czaplinska, U. (2003). *Trypanosoma (Herpetosoma) microti* Laveran et Pettit, 1909 in the social vole, *Microtus socialis* (Pallas, 1771) from Ukraine. *Acta Parasitologica*, *48*(3), 155–162.

978. Wita, I., Karbowiak, G., & Jezewski, W. (2001). *Wystepowanie swidrowcow z rodzaju Trypanosoma* u leszcza *Abramis brama* w jeziorach Goslawskim i Goplo. *Wiadomości Parazytologiczne*, *47*, 383–387.

979. Wita, I., Kingston, N., Karbowiak, G., & Czaplinska, U. (1998). The occurrence of *Megatrypanum* trypanosomes in wild ruminants in Poland. *Wiadomości Parazytologiczne*, *44*, 492.

980. Wolf, K. (1974). Whirling disease research: A delineation of needs. *Fish Health News*, *3*, 20–27.

981. Woo, P. T. K. (1969). The haematocrit centrifuge technique for the detection of trypanosomes in blood. *Canadian Journal of Zoology*, *47*, 921–923.

982. Woo, P. T. K. (1978). The division process in *Trypanoplasma salmositica* in experimentally infected rainbow trout (*Salmo gairdneri*). *Canadian Journal of Zoology*, *56*, 1514–1518.

983. Woo, P. T. K. (1979). *Trypanoplasma salmositica*: Experimental infections

in rainbow trout, *Salmo gairdneri*. *Experimental Parasitology*, *47*, 36–48.

984. Woo, P. T. K. (1981a). *Trypanosoma danilewskyi*: A new multiplication process for *Trypanosoma* (Protozoa: Kinetoplastida). *Journal of Parasitology*, *67*, 522–526.

985. Woo, P. T. K. (1981b). Acquired immunity against *Trypanosoma danilewskyi* in goldfish, *Carassius auratus*. *Parasitology*, *83*, 343–346.

986. Woo, P. T. K. (1987a). Cryptobia and cryptobiosis in fishes. *Advances in Parasitology*, *26*, 199–237.

987. Woo, P. T. K. (1987b). Immune response of fish to parasitic protozoa. *Parasitology Today*, *3*, 186–188.

988. Woo, P. T. K. (1990). MISET, an immunological technique for the serodiagnosis of *Cryptobia salmositica* (Sarcomastigophora, Kinetoplastida) infection in *Oncorhynchus mykiss*. *Journal of Parasitology*, *76*, 389–393.

989. Woo, P. T. K. (1994). Flagellate parasites of fishes. In J. P. Kreier (Ed.), *Parasitic Protozoa* (2nd ed., Vol. 8, pp. 1–80). Academic Press.

990. Woo, P. T. K. (1995). *Fish Diseases and Disorders* (Vol. 1). CAB International.

991. Woo, P. T. K. (1998). Protection against *Cryptobia* (*Trypanoplasma*) *salmositica* and salmonid cryptobiosis. *Parasitology Today*, *14*(7), 272–277.

992. Woo, P. T. K. (1999). The use of monoclonal antibody probes in salmonid cryptobiosis. *Aquaculture*, *177*(1–4), 311–324.

993. Woo, P. T. K. (2003). *Cryptobia* (*Trypanoplasma*) *salmositica* and salmonid cryptobiosis. *Journal of Fish Diseases*, *26*, 627–646.

994. Woo, P. T. K. (2005). Salmonid cryptobiosis and strategies against the disease. *Acta Zoologica Sinica*, *51*(4), 546–549.

995. Woo, P. T. K. (2006). Diplomonadida (Phylum Parabasalia) and Kinetoplastea (Phylum Euglenozoa). In P. T. K. Woo (Ed.), *Fish Diseases and Disorders* (2nd ed., Vol. 1, pp. 46–115). CAB International.

996. Woo, P. T. K. (2010). Immunological and therapeutic strategies against salmonid cryptobiosis. *Journal of Biomedicine and Biotechnology*, 341783, 9. https://doi.org/10.1155/341783.

997. Woo, P. T. K., & Black, G. A. (1984). *Trypanosoma danilewskyi*, host specificity and host's effects on morphometrics. *Journal of Parasitology*, *70*, 788–793.

998. Woo, P. T. K., & Kobayashi, A. (1975). Studies on the anaemia in experimental African trypanosomiasis: I. A preliminary communication

on the mechanism of anaemia. *Annales de la Société Belge de Médecine Tropicale*, *55*, 37.

999. Woo, P. T. K., & Li, S. (1990). *In vitro* attenuation of *Cryptobia salmositica* and its use as a live vaccine against cryptobiosis in *Oncorhynchus mykiss*. *Journal of Parasitology*, *76*(5), 752–755.

1000. Woo, P. T. K., & Poynton, S. L. (1995). Diplomonodida, Kinetoplastida and Arnoebida (Phylum Sarcomastigophora). In P. T. K. Woo (Ed.), *Fish Diseases and Disorders* (2nd ed., Vol. 1, pp. 27–96). CAB International.

1001. Woo, P. T. K., & Thomas, P. T. (1991). Polypeptide and antigen profiles of *Cryptobia salmositica*, *C. bullocki* and *C. catostomi* (Kinetoplastida, Sarcomastigophora) isolated from fishes. *Diseases of Aquatic Organisms*, *11*, 201–205.

1002. Woo, P. T. K., & Thomas, P. T. (1992). Comparative *in vitro* studies on virulent and avirulent strains of *Cryptobia salmositica* Katz, 1951 (Sarcomastigophora, Kinetoplastida). *Journal of Fish Diseases*, *15*, 261–266.

1003. Woo, P. T. K., & Wehnert, S. D. (1983). Direct transmission of a haemoflagellate, *Cryptobia salmositica* (Kinetoplastida: Bodonina) between rainbow trout under laboratory conditions. *Journal of Protozoology*, *30*, 334–337.

1004. Woo, P. T. K., & Wehnert, S. D. (1986). *Cryptobia salmositica*: Susceptibility of infected rainbow trout, *Salmo gairdneri*, to environmental hypoxia. *Journal of Parasitology*, *72*(3), 392–396.

1005. Woo, P. T. K., Wehnert, S. D., & Rodgers, D. (1983). The susceptibility of fishes to hemoflagellates at different ambient temperatures. *Parasitology*, *87*, 385–392.

1006. Woodcock, H. M. (1909). The haemoflagellates and allied forms. In E. Ray (Ed.), *Lankester* (Vol. 1, Fascicle I, 193 pp.).

1007. Woodruff, A. W. (1975). Trypanocidal activity of antitumour antibiotics and other metabolic inhibitors. *Transactions of the Royal Society of Tropical Medicine and Hygiene*, *69*(1), 1, 2.

1008. Yadav, P. (1990). Biochemical and pathological changes in fish parasitized by haemoflagellates (Doctoral dissertation, M.J.P. Rohilkhand University, Bareilly).

1009. Yakimoff, W. L. (1912). Trypanosomes parasites du sang des poisons marins. *Archiv für Protistenkunde*, *27*, 1–8.

1010. Yamasaki, M., Hossain, M. A., Jeong, J. R., Chang, H. S., Satoh, H., Yamato, O., & Maede, Y. (2003). *Babesia gibsoni* specific isoenzymes related to energy metabolism of the parasite in infected erythrocytes. *Journal of Parasitology*, *89*(6), 1142–1146.

1011. Yatindra, & Mathur, B. N. (1986). Studies on *Trypanosoma seenghali* var. nov. sophorae from *Puntius sophore*. *Current Science*, *55*, 471–473.

1012. Yeld, E. M., & Smit, N. J. (2006). A new species of *Trypanosoma* (Kinetoplastida: Trypanosomatidae) infecting catsharks from South Africa. *Journal of the Marine Biological Association of the United Kingdom*, *86*, 829–833.

1013. Zajicek, P. (1991). Enzyme polymorphism of freshwater fish trypanosomes and its use for strain identification. *Parasitology*, *102*, 221–224.

1014. Zajicek, P., & Peckova, H. (1995). A comparison of morphology, growth rate and selected enzyme activities among freshwater fish trypanosomes and the marine species *Trypanosoma boissoni*. *Archiv für Protistenkunde*, *146*, 95–100.

1015. Zar, J. H. (1974). *Biostatistical Analysis*. Englewood Cliffs, NJ: Prentice Hall.

1016. Zatkis, A., Zak, B., & Boyl, A. J. (1953). A new method for the direct determination of serum cholesterol. *Journal of Laboratory and Clinical Medicine*, *41*, 486–492.

1017. Zelmer, D. A., & Arai, H. P. (1998). The contributions of host age and size to the aggregated distribution of parasites in yellow perch, *Perca flavescens*, from Garner Lake, Alberta, Canada. *Journal of Parasitology*, *84*, 24–28.

1018. Zhai, L., Blom, J., Chen, M., Brogger, S., & Kharazmi, A. (1995). The anti-leishmanial agent licochalcone A interferes with the function of the parasite mitochondria. *Antimicrobial Agents and Chemotherapy*, *12*, 2742–2748.

1019. Zhao, Y. J. (1998). Parasitic flagellates of freshwater fishes from Sichuan Province of China. *Acta Hydrobiologica Sinica*, *22*, 114–119.

1020. Zintl, A., Voorhels, H. P., & Holland, V. (2000). Experimental infections of farmed eels with different *Trypanosoma granulosum* life-cycle stages and investigations on pleomorphism. *Journal of Parasitology*, *86*, 56–59.

INDEX

I

K

L

P

R

S

W

Z

Kolibri 360 GmbH
Ferdinand-Jühlke-Straße 7
99095 Erfurt, Deutschland
produktsicherheit@kolibri360.de